Birds of
ARGENTINA
AND THE SOUTH-WEST ATLANTIC

Dedicated to Alejandra, Ian and Luca, and Mecky, Simona and Ainara,

and to all Argentine ornithologists and birders

In memory of Harry Pearman and Pacho Centeno

HELM FIELD GUIDES

Birds of
ARGENTINA
AND THE SOUTH-WEST ATLANTIC

Mark Pearman and Juan Ignacio Areta

Illustrated by
Aldo Chiappe, Jorge Rodríguez Mata,
Richard Johnson and Alan Harris

H E L M
LONDON • OXFORD • NEW YORK • NEW DELHI • SYDNEY

HELM
Bloomsbury Publishing Plc
50 Bedford Square, London, WC1B 3DP, UK
29 Earlsfort Terrace, Dublin 2, Ireland

BLOOMSBURY, HELM and the Helm logo are trademarks of Bloomsbury Publishing Plc

First published in the United Kingdom in 2020

ISBN: HB: 978-1-4729-8432-6
PB: 978-0-7136-4579-8
ePub: 978-1-4729-7310-8

2 4 6 8 10 9 7 5 3

Project manager and Editor: Nigel Redman
Design by Julie Dando, Fluke Art
Printed and bound in India by Replika Press Pvt. Ltd.

MIX
Paper from
responsible sources
FSC® C016779
www.fsc.org

Cover Artwork
Front: Strange-tailed Tyrants (Aldo Chiappe)
Back (top to bottom): Black-crowned Monjita (Jorge Rodríguez Mata), Crested Gallito (Aldo Chiappe),
Tucumán Amazon (Alan Harris), South American Painted-snipe (Richard Johnson)

To find out more about our authors and books visit www.bloomsbury.com and sign up for our newsletters

CONTENTS

ACKNOWLEDGEMENTS

First and foremost, the authors and art team greatly acknowledge the principal reviewers of this book for their diligent, constructive and often illuminating comments and criticism, which for many were provided throughout the compilation of this book. We are therefore indebted to the immense help provided by Alejandro Bodrati, Flavio Moschione, Juan Mazar Barnett, Santiago Imberti, Ignacio Roesler, Hernán Casañas, Germán Pugnali, Sergio Seipke and Christian Savigny. The book would be far from complete without the important input by these dedicated field ornithologists who have helped to set a new benchmark in Argentine ornithology.

We are also particularly indebted to various provincial and family specialists for sharing their intimate knowledge: Alejandro Di Giacomo and Fabricio C. Gorleri (Formosa); Alejandro Bodrati (Chaco, Misiones and Buenos Aires); Miguel Castelino and Ernesto Krauczuk (Misiones); Freddy Burgos and José Segovia (Jujuy); Emilio A. Jordan (Entre Ríos); Martín Manassero (Santa Fé); Hernán Casañas and Juan Klavins (Córdoba); Pedro Blendinger and Jorge Gonnet, Raúl Ábalos (Mendoza); Jorge Veiga (Neuquén), Miguel Christie, Eduardo Ramilo, Marcelo Bettinelli and Ignacio Hernández (Neuquén and Río Negro); Lorenzo Sympson, Guillermo Amico and Hernán Povedano (Río Negro), Felix Vidoz, Quillen Vidoz and Luis Segura (Chubut), Santiago Imberti (Santa Cruz), Marcelo de Cruz (Tierra del Fuego); Alan Henry (Falklands); Santiago Imberti, Tony Palliser and Christian Savigny (seabirds), Tony Prater (shorebirds), Pim Edelaar (steamer-ducks), Sergio Seipke and Matías A. Juhant (diurnal raptors), Emiliano A. Depino (rallids), Manuel Marín (swifts), Jon Curson (warblers) and Rosendo Fraga (icterids).

For the extraordinary variety of help and special requests during the many years of museum work, MP is indebted to Jorge R. Navas and Pablo Tubaro with the assistance of Giovanna Crespo and Yolanda Davies at the Museo Argentino de Ciencias Naturales "Bernardino Rivadavia", Buenos Aires (MACN), without whose help this book would still be in its infancy. We are also extremely grateful to all the help received at other important collections: Aníbal Camperi, Carlos Darrieu, Nelly Bó, and Luis Pagano at Museo de La Plata, Buenos Aires (MLP); Estella Alabarce, José Luis Ciamarello, Sebastián Aveldaño, Ada Echevarría and Sara Bertelli at Fundación Miguel Lillo, Tucumán (FML); Ana Maria Scollo and Elba Pescetti at Instituto Argentino de Zonas Aridas, Mendoza (IADIZA); Robert Prys-Jones and Michael Walters at the Natural History Museum, Tring (BMNH); Julio Milat at Museo Ornitológico Municipal de Berisso (MOMB), and Luis Benegas at Museo Municipal de Río Grande, Tierra del Fuego (MMRG); Kim Bostwick, Kevin Winkler and Charles Dardia at Cornell University Museum of Vertebrates (CUMV); Jorge de León at Yale Peabody Museum (YPM); Paul R. Sweet, Peter Capainolo and Tom Trombone at the American Museum of Natural History (AMNH). Robert Payne kindly provided a database of Argentine specimens held at the University of Michigan, Museum of Zoology (UMMZ), as did Brian K. Schmidt at the United States National Museum (of Natural History), Washington D.C. (USNM). Hernán Casañas and Martín Toledo provided some important data on specimens held at the Museo Zoológico de Universidad Nacional de Córdoba (MZUNC), and Hernán Povedano commented on specimens held at the Museo Provincial de Ciencias Naturales "Florentino Ameghino", Santa Fe (MFA) and Museo de Ciencias Naturales y Antropólogicas "Prof. Antonio Serrano", Paraná, Entre Ríos (MAS). Alice Cibois of the Natural History Museum of Geneva (MHNG) and Pam Rasmussen of the USNM kindly responded to specific requests. MP thanks Graeme and Carole Green for their kind hospitality during numerous visits to the Natural History Museum at Tring.

The following people have unreservedly contributed their own sound recordings including various previously unknown voices, or have sourced published material for the project: Raúl Abalos, Freddy Burgos, Hernán Casañas, Miguel Castelino, Dave Farrow, Gabriela Ibarguchi, Santiago Imberti, Frank Lambert, Bernabé López-Lanús, James Lowen, Sjoerd Mayer, Juan Mazar Barnett, Diego Monteleone, Germán Pugnali, Ignacio Roesler, Mark Sokol, Myriam Velázquez, Barry Walker, Andrew Whittaker and Dave Willis. The project has also benefited from the use of additional recordings from the National Sound Archive (NSA) of the British Library with particular thanks to Richard Ranft and Paul Duck, and from the Macaulay Library of Natural Sounds (MLNS) with special thanks to Greg F. Budney and Eduardo Iñigo-Elías.

Both the authors and artists express their deep thanks for the photographic material received from Raúl Abalos, Danny Aeberhard, Daniel Almirón, Anders Andersson, Phil Benstead, Marcelo Bettinelli, Alejandro Bodrati, Norberto Bolzón, Adriana H. Centeno, Juan Claver, Kristina Cockle, Freddy Burgos, Ruben Cargnelutti, Marcelo de-Cruz, Rubén Dellacasa, Alejandro Di Giacomo, Ignacio Hernández, James Lowen, Ricardo Doumecq Milieu, Pim Edelaar, Rainer

Ertel, Roberto Güller, Alberto Gurni, Alan Henry, Jon Hornbuckle, Steve Howell, Santiago Imberti, Alvaro Jaramillo, Ramón Moller Jensen, Juan Klavins, Frank Lambert, James Lowen, Ricardo Matus, Andrew Moon, Juan Mazar Barnett, Norberto Oste, Luis Pagano, Tony Palliser, Mario Peña, Germán Pugnali, Christian Savigny, Julio Schindler, Federico Schulz, Bill Simpson, Ken Simpson, Dave Stejskal, Paul Smith, Pablo Sturzenbaum, Marcus Sulley, Joe Tobias, Jacob and Tini Wijpkema, and Dave Williamson. The artists especially thank Jorge Anfuso (Guira Oga) for permission to photograph raptors in his care and for his help with video footage of raptors in flight. In particular, we greatly appreciate the help of Marcelo Bettinelli and Roberto Güller in tracking down many of our photographic requests. Richard Johnson thanks Joe Tobias, James Lowen, Georgina Willmot and Roger and Margaret Johnson. Alan Harris thanks the staff at the Natural History Museum, Tring, especially Robert Prys-Jones and Mark Adams.

Not least, these acknowledgements are principally dedicated to the huge variety of help and information in the form of personal observations, behavioural data, logistical help and distributional data without which this book would be incomplete, and as such we are indebted to the following people, institutions and companies: Esteban Abadie, Rafael and Raúl Abalos, Daniel Almirón, Alparamis S.A., Guillermo Amico, Mirko Avedano Schaler, Björn Anderson, Emilio Arauz, Aves Argentinas, Marcos Babarskas, Pedro Babsia, Javier Orlando (El Colo) Baez, Chris Balchin, Juan Mazar Barnett, Luis Benegas, Elida and Marcelo Bettinelli, Birdquest Ltd., Pedro Blendinger, Alejandro Bodrati, Andres Bosso, Neil Bostock, Oscar Braslavsky, Freddy Burgos, Angel Caradonna, Ruben Cargnelutti, Hernán Casañas, Miguel Castelino, Juan Carlos Chebez, Mariano Chudy, Ricardo Clark, Rob Clay, Kristina Cockle, Eugenio (Coco) Coconier, Daphne Colcombet, Roberto Comparín, Paul Coopmans, Patricio Cowper Coles, Victor Cueto, Esteban Daniels, William Davidson, Marcelo de Cruz, Marco della Zeta Somi, Adrián S. Di Giacomo, Alejandro G. Di Giacomo, Cristóbal Doiny Cabré, Ricardo Doumecq Milieu, Alan Eardley, Alec Earnshaw, Pim Edelaar, Gunnar Engblom, Graciela Escudero, Carlos and Silvia Ferrari, Daniel Fernández, Eduardo Fernández-Duque, Miguel Fiameni, Rosendo Fraga, Marcelo Gallegos, Carlos Galliari, Marcos García Rams, Nick Gardner, Mariano Gelain, Guillermo Gil, Jorge Gonnet, Hernán Rodríguez Goñi, Carlos and Fabricio Gorleri, Alejandra Grigoli, Pablo Grilli, Roberto Güller, Sharon Halford, Peter Hayman, Alan and Bob Henry, Christian Henscke, Paul Hilder, Steve Hilty, Ingrid Holzmann, Steve N. G. Howell, Judith and Michael Hutton, Santiago Imberti, Ingleby Farms & Forests, Alvaro Jaramillo, Roberto Jensen, Anders Jihmanner, Guy Kirwan, Juan Klavins, Santiago Krapovickas, Ernesto Krauczuk, Hernán Laita, Frank Lambert, Reginaldo (Chino) Lejarraga, José Lieberman, Huw Lloyd, Bernabé López-Lanús, James Lowen, Niceforo Luna, Niall Machin, Claudio Maders, Marcos Malaspina, Mauricio Manzione, Carlos Marchisio, Manuel Marín, Victor Matuchaka and park rangers from central Misiones, Gabriel Maugeri, Carlos Maza, Emilse Merida, Isabel Merle, Rodolfo Miatello, Pablo Michelutti, Diego Monteleone, Flavio Moschione, Maxi Navarro, Liliana Olveira, Norberto Oliveira (Moconá), Ulises Ornstein, Norberto Oste, Luis Pagano, Tony Palliser, Pablo Petracci, Gabriel Piloni, Tony Prater, Pablo Pratts, Germán Pugnali, Fernando Raffo, Juan Raggio, Colón Rivero, Gabriel Rocha, Ignacio Roesler and family, Román Ruggera, Charlie Sandoval, Christian Savigny, Rosemary Scoffield, Luis Segura, Sergio Seipke, Jenny Sherriff, Bill Simpson, Paul Smith, Mark and Elaine Sokol, María Jose Solís, Dave Stejskal, Gary Stiles, Marcus and Simon Sulley, Lorenzo Sympson, Ryan Terrill, Dave Thorns, Phil Tizzard, Joe and Nat Tobias, James Van Remsen, Julio Vázquez, Jorge Veiga, Félix and Quillén Vidoz, Barry Walker, Dave Willis, Project Yacutinga '95 team, Emilio White, Bret Whitney, Jacob and Tini Wijpkema, Martin Young, Marcelo Zambrano, Mateo Zelich and to anyone inadvertently omitted.

The seemingly endless review of published material was alleviated by help, direction and special requests gratefully received from Danny Aebehard, Thomas Arndt, Marcelo Bettinelli, Daniel Blanco, Alejandro Bodrati, Andres Bosso, Hernán Casañas, Eugenio Coconier, Adrián and Alejandro Di Giacomo, Pim Edelaar, Rosendo Fraga, Graeme Green, Jeremy Hatch, Alan and Trish Henry, Baz Hughes, Guy Kirwan, Adrian Long, James Lowen, Gabriel Maugeri, Juan Mazar Barnett, Diego Monteleone, Robert Prys-Jones, Ignacio Roesler, Christian Savigny, Sergio Seipke, Francois Vuilleumier, Effie Warr and David Wege.

Our botanical direction in this project was greatly aided by the help of Rosemary Scoffield who also kept us up to date with botanical taxonomy.

Finally, we owe special thanks to the editors Robert Kirk, Jim Martin and in particular Nigel Redman for their patience, understanding and diligence during the project, and extend our appreciation to the designer Julie Dando for her considerable layout and map expertise.

Last, but not least, we thank Alejandra and Mecky for their patience and support throughout the project, especially during long field trips and tours.

BOOK PLAN AND AIMS OF THE BOOK

The book consists of two volumes: **Volume 1 (Field guide)** and **Volume 2 (Identification, natural history, distribution and taxonomy)**. Volume 1 is a light, field-friendly guide that contains illustrated plates and accompanying succinct texts focused on bird identification, as well as distribution maps. Volume 2 forms the bulk of the work and is a detailed account of identification at subspecific level, describing the differences between similar species and providing behavioural and vocal data, as well as giving the status and distribution of every taxon, the worldwide species range, the global conservation status, and taxonomic notes.

The main aims of these two volumes are:

- To provide detailed descriptions of all species and subspecies occurring in the region, combined with a user-friendly synthesis of diagnostic features and comparisons of potential confusion species.
- To describe, define and depict the current and historical distribution, altitudinal range and relative abundance of all bird taxa known in the region, on a provincial and maritime scale, with special emphasis on the complex migratory patterns.
- To provide accurate illustrations of all species together with subspecies, colour morphs and age-related plumages which differ to the extent that they might cause confusion or promote certain interest.
- To provide relevant taxonomic notes with sourced references and a synthesis of possible future updates.
- To provide a comprehensive dataset on behaviour and voice, partially drawn from published material and partially from the authors' own field research.
- To describe the habitat preferences for every species, and provide a detailed illustrated account of ecosystems and habitats.
- To provide the first detailed synthesis of breeding data for all species in the region.

AREA COVERED BY THE BOOK

This book has been compiled from a biological, ornithological and non-political, perspective and covers a single biogeographical unit, namely continental Argentina, associated Fuegian and subantarctic islands (including those belonging to other countries) and the maritime Exclusive Economic Zones associated with these regions. The EEZs extend 200 nautical miles from land, their boundaries entering the Drake Passage in the far south. The foremost region covered includes the twenty-three provinces of the Republic of Argentina (Map 1), representing the eighth largest country in the world.

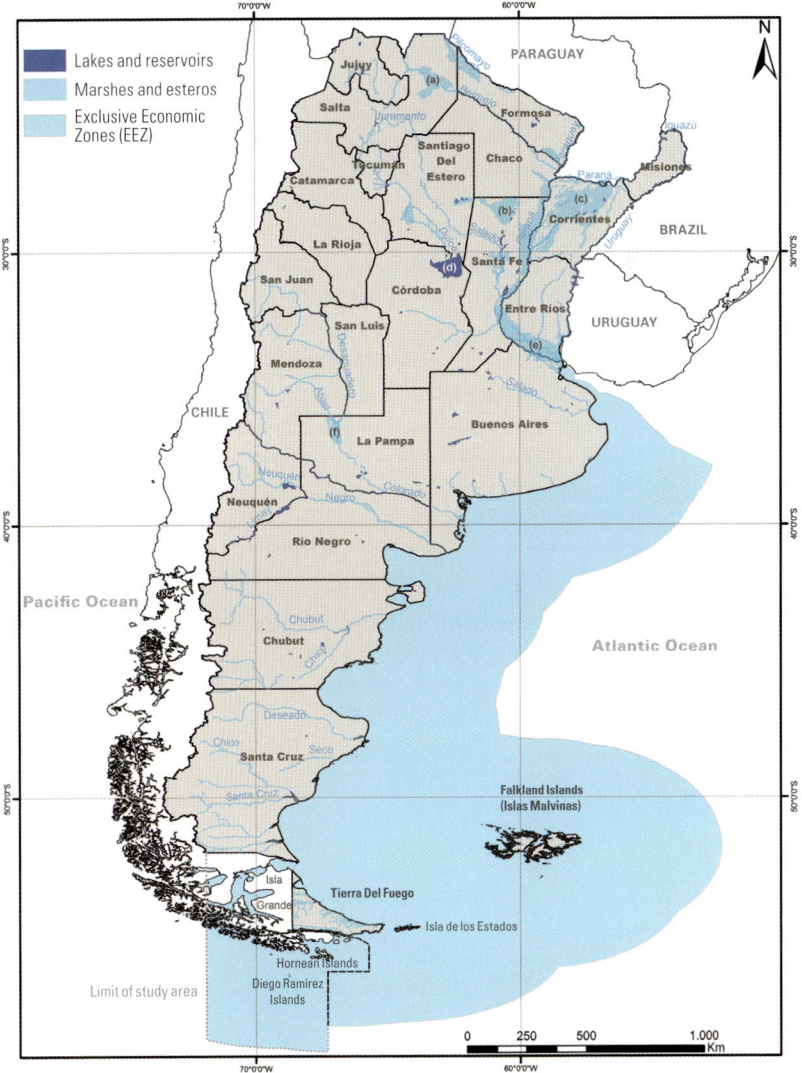

Map 1. Boundaries and hydrography
This map delimits the area covered by the book, the international and provincial boundaries of Argentina, and shows the major rivers and water bodies. **Key**: (a) Bañados del Quirquincho, (b) Bajos Submeridionales, (c) Esteros del Iberá, (d) Laguna de Mar Chiquita and Bañados del Río Dulce, (e) Paraná Delta and (f) Bañados del Atuel.

The subantarctic islands included are all adjacent to the continent and either comprise undisputed Argentine territory, or are otherwise claimed by, inhabited by or disputed between Argentina, the United Kingdom or Chile. The Islas de los Estados (or Staten Islands) lying south-east of Isla Grande represent an outstanding and ornithologically important inclusion. The remaining Atlantic islands of Tierra del Fuego, to the south of Isla Grande and belonging to Chile, are also covered. These include all islands in the Beagle Channel and, in particular, the Hornean islands complex including Wollaston, Amarilla, Hoste and Cabo de Hornos (Cape Horn) islands; which extend the Chilean EEZ further south as a consequence. These islands represent an intrinsic component of the Fuegian avifauna in that some species are largely restricted to the Hornean island group, and only occasionally reach Isla Grande. Other Atlantic islands such as Evout and Barnevelt and the remote Diego Ramírez (Chile) have received little attention, but are covered here (Map 1) because of their important breeding seabird populations, details of which have been published in little-known journals, and have been rarely documented in the mainstream literature.

Within the defined oceanic boundary, the only offshore island group in addition to the several more remote islands mentioned above, are the Falkland Islands (Map 1) which lie 490 km east of southern Patagonia. The name given to this archipelago differs in the Argentine Spanish language in that they are known as the *Islas Malvinas* or more simply *Las Malvinas*. Since this is an English language edition, we considered it appropriate to use the English name for these islands and localities within the island group. A forthcoming Spanish edition will use Argentine names. While there is an indisputable link between the avifaunal and botanical composition between Tierra del Fuego, Isla de los Estados, the Hornean group and the Falkland Islands at family and generic level, other islands to the east and south, e.g. South Georgia, South Sandwich, South Orkney and South Shetland, have extremely weak ornithological links and share virtually no botanical links. These are devoid of all but one land bird: South Georgia Pipit *Anthus antarcticus*, a typical representative of a cosmopolitan genus that recent molecular phylogenetic work suggests may be just a subspecies of the widespread Correndera Pipit *Anthus correndera*, and it is excluded here. Numerous species of seabirds breed on islands to the south of the Antarctic convergence, with a few breaching both sides, and as such this is usually regarded as a natural biological barrier. The oceanic and island coverage in this book therefore relates exclusively to a single biogeographic area, which is explicitly related to the continental avifauna of Argentina and its adjacent maritime region.

It should also be noted that the Antarctic islands and the Antarctic Peninsula have already received much ornithological exploration resulting in many detailed books and publications, not to mention their location in a completely distinct continent. Their inclusion here would have meant considerable additional research, deferring from the title of this work and its original concept. Nevertheless, many Antarctic species have occurred as vagrants inside our study area and hence their inclusion. Ultimately, the species coverage in this book includes almost 100% of bird species from Uruguay, in addition to approximately 86% of the Chilean avifauna and 93% of the Paraguayan avifauna.

GEOGRAPHY AND HYDROGRAPHY

The Republic of Argentina ranges from sea-level to the highest peak in the Americas, Aconcagua with its 6961 m peak also representing the highest mountain in the Western Hemisphere and Southern Hemisphere. Two main geographical features characterise the Argentine landscape; the **Andean Cordillera** and the **Chaco/Pampean plain** (Map 2). The Cordillera de los Andes extends throughout the western flank of Argentina from north to south for more than 3700 km. Numerous peaks extend over 6300 m above sea-level, from Jujuy in the north to Mendoza (e.g. Volcán Llullaillaco, Nevado de Cachi, Volcán Ojos del Salado, Monte Pissis, Cerro Bonete, Cerro Mercedario), while several active volcanoes are found from northern Neuquén

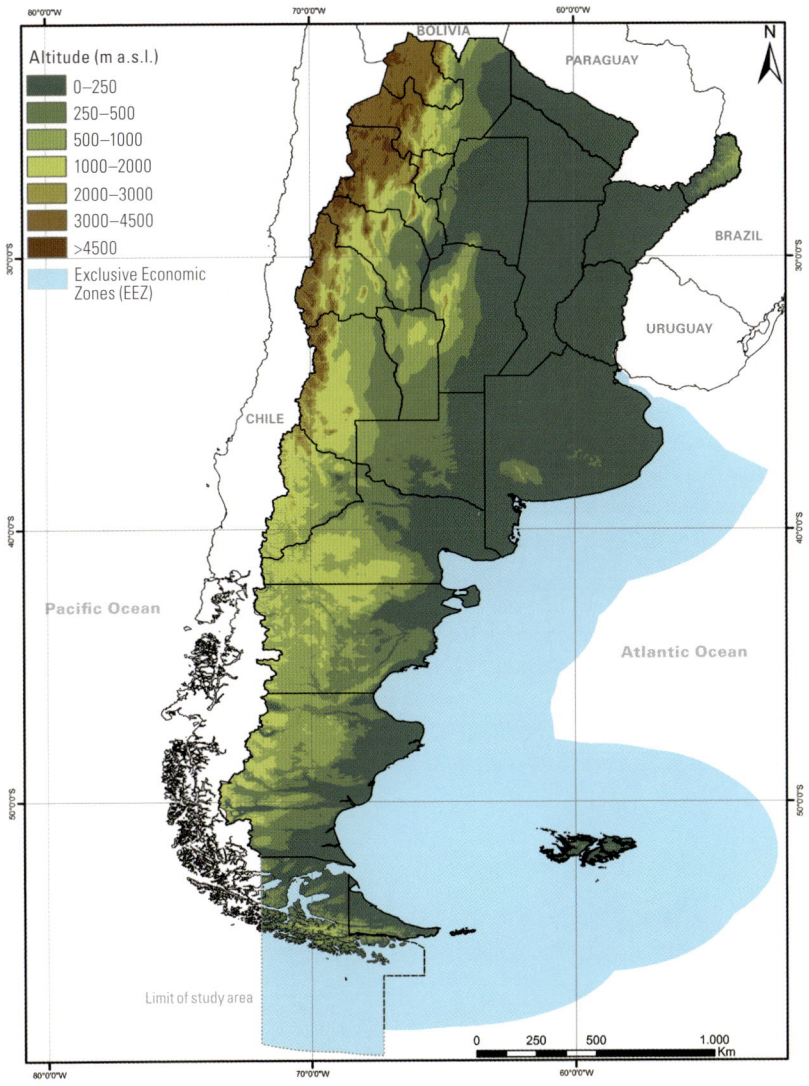

Map 2. Geography
This map shows the relief of Argentina and the study area.

to the south (Volcán Domuyo, Volcán Copahue and Volcán Lanín), from where the Cordillera decreases progressively in height as it crosses Tierra del Fuego and terminates on Isla de los Estados. To the east of the southern Andes buttress, the landscape is dominated by desertic *mesetas* (table-top mountains), of which the Meseta de Somuncurá in central Río Negro and the mesetas of Santa Cruz province should be highlighted. Several extra-Andean ranges rise either from near the Andes (the Subandean ranges) or further inland (the Pampean ranges). The **Subandean ranges** are of biogeographic importance for birds, as it is here that some endemic species occur (Sierra de Santa Bárbara/Cresta del Gallo, Sierras de Medina/Castillejos, Serranía del Aconquija, Serranía de Ambato/Ancasti and Sierra de Velasco). The **Chaco/Pampean plain** inclines mostly to the southeast spanning 1.2 million km² encompassing the chaco, Pampas, Espinal, Esteros del Iberá and Campos ecoregions. In this vast flat area, the **Pampean ranges** of the Sierras de Córdoba together with the Sierras de San Luis (the Central Sierras) harbour several endemic species and subspecies. Other important ranges are the Sierra de Guasayán, rising in the western portion of the dry chaco, and the Sierra de Ventania in the southern Pampas. The northeast and central portions of Paraná forest also experience drastic avifaunal changes despite the relative low altitude (maximum of 1000 m) of local mountain ranges.

The complex topography of Argentina results in a similarly complex hydrography (Map 1). The **Rio de la Plata basin** is the second largest in South America; its main tributaries, the Paraguay, Paraná and Uruguay rivers run essentially from north to south. The Paraná and Uruguay rivers constitute important biogeographic corridors and biogeographic breaks, and jointly delimit the region known as **Mesopotamia** which includes the provinces of Misiones, Corrientes and Entre Ríos, from north to south. The Delta of the Paraná river extends for 14,000 km² and for some 320 km from Entre Ríos province to the Rio de la Plata estuary (see Pampas ecoregion below). The Bermejo river traverses the chaco region from west to east, while the parallel Pilcomayo is interrupted in the middle portion; these sinuous, meandering rivers flow into the Paraguay river and are characterised by the presence of oxbow lakes known as *madrejones*. An important tributary of the Paraná river is the Río Salado (Salado del Norte or Juramento), beginning in the high Andes above 5000 m in the Calchaqui valley, Salta and representing the longest river in the country. In central Argentina, the **Rio Desaguadero basin** collects water from the San Juan, Mendoza, Tunuyán and Atuel rivers, reaching the sea only intermittently during exceptional floods. The flooding of the Atuel and Desaguadero rivers form the *bañados del Atuel*. Further south, the **Patagonian system** includes several independent rivers that run west to east into the sea, and include the Colorado, Río Negro, Chubut and Santa Cruz rivers. Most basins drain to the Atlantic Ocean, but a few Patagonian rivers (e.g. Hua Hum, Manso, Puelo and Futaleufú) drain through Chile into the Pacific slope. **Endorheic basins** give rise to extensive salt flats or large saltwater bodies. Among the latter, the Laguna de los Pozuelos (Jujuy province) and the vast Laguna de Mar Chiquita (Córdoba province) covering 2000 km², deserve special mention for the number of waterbirds, and especially waders, that they harbour in spring and autumn. The Río Dulce is the main affluent of Laguna de Mar Chiquita and forms the *bañados del Río Dulce* to the north of the lake.

ECOREGIONS AND HABITATS

The distribution of birds in Argentina is intrinsically linked to the major ecoregions on the mainland and subantarctic islands in the south-west Atlantic (Map 3). The distribution of these ecoregions is heavily influenced by mountainous regions and climatic factors. Most of these ecoregions contain a variety of distinct habitats. Floristic composition varies from a relatively low diversity in the Puna and Patagonian steppe, to a highly complex one in the chaco, Yungas and Paraná forests, but this diversity does not strictly reflect the diversity of the avifauna. Most ecoregions and habitats grade into one another forming ecotones, and some even contain flora from disjunct regions. A schematic representation of the distribution of ecoregions

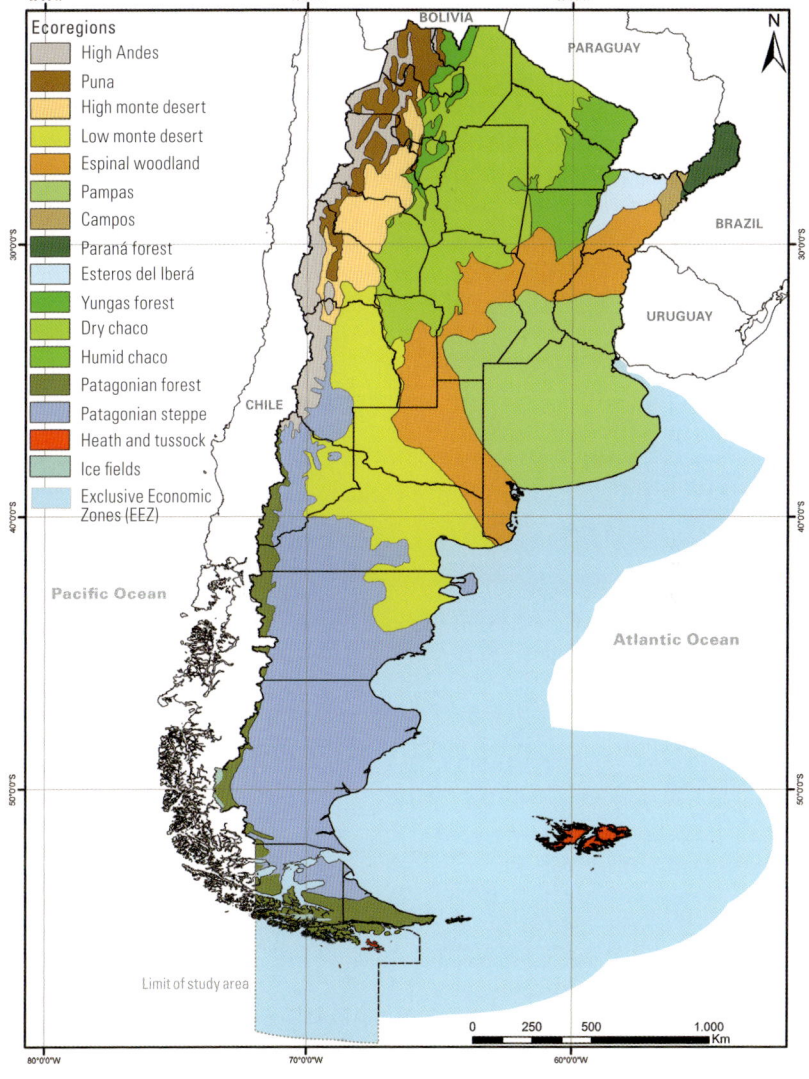

Ecoregions
- High Andes
- Puna
- High monte desert
- Low monte desert
- Espinal woodland
- Pampas
- Campos
- Paraná forest
- Esteros del Iberá
- Yungas forest
- Dry chaco
- Humid chaco
- Patagonian forest
- Patagonian steppe
- Heath and tussock
- Ice fields
- Exclusive Economic Zones (EEZ)

Limit of study area

BOLIVIA
PARAGUAY
BRAZIL
URUGUAY
CHILE
Pacific Ocean
Atlantic Ocean

0   250   500   1.000 Km

Map 3. Ecoregions
This map defines the main ecoregions of Argentina and the study area.

(Map 3) showcases the general distributional patterns. At a more local scale, patches of habitat from some ecoregions may occur locally embedded in others. Thus, the distribution of ecoregions should be used as a general aid to birdwatching, bearing in mind that local settings may vary substantially from the general map, while taking into account the size of the area that is covered. The Southern Patagonian ice field covering 19,500 km² is devoid of vegetation and is not further discussed (see Map 3).

After describing the principal floral composition of each ecoregion and habitat, the endemic bird species and genera are mentioned for each habitat, together with any other noteworthy species. Note that a large variety of native flora is illustrated on the plates (see Appendix 5). The flora taxonomy employed follows that of *Flora Argentina* (1997–2017) published by the Instituto de Botánica Darwinion.

PUNA AND HIGH ANDES

(**Habitats**: Pre-puna, Puna brush-steppe, Tola heath, Puna grasslands, Andean grass-steppe, sierran grasslands)

The dry Andean slopes above the Yungas forest range from 2000 to 3400 m in Jujuy and Salta south to La Rioja (descending to 1000 m) and are classed as **Pre-puna**, becoming intermixed and grading into the Monte Desert. The dominant vegetation of Pre-puna is formed by leguminous bushes such as Pichana *Senna crassirame*, and composite shrubs such as *Aphyllocladus spartioides* and *Gochnatia glutinosa*. Certain sectors are dominated by cacti including stands of giant columnar Cardón del Valle *Trichocereus terschekii* reaching 12 m in height, and the smaller Pasacán *T. pasacana*, Airampu *Tunilla soehrensii*, and a variety of other cacti including *Opuntia tilcarensis*, *Parodia maassii* and the low spherical *Lobivia formosa*. Some rivers harbour low woodlands of Visco *Acacia visco*, thorny Churqui *Prosopis ferox*, Chilca *Baccharis salicifolia* and Molle *Schinus areira*. The highest slopes only have a sparse cover of bromeliads (*Deuterocohnia brevifolia*, *D. lorentziana*) and *Tillandsia bryoides*. Typical bird species include Brown-backed Mockingbird *Mimus dorsalis* and Rusty-vented Canastero *Asthenes dorbignyi*, while the endemic Moreno's Ground Dove *Metriopelia morenoi* is almost exclusively found in rocky areas in Pre-puna.

Above the Pre-puna is a flat desert known as Puna or 'altiplano', extending from southern Peru across northern Chile and western Bolivia into northwest Argentina. This arid landscape alternates between brush- and grass-steppe. **Puna brush-steppe** is dominated by Tolilla *Fabiana densa*, Añagua *Adesmia horrida*, Chijúa *Baccharis boliviensis* and Tola *Parastrephia* spp. **Puna grasslands** are dominated by *Festuca orthophylla* bunchgrass and climax grasses of the genera *Festuca*, *Pennisetum*, *Bouteloua*, *Eragrostis* and *Aristida*. Bird species diversity is relatively low but a number of exclusive species include Puna Tinamou *Tinamotis pentlandii*, Puna Miner *Geositta punensis* and Puna Yellow Finch *Sicalis lutea*. Boulders and rocks covered with cushion-plants, especially *Frankenia triandra*, form a microhabitat for Red-backed Sierra Finch *Idiopsar dorsalis*. In striking contrast, lakes and wetlands in the Puna region are lined with *Zameioscirpus atacamensis*, *Juncus stipulatus* and *Eleocharis quinqueflora*, and support a large and varied avifauna including many Puna specialties e.g. Andean Goose *Oressochen melanopterus*, Puna Ibis *Plegadis ridgwayi*, Puna and Andean Flamingos *Phoenicoparrus jamesi* and *P. andinus*, Horned, Giant and Andean Coots *Fulica cornuta*, *F. gigantea*, *F. ardesiaca*, and Andean Avocet *Recurvirostra andina*. From San Juan to Santa Cruz and above the tree-line in Tierra del Fuego, vegetation is more sparse although **Andean grass-steppe** dominates in certain areas with several compact bushes.

Above the tree-line from 1500 to 3000 m in the sierras of Córdoba, San Luis, Catamarca, La Rioja and western Santiago del Estero, the **sierran grasslands** are composed of a variety of grasses of the genera *Stipa*, *Festuca*, *Digitaria*, *Diplachne*, *Trichloris*, *Pappophorum* and *Elionurus*. These grasslands also have sparse bush cover with Chilca *Eupatorium buniifolium* and Piquillín *Condalia microphylla*, also *Senna*, *Berberis* and *Baccharis* spp. among others. In the Central Sierras, the habitat supports the endemic Córdoba Cinclodes *Cinclodes comechingonus* and Sierran Meadowlark *Leistes* [*loyca*] *obscura*, and another 11 endemic subspecies accepted here.

MONTE DESERT

(**Habitats**: low monte desert, high monte desert)

The Monte desert is the only ecosystem which is strictly endemic to Argentina, and this is reflected by a number of endemic bird species, endemic breeders and other species shared only between the monte and the adjacent dry chaco or Patagonian steppe. Monte desert is an arid steppe ecosystem with very low rainfall, sandy soils and is dominated by thorn bushes. It forms a band which commences in the arid northern intermontane Andean valleys and is bordered by Pre-puna and dry chaco woodlands, and in the south intergrades with Espinal woodland and Patagonian steppe. The Monte desert strip extends from Salta through Tucumán, Catamarca, La Rioja, San Juan, north-west San Luis, Mendoza, western La Pampa, eastern Neuquén, Río Negro and north-east Chubut. The dominant flora comprises three species of low thorny Creosote bushes (*Larrea divaricata*, *L. nitida* and *L. cuneifolia*) which have small yellow flowers. These are usually intermixed with other xerophytic bushes and stunted trees which rarely exceed three metres in height, and include the green-barked Brea *Cercidium praecox*, Retamo *Bulnesia retama*, various algarrobos including *Prosopis flexuosa* and *P. alpataco*, Chañar *Geoffroea decorticans*, Molle *Schinus fasciculatus*, Piquillín *Condalia microphylla*, Mata Sebo *Monttea aphylla*, Monte Negro *Bougainvillea spinosa*, Pichana *Senna aphylla*, and Chirriadora *Chuquiraga erinacea*. A large variety of cacti are found (e.g. *Opuntia sulphurea*, *O. anacantha*, *Pterocactus*, *Tephrocactus* and *Trichocereus* spp.) while grassland communities are represented by *Aristida*, *Bouteloua*, *Pappophorum*, *Eragrostis*, *Stipa*, *Hordeum*, *Bromus*, etc. Sand dune formations are frequent throughout the monte. In the south, the **low monte desert** extends mostly over a vast plain while in the north the **high monte desert** is mostly distributed as narrower strips of habitat surrounded by mountains.

Bird species diversity is low, compared to all other wooded or forested habitats in Argentina. However, five species are completely endemic to Argentina: White-throated Cacholote *Pseudoseisura gutturalis*, Steinbach's Canastero *Pseudasthenes steinbachi*, Sandy Gallito *Teledromas fuscus*, Carbonated Sierra Finch *Rhopospina carbonaria* and Monte Yellow Finch *Sicalis mendozae*. There are another five endemic breeding species: Black-crowned Monjita *Neoxolmis coronata*, Hudson's Black Tyrant *Knipolegus hudsoni*, Straneck's Tyrannulet *Serpophaga griseicapilla*, White-banded Mockingbird *Mimus triurus* and Cinnamon Warbling Finch *Poospiza ornata*. Finally, three other species are almost exclusively found in the Monte desert, but may also breed in adjoining Patagonian steppe: Patagonian Canastero *Pseudasthenes patagonica*, Rusty-backed Monjita *Neoxolmis rubetra* and Lesser Shrike-Tyrant *Agriornis murina*.

YUNGAS

(**Habitats**: *Polylepis* woodlands, *Podocarpus* forest, alder forest, cloud forest, transitional foothill forest)

The Yungas includes a complex and varied number of montane forest types ranging mostly from 400 to 2600 m on the Andean slopes of Jujuy, Salta, Tucumán, Catamarca and adjacent La Rioja province with an outlying sector on the Sierra de Guasayán, Santiago del Estero, where Yungas forest grades into sierran chaco woodlands. There is a marked decline in floral and faunal diversity from north to south, and also dramatic variation in composition by altitude. The forest is contiguous only with similar forests in southern Bolivia. *Polylepis* **woodlands** are representative of the highest altitude trees in the world, occurring at and above the usual tree-line, mainly along river gorges, and also forming isolated habitat within the Puna grass-steppe. *Polylepis australis* extends sparsely in northern Salta and Jujuy (at 1900–3000 m) and reappears in the sierras of Córdoba and San Luis, while *P. tomentella* occurs at 2400–4300 m in Jujuy only. Exclusive bird species include Tawny Tit-Spinetail *Sylviorthorynchus yanacensis* and Giant Conebill *Conirostrum binghami*, although many other bird species have strong ties. Two other high altitude Yungas trees have a somewhat localised distribution and each may be dominant in certain areas, forming monospecific woodlands. **Podocarpus forest** *P. parlatorei* occurs from 1500 to 2000 m, providing a seasonal food source for Tucumán Amazon *Amazona tucumana*, and is also inhabited by local species such as Smoky-brown Woodpecker *Dryobates fumigatus*, Fulvous-headed Brushfinch *Atlapetes fulviceps*, and the fairly specialised Buff-banded Tyrannulet

Mecocerculus hellmayri. **Alder forest** *Alnus acuminata* is somewhat more widespread (from 1300 to 2500 m) and is inhabited by Tucumán Amazon, and forms typical, but not exclusive, habitat for the near-endemic Tucumán Mountain Finch *Poospiza baeri* and the endemic White-browed Tapaculo *Scytalopus superciliaris* and Yellow-striped Brushfinch *Atlapetes citrinellus*. It also lines the rivers on which Rufous-throated Dipper *Cinclus schulzi* can most frequently be encountered.

The floristic composition becomes subtropical, complex and dense below 1600 m and descends the Andean slope mainly to 550 m. The mid- and upper elevation forest is classed as **cloud forest** because it is shrouded in cloud cover in summer and autumn. The dominant trees are the endemic 30–40 m-high Horco Molle *Blepharocalyx salicifolius*, the 25 m-high endemic Laurel de la Falda *Cinnamomom porphyria* (usually heavily laden with epiphytes), the 15–30 m-high Nogal Criollo *Juglans australis* and, among others, the Cedro Tucumano or Coya *Cedrela lilloi*, Cedro Salteño *C. balansae*, Saúco *Sambucus peruviana*, Horco-Cebil *Parapiptadenia excelsa*, Lapacho Amarillo *Tabebuia lapacho* and Carnaval *Senna spectabilis*. Humid areas support extensive stands of *Chusquea lorentziana* bamboo and *Alsophila odonelliana* tree-ferns. This sector of Yungas forest holds the highest bird species diversity. The entire Yungas forest system is home to some 18 species which can be considered endemic to Bolivia and Argentina. However, species distribution is by no means uniform since the forest is divided into a number of semi-connected and unconnected blocks which follow major Andean and extra-Andean chains and sierras. Representative Yungas birds include Red-faced Guan *Penelope dabbenei*, Yungas Guan *Penelope bridgesi*, Blue-capped Puffleg *Eriocnemis glaucopoides*, Slender-tailed Woodstar *Microstilbon burmeisteri*, Hoy's Screech Owl *Megascops hoyi*, Spot-breasted Thornbird *Phacellodomus maculipectus*, Sclater's Tyrannulet *Phyllomyias sclateri*, Rufous-browed Warbling Finch *Microspingus erythrophrys* and Fulvous-headed Brushfinch *Atlapetes fulviceps*.

The lower slopes of the Yungas from 400 to 700 m are mostly covered with **foothill forest** which shares floristic characters with the Caatinga of Brazil and the guajira of NE Colombia and NW Venezuela. Here, the dominant trees include the Tipa Blanca *Tipuana tipu*, a Yungas endemic reaching 40 metres, the 30 m-high Palo Blanco *Calycophyllum multiflorum*, the endemic 20–30 m-high Lapacho Rosado *Tabebuia impetiginosa*, Urundel *Astronium urundeuva*, Roble Criollo *Amburana cearensis*, Jacarandá *Jacaranda mimosifolia* and Quina *Myroxylon peruiferum*. This basal forest is not a transition between dry chaco woodlands and Yungas forest but has instead been invaded by some chaco flora only in recent times, e.g. the Palo Amarillo *Phyllostylon rhamnoides* and the 30 m-high Pacará *Enterolobium cortotisiliquum* which are widespread chaco trees. Virtually all of the foothill forest from 400 to 550 m has been clear-felled for sugar cane, citrus and tobacco plantations. The avifauna mostly comprises Yungas forest species, but some typical lowland chaco species, such as Cream-backed Woodpecker *Campephilus leucopogon* and Great Rufous Woodcreeper *Xiphocolaptes major*, also occur and ascend the Andean slopes.

INTER-ANDEAN DRY VALLEYS

(**Habitats**: *Polylepis* forest, dry shrub, low cactus beds, Aguaribay woodland)

An impoverished version of the Inter-Andean Dry Valleys of Bolivia extends into a very short and narrow section of NW Argentina in extreme nc. Salta, centred around the town of Santa Victoria, north to the Bolivian border and south possibly slightly beyond Punco Viscana. These valleys range from 2200 to 3200 m and are dry by virtue of the rain shadow created from the west by the high Sierra de Santa Victoria range and from the east by lower mountain chains with luxuriant Yungas cloud forest. The deep gorges are characterised by sparse forests of *Polylepis tomentella*, while the drier valley slopes harbour a diversity of small cacti, sparse shrub cover and generally isolated Aguaribay *Schinus areira* trees. Small-scale farming is widespread. Despite its restricted distribution in Argentina, this ecoregion is ornithologically relevant. The very sparse woods that grow alongside rivers harbour a small population of Bolivian Woodpeckers *Dryobates* [*lignarius*] *puncticeps*, sparse shrubs on steep rocky slopes are inhabited by Bolivian Earthcreeper *Tarphonomus harterti*, while Bolivian Warbling Finch *Poospiza boliviana* and a white-browed form of Azara's Spinetail *Synallaxis azarae* are also found. The only reports of Cliff Parakeet *Myiopsitta* [*monachus*] *luchsi* come from this area.

CHACO

(**Habitats**: Dry chaco woodlands, sierran chaco woodlands, humid chaco woodlands, forest islands in savanna, palm savanna, gallery forest)

One of the largest ecoregions in Argentina are the mainly low-lying thorn woodlands known as the chaco (of which Chaco province itself only forms a small part) and which also extends over western Paraguay and south-eastern Bolivia, covering an area of 1,000,000 km². As such, this is one of the largest and most continuous areas of arid woodland on the planet. Throughout the text we use the term 'chaco' to indicate this habitat, while 'Chaco' is used to indicate the province of Chaco. The chaco can be subdivided into two main habitats, humid chaco in the east and dry chaco in the west. The dry chaco is often subdivided again with 'sierran chaco' on higher ground. In general terms, chaco woodlands are characterised by four currently recognised species of quebracho (*Aspidosperma* and *Schinopsis* spp.) which represent the tallest trees, and which have undergone severe exploitation. The majority of trees are thorny and widespread throughout the life-zone, and include the Algarrobo Negro *Prosopis nigra*, Algarrobo Blanco *Prosopis alba*, Mistol *Ziziphus mistol*, Guayacán *Caesalpinia paraguariensis*, Guaraniná *Sideroxylon obtusifolium*, Chañar *Geoffroea decorticans* and Espinillo *Acacia caven*. Cacti are also dominant, especially in the dry chaco, and some take on the proportion of large trees, e.g. the Cardón *Stetsonia coryne*. Bird species diversity is high although many species range into adjacent Monte desert. Several species are exclusive, or almost exclusively restricted to the chaco in Argentina and these include Brushland Tinamou *Nothoprocta cinerascens*, Spot-backed Puffbird *Nystalus maculatus*, Red-billed Scythebill *Campylorhamphus trochilirostris*, Stripe-backed Antbird *Myrmorchilus strigilatus*, Chaco Warbling Finch *Microspingus* [*torquatus*] *pectoralis*, and Chaco Sparrow *Rhynchospiza strigiceps*. Endemic species to each of the chaco habitats are mentioned below.

The **dry chaco** of eastern Jujuy and eastern Salta across western Formosa and western Chaco, south through eastern Tucumán, much of Santiago del Estero, and eastern Catamarca and La Rioja to north-west Córdoba is the largest area of chaco and is dominated by the narrow-leaved Quebracho Colorado Santiagueño *Schinopsis lorentzii* reaching a height of 24 m, and the more widespread Quebracho Blanco *Aspidosperma quebracho-blanco*. Other characteristic trees include the Palo Santo *Bulnesia sarmientoi* with its distinctive pendulous fruits, Palo Cruz *Tabebuia nodosa*, Yuchán or Palo Borracho *Ceiba chodatii* with its distinctive spiky and swollen trunk, and the thorny Itín *Prosopis kuntzei*. The Quimil *Opuntia quimilo* and Ucle *Cereus forbesii* are two of the most dominant species of cactus. The understorey is dense and characterised by large terrestrial bromeliads of the genus *Bromelia*, collectively known as Chaguars. Two endemic bird species of this habitat are the Quebracho Crested Tinamou *Eudromia formosa* and Cinereous Tyrant *Knipolegus striaticeps*, while shrublands of Aliso de Río *Tessaria integrifolia* and *Baccharis* spp. harbour a chaco endemic breeder, Dinelli's Doradito *Pseudocolopteryx dinelliana*. The southern portion of the dry chaco holds two vast areas of salt flats known as salinas: the Salinas Grandes in north-west Córdoba and south-east Catamarca, and the Salinas de Ambargasta in south-west Santiago del Estero. Their periphery is dominated by a succulent shrub ecotone reaching 80 cm in height (*Heterostachys*, *Allenrolfea*, *Atriplex*, *Suaeda* spp. and *Prosopis reptans*) supporting the principal populations of the endemic Salinas Monjita *Neoxolmis salinarum*.

Hill ranges known as sierras in the provinces of Córdoba, San Luis, Catamarca, La Rioja and the Sierra de Guasayán in west Santiago del Estero are covered with **sierran chaco woodlands** between 700 and 1300 m, with bush cover up to 1700 m. Dominant species include the Horco-Quebracho *Schinopsis marginata*, Molle de Beber *Lithraea molleoides*, Cochucho *Zanthoxylum coco*, Maitén *Maytenus boaria* (also characteristic of the Patagonian forest) and areas of *Polylepis australis* at and above the tree-line. The shrub-zone is dominated by the Sierran Romerillo bush *Heterothalamus alienus*, other bushes such as *Eupatorium*, *Baccharis* and *Colletia* spp., abundant mistletoe-like semi-parasites (*Phoradendron*, *Tripodanthus* and *Ligaria* spp.), and some sectors are dominated by stunted Caranday Palms *Trithrinax campestris* showing a strong link with the humid chaco (see below). The avifauna is diverse and contains elements of Andean and dry chaco species. It supports the world's most important population of Black-bodied Woodpecker *Dryocopus schulzi*, possibly the most threatened woodpecker in South America.

The **humid chaco** spans central and eastern Formosa and Chaco provinces as well as north-east Santiago del Estero and northern Santa Fe provinces, and is dominated by the Quebracho Colorado Chaqueño *Schinopsis balansae*, Urunday *Astronium balansae*, Marmelero *Ruprechtia laxiflora* and Guayaibí *Patagonula americana*, in addition to many widespread dry and sierran chaco trees (listed above), bushes such as Vinal *Prosopis ruscifolia*, Garabato Negro *Acacia praecox* and Colquiyoyo *Maytenus vitis-idaea*, and an understorey of terrestrial bromeliads (*Bromelia, Dyckia, Aechmea* spp.). Humid chaco woodlands often take on the form of **forest islands in savanna** while oxbow lakes are usually dominated by heron colonies. The entire area is interspersed by a mosaic of lakes and open water bodies known as **esteros**, lined with *Schoenoplectus californicus* sedges and *Typha* spp. cattails, floating mats of water hyacinths *Eichhornia* spp., giant flowering lily pads *Victoria cruziana*, stands of the large-leaved Pehuajó *Thalia* spp. and dense stands of the papyrus-like Pirí sedge *Cyperus giganteus*, intersected with periodically inundated *Spartina* and *Elionurus* grasslands. Extensive areas of **palm savanna** are dominated by stands of Palma Blanca *Copernicia alba* with sectors of Pindó Palm *Syagrus romanzoffiana*, Yatay *Butia yatay*, the stunted Carandillla *Trithrinax schyzophylla*, and the thorny Mboyaca Palm *Acrocomia aculeata*. Bird species diversity is far greater in the humid chaco than in the dry chaco, and contains notable elements of the Mesopotamian grasslands and Paraná forest. Bird species that are exclusively found in Argentina in the humid chaco include the Undulated Tinamou *Crypturellus undulatus* and Peach-fronted Parakeet *Eupsittula aurea*.

Gallery forests, a forest type that is typical of rivers and streams in many biomes, are well distributed along the Bermejo and Pilcomayo rivers from eastern Salta across northern Formosa, and along the border of Formosa and Chaco provinces. The western sectors act as a corridor of dispersal of lower Yungas flora and birds with species such as Blue-crowned Trogon *Trogon curucui* and Black-banded Woodcreeper *Dendrocolaptes picumnus* reaching gallery forests as far as extreme north-east Formosa province. Eastern gallery forests are more homogenous with those found in the Mesopotamian savannas (described below) and allow movements of birds such as Green Ibis *Mesembrinibis cayennensis* from the Paraná rainforest. They also hold a number species which are local in Argentina, being mostly restricted to eastern Formosa and Chaco such as Grey-lined Hawk *Buteo nitidus*, Bare-faced Curassow *Crax fasciolata* and Flavescent Warbler *Myiothlypis flaveola*.

ESTEROS DEL IBERÁ AND CAMPOS

(**Habitats**: esteros, dry grasslands, Campos, palm savanna, gallery forest, Paraná Delta)

The **Esteros del Iberá** is a globally unique ecosystem which comprises a mosaic of higher-lying dry grasslands and low-lying humid climax grasslands with a vast network of subterranean water channels that feed numerous open lakes. Much of the low-lying areas comprises floating vegetation-covered water bodies known as esteros. The Esteros del Iberá system covers 20,000 km², harbours some 350 species of birds and boasts a biodiversity comparable only with the Llanos of Venezuela or the Pantanal of Brazil. Some of the most noteworthy species include Black-and-white Monjita *Heteroxolmis dominicanus*, Strange-tailed Tyrant *Alectrurus risora* (the global stronghold), Marsh Seedeater *Sporophila palustris*, Rufous-rumped Seedeater *Sporophila hypochroma*, Chestnut Seedeater *Sporophila cinnamomea*, Yellow Cardinal *Gubernatrix cristata* and Saffron-cowled Blackbird *Xanthopsar flavus*. Elevated areas in southern Misiones, northwest Corrientes and central Entre Ríos have rolling hill formations of **dry grasslands**. The principal grasses include *Paspalum notatum, Aristida jubata, Andropogon lateralis, Elionurus tripsacoides, Setaria* and *Stipa* spp., while inundated areas support giant stands of *Panicum prionitis* and *Cortaderia selloana*.

Rolling grasslands in the red soil region of north-east Corrientes and southern Misiones cover around 1.1 million hectares and are defined as **Campos**, covered principally by *Aristida pallens* and *Paspalum* *quadrifarium* grasses, over a basaltic bedrock with islands of Urunday (*Astronium balanse* and *Acosmium subelegans*) woodland. This area was formerly characterised by the dwarf palm Yatay Poní *Allagoptera leucocalyx*, although very few pockets now remain. The Campos support the main populations of Sickle-winged Nightjar *Eleothreptus anomalus*, Streamer-tailed Tyrant *Gubernetes yetapa*, Ochre-breasted Pipit *Anthus nattereri*, and Pearly-bellied Seedeater *Sporophila pileata*, and the ever-decreasing Saffron-cowled Blackbird *Xanthopsar flavus* and Black-masked Finch *Coryphaspiza melanotis*, as well as the only populations of Least Nighthawk *Chordeiles pusillus*. Cock-tailed Tyrant *Alectrurus tricolor* and hypothetical Collared Crescentchest *Melanopareia torquata* formerly occurred here but now appear to be extinct. Elsewhere, more characteristic **palm savanna** is represented by areas of Palma Blanca *Copernicia alba* in Corrientes, and the last remaining extensive Yatay Palm forest *Butia yatay* is protected in El Palmar National Park in central-eastern Entre Ríos. Grass communities in these savannas are dominated by *Andropogon lateralis*, *Paspalum notatum*, *Stipa* (*Nassella*) *megapotamica* and *Setaria parviflora*. Bird species diversity is very high and there are a large number of species which are threatened on a global scale. Although the entire Mesopotamian life-zone has traditionally been used for cattle-raising, much of the elevated areas are being converted to pine and *Eucalyptus* plantation causing a dangerous threat to many of the grassland species. The Mesopotamian life-zone is interspersed with large areas of Espinal woodland (see below), and bordered by **gallery forest** along the rivers Uruguay and Paraná, which also penetrates some internal river systems. This forest has strong affinities with the Paraná rainforest and acts as a corridor for the dispersal of birds and plants from that region. The forests comprise Curupí *Sapium haematospermum*, Timbó Blanco *Albizia inundata*, Seibo *Erythrina crista-galli*, Sauce Criollo *Salix humboldtiana*, Sangre de Drago *Croton urucurana*, Ubajay *Hexachlamys edulis*, Aliso del Río *Tessaria integrifolia*, Ibapoy *Ficus luschnathiana*, Anchicó Colorado *Parapiptadenia rigida*, Ingá *Inga uraguensis*, Ombú *Phytolacca dioica*, Bugre *Lonchocarpus nitidus* and stands of Picanilla Bamboo *Guadua paraguayana*.

PARANÁ FOREST

(**Habitats**: Paraná rainforest, riverine forest, bamboo stands, *Araucaria* forest)
This subtropical evergreen rainforest, also known as Interior Atlantic Forest, was formerly connected to the Coastal Atlantic forest, or *Mata Atlantica* of coastal Brazil, and is so-named because of its exclusive distribution in the Paraná watershed. In Argentina this forest is restricted to Misiones and extreme north-east Corrientes province. Elsewhere, it extends rather patchily in eastern Paraguay and in Paraná state, Brazil with virtually all of the Paraná forest having been cleared in the adjacent Brazilian states of Santa Catarina and Rio Grande do Sul. **Paraná rainforest** is characterised by more than 200 species of trees and over 2000 species of vascular plants. The tree canopy is generally 20–30 m above the ground with emergents reaching 40 m, and there is an abundance of lianas, epiphytes and tree orchids. Some of the most representative trees include the Cedro Misionero *Cedrella fissilis*, Lapacho Rosado *Tabebuia impetiginosa*, Peteribí *Cordia trichotoma*, Laurel Negro *Nectandra megapotamica*, Guatambú Blanco *Balfourodendron riedelianum*, Timbó or Pacará *Enterolobium contortisiliquum*, Caña Fístula *Peltophorum dubium*, Incienso *Myrocarpus frondosus*, Grapia *Apuleia leiocarpa* and the Palo Rosa *Aspidosperma polyneuron*, the tallest of the emergents reaching 42 metres. The principal fruiting trees include the Pindó Palm *Syagrus romanzoffiana* reaching 20 m, Ubajay *Hexachlamys edulis*, Cerella *Eugenia involucrata*, Ñangapirí *E. uniflora*, Aguay *Chrysophyllum gonocarpum* and Cocú *Allophylus edulis*. Some 400 species of birds inhabit this ecosystem, from the forest floor through all strata to the canopy. Species such as the Brazilian Merganser *Mergus octosetaceus* were formerly restricted to river corridors within the Paraná forest, and the associated **riverine Paraná forest** supports large avifaunal communities including the threatened Black-fronted Piping Guan *Pipile yacutinga*. The dominant tree species in this habitat are Laurel Blanco *Ocotea acutifolia*, Laurel del Río *Nectandra angustifolia*, Mata Ojo *Pouteria salicifolia* and Ambaí *Cecropia pachystachya*.

The understorey of Paraná forest may be open or dense, and in addition to sectors of giant tree ferns (*Alsophila* spp. and *Dicksonia sellowiana*), **bamboo stands** represent an important vegetation community. The five common bamboo species involved are the giant spiny Tacuaruzú *Guadua chacoensis* which reaches 30 metres in height and forms dense stands; the slightly smaller, and more curved, spiny Yatevó *Guadua trinii*; the medium-sized Tacuapí *Merostachys claussenii*; the low, slender Tacuarembó *Chusquea ramosissima*, and the local Pitinga *Chusquea tenella*. The twelve bird species which can be classed as strict bamboo specialists in Argentina are Purple-winged Ground Dove *Paraclaravis geoffroyi*, White-bearded Antshrike *Biatas nigropectus*, Bertoni's Antbird *Drymophila rubricollis*, Rufous-tailed Antthrush *Chamaeza ruficauda*, Spotted Bamboowren *Psilorhamphus guttatus*, Yellow Tyrannulet *Capsiempis flaveola*, Large-headead Flatbill *Ramphotrigon megacephala*, Brown-breasted Bamboo Tyrant *Hemitriccus obsoletus*, Temminck's Seedeater *Sporophila falcirostris*, Buffy-fronted Seedeater *Sporophila frontalis*, Unicolored Finch *Haplospiza unicolor* and Blackish-blue Seedeater *Amaurospiza moesta*, of which the ground dove and the *Sporophila* seedeaters depend largely on *Guadua* bamboo seeding which occurs in cycles of up to thirty years. Other species are frequently but not exclusively found in bamboo stands, including Ochre-collared Piculet *Picumnus temminckii* and Tufted Antshrike *Mackenziaena severa* among other birds.

In north-east Misiones two hill ranges, the Sierra de Santa Victoria and the Sierra de Misiones, reach 1000 m and are flanked by relictual stands of ***Araucaria*** forest where less than 1% of the original 210,000 hectares of this forest type survives today. Here, the Paraná Pine *Araucaria angustifolia*, which is otherwise known only in the states of Rio Grande do Sul and Santa Catarina, Brazil, forms a few pure forest areas and many isolated stands, but is usually mixed with Paraná rainforest, and some trees can reach heights of 40 m. Although numerous species of birds inhabit this forest type, Araucaria Tit-Spinetail *Leptasthenura setaria* and Vinaceous-breasted Amazon *Amazona vinacea* are the only exclusive inhabitants, while Azure Jay *Cyanocorax caeruleus* and a recently discovered small population of Black-capped Piprites *Piprites pileata* also have important links with this forest type.

ESPINAL

(**Habitats**: Espinal and Caldén woodlands)

The **Espinal** is a thorn woodland which also occupies parts of western Uruguay and the extreme south-western tip of Rio Grande do Sul State, Brazil where it is known as *Espinillo*. However, Espinal is virtually an endemic biome to Argentina and expands extensively across southern Corrientes and Entre Ríos, forming a mosaic of hot, humid woodlands over the Mesopotamian savanna, and continuing west through central Santa Fe. They reappear in southern Córdoba, south-east San Luis and cut a broad swathe through La Pampa province, reaching the coast in the extreme southern Buenos Aires province. The western and southern region is dry and xerophytic in comparison with that found in Mesopotamia. Espinal woodlands in Mesopotamia are sometimes referred to as Mesopotamian Parkland because they are reminiscent of parkland on the plains of East Africa and have an open understorey. The dominant species from east to west are the Ñandubay *Prosopis affinis*, Algarrobo Blanco *P. alba*, Algarrobo Negro *P. nigra*, Espinillo *Acacia caven*, Cina-cina *Parkinsonia aculeata*, Quebracho Blanco *Aspidosperma quebracho-blanco*, Caranday Palm *Trithrinax campestris*, Tembetarí *Zanthoxylum fagara*, Sombra del Toro *Jodina rhombifolia* with distinctive holly-shaped leaves, Coronillo *Scutia buxifolia*, Chañar *Geoffroea decorticans* and Molle *Schinus longifolius*. Several bird species are shared between the Espinal and chaco, including Lark-like Brushrunner *Coryphistera alaudina*, Short-billed Canastero *Asthenes baeri* and Little Thornbird *Phacellodomus sibilatrix*.

In the extreme south of the Espinal woodlands in south-east San Luis, La Pampa and southern Buenos Aires, **Caldén woodlands** *Prosopis caldenia* are intermixed with other typical Espinal trees and shrubs mentioned above, and with grassland communities of the genera *Trichloris*, *Stipa*, *Elionurus*, *Digitaria*, *Poa* and *Aristida*. Bird species diversity is fairly high and contains components of chaco woodlands and Monte desert. One extremely threatened Espinal species is the Yellow Cardinal *Gubernatrix cristata*.

PAMPAS

(**Habitats**: Pampas grassland, Pampas marshes, Tala woodlands, Paraná Delta)

The Pampas region, of which La Pampa province only forms a small part, was formerly a vast plain of flat low-lying grasslands covering an area of 430,000 km² in the provinces of Buenos Aires excluding the southern 'pan handle', southern Santa Fe, southern Córdoba, adjacent parts of San Luis and extreme north-west La Pampa. Today, natural Pampas grassland has been reduced to small disjunct areas, one of the largest pristine areas being in southern Buenos Aires province to the west of Bahía Blanca. The dramatic change is the result of major agricultural advances between 1890 and 1950 which have almost transformed the entire area to agriculture and cattle-raising. Climax **Pampas grasslands** includes Cebadilla Criolla *Bromus catharticus*, Cortadera or Pampas Grass *Cortaderia selloana*, Pasto Miel *Paspalum dilatatum*, Paja Colorada *P. quadrifarium*, Pelo de Chancho *Distichlis scoparia* and *D. spicata*, flechillas *Piptochaetium montevidense* and *Stipa* (*Nassella*) *neesiana*, and espartillos *Spartina densiflora* and *Elionurus muticus*. Coastal sand dunes are covered with *Poa lanuginosa* grass and *Adesmia incana*. Specialist grassland birds such as Pampas Meadowlark *Leistes defilippi* are now almost exclusively restricted to small areas near Bahía Blanca, and the Black-and-white Monjita *Heteroxolmis dominicanus* to a small area in central-eastern Buenos Aires. The enigmatic Pampas Pipit *Anthus chacoensis* is an endemic breeder in this habitat. It is noteworthy that several other species including Strange-tailed Tyrant *Alectrurus risora* and Saffron-cowled Blackbird *Xanthopsar flavus* are now extinct in the Pampas grasslands. Numerous freshwater and brackish lakes, and a mosaic of rivers, combine to form the **Pampas marshes** which still dominate certain sectors of the Pampas landscape. These support some of the most important populations of Dot-winged Crake *Laterallus spilopterus*, Red-and-white Crake *Laterallus leucopyrrhus*, Freckle-breasted Thornbird *Phacellodomus striaticollis*, Bay-capped Wren-Spinetail *Spartonoica maluroides*, Sulphur-bearded Reedhaunter *Limnoctites sulphuriferus*, Curve-billed Reedhaunter *Limnornis curvirostris*, Hudson's Canastero *Asthenes hudsoni* and Warbling Doradito *Pseudocolopteryx flaviventris*, which also extend locally into the Mesopotamian grasslands.

Two mountain ranges, Ventania and Tandilia, in southern Buenos Aires province are dominated by *Stipa* and *Piptochaetium* grasses. The higher of these two ranges, Sierra de la Ventana, supports isolated populations of Cordilleran Canastero *Asthenes modesta*, Black-billed Shrike-Tyrant *Agriornis montana* and Greater Yellow Finch *Sicalis auriventris*.

A narrow strip of **Tala woodlands** (*Celtis ehrenbergiana*) extends along the littoral region of the Río de la Plata estuary, from the Paraná Delta south-east to Punta Rasa at the mouth of this, the world's widest river, and these woodlands support a variety of typical Espinal bird species.

In the northern sector of the Pampas, floral composition changes as one descends into the deep depression of the **Paraná Delta**, often described as a separate ecoregion or life-zone in itself. This low-lying and periodically flooded region with limited access is dominated by marsh grasses including Paja Brava *Scirpus giganteus*, Espartillo *Spartina densiflora*, Paja Mansa *Paspalum quadrifarium*, Totora *Typha latifolia* and various sedges and rushes (*Schoenoplectus*, *Juncus* spp.). Sparse tree and bush cover includes the Seibo *Erythrina crista-galli*, Acacia Mansa *Sesbania punicea*, Carpinchera *Mimosa pellita*, Espinillo Manso *M. pilulifera*, Sarandí Blanco *Phyllanthus sellowianus*, Palo Amarillo *Terminalia australis* and Aliso del Río *Tessaria integrifolia*. Among the high bird species diversity, which also includes a variety of gallery forest, Pampas and Paraná Forest species, the threatened Marsh Seedeater *Sporophila palustris* appears with some regularity, along with several poorly known species such as the Dot-winged Crake *Laterallus spilopterus* and Straight-billed Reedhaunter *Limnoctites rectirostris*.

PATAGONIAN STEPPE

(**Habitats**: Patagonian grass-steppe, Fuegian grass-steppe, Patagonian marshes, Patagonian steppe-lakes, Patagonian brush-steppe, scree slopes, Fuegian bogs)

The Patagonian steppe is an arid desert with little rainfall and is the largest ecoregion in Argentina covering *c.*798,000 km², or approximately 70% of Patagonia which in itself is approximately the size of Colombia. Cold winds from the Pacific Ocean drop most of the rain on the Andean cordillera and otherwise cause rapid evaporation across the steppe zone. The landscape is barren and rocky, covered with grasslands and low thorn scrub, and dotted with seasonal lakes. Three types of steppe prevail. In Neuquén, a narrow strip of grasslands, juxtaposed to the Patagonian forests, extends along Andean slopes from 750 to 1400 m and expands south of 51°S over Santa Cruz province. This **Patagonian grass-steppe** is dominated by three grasses: Coirón Dulce *Festuca pallescens* on the highest slopes, and Coirón Amargo *Stipa* (*Jarava*) *speciosa* and *J. humilis* on lower slopes. Other characteristic flora includes Neneo *Mulinum spinosum*, *Acaena* spp., *Euphorbia portulacoides*, *Viola maculata*, etc. In the south, this habitat supports the poorly known near-endemic Patagonian Tinamou *Tinamotis ingoufi*. **Fuegian grass-steppe**, found on northern Isla Grande, is somewhat different, being dominated by *Festuca gracillima*, *Poa*, *Hordeum*, *Agrostis* and *Bromus* spp., where Short-billed Miner *Geositta antarctica* breeds. The Mata Negra *Chiliotrichum diffusum* shrubs dominating areas of rolling hills provide habitat for Austral Canastero *Asthenes anthoides*. The southernmost sector of Patagonian grass-steppe, together with the Fuegian grass-steppe, support the precarious continental migratory population of Ruddy-headed Goose *Chloephaga rubidiceps*, together with an endemic subspecies of White-bridled Finch *Melanodera melanodera princetoniana,* which occupies a similar range. Certain sectors from Chubut to Tierra del Fuego are also interspersed with **Patagonian marshes** dominated by *Schoenoplectus californicus* sedges. These provide habitat for numerous marshbirds, waterfowl and, mainly in Santa Cruz province, the poorly known Austral Rail *Rallus antarcticus* which was rediscovered in 1998. Sub-Andean **Patagonian steppe-lakes** are temporary fresh or brackish waterbodies which dominate western zones and spread across northern Tierra del Fuego. Often, the only vegetation is the submergent water milfoil known as Vinagrilla *Myriophyllum quitense*, and in Santa Cruz this provides the exclusive nesting material for the endemic breeding Hooded Grebe *Podiceps gallardoi*. The poorly known Magellanic Plover *Pluvianellus socialis* also breeds along these lake shores, before migrating north along the coast. Some of the larger steppe-lakes such as Laguna Blanca in Neuquén province support a huge number of waterbirds and provide wintering grounds for long-distance migratory waders.

A mix of grass-steppe and shrub cover beginning in northern Neuquén extends southward and expands across Patagonia from central Río Negro through much of Santa Cruz, but also appears on the Valdés Peninsula. This **Patagonian brush-steppe** is connected in the north to Patagonian grass-steppe and to Monte desert and occupies the driest region of Patagonia with only 100–150 mm annual rainfall. It is dominated by a variety of low bushes and dwarf trees including the rounded Quilembai bush *Chuquiraga avellanedae*, together with Colapiche *Nassauvia glomerulosa*, Molle Patagónico *Schinus johnstoni*, Malaspina *Retanilla patagonica*, Mata Negra *Junellia tridens* and Mataguanaco *Anarthrophyllum rigidum*, etc. Occasionally the landscape is interrupted by basaltic outcrops. Noteworthy birds include the endemic Patagonian Canastero *Pseudasthenes patagonica* and the near endemic Band-tailed Earthcreeper *Ochetorhynchus phoenicurus*. Above the tree-line, where Ñire *Nothofagus antarctica* becomes stunted ('krumholz'), **scree slopes** only support a sparse covering of moss and lichens and in spite of the lack of vegetation, this habitat supports two Patagonian endemics; White-bellied Seedsnipe *Attagis malouinus* and Yellow-bridled Finch *Melanodera xanthogramma*, both of which descend in winter to the lowlands and coastal regions.

In south-east Tierra del Fuego, Patagonian grass-steppe gives way to **Fuegian bogs** which are intermixed with stunted Patagonian forest. This tundra-like landscape is home to one of the world's least known waders, the Fuegian Snipe *Gallinago stricklandii,* although for unknown reasons it is chiefly found on higher ground on uninhabited islands.

PATAGONIAN FOREST

(**Habitats:** *Araucaria* forest, Valdivian forest, Coihue forest, steppe-forest, Magellanic forest)

The Patagonian forest is a subantarctic Andean temperate evergreen and deciduous forest, in a glacial landscape with numerous lakes and rivers. Tree composition varies with latitude and altitude and is mostly dominated by 'southern beech' trees of the genus *Nothofagus*. The main forest block extends from 39°S in south-west Neuquén to Santa Cruz and reappears on Isla Grande and Isla de los Estados in Tierra del Fuego. In Chile the forest extends almost 600 km further north; however, an extensive remnant forest of Roble Pellín *Nothofagus obliqua*, of Chilean affinity, survives in north-west Neuquén in an area of arid steppe grasslands, and holds Argentina's only population of Chestnut-throated Huet-huet *Pteroptochos castaneus*. In central-west Neuquén at 37°45'S, the first areas of 45 m-high Pehuén ('monkey-puzzle') *Araucaria araucana* forest dominate the skyline. These extend southwards in patches to 40°S, intermixing with Lenga *Nothofagus pumilio* around 38°S. The two most widespread trees in the entire Patagonian forest are the Lenga, ranging from sea-level to 1800 m, but stunted above 1400 m, and the small-leaved Ñire *N. antarctica*, ranging from sea-level to 1550 m and occurring mostly in humid areas. These species are sometimes mixed in the northern sector with *N. obliqua* and Raulí *N. nervosa*, reaching a height of 35 metres. The term **Valdivian forest** is usually applied to Patagonian forests in Chile, but typical Valdivian forest also straddles the Andean chain from south-west Neuquén to north-west Chubut, where annual rainfall is over 3000 mm. The three dominant trees are all evergreen and include the Coihue *Nothofagus dombeyi*, mainly 25–30 m and occasionally 45 m high; the formidable Alerce *Fitzroya cupressoides* reaching 30–70 m and living up to 3,600 years; Mañiú Macho *Podocarpus nubigenus*, Mañiú Hembra *Saxegothaea conspicua*, Fuique *Lomatia ferruginea*, Huahun *Laureliopsis philippiana* and Tineo *Weinmannia trichosperma*. To the east of the Valdivian Forest, **Coihue forest** occurs from lake level to 900 metres and is dominated by Coihue *N. dombeyi*, often mixed with Ñire and Lenga, Radal *Lomatia hirsuta*, and the pyramidal-shaped Cordilleran Cypress *Austrocedrus chilensis*. Lakes are usually bordered by Arrayán *Luma apiculata* with its distinctive cinnamon-coloured bark, and Patagua *Myrceugenia exsucca*. The understorey of Valdivian and Coihue Forest, and *N. obliqua* forest in the extreme north of this life-zone, is always dominated by stands of Coligüe bamboo *Chusquea culeou* which can reach a height of 4 metres and sometimes extends outside of the tree-cover along rivers and streams. Dominant understorey bushes include Barberry *Berberis* spp., Wild Currant *Ribes* spp. and Canelo *Drymis winteri*.

The lower slopes of the northern sector of forest from south-west Neuquén to Chubut also have **steppe-forest** on the lower slopes, which constitutes an ecotone with the Patagonian Steppe. Here, dominant trees include the Cordilleran Cypress, which often forms pure forests below 1000 m on drier slopes, Radal, Laura *Schinus patagonicus*, Retama *Diostea juncea*, Espino Negro *Colletia spinosissima* and Palo Piche *Fabiana imbricata*. Maitén *Maytenus boaria* and Chacay *Discaria* spp. occupy the lowest slopes along rivers.

In western Santa Cruz and Tierra del Fuego, including Isla de los Estados and Isla Navarino (Chile), the forest changes composition again and forms **Magellanic forest**. In addition to the omnipresent Ñire and Lenga, Magellanic Forest is characterised by the 25 m-high Guindo *Nothofagus betuloides*, the 20 m-high Ten *Pilgerodendron uviferum* (mostly in *Sphagnum* bogs), and Canelo *Drymis winteri* which takes on tree form and reaches 7 m. The understorey is dominated by *Blechnum*, *Asplenium* and *Lophosoria* ferns with a huge diversity of mosses and hepatic plants.

Bird species are remarkably widespread through the ecoregion. Patagonian forests are home to some 24 endemic bird species including five endemic genera (*Enicognathus*, *Pygarrhichas*, *Eugralla*, *Colorhamphus* and *Curaeus*), and species such as Rufous-tailed Hawk *Buteo ventralis*, Rufous-legged Owl *Strix rufipes*, Magellanic Woodpecker *Campephilus magellanicus*, Thorn-tailed Rayadito *Aphrastura spinicauda*, Des Murs's Wiretail *Sylviorthorhynchus desmursii*, Patagonian Forest Earthcreeper *Upucerthia saturatior*, Fire-eyed Diucon *Pyrope pyrope* and Patagonian Sierrra Finch *Phrygilus patagonicus*. Also noteworthy are the wide diversity of tapaculos (Rhinocryptidae), ranging from the small plain *Scytalopus* tapaculos to the giant huet-huets (*Pteroptochos* spp.), all with far-carrying vocalisations which reverberate through the forest. Endemic waterbirds include Spectacled Duck *Speculanus specularis* and an endemic subspecies of Great Grebe *Podiceps major navasi*.

ATLANTIC SHORE AND OCEANIC WATERS

(**Habitats**: mudflats, sandy beaches, saltmarsh, cliffs, pebble beaches, kelp forest, temperate inshore and offshore ocean, subantarctic inshore and offshore ocean)

The Atlantic coastline of Argentina covers more than 4500 km, from the Río de la Plata estuary to Tierra del Fuego, and comprises two main shore types. From Punta Lara to Bahia Blanca shores comprise a mix of tidal **mudflats** and **sandy beaches** with adjacent **saltmarshes,** usually dominated by *Spartina densiflora* and sand-dunes and where marine vegetation is dominated by green algae (*Ulva, Enteromorpha* and *Chaetomorpha* spp.) and red algae (*Porphyra, Condria* etc). Between Bahía Blanca and Bahía Anegada, flat **sandy islands** serve as the unique, and limited, breeding grounds for Olrog's Gull *Larus atlanticus*.

From the mouth of the Río Negro south to Río Gallegos, in southern Santa Cruz, the seaboard includes vertical **cliffs**, rocky islets, noteworthy capes and gulfs, usually with **pebble beaches** which continue into northern Tierra de Fuego, until they are interrupted by the extensive tidal **mudflats** of Bahia San Sebastián. The Patagonian shoreline, and in particular the Fuegian shoreline, varies from having a wide to extreme intertidal zone. Here, the marine vegetation is characterised by beds of giant **kelp forest** *Macrocystis pyrifera* and other seaweeds (*Lessonia, Durvillea* spp. etc). The endemic Chubut Steamer Duck *Tachyeres leucocephalus*, described new to science in 1974, is restricted to rocky shorelines with kelp beds.

To the south-east of Río Grande in Tierra del Fuego, coasts are flanked by Magellanic forest, mixed with tundra bogs, cliffs, sandy and pebble beaches and capes with important sea-lion and shag colonies. The Mitre Peninsula supports a small breeding population of Striated Caracara *Phalcoboenus australis* and Fuegian Cinclodes *Cinclodes* [*antarcticus*] *maculirostris*.

Some outstanding sites along the Atlantic shore are Punta Rasa (wintering and staging ground for migrant shorebirds); the Valdés Peninsula (breeding colonies of Imperial Shag *Phalacrocorax atriceps* and Rock Shag *Phalacrocorax magellanicus*, Kelp Gull *Larus dominicanus*, South American Tern *Sterna hirundinacea* and Magellanic Penguin *Spheniscus magellanicus*); Punta Tombo (largest Magellanic Penguin colony); Deseado Estuary (breeding colonies of Red-legged Cormorant *Phalacrocorax gaimardi*, Rock Shag, Brown Skua *Stercorarius antarcticus* and Dolphin Gull *Leucophaeus scoresbii*); Coig Estuary (wintering grounds of Hooded Grebe *Podiceps gallardoi*); San Sebastián Bay (main wintering grounds of Hudsonian Godwit *Limosa haemastica*, Red Knot *Calidris canutus*, White-rumped Sandpiper *Calidris fuscicollis* and Sanderling *Calidris alba*); the Beagle Channel (Imperial Shag, Magellanic Oystercatcher *Haematopus leucopodus*, Fuegian Steamer Duck *Tachyeres pteneres*, Magellanic Diving Petrel *Pelecanoides magellani*, small numbers of breeding Gentoo Penguin *Pygoscelis papua*); and Isla de los Estados (breeding colonies of Rockhopper Penguin *Eudyptes chrysocome*, Southern Giant Petrel *Macronectes giganteus* and Blackish Oystercatcher *Haematopus ater,* with recent proven breeding of King Penguin *Aptenodytes patagonicus*).

Three main feeding areas are known for seabirds in waters adjacent to the continent. The first lies a short distance off Mar del Plata (central-east Buenos Aires) where the surface water has a temperature range of 18–20°C being strongly influenced by the warm Brazilian Current, but underlain by cold subantarctic water of the Falkland Current with a temperature range of 4–14°C. The relatively high phosphate levels and high salinity compared to offshore waters enhance biological productivity. This area forms an important feeding zone for several warm water species such as Yellow-nosed Albatross *Thalassarche chlororhynchos* and Cory's Shearwater *Calonectris diomedea* among others, but holds its highest diversity of pelagic species in mid-winter with the regular movement of many subantarctic and Antarctic breeders into these temperate waters. To the south, another biologically rich area lies 200 km east of the Valdés Peninsula, Chubut and extends southwards to the Falklands with a sea temperature of 14°C and high levels of phosphate, nitrate and chlorophyll. This is the main feeding area for many seabirds which breed on the Falklands. In contrast, many breeding species from South Georgia tend to move to waters off South Africa. The third important area lies off the east coast of Tierra del Fuego where the coastal upwelling provides nutrient-rich waters. The other areas are rich in seabirds year-round, with notable absences during the breeding season which varies greatly between species.

HEATH AND TUSSOCK (SOUTH-WEST ATLANTIC ISLANDS)

(**Habitats**: subantarctic heath, tussock grass, kelp forest)

A variety of islands in the south of the region exhibit a floral composition distinct from the Patagonian steppe and Patagonian forest life-zones. This is perhaps most notable on the Falkland Islands, outer islands in the Isla de los Estados group, in the Hornean island complex and at the Diego Ramírez islands.

The Falkland Islands comprise two large islands and some 700 comparatively small islands, some of which are inhabited. The main islands show sinuated and gently sloping coasts in the north-west and steep cliffs in the south-east. The landscape is treeless and is mainly dominated by **subantarctic heath** comprising grasses, principally White Grass *Cortaderia pilosa*, and Fachine *Chiliotrichum diffusum* as well as brush-steppe of which the main component is Diddle-dee *Empetrum rubrum*, Mountain Berry *Gaultheria pumila* and 'Christmas Bush' *Baccharis magellanica*, with certain areas dominated by Balsam Bog *Bolax gummifera* and *Valeriana sedifolia*, more humid areas by *Blechnum* fern species, while on higher ground cushion plants such as *Azorella monantha* are notable. The Tussacbird *Cinclodes antarcticus* is a widespread endemic of the heath and shoreline, mostly on rat-free islands. By far the most outstanding microhabitat of the islands, and one upon which numerous species of birds are dependent, is represented by the giant stands of **tussock grass** *Poa flabellata* which reach a height of 3 metres and dominate some coastal regions, especially in the east and south. Grazing and burning of this grass has resulted in the destruction of 80% of this habitat, although new plantation programmes are in progress. This habitat forms an important refuge for the entire world population of the distinctive Cobb's Wren *Troglodytes cobbi*, as well as for Grass Wren *Cistothorus platensis falklandicus*. It also provides the breeding substrate for Grey-backed Storm Petrel *Garrodia nereis* and shelter for a variety of pinnipeds. The Falkland Steamer Duck *Tachyeres brachypterus* is another endemic species, thriving in all coastal regions. This and other marine waterfowl and several pinnipeds are dependent or semi-dependent upon **kelp forest** comprising the *Macrocrystis* and *Lessonia* Giant Kelp species. Of the remarkable number of endemic bird taxa, some twelve other endemic subspecies found in the archipelago are mostly representatives of Patagonian steppe and Patagonian forest birds. Up to 27,000 pairs of Ruddy-headed Goose *Chloephaga rubiceps*, are resident on the Falklands, in comparison with the small migratory mainland population. Among the breeding seabirds, the Falklands support around 75% of the world's breeding population of Black-browed Albatross *Thalassarche melanophris* including the world's largest single colony of 157,000 pairs on Steeple Jason, in addition to breeding colonies of Southern Giant Petrel *Macronectes giganteus*, Slender-billed Prion *Pachyptila belcheri*, Fairy Prion *P. turtur*, White-chinned Petrel *Procellaria aequinoctialis*, Great Shearwater *Ardenna gravis*, Sooty Shearwater *A. grisea*, Wilson's Storm Petrel *Oceanites oceanicus*, Grey-backed Storm Petrel *Garrodia nereis*, Common Diving Petrel *Pelecanoides urinatrix*, Rock Shag *Phalacrocorax magellanicus* and Imperial Shag *P. atriceps*. Five species of penguin also breed regularly which, in order of abundance, are Rockhopper *Eudyptes chrysocome*, Magellanic *Spheniscus magellanicus*, Gentoo *Pygoscelis papua*, King *Aptenodytes patagonicus* and Macaroni Penguins *Eudyptes chrysolophus*.

The Islas de los Estados (Staten Island) are separated by the 24 km-wide Le Maire Straits from the Península Mitre of Isla Grande, the largest Fuegian island. These islands extend 65 km from west to east and are indented by broad bays and deep, narrow fiords. The islands are covered with numerous glacial lakes, and Magellanic forest (described above). **Subantarctic heath** covers a large proportion of the islands with peat turf and coastal grasslands; it is also found on higher ground. **Tussock grass** (*Poa flabellata*) also covers certain coastal regions on the main island, but is abundant on the treeless off-islands in the north of the archipelago: Alferéz Goffre, Zeballos, Elizalde and Observatorio. Giant **kelp forest** *Macrocystis pyrifera* characterises the inshore. In many ways the flora is rather intermediate between that found on Isla Grande and the Falkland Islands, comprising luxuriant Magellanic forest but also subantarctic heath and tussock grass. The islands have a typical Fuegian avifauna, but due to the presence of tussock grass and pinniped hauling grounds they also support Striated Caracara and Fuegian Cinclodes *Cinclodes* [*antarcticus*] *maculirostris*. Rockhopper and Magellanic Penguins breed and one pair of King Penguins was found breeding in 2004, although the species was formerly abundant on the islands. Breeding seabirds include Southern Giant Petrel, Rock Shag and Imperial Shag while a number of other species presumably breed but there has been little investigation.

Similar subantartic heath and tussock grass is also found on the Hornean Islands complex south of Isla Navarino, and Snipe Island in the western Beagle Chanel also has some tussock grass. Fuegian Cinclodes and Striated Caracara have been found in all of these places where tussock grass is located close to pinniped or shag colonies. The Diego Ramírez islands lying 100 km south-west of Cape Horn in the Drake Passage are covered in *Poa flabellata* tussock grass, where surprisingly the Thorn-tailed Rayadito *Aphrastura spinicauda* is the commonest species. This small island group is the only breeding site for Grey-headed Albatross *Thalassarche chrysostoma* in South America.

BIBLIOGRAPHY AND FURTHER READING ON ECOREGIONS AND HABITATS OF ARGENTINA

Biloni, J.S. (1990) *Arboles autóctonos argentinos*. Tipográfica Editora Argentina, Buenos Aires.

Boelcke, O. (1981) Plantas Vasculares de la Argentina – nativas y exóticas. FECIC, Buenos Aires.

Burkhart, R., Bárbaro, N.O., Sánchez, R.O. & Gómez, D.A. (undated) *Eco-regiones de la Argentina*. Admin. Parques Nacionales, Sec. Recursos Nats y Des. Sust.

Cabrera, A.L. (1976) *Regiones fitogeográficas argentinas. Enciclopedia argentina de agricultura y jardinería*. Vol. 2 (parte 1). Editorial ACME, Buenos Aires.

Cabrera, A.L. & Willink, A. (1980) *Biogeografía de America Latina*. Monog. 13. Serie de Biología. Org. De Estados Americanos, Wash. D.C.

Cabrera, A.L. & Zardini, E.M. (1993) *Manual de la Flora de los alrededores de Buenos Aires*. Editorial ACME, Bs As.

Cozzo, D. (1990) Ubicación y riqueza de los bosques espontáneos de "pino" Paraná (*Araucaria angustifolia*) existentes en Argentina. *Rev. Forestal Arg.* 4: 46–54.

Erize, F. ed. (1997) *El nuevo libro del árbol*. Vols 1 & 2. El Ateneo, Buenos Aires.

Erize, F., Canevari, M., Canevari, P., Costa, G. & Rumboll, M. (1981) *Los Parques Nacionales de la Argentina y otras de sus áreas naturales*. INCAFO, Madrid.

Mandelli, E.F. & Oralando, A.M. (1966) La producción orgánica primaria y las características físico-quimicas de la Corriente de Malvinas. *Bol. Serv. Hidrog.* Naval 3: 185–196.

Olson, D. M., Dinerstein, E., Wikramanayake, E. D. *et al.* 2001. Terrestrial ecoregions of the world: a new map of life on Earth. Bioscience 51(11): 933–938.

Shirihai, H. (2007) *A Complete Guide to Antarctic Wildlife: The Birds and Marine Mammals of the Antarctic Continent and the Southern Ocean*. Second Edition. A & C Black, London.

Thomsen, H. (1962) Masas de agua características del océano Atlántico (parte sudoeste). *Serv. Hidrog. Naval*, H. 632.

TAXONOMY AND NOMENCLATURE

Taxonomy is in constant flux. Keeping up with scientific changes is challenging but necessary, as these changes place birders and biologists in the frontier of knowledge on the phylogenetic relationships and species limits in birds. With the wealth of new studies and new information, the taxonomy presented here, albeit refined in present-day terms, can only reflect what is currently known and is by no means the last word on the subject. The nomenclature and species level taxonomy adopted here mainly follow those of the **South American Classification Committee (SACC)** (Remsen *et al.* 2020). Specific and subspecific taxonomy has recently become a hotbed of debate and, as elsewhere in the world, there is some controversy over the validity of certain species found in the region. Changes involve the elevation of subspecies to species status (colloquially known as 'splitting'), relegation of species to subspecific rank ('lumping'), invalidation of subspecies, or re-allocation of species to different genera. Subspecific taxonomy is a complex area by virtue of the numbers of taxa involved, conceptual matters and the general lack of representative specimen series. In numerous cases, subspecies were described from very few specimens which were originally only compared to a very small series of conspecifics or rarely also with congeners. In cases in which enough information was available to us to be certain of a future species-level split, we have signalled this in the text by putting the current specific epithet in square parentheses. For example, the distinctive northern (*maculatus*) and southern (*solitarius*) subspecies of Streaked Flycatcher *Myiodynastes maculatus* are firm candidates for a split. In this case, the bird inhabiting Argentina is given as *Myiodynastes* [*maculatus*] *solitarius*. This means that a very likely future taxonomic change would result in the bird in our study area being known by the name *Myiodynastes solitarius*. In many instances, we or others are actively working on taxonomic papers that should soon be published. In this way, we have avoided creating further 'field-guide taxonomy', while providing readers with critical indications of the most likely future changes. Detailed comments on the basis for these and other taxonomic decisions appear in Volume 2 (but see brief **Taxonomic notes** in Appendix 6).

MIGRATION AND MOVEMENTS

The number of migration patterns and types of movements in the Argentine avifauna is overwhelmingly complex. Migration is a very important factor as the country is host to numerous species which depart north after breeding, others which displace north after breeding but stay within the country, long distance non-breeding **boreal migrants** from the Northern Hemisphere, and yet others which even wander southwards or northwards in the austral winter. These are the basic migratory patterns, but it is commonplace to find that within a single species there may be populations or subspecies which are resident, altitudinal migrants and partial austral migrants within Argentina (e.g., Patagonian Tyrant *Colorhamphus parvirostris*), and yet others with breeding populations and non-breeding boreal populations (e.g., Barn Swallow *Hirundo rustica*). In order to begin to understand migratory patterns in Argentina it is imperative that one is aware that the seasons are the reverse of those in the Northern Hemisphere. Therefore, any mention of 'summer' in the text refers to the **austral (or southern) summer** which, including spring, roughly spans the period from September to March, and corresponds to autumn and winter in the Northern Hemisphere. Note that many raptors, waders and several shearwaters, flycatchers, ducks, nighthawks and passerines that are often described as 'wintering' in Argentina, actually occur during the austral spring and summer and not therefore during the **austral (or southern) winter** that roughly spans April to August. The small number of birds which fail to make the return migration to boreal breeding grounds are here termed as 'over-summering'.

Resident species are sedentary and remain throughout the year in their home ranges, showing only insignificant movements at best. Some 680 species are here considered to be resident in Argentina and the south-west Atlantic. Virtually all of the resident species are landbirds and some coastal breeders, while the vast majority of tubenose seabirds found in Argentina are non-breeders with wide-ranging migratory patterns. A further six species appear to be extinct in Argentina: Brazilian Merganser *Mergus octosetaceus*, Eskimo Curlew *Numenius borealis* (a boreal migrant), Glaucous Macaw *Anodorhynchus glaucus*, Red-and-green Macaw *Ara chloropterus* (possibly extralimital in Argentina), Blue-winged Macaw *Primolius maracana* and Cock-tailed Tyrant *Alectrurus tricolor*. The current status of Black-masked Finch *Coryphaspiza melanotis* and Black-collared Swallow *Pygochelidon melanoleuca* requires elucidation, as both appear to have declined precipitously and lack documented records in the last 15–20 years.

Austral migrants behave in the reverse manner of boreal migrants in that they breed in the region covered by the book during the austral spring and summer, and then migrate northwards. Broadly defined, austral migrants do not migrate to the Northern Hemisphere, and instead 'winter' in warmer climates in northern South or even Central America, while many 'winter' a relatively short distance to the north of Argentina, in Bolivia, Paraguay and Brazil. Almost 25% of the species recorded in Argentina are austral migrants, making them a very important component of the avifauna. Austral migrants fall into three broad categories. Austral migrants *sensu stricto* are species that breed in Argentina and then leave the country. Partial austral migrants migrate northwards after breeding (i.e. chiefly in March–April), although their migratory distance is much shorter and they 'winter' wholly or partially within Argentina, at more northerly latitudes than their breeding range. Given the length of Argentina, some Patagonian partial austral migrants actually achieve rather long migrations. In spring (September–November) these species migrate back to more southerly latitudes in order to breed. Not all populations of partial austral migrants are migratory, and some populations are resident in certain parts of the country. In many cases different subspecies of the same species can be resident or migratory, adding to the complexity of migration patterns in the region.

It must be noted that the term 'partial migration' has also been used with a different meaning, referring to the differential migration of individuals of a single population, in which some individuals may migrate while others may stay. Some migratory species are nomadic austral migrants that exhibit complex and unpredictable seasonal patterns of migration. For example, the migration dates and seasonal occurrence of Carbonated Sierra Finch and Cinnamon Warbling Finch (both endemic breeders in Argentina) are very erratic, depending on patterns of rainfall and food abundance. However, their usual seasonal migration has some degree of nomadism, meaning that they could be abundant in one year at a certain locality, but absent the next year during the same dates.

Bamboo-seed specialists are nomadic and track specific food resources over large areas. These species may be present for several consecutive years in an area or region, without performing any seasonal movement, and then vacate the area for many years when the resources needed for their subsistence are not available. Altitudinal migrants include montane species that experience partial or complete minor-scale altitudinal movements, typically descending during the winter. Note that some species that are not regular altitudinal migrants may still displace to lower altitudes when facing extreme conditions (e.g. Scribble-tailed Canasteros *Asthenes maculicauda* may descend when their high-Andean habitats are snow-covered). Finally, a group of species appears to visit the Paraná forest of Argentina during the winter, moving westwards and descending in altitude after breeding in the Atlantic forests of southern Brazil. This system is known as the Southern Atlantic Forest longitudinal migratory system and includes species such as Golden-rumped Chlorophonia *Chlorophonia cyanocephala aureata*, Shear-tailed Grey Tyrant *Muscipipra vetula*, Swallow-tailed Cotinga *Phibalura flavirostris* and Black Jacobin *Florisuga fusca*. The bewildering complexity of South American migratory patterns is awaiting a continent-wide analysis to be properly described. Until then, the terminology used to describe migration patterns will only remain partially accurate, and is likely to experience a major overhaul in the next decade.

OVERVIEW OF THE ARGENTINE AND SOUTH-WEST ATLANTIC AVIFAUNA

Here, we treat some 1,085 naturally occurring species and an additional nine introduced species (up to July 2020) reported from Argentina and the south-west Atlantic region. An amazing 88% of species recorded in Argentina are believed to breed in the region. Among these breeders 72% are sedentary and 28% are austral migrants (see Migration and movements). Only 6% are non-breeding boreal migrants. A deeper analysis on the composition of the avifauna of Argentina and the south-west Atlantic will be found in Volume 2.

ENDEMIC SPECIES

Endemism is fairly high in Argentina and the south-west Atlantic: 25 species breed only in continental Argentina and three species are restricted to the Falkland Islands, while the ranges of an additional 17 species are primarily within continental Argentina. **Species endemic to continental Argentina (13)** and **endemic to the Falklands (3)** are breeding year-round residents restricted to these areas. **Endemic breeding species in continental Argentina (12)** breed exclusively in continental Argentina but regularly or occasionally reach other countries during the austral winter. **Near-endemic species in continental Argentina (17)** have approximately 90% or more of their ranges within the country.

Common English name	Scientific name	Other countries of occurrence
Endemic to continental Argentina (13)		
Chubut Steamer Duck	*Tachyeres leucocephalus*	
Moreno's Ground Dove	*Metriopelia morenoi*	
Sandy Gallito	*Teledromas fuscus*	
White-browed Tapaculo	*Scytalopus superciliaris*	
Córdoba Cinclodes	*Cinclodes comechingonus*	
White-throated Cacholote	*Pseudoseisura gutturalis*	
Steinbach's Canastero	*Pseudasthenes steinbachi*	
Patagonian Canastero	*Pseudasthenes patagonica*	
Salinas Monjita	*Neoxolmis salinarum*	
Yellow-striped Brushfinch	*Atlapetes citrinellus*	
Carbonated Sierra Finch	*Rhopospina carbonaria*	
Monte Yellow Finch	*Sicalis mendozae*	
Sierran Meadowlark	*Leistes [loyca] obscura*	
Endemic breeding species to continental Argentina (12)		
Hooded Grebe	*Podiceps gallardoi*	Chile [o]
Olrog's Gull	*Larus atlanticus*	Uruguay [r], Brazil [o]
Straneck's Tyrannulet	*Serpophaga griseicapilla*	Brazil [o], Paraguay [r], Uruguay [r]
Dinelli's Doradito	*Pseudocolopteryx dinellianus*	Paraguay [o]

Lesser Shrike-Tyrant	*Agriornis murina*	Bolivia [r], Paraguay [r], Uruguay [r]
Rusty-backed Monjita	*Neoxolmis rubetra*	Brazil [o], Uruguay [o]
Black-crowned Monjita	*Neoxolmis coronata*	Bolivia [r], Paraguay [r], Uruguay [r], Brazil [o]
Hudson's Black Tyrant	*Knipolegus hudsoni*	Bolivia [r], Brazil [o], Peru [o]
White-banded Mockingbird	*Mimus triurus*	Chile [o], Bolivia [r], Peru [o]
Pampas Pipit	*Anthus chacoensis*	Paraguay [o]
Chaco Sparrow	*Rhynchospiza strigiceps*	Bolivia [r], Paraguay [r]
Cinnamon Warbling Finch	*Poospiza ornata*	Uruguay [r]
Near-endemic species in continental Argentina (17)		
Elegant Crested Tinamou	*Eudromia elegans*	Chile
Patagonian Tinamou	*Tinamotis ingoufi*	Chile
Austral Rail	*Rallus antarcticus*	Chile
Dot-winged Crake	*Laterallus spilopterus*	Chile, Uruguay, Brazil
Burrowing Parrot	*Cyanoliseus patagonus*	Chile, Uruguay
Creamy-rumped Miner	*Geositta isabellina*	Chile
Band-tailed Earthcreeper	*Ochetorynchus phoenicurus*	Chile
Tufted Tit-Spinetail	*Leptasthenura platensis*	Paraguay, Uruguay, Brazil
Bay-capped Wren-Spinetail	*Spartonoica maluroides*	Uruguay, Brazil
Chocolate-vented Tyrant	*Neoxolmis rufiventris*	Chile, Uruguay, Brazil
Patagonian Mockingbird	*Mimus patagonicus*	Chile
Rufous-throated Dipper	*Cinclus schulzi*	Bolivia
Yungas Sparrow	*Rhynchospiza dabbenei*	Bolivia
Yellow Cardinal	*Gubernatrix cristata*	Uruguay, Brazil
Tucumán Mountain Finch	*Poospiza baeri*	Bolivia
Patagonian Yellow Finch	*Sicalis lebruni*	Chile
Pampas Meadowlark	*Leistes defilippii*	Uruguay, Brazil
Species endemic to the Falklands (3)		
Falkland Steamer Duck	*Tachyeres brachypterus*	
Tussacbird	*Cinclodes antarcticus*	
Cobb's Wren	*Troglodytes cobbi*	

o = occasional in winter
r = regular in winter

HOW TO USE THE PLATES AND FACING TEXTS

The plates were designed to include as many different plumages as possible, including sexually dimorphic plumages, subspecies, commonly seen juvenile, immature, subadult or polymorphic plumages which differ significantly to the extent that they either create specific interest or could cause confusion with other species. Birds are mostly illustrated to scale with one another, apart from the obvious decreased size of flying or more distant birds. When this was not possible, a change in scale within a plate is indicated by a line across the plate.

Each illustration is labelled on the plate and described in the facing text. Readers are urged not to make hasty identifications using the plates alone, but to refer to the text to confirm the identification. Note that there is nearly always some vital piece of information in the brief identification texts which makes for crucial reading. For polytypic species (i.e. species which have two or more subspecies), the trinomial scientific name on the plate identifies which subspecies is illustrated, even if only one subspecies occurs in our study area.

Birds are illustrated perched and in flight when it was deemed useful for field identification, e.g. ducks, geese, herons, storks, vultures, raptors, waders, gulls and terns, pigeons and doves, nightjars, large woodpeckers, a number of tyrant-flycatchers, swallows etc., while those families which are more likely to be seen only in flight, e.g. tubenoses, storm petrels, skuas and swifts etc. are only illustrated in flight.

Raptors are divided into open country species followed by forest species, and the species texts for raptors are divided into descriptions of perched birds followed by descriptions of flying birds.

What is illustrated

Our aim has always been to try to illustrate all the breeding species in the study area (see Area covered by the book) and all those which regularly occur as non-breeding visitors, as well as scarce and vagrant visitors. In total 1,070 species have been illustrated on the colour plates. The number of vagrants, splits and even invading species has been steadily increasing during the past two decades and the compilation of this book, but all have been illustrated apart from a very few recent and extreme vagrants.

In recent years we have been designated as the list-keepers for Argentina by the South American Classification Committee of the American Ornithological Society. The committee has a strict code for inclusion of species onto country lists and only accepts those with tangible archived evidence. In this book, however, we were aware of many additional unpublished species records. Within reason, these species are incorporated and illustrated here, but are termed as 'unproven' in Volume 2. There are also many historical records from the region which cannot now be proven; some seem possible while others very unlikely. We have illustrated some of these, choosing those which are most likely and which we define as being 'hypothetical' in Volume 2. A few confusion species which occur close to the borders of the region but which have not been certainly recorded in Argentina are also illustrated. Thus, female Frilled Coquette *Lophornis magnifica*, which is an obvious confusion species with female Festive Coquette *Lophornis chalybeus*, is illustrated.

A large variety of native flora is painted to scale with the birds, and is named at species or generic level in Appendix 5, with a key to their relative importance. This will allow users to recognise specific floral requirements of certain birds.

What is not illustrated

Four **cryptic species** which were all split in recent years, and which are virtually identical in plumage to other species, are not illustrated but are referred to under their look-alike sister species. These are Bolivian Woodpecker *Dryobates* [*lignarius*] *puncticeps*, Zimmer's Tapaculo *Scytalopus zimmeri*, Ticking Doradito *Pseudocolopteryx citreola* and Puna Pipit *Anthus* [*furcatus*] *brevirostris*. It was felt that in these cases illustrations would add nothing. However, we highlight the importance of biogeographical and vocal differences in these species (see sonograms in Appendix 4).

A few **extreme rare vagrants** (usually only a single occurrence in the region) are also not illustrated but are mentioned in the texts facing the plates, and are indexed. These are Northern Shoveler S*patula clypeata*, Amsterdam Albatross *Diomedea amsterdamensis*, Yellow-crowned Night Heron *Nyctanassa violacea*, Curlew Sandpiper *Calidris ferruginea*, Peruvian Booby *Sula variegata*, Belcher's Gull *Larus belcheri*, Lesser Black-backed Gull *Larus fuscus*, White-winged Tern *Chlidonias leucopterus*, Silver-beaked Tanager *Ramphocelus carbo*, Scarlet Tanager *Piranga olivacea* and Black-backed Tanager *Stilpnia peruviana*. Four other vagrants are illustrated only with line drawings in the text: Peruvian Pelican *Pelecanus thagus*, Sooty Tern *Onychoprion fuscatus*, Bridled Tern *Onychoprion anaethetus* and Black Noddy *Anous minutus*. We have steered a wide berth of several purported pseudo-published inclusions which lack any kind of field description, and have excluded a few others which had bizarre published descriptions and had been previously rejected from peer-reviewed journals.

We have not illustrated most of the nine **introduced species** which have viable populations, but these are described at the end of the plates in Appendix 1.

Finally, we have decided to remove Black Rail *Laterallus jamaicensis* from the Argentine list since we consider that all putative records of this species belong instead to Dot-winged Crake *Laterallus spilopterus*. Also, although piculets are currently in a state of taxonomic turmoil, we found that the traditional inclusion of Ocellated Piculet *Picumnus dorbignyanus* for Argentina appears to relate to White-wedged Piculet *P. albosquamatus*, based on plumage and altitude data from specimen evidence, and Ocellated Piculet is not included in the book.

Plate sequence

The sequence of plates loosely follows the taxonomic sequence of the family and species accounts of Volume 2 (which is based on the SACC taxonomic sequence), as far as this was possible, but mostly follows a more traditional family sequence in order to be more user-friendly in the field. In other cases, the sequence was compromised to allow the direct comparison of confusion species or to compare similar but not closely related species on the same plate.

Texts facing the plates

Some plate texts provide introductory summary notes of each genus or species group, detailing characters common to each species in the group. Individual species accounts include details of many subspecies, age-related plumages and colour morphs. These texts give a mean average total length and sometimes wingspan (see Measurements below). Species accounts begin with a few words on the principal range or habitat, followed by the following subheadings: **ID** (Identification) A brief summary of salient identification features for each plumage illustrated (and a few that are not illustrated). Confusion species are cross-referenced. **Voice** Descriptions and transcriptions of the most frequent vocalisations include comparisons to confusion species. **Tax note** Brief taxonomic notes on recent or likely future taxonomic changes have been included for selected cases (see Appendix 6). Emphasis is placed on taxa that occur on a regular basis, or that have a certain importance in our study area, rather than well-known taxonomic problems of wide-ranging species such as *Diomedea* and *Thalassarche* albatrosses. More extensive and detailed notes on taxonomy, including long-standing taxonomic problems, phylogenetic relationships and generic treatment of the avifauna are fully covered in Volume 2. A numbered taxonomic note (e.g., **Tax note 1**) at the end of a species account refers to the relevant entry in Appendix 6. **Alt** Some alternative names are given where relevant.

Argentine **Spanish names** are provided in square brackets at the end of each species entry.

English vernacular names

The English vernacular names of Neotropical birds have been subject to moderate debate. We have opted to maintain as many traditional names as possible, accepting some logical changes or the revalidation of a few older names; the names largely follow those used by the International Ornithological Committee (www.worldbirdnames.org/ioc-lists/) barring a few minor exceptions. Importantly, spelling and the usage of hyphens follows the IOC recommendations with all hyphenated compound names restricted to bird group names, thus massively reducing the usage of hyphens in compound names. Some alternative English names are listed at the end of a species account, but all the alternatives are mentioned in Volume 2 in the species footnotes. We have also chosen to use accents in English names when they relate to an Argentine place name that has an accent, not only because it is the correct and only legitimate spelling but because anyone with a minimum understanding of Spanish would be able to pronounce these names correctly rather than having to guess how to pronounce them. Several tentative novel names are included when none was available due to probable future splits (see Taxonomy), and these are just suggestions for interim usage.

Scientific names

The Scientific names of species are always written in italics, and are given as binomials next to the English vernacular name. For polytypic species, the subspecies illustrated is labelled on the plate. Illustrations depict representative subspecies, but when two or more distinctive subspecies are illustrated, their names are always indicated. Full descriptive, distributional and taxonomic texts are provided in Volume 2 for every subspecies.

Measurements

An average **total length** measurement, from the tip of the bill to the tip of the tail, is given in centimetres for each species, and for some sexes and some subspecies where these differ appreciably. When available measurements were not accurate, we indicate this by a *circa* abbreviation. The **wingspan** measurement is indicated by the abbreviation **WS**, but is only given when deemed useful, principally for seabirds and diurnal raptors. In general terms, total length and, in certain families, wingspan can give a good impression of a bird's overall size.

Altitude ranges

Given the complex topography of Argentina, **altitudinal ranges** are key to understanding bird distribution in the country. Some species range only between or above certain altitudes in the Andes and Central Sierras. Upper and lower altitudinal limits are given in metres (abbreviated to m) for species which occur above 600 m above sea level. Therefore, the term 'lowlands' implies a range below and up to 600 m. Only the upper altitudinal limit is mentioned for species which range in both the lowlands and the Andes or Central Sierras. All the ranges given apply only to species' ranges within Argentina and not to altitudinal ranges in other countries which are often different and usually higher (many species of north-west Argentina tend to occur at lower altitudes than they do in Bolivia or Peru). These data were taken from the authors' own data and other observers' records, and are supplemented by museum data, and published and reliable unpublished sight records from a known locality. Lower limits are also given for altitudinal migrants which descend in winter using the same methodology.

Sonograms

Sonograms are graphic representations of sound in which frequency (pitch) is represented on the vertical axis, time on the horizontal axis and 'loudness' (amplitude) is represented by the intensity (darkness) of the graph. A loud, high-pitched, pure whistle will appear as a dark line and occupy the upper region of the graph, while a soft harsh sound will be paler, and occupy a wider area in the lower region of the sonogram. Visualising sounds helps one learn to listen to them, and the experience of listening and studying sonograms can increase awareness of how birds sound and their identification. We have provided sonograms of 23 species that represent particularly complex identification challenges (Appendix 4). Examples include several similar tyrannulets, all pipits, and cryptic species which cannot be otherwise safely identified in the field except by vocalisations (Zimmer's Tapaculo, Ticking Doradito and Puna Pipit, none of which are illustrated). Sonograms have been grouped to include the most likely confusion species to simplify comparisons. Descriptions and details on what to look for and listen to can be found in the corresponding texts facing the plates.

Distribution maps

Information used for the distribution maps came from multiple sources, including museum specimens, sound recordings, publications, a selection of trustworthy records from citizen science initiatives, and our own unpublished and third-party records. Depicting seasonal and dynamic patterns in a printed map is challenging. Nonetheless, we have tried to convey as much information as possible in the maps to help understand movements and the chances of seeing a bird at a certain time and place. The maps are working hypotheses that need to be tested in the field and refined with information that is currently missing. Identification of birds based solely on distribution should be avoided.

The colours and symbols indicate to the best of our knowledge what the majority of birds in a given area are expected to do, but the nature of bird movements is diverse and difficult to categorise. Individuals of some migratory breeders (e.g., hummingbirds and tyrant-flycatchers; mapped in red) frequently overwinter, yet the seasonal change in abundance is so drastic that it is clear that most members of a population regularly leave their breeding grounds every year. Some partial migrants experience good years (in which many birds stay) and bad years (in which many birds leave); depending on the general overall trends that we were able to detect, some such species have been mapped in green and others in red. Also, in many partial austral migrants, birds can depart from their breeding range only to be replaced by individuals of the same species coming from other breeding areas, creating an effect of residency (sometimes shown in the maps as green areas separating red and blue ones). In some cases, partial migration can be easily recognised in the maps because birds appear only in winter in some areas while they are mapped as resident elsewhere (green and blue maps). Altitudinal migration is seldom complete, and is typically represented by green on the montane breeding grounds and blue in adjacent low-lying areas.

Ten basic templates were designed, based on general distribution patterns to allow detailed mapping as far as possible. The bulk of the maps belong to these templates, although occasional departures from this plan occur to show specific records.

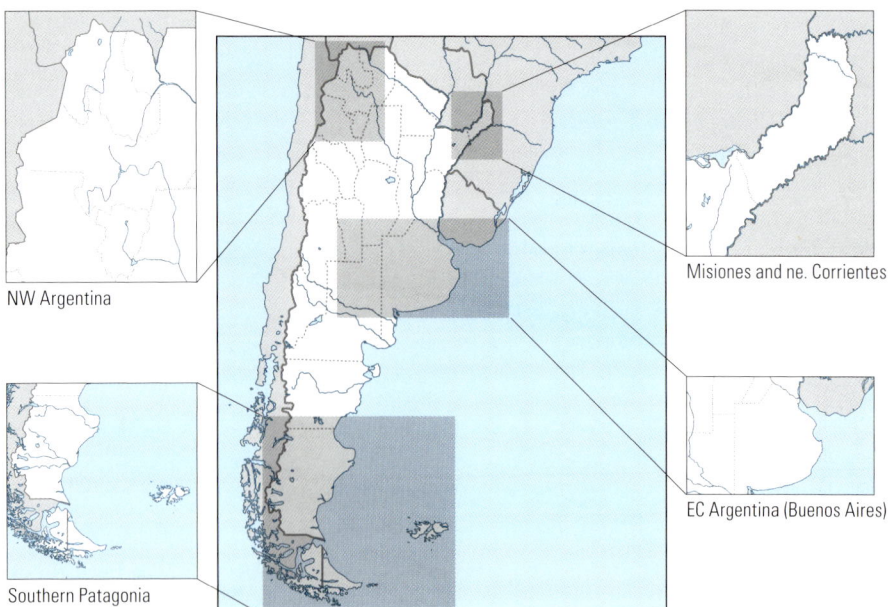

Map template used for the distribution maps, either whole for wide-ranging species or cropped into smaller discrete regions (e.g. Misiones and ne. Corrientes or NW Argentina) to show ranges at a greater scale.

Key to map colours and symbols

Year-round resident. Birds that breed and stay year-round in the same area.

Spring–summer resident. Birds that breed and then wholly or mostly vacate the breeding grounds. Note that a small proportion may overwinter.

Seabirds close to their breeding range during spring–summer.

Winter visitor.

Scarce winter visitor (seabirds only).

Seasonal non-breeding visitor. Used mainly for spring–summer boreal migrant landbirds, and for seabirds breeding elsewhere and visiting the South-west Atlantic mostly during spring–summer.

Scarce seasonal non-breeding visitor (seabirds only).

Passage migrant. Birds that are found only while on passage. Also used for **nomadic breeders** which may be present/absent for many years without any seasonal pattern (mostly bamboo seed-eating birds). It is also used for Saffron-cowled Blackbird whose colonies are seldom in the same places, thereby suggesting some degree of inter-seasonal movements.

Year-round non-breeding range at sea or along coasts.

Sparse occurrence. Frequently used for vagrants that have occurred repeatedly and which are likely to occur again. Also used for some birds that are in expansion and, in a few cases, for areas for which information did not allow us to decide whether a bird breeds, overwinters or is merely on passage. Overall, pink indicates situations of low-density, low chances of seeing the bird and uncertainty on status.

—— Bold lines indicate the former area occupied by birds that have experienced drastic range retractions (e.g., Yellow Cardinal), or that are currently considered extinct in the study area (e.g., Glaucous Macaw).

✕ Accidental, unique or extremely sparse records.

○ Confirmed specimen record when other evidence is lacking, or a species has a restricted range or is rare in the study area.

? Uncertain or inconclusive records.

⟵ Small arrows are used to point to small distribution areas and seabird breeding colonies. Longer or ⟷ double-ended arrows indicate migration routes.

Abbreviations and conventions

♂/♂♂: male/males
♀/♀♀: female/females
ad: adult
imm: immature
juv: juvenile
br: breeding
non-br: non-breeding
1st-yr: first-year etc
sp: species, used to denote an unidentified species, e.g. *Spinus* sp.
spp: species (plural)

ssp: subspecies (singular and plural)
WS: wingspan
c.: *circa*
n., s., e., w.: compass directions used for provinces
N, S, E, W, C: compass directions used for general regions of Argentina (C = Central)
Chaco: relates to the province of Chaco.
chaco: relates to the chaco ecoregion.
C Sierras: refers to the sierras of Cordoba and San Luis in central Argentina.

BIRD TOPOGRAPHY

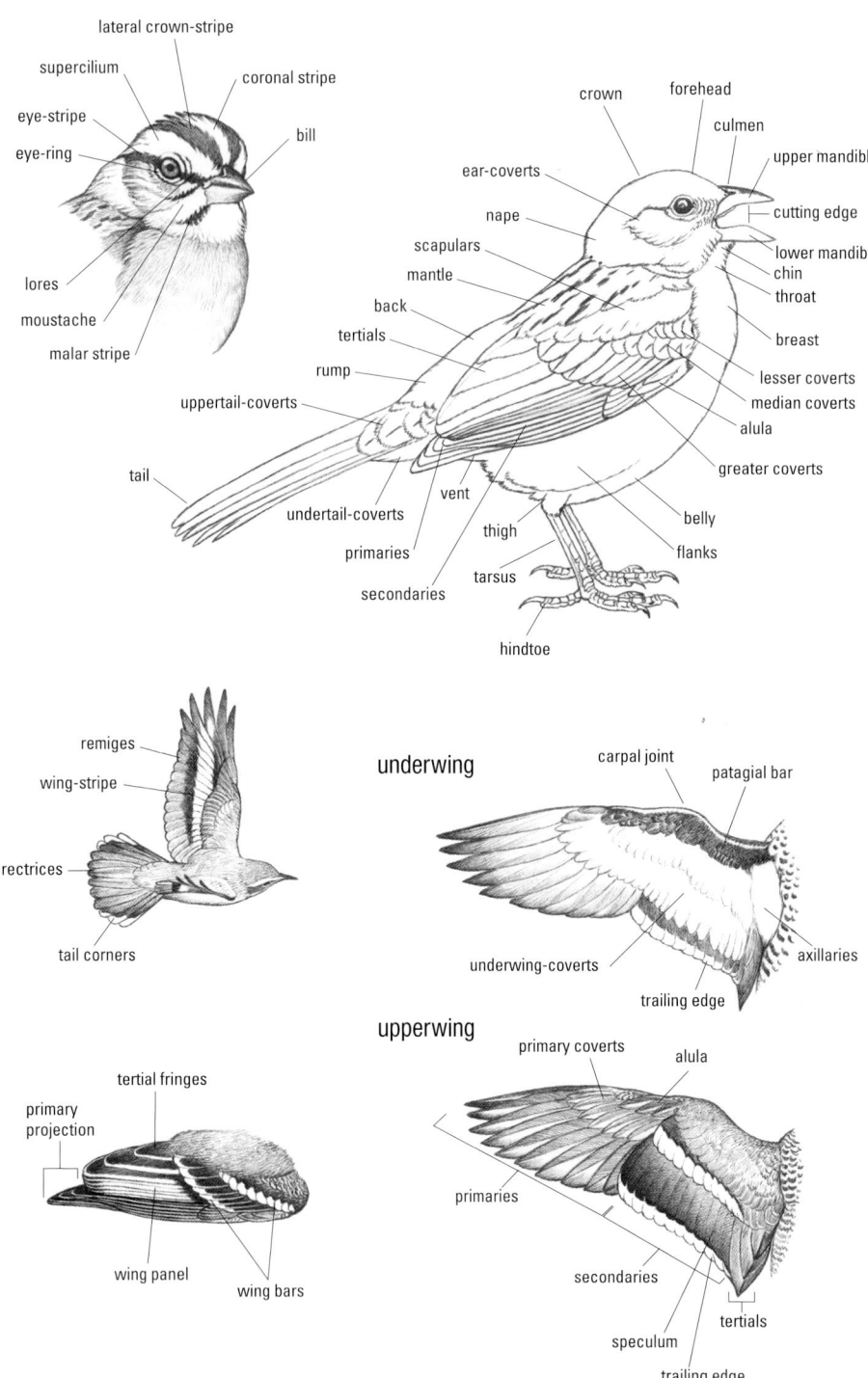

lateral crown-stripe
supercilium
coronal stripe
eye-stripe
eye-ring
bill
lores
moustache
malar stripe

crown
forehead
culmen
upper mandible
ear-coverts
cutting edge
nape
lower mandible
scapulars
chin
mantle
throat
back
breast
tertials
lesser coverts
rump
median coverts
uppertail-coverts
alula
greater coverts
tail
vent
belly
undertail-coverts
flanks
primaries
thigh
tarsus
secondaries
hindtoe

remiges
wing-stripe

underwing
carpal joint
patagial bar

rectrices
tail corners

underwing-coverts
axillaries
trailing edge

upperwing
tertial fringes
primary coverts
alula
primary
projection
primaries
wing panel
secondaries
wing bars
speculum
tertials
trailing edge

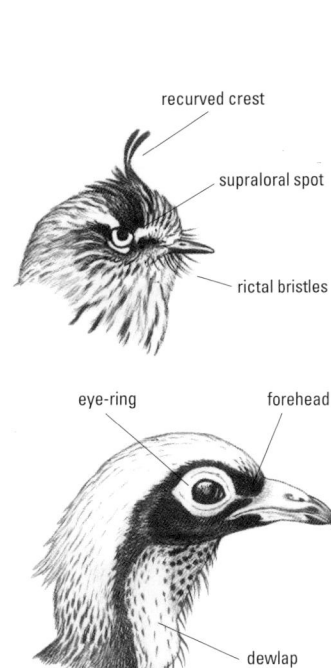

recurved crest

supraloral spot

rictal bristles

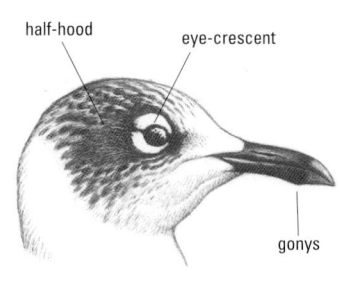

half-hood

eye-crescent

gonys

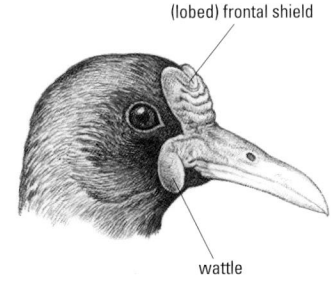

eye-ring

forehead

dewlap

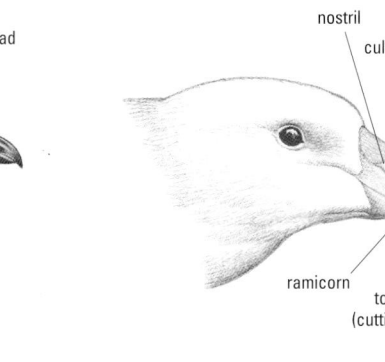

nostril

culminicorn

latericorn

maxilliary unguis (nail)

ramicorn

tomium (cutting edge)

mandibular unguis

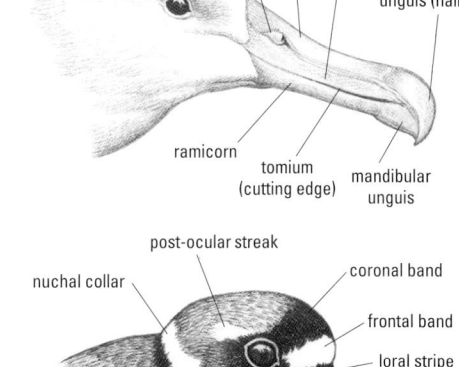

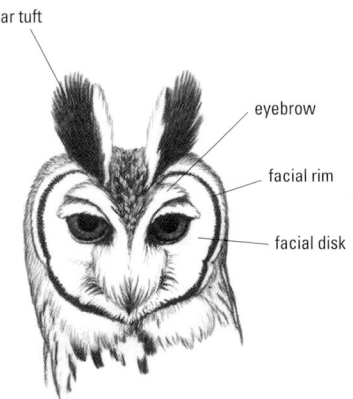

(lobed) frontal shield

wattle

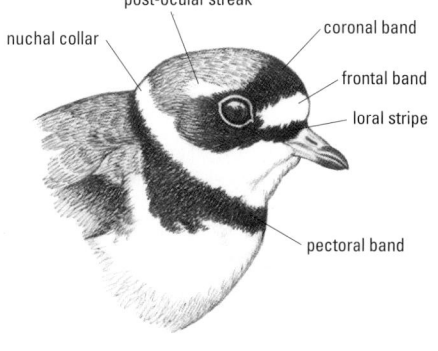

post-ocular streak

nuchal collar

coronal band

frontal band

loral stripe

pectoral band

ear tuft

eyebrow

facial rim

facial disk

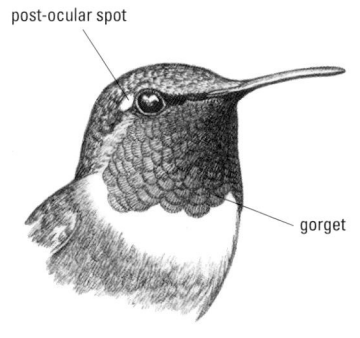

post-ocular spot

gorget

Rhea Unmistakable long-necked and long-legged flightless birds. Males larger, especially in Greater Rhea. Lesser Rhea was formerly placed in the genus *Pterocnemia*.

Lesser Rhea *Rhea pennata*   90 cm

Puna, NW Andean and Patagonian steppe. **ID Adult** *garleppi* (Jujuy to Mendoza): Breeding plumage is rather brown, thickly spotted with white, especially on the shoulder. **Adult** *pennata* (Patagonia north to w. Mendoza): Considerably greyer than *garleppi* with smaller spots. Non-breeding birds of both ssp. lack or show reduced spotting. Chicks have a complex black, brown and white-striped pattern. [Suri, Choique]

Greater Rhea *Rhea americana*   140 cm

Widespread in N and C lowlands. **ID Adult** ♂: Black crown, base of neck and sides of breast with unmarked grey back. **Adult** ♀: Smaller than ♂ with black restricted to sides of neck base. **Juvenile** Smaller still, with brown plumage and base of neck. Chicks are pale brown and white with striped backs. **Voice** Loud, extremely low-pitched, booming calls are ventriloquial. [Ñandú]

Cariama and *Chunga* Extremely large, terrestrial and arboreal, open-country carnivorous predators with long neck and tarsus. Rarely fly except when pressed; both species exhibit thickly barred black-and-white raptor-like wings.

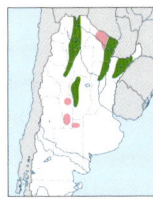

Red-legged Seriema *Cariama cristata*   92 cm

N Mesopotamia, humid and sierran chaco. **ID Adult**: Unmistakable with well-developed frontal crest, red bill and tarsus. Overlaps locally with Black-legged Seriema. **Voice** Loud and slow duetting calls are more bubbly and descend in pitch rapidly, unlike Black-legged Seriema. [Chuña Patas Rojas]

Black-legged Seriema *Chunga burmeisteri*   81 cm

Dry chaco and monte desert. **ID** Similar to Red-legged Seriema in shape and habits but notably smaller and lacks the frontal crest. **Adult**: Blackish bill and dark grey tarsus. Note conspicuous white supercilium. Overlaps locally with Red-legged Seriema. **Voice** Duets are similar to Red-legged and difficult to differentiate, but notes are harsher, less musical and more evenly pitched. Short song is more cackling and kookaburra-like. [Chuña Patas Negras]

Lesser Rhea

pennata

garleppi

ad br

ad br

chicks

Red-legged Seriema

ad

ad

Black-legged Seriema

Greater Rhea

chicks

ad ♂

albescens

ad ♀

juv

Crypturellus Small to large tinamous of forest and light woodland. All species are best located by their distinctive voices.

Small-billed Tinamou *Crypturellus parvirostris* 23 cm

Forest, scrub and plantations in Misiones; rare in Yungas foothills. **ID** Overlaps with somewhat larger Tataupa Tinamou. **Adult:** Differs by fairly bright reddish-pink tarsus, duller upperparts and greyish-olive or olive breast. **Voice** A few accelerating trills followed by longer descending trills, sometimes in an extended series. [Tataupá Chico]

Tataupa Tinamou *Crypturellus tataupa* 25.5 cm

Widespread in N woodlands and foothill forest. **ID Adult:** Plumbeous-grey crown and neck, becoming blue-grey on breast. Rich purplish-rufous upperparts. Bill longer and brighter than Small-billed Tinamou, being plastic orange-red, but tarsus much duller pinkish, brown or grey. Compare with Brown Tinamou in Paraná forest and Undulated Tinamou in humid chaco woodlands. **Voice** Most frequent song comprises 4–6 short, explosive, stuttered, mostly descending trills; more rarely, a long series of slowly descending trills that accelerates towards the end. [Tataupá Montaraz]

Undulated Tinamou *Crypturellus undulatus* 33 cm

Very local in humid chaco woodlands of e. Chaco and e. Formosa. **ID Adult:** Rather uniform brown above but finely vermiculated black at close range. Contrasting whitish throat and rufescent lower foreneck and upper breast. Creamy belly and cinnamon wash over brown-barred flanks. Creamy tarsus. Overlaps with Tataupa Tinamou. **Voice** Song comprises 4 deep slurred whistles, inflected upwards at end. [Tataupá Listado]

Brown Tinamou *Crypturellus obsoletus* 29.5 cm

Paraná forest. **ID Adult:** Grey head and neck with paler throat. Rich russet-brown mantle. Vinaceous breast, becoming paler on belly and flanks which show blackish crescents. Pale olive tarsus. Overlaps with Solitary, Small-billed and Tataupa Tinamou. **Voice** Song is a long series of frenetic accelerating trills. [Tataupá Rojizo]

Rhynchotus Large long-necked grassland tinamous with short hind toe and long decurved bill used for digging. Striking rufous primaries visible in flight. Both species are best located by their distinctive voices.

Huayco Tinamou *Rhynchotus maculicollis* 41 cm

Grasslands in the NW Andes. **ID Adult:** Resembles Red-winged (no definite overlap) but more spotted, rather than barred, on upperparts and black-streaked foreneck. Rufescent breast helps to distinguish from Ornate Tinamou. **Voice** Two-note, onomatopoeic ascending-descending whistle. [Guaipo]

Red-winged Tinamou *Rhynchotus rufescens* 41 cm

Grasslands in the NE and C lowlands. **ID Adult:** Upperparts barred brown and cream. Grey breast. Birds from campos region lack grey on the breast and have richer, more ochraceous neck. **Voice** Four-note descending whistled call, with longer pause between first and second notes. [Colorada]

Small-billed Tinamou

ad

Tataupa Tinamou

ad

tataupa

Undulated Tinamou

ad

undulatus

Brown Tinamou

ad

obsoletus

Huayco Tinamou

ad

Red-winged Tinamou

ad

pallescens

Nothura Small tinamous of open grassland, semi-desert and agricultural land. Best located by their confusingly similar voices.

Darwin's Nothura *Nothura darwinii*   26.5 cm

Grasslands and crops in Andean foothills and W lowlands south to N Patagonia. **ID Adult:** Generally paler than more richly coloured Spotted Nothura (some areas of overlap) with shorter tarsus, shorter neck, more linear breast streaking, separated by whitish intermediate streaks, and fine white dorsal streaking. Compare with Brushland Tinamou. **Voice** Trill (10–11 notes/sec) is much slower than Spotted Nothura. Slow whistled song often continues into a trill and finishes in a short series of slower notes. Both species can give very similar songs without the distinctive ending. [Inambú Pálido]

Spotted Nothura *Nothura maculosa*   26.5 cm

Widespread in N and C lowlands. **ID Adult:** Highly variable small grassland tinamou with a spotted breast. Generally more richly coloured than Darwin's Nothura and with a longer tarsus and neck. **Voice** Trill (16–17 notes/sec) is much faster than Darwin's Nothura. Slow whistled song finishes with short series of faster, distinctive, mildly accelerating notes. Both species can give very similar songs without the distinctive ending. [Inambú Campestre]

Nothoprocta Medium-sized and large tinamous of grasslands, light woodlands and scrub. All species are best located by their distinctive voices.

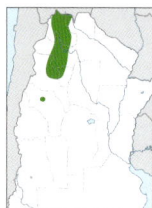

Ornate Tinamou *Nothoprocta ornata*   36 cm

Pre-puna and puna grasslands. **ID Adult:** Erectile blackish crown. Broad, often puffed out, cinnamon flanks are diagnostic. Limited overlap with rufous-winged Huayco Tinamou, but mainly overlaps with smaller, grey-breasted Andean Tinamou and very different, larger Puna Tinamou. **Voice** Loud ascending drawn-out whistle, sometimes followed by a soft low-pitched grunt, is flatter than Andean Tinamou. [Inambú Serrano]

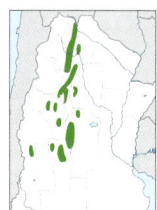

Andean Tinamou *Nothoprocta pentlandii*   29 cm

Scrub and light woodlands in the NW Andes and C Sierras. **ID Adult** *doeringi* (C Sierras): Decurved pinkish bill, yellowish tarsus, bushy crown and dark grey face and underparts, finely spotted white. **Adult** *pentlandii* (NW Andes): Paler grey foreneck and breast. Browner back and wings. **Voice** Short sudden loud whistle ascends faster than Ornate Tinamou. Compare with Ornate and Huayco Tinamous. [Inambú Silbón]

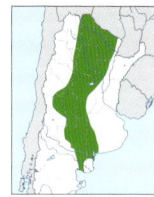

Brushland Tinamou *Nothoprocta cinerascens*   34 cm

Monte desert and farmland in dry chaco. **ID Adult:** Bushy erectile black crown. Greyish upperparts heavily streaked white, and breast spotted white. Overlaps broadly with Darwin's Nothura, and with Andean Tinamou in the sierran foothills. **Voice** Loud series of 6–10 clear whistles, rising slightly in pitch and often becoming disyllabic towards the end. [Inambú Montaraz]

Taoniscus Diminutive savanna tinamou with distinctive voice.

Dwarf Tinamou *Taoniscus nanus*   16 cm

Historical records from the Río Bermejo region of Formosa or Chaco. **ID Adult:** Rather prominent black checkered pattern on the mantle and wing-coverts with contrasting white throat. Breast and flanks barred with black. **Voice** Insect-like voice, consisting of a long series of short, evenly spaced high-pitched whistles. [Inambú Enano]

Darwin's Nothura

ad

darwinii

Spotted Nothura

ad

annectens

Ornate Tinamou

ad

rostrata

Andean Tinamou

pentlandii

ad

ad

doeringi

Brushland Tinamou

ad

Dwarf Tinamou

ad

cinerascens

ad

Eudromia Large tinamous with distinctive recurved crests and double white head and neck stripes. No hind toe. No other crested tinamou occur in the range of each species. Often seen in trios during breeding season, otherwise in flocks.

Elegant Crested Tinamou *Eudromia elegans*   41 cm

Widespread with 8 ssp. in the lowlands west of the Río Paraná; mainly in Andean scrub and Patagonian steppe. **ID** Adult *elegans* (C Argentina): Dark brown upperparts, finely spangled with small buff spots. Extensive flank barring. Adult *intermedia* (NW Andes to 2850 m): From *elegans* by paler brown upperparts with larger buff spotting, and clearer belly with diffuse flank markings. **Voice** Low-pitched whistle *WE-see-u* with final note slightly lower-pitched. [Martineta Copetona]

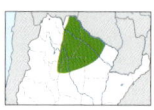

Quebracho Crested Tinamou *Eudromia formosa*   41 cm

Dry chaco. **ID** Adult: From Elegant Crested (no definite overlap) by presence of black ladder-barring on upperparts, with pairs of buff spots, and thick black chevrons on breast and flanks. **Voice** Two or three whistled ascending liquid notes with last one often lower-pitched. [Martineta Chaqueña]

Tinamotis Large tinamous of desert scrub and grasslands, with short tarsus, no hind toe and broad toe pads. Most often seen in trios. Both species show diagnostic triple white stripes on the head and neck, but lack crests. Both species are best located by their distinctive voices.

Puna Tinamou *Tinamotis pentlandii*   42 cm

Puna grasslands. **ID** Adult: Rufous belly and vent. Overlaps only with Ornate Tinamou. **Voice** Medium-pitched whistled *TOO-WHA* repeated incessantly and usually given in chorus. [Quiula Puneña]

Patagonian Tinamou *Tinamotis ingoufi*   37 cm

Patagonian steppe; mainly Santa Cruz. **ID** Adult: Rufous primaries and cinnamon belly and vent. Overlaps only with Elegant Crested Tinamou. **Voice** Medium-pitched, melancholic and descending whistled *PEE-WOOO* often given in chorus. [Quiula Patagónica]

Tinamus Very large forest tinamou with proportionately very small head. This genus roosts in trees.

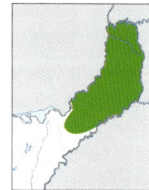

Solitary Tinamou *Tinamus solitarius*   46 cm

Paraná forest. **ID** Adult: Rusty neck stripe, which can appear creamy in poor light. Brown upperparts with fine wavy black barring. Whitish throat and brownish-grey breast. **Voice** Low-pitched tremulous crepuscular whistle resonates in the forest. [Macuco]

Spot-winged Wood Quail *Odontophorus capueira*   28 cm

Paraná forest. **ID** Gregarious forest quail with loud crepuscular voice. Adult: Pink skin around eye, erectile brown crest, rufous forehead and tawny supercilium. Wings spangled with white spots. Slate-grey underparts. **Voice** Loud crepuscular voice is a monotonous series of rich disyllabic *OO-Rúk* calls with the second note higher-pitched. [Urú]

Elegant Crested Tinamou

intermedia

ad

elegans

ad

Quebracho
Crested Tinamou

ad

Puna Tinamou

ad

Patagonian Tinamou

ad

ad

Solitary Tinamou

ad

capueira

Spot-winged Wood Quail

Penelope Large brown, monogamous, forest-dwelling cracids with hanging red gular sac. Spectacular wing-whirring displays at dawn and dusk. Alarm voices are similar to calls, but generally, louder, quicker and often hysterical when startled.

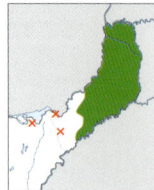

Rusty-margined Guan *Penelope superciliaris* 61 cm

Paraná forest. **ID Adult**: Grey-brown breast with paler greyish fringes creating scaly effect. Tertials, secondaries and greater-coverts fringed rufous when fresh. Often looks darker in forest interior. Unconfirmed overlap with Dusky-legged Guan in Misiones. **Voice** Low-pitched, fairly well spaced, coarse grating *kargh, kargh, kargh* etc. [Yacupoí]

Dusky-legged Guan *Penelope obscura* 58 cm

Gallery forest in the NE. **ID Adult**: Bronze-olive with sparse white streaking on breast, foreneck, scapulars and wing-coverts. Unconfirmed overlap with Rusty-margined Guan in Misiones. **Voice** Medium-pitched and disyllabic with or without terminal notes, lacking guttural rasps e.g. *Wikik Wikik-u, Wikik-u-u, Wikik, Wikik-u* etc. [Pava de Monte Ribereña]

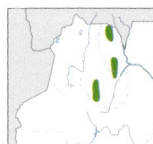

Red-faced Guan *Penelope dabbenei* 63 cm

N Yungas forest at 1300–2700 m; rarely below 1650 m. Prefers tree-line alder forest. **ID Adult**: Short-legged guan with pink skin around the eye, narrow black forehead and contrasting frosty white forecrown or supercilium. Indistinct white streaking on the mantle, scapulars, wing-coverts and breast. Warm chestnut belly and vent. Bronzy tail. Overlaps with Yungas Guan below 2000 m. **Voice** Distinctive *Kee-Wok* calls. [Pava de Monte Alisera]

Yungas Guan *Penelope bridgesi* 79 cm

Yungas forest to 2000 m. **ID** Largest guan. **Adult**: Prominent white streaks on mantle, scapulars and wing-coverts, and sparser streaking on the breast. Dark grey facial skin. Very long and broad black tail. Overlaps with Red-faced Guan and Chaco Chachalaca. **Voice** Low-pitched whistles mixed with guttural rasps e.g. *SWig SWI-u, SWig-u KREG, swig-u, u, KREG KREG KREG* etc. **Tax note 1**. [Pava de Monte Yungueña]

Chaco Chachalaca *Ortalis canicollis* 60 cm

Humid and dry chaco, ranging into Yungas foothill forest. **ID** Relatively small, noisy gregarious cracid. Arboreal and terrestrial. **Adult**: Red facial skin and small dewlap. Greyish head and neck. Rufous vent and chestnut tips to outer tail feathers; best seen in flight. **Voice** At dawn and dusk, a loud far-carrying cacophony of hoarse grating *char.... CHaRa-RaTá, CHaRa-RaTá, CHaRa-RaTá* etc. [Charata]

Black-fronted Piping Guan *Pipile jacutinga* 71 cm

Rare in riverine Parana forest, especially with *Euterpe* palms. **ID** Slender-necked, black and white, mostly arboreal forest cracid with a mainly red dewlap. **Adult**: Striking white crown, nape, eye-ring and wing-coverts contrasting with predominantly black plumage. **Voice** Ascending series of 4–6 shrill high-pitched whistles, each louder than the previous. Wing rattle crepuscular display like *Penelope* guans. [Yacutinga]

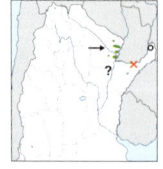

Bare-faced Curassow *Crax fasciolata* ♂ 94 cm; ♀ 89 cm

Rare in gallery forest of the humid chaco. Historical records from n. Corrientes and w. Misiones. **ID** Unmistakable large cracid, terrestrial and arboreal, sexually dimorphic, with erectile curly crest. Very reclusive. **Adult ♂**: Plumage mainly black with white lower belly and vent. Broad yellow cere around bill base. **Adult ♀**: Head and neck black. Upperparts black heavily barred white, extending onto breast. Rest of underparts cinnamon. Black tail, narrowly barred white and tipped cinnamon. **Voice** Best detected by hollow booming at dawn. Shrill whistled alarm calls. [Muitú]

Rusty-margined Guan

ad

major

Dusky-legged Guan

ad

obscura

Red-faced Guan

ad

Yungas Guan

ad

Chaco Chachalaca

canicollis

ad

ad ♂

ad ♀

fasciolata

Black-fronted Piping Guan

ad

Bare-faced Curassow

Coscoroba Swan *Coscoroba coscoroba*   98 cm

Lowlands, Andes to 2000 m and the Falklands. **ID** Unusual white 'swan' with long legs and a duck-like bill. **Adult:** Entirely white with a bright waxy red bill. In flight, white with black primary tips. **Juvenile:** Grey bill, blackish cap and grey-brown wash on back. **Voice** Far-carrying bugle-like *Cos-Co-roba*. [Coscoroba]

Black-necked Swan *Cygnus melancoryphus*   ♂ 119 cm, ♀ 106 cm

Lowlands, Andes to 1450 m and the Falklands. **ID** A large, short-legged swan which undertakes partial austral migration. Sexes differ by size and number of bill caruncles. **Adult** ♂: Black head and neck contrast with white body and wings. In flight, long black head and neck, and white wings. Noisy wing-beats and whooshes audible from a distance. **Juvenile:** Greyish head and neck. Primaries tipped blackish like Coscoroba Swan. [Cisne Cuello Negro]

Southern Screamer *Chauna torquata*   81 cm

N lowlands and foothills. **ID** Corpulent marsh bird with a comparatively small head, unwebbed toes and two carpal spurs on each wing. Often perches in bare trees. Monogamous, but forms huge post-breeding flocks. **Adult:** Essentially grey with a paler head and wispy crest, large red legs, and a black collar. In flight, broad wings with white underwing-coverts and short tail; recalls a vulture when soaring. **Voice** Loud, far-carrying, explosive *Ker-Ahhh*, sometimes duetting. [Chajá]

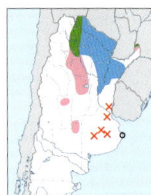

Comb Duck *Sarkidiornis sylvicola*   ♂ c.76 cm, ♀ 67 cm

Local in the N lowlands and foothills. **ID** Large aberrant goose-like duck. Laboured flight often in line or V formation. **Adult** ♂: Large blackish comb on bill. Black crown and back, with mostly white face, neck and breast, and very broad black flanks. **Adult** ♀: Slightly less green gloss on inner wings than male and lacks the bill comb. In flight, dark underwing and flanks. **Juvenile:** Dark brown crown, back and wings. White spot at base of bill. Buffy head and neck with a dark eye-stripe. Mottled flanks and variable white to creamy vent. **Tax note 2.** [Pato Crestudo]

Muscovy Duck *Cairina moschata*   ♂ 75 cm, ♀ 68 cm

Uncommon in the N lowlands. **ID** Robust dabbling duck of forested regions with peaked crown and nuchal mane in ♂. Wary of man, quickly undertaking a strong but laboured escape flight. Caruncles on base of bill. Facial pattern becomes enhanced during the breeding season. **Adult** ♂: Black, glossed green and purple above. White wing-coverts visible on folded wings. **Adult** ♀: Reduced caruncles, lacks nuchal ruff, browner below than ♂. In flight, contrasting white wing-coverts on upper- and underwing. **Juvenile:** Lacks bill caruncles. Sooty blackish-brown throughout without gloss. White wing-coverts develop after first year. [Pato Real]

Coscoroba Swan

juv

ad

ad

juv

ad ♂

Black-necked Swan

Southern
Screamer

ad

Comb Duck

ad ♂

ad ♀

juv

juv

ad ♂

Muscovy Duck

ad ♀

ad ♀

Andean Goose *Oressochen melanopterus*   75–80 cm

High NW Andes. **ID** Large thick-necked, grazing goose of meadows, lake shore and cushion-plant bogs. Sexes alike but ♂♂ are considerably larger than ♀♀. **Adult:** Predominantly black and white. Nothing similar in range. In flight, contrasting blue-black primaries and unconnected short purple-black bar across greater coverts. [Guayata]

Orinoco Goose *Neochen jubata*   55–66 cm

Rare and local in e. Salta – w. Formosa; vagrant to Catamarca and Corrientes. **ID** A robust goose of sandy rivers and marshes in the dry chaco. Has unique grooves between feather tracts on the sides of neck, creating a ruffled appearance. ♂♂ are noticeably larger than ♀♀. **Adult:** Reddish tarsus, creamy head and neck with chestnut back and flanks, and a contrasting white vent. In flight, has broad blackish wings with a small oval or squarish white speculum in the inner secondaries. [Ganso de Monte]

Chloephaga Large, thick-necked, grazing geese of open grasslands, wetlands and shore. Wing patterns identical: white, with black primaries connecting with a glossy green-black bar across greater coverts. ♂♂ are generally larger than ♀♀. Upland and Kelp are sexually dimorphic.

Ashy-headed Goose *Chloephaga poliocephala*   50–58 cm

S Andes and Patagonia; winters in s. Buenos Aires. Vagrant to the Falklands. **ID Adult:** Grey head and neck, rufous breast and white belly with heavily barred flanks. In flight, rufous breast contrasts with pale head and body. **Juvenile:** Breast and back mixed with black-and-white barring. [Cauquén Real]

Upland Goose *Chloephaga picta*   66–76 cm (*picta*); 72–78 cm (*leucoptera*)

Widespread in Patagonia; winters in s. Buenos Aires (*picta*). Resident on the Falklands (*leucoptera*). **ID Adult** ♂ *picta* **white morph:** White with mainly grey mantle and extensive black barring on flanks. **Adult** ♂ *picta* **barred morph:** Breast, flanks and most of belly barred with black. **Adult** ♀ *picta*: Like adult Ruddy-headed Goose but larger, proportionately longer-necked with blacker flanks and at least some white (not chestnut) on the vent. **Adult** ♂ *leucoptera* **white morph:** Differs from *picta* only by its larger size. No overlap with Andean Goose. **Adult** ♀ *leucoptera*: Brighter head than *picta* and larger. From Ruddy-headed Goose by longer neck, blacker flanks and white on the undertail-coverts. **Juvenile** ♂: Resembles adult ♂ barred morph in both ssp. but with finer barring and a grey wash over the head and neck. **Juvenile** ♀: Duller and more washed out than adult ♀, with less well-defined barring. Darker overall than juvenile Ruddy-headed Goose. [Cauquén Común]

Ruddy-headed Goose *Chloephaga rubidiceps*   45–50 cm

Extremely rare in S Patagonia; winters in s. Buenos Aires. Abundant on the Falklands. **ID Adult:** Narrow white eye-ring. Short-necked with a slight nuchal ruff. Rounded chestnut head and uppermost neck only, sharply demarcated from barred foreneck. Contrasting pale forehead. Chestnut centre of belly and vent. Falklands birds appear to be slightly paler-headed. In flight, very like ♀ Upland Goose but note shorter neck and rusty undertail-coverts. **Juvenile:** Pale brown head and upper neck with finer ventral barring. [Cauquén Colorado]

Kelp Goose *Chloephaga hybrida*   55–65 cm (*hybrida*); 66–74 cm (*malvinarum*)

Resident on rocky shores of S Patagonia (*hybrida*) and Falklands (*malvinarum*). **ID Adult** ♂: All white with black bill and yellow-orange legs. **Adult** ♀: Chocolate-brown above with narrow white eye-ring and pink bill. Underparts with coarse black-and-white barring, and white belly and tail. In flight, note contrasting white rump and tail. **Juvenile** ♂ (not illustrated): As ♀ but with white head and spotted mantle. **Second-year** ♂ (not illustrated): White with dusky primary tips. [Caranca]

Andean Goose

Orinoco Goose

ad

ad

ad

Ashy-headed Goose

ad

juv ♀

Upland Goose

ad ♀

ad

juv

ad

picta

Ruddy-headed Goose

picta

juv ♂

Upland Goose

ad ♂ barred morph

ad ♀

leucoptera

picta

ad ♂ white morph

ad ♂ white morph

picta

leucoptera

ad ♀

ad ♂

hybrida

Kelp Goose

juv ♂

Tachyeres Large robust diving ducks with three adult plumages. ♂♂ are generally larger. Head and bill pattern/shape are the most important features. The flightless species (Fuegian, Chubut and Falkland) are allopatric and have relictual primaries, but beware of moulting Flying. Flightless species can become slightly airborne when 'steaming'.

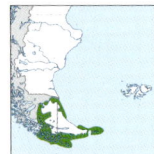

Fuegian Steamer Duck *Tachyeres pteneres*   79 cm

Fuegian shores. **ID** Breeding ♂: Large orange bill with a swollen base; rarely yellower distally. White head. Darker grey body than congeners. **Breeding ♀**: Bill as ♂. Grey to grey-brown head, sometimes with a very indistinct whitish or pale grey post-ocular streak. **Non-breeding** ♂: Bill as breeding ♂. Pale blue-grey head with a short, indistinct greyish post-ocular streak. **Non-breeding ♀**: The only ♀ steamer duck with an entirely orange bill. Head similar to non-breeding ♂ but darker grey. **Eclipse** ♂: Similar to non-breeding ♂, with darker lores and a darker grey-brown head, thus the post-ocular streak tends to be more prominent. **Eclipse ♀**: Similar to eclipse ♂ but head slightly browner and foreface warmer, with a more prominent post-ocular streak. **Juvenile**: Dark grey head with a darker foreface and lores. Lacks a post-ocular streak but may show a white eye-ring. [**Quetro Austral**]

Chubut Steamer Duck *Tachyeres leucocephalus*   68 cm

Endemic. Only on rocky shores. **ID** Breeding ♂: Yellowish-olive or grey bill with a contrasting swollen orange base. White head, often with some grey. Less distinct dorsal spotting than breeding ♂ flying. **Breeding ♀**: Olive bill with a paler or greyer tip, and swollen orangey or yellowish base. Dark brown head and dark greyish crown. Long, uninterrupted, broad whitish post-ocular streak, extending down the sides of the neck into an obvious pale pectoral collar. Cinnamon speckling on breast. **Non-breeding** ♂: Bill as breeding ♂. Grey crown and grey to reddish cheeks. White post-ocular streak reaches the sides of the neck. **Non-breeding ♀**: Dull grey or olive bill usually lacks a brighter base. Entirely dark brown head with a short white post-ocular streak. Dorsal and flank plumage darker than non-breeding ♂, and grey spotting rarely prominent. **Eclipse** ♂: Bill as breeding ♂ but duller and greyer; usually with some orange at the base. Brown head with a larger white eye-ring, and chestnut cheeks and throat. **Eclipse ♀**: Not illustrated; possibly like non-breeding ♀. **Juvenile**: Dark brown head with a paler foreneck. Indistinct buffish post-ocular streak. [**Quetro Cabeza Blanca**]

Flying Steamer Duck *Tachyeres patachonicus*   67 cm

Widespread. **ID** Only species capable of sustained flight with changes of direction. Bill not swollen at the base. Black tail slightly longer and more recurved than flightless species. **Breeding** ♂: Grey or olive-yellow bill with an orange base; sometimes entirely orange in coastal birds. White head, with variable grey forecrown, eye-stripe joining in a V on nape and line down the hindneck. Cinnamon throat patch. Brownish back and flanks, more heavily spotted pale grey than Chubut. Grey breast, tinged pink in centre. **Breeding ♀**: Bill grey, olive or dull yellow, sometimes with a yellow base and culmen. Dark brown head with a dark greyish crown. From ♀ Chubut by much narrower post-ocular streak; continues down sides of neck and can form a very indistinct pectoral band. **Non-breeding** ♂: Bill as breeding ♂. Grey head, often browner on cheeks. White eye-ring and broad white post-ocular streak. Some show a cinnamon throat. **Non-breeding ♀**: As breeding ♀, but post-ocular streak usually broken at the rear, continuing faintly onto sides of neck. Darker dorsal plumage than non-breeding ♂. **Eclipse** ♂/♀: Entirely brown head. White eye-ring and very short, indistinct whitish post-ocular streak; sometimes lacking. Dark chestnut throat, perhaps only in the ♂. **Juvenile**: Brown head, darker on the crown, chestnut throat patch. White eye-ring and narrow indistinct post-ocular streak, extending onto the sides of the neck. [**Quetro Volador**]

Falkland Steamer Duck *Tachyeres brachypterus*   69 cm

Falklands shores. **ID** Breeding ♂: Bill entirely bright orange with a swollen base, sometimes paler towards tip. White head with variable grey wash on lores, forehead, crown and cheeks. Chestnut throat usually extending onto lower cheeks, unlike congeners. Vinaceous breast speckling. **Breeding ♀**: Olive bill with a yellowish or orangey base. Dark brownish-grey head with a greyer crown and chestnut throat. Unbroken, distinct white post-ocular streak, arching down sides of neck, fading into an ill-defined pectoral band. **Non-breeding** ♂: Pale orange bill; yellower towards the tip. Grey crown and hindneck with contrasting dark brown or deep chestnut lores and cheeks. Broad, white, flared post-ocular streak. **Non-breeding ♀**: Olive bill with a yellow or orange base. Head as breeding ♀, but white eye-ring and indistinct whitish post-ocular streak, broken behind the eye, and continuing down sides of the neck. **Eclipse** ♂/♀: Dark brown lores, foreface and crown with greyish hindcrown. Dark chestnut rear portion of ear-coverts and throat. White eye-ring and short white post-ocular streak. **Juvenile**: Entirely brown head. White eye-ring and short white post-ocular streak, becoming indistinct while curving onto sides of neck. [**Quetro Malvinero**]

Fuegian Steamer Duck

♂ non-br

juv

♀ eclipse

♂ eclipse

♀ non-br

♀ br

♂ br

♂ non-br

♀ non-br

Chubut Steamer Duck

♀ br

juv

♂ br

♂ eclipse

Flying Steamer Duck

♂ non-br

♀ non-br

♂ non-br

♂ non-br steaming

♀ br

juv

♂ br

♂/♀ eclipse

♂ non-br

♀ non-br

Falkland Steamer Duck

♀ br

juv

♂ br

♂/♀ eclipse

Dendrocygna Fairly large, gregarious ducks with long legs that project beyond the tail in flight. All show dark underwings. Relatively slow flight. Distinctive whistled calls. Nocturnal feeding.

Fulvous Whistling Duck *Dendrocygna bicolor* 48 cm

Widespread in the N lowlands. **ID** Adult: Grey bill and legs. Face and body rich buff with white flank stripes. In flight, shows blackish wings and creamy-white horseshoe over uppertail-coverts. **Juvenile:** Like washed-out version of adult with grey uppertail-coverts. **Voice** Call is a whistled nasal *who-TEER, who-TEER.* [Sirirí Colorado]

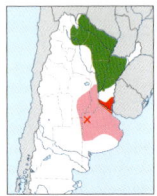

Black-bellied Whistling Duck *Dendrocygna autumnalis* 49 cm

N lowlands. **ID** Adult: Red bill and pink legs. Grey face and throat contrasts with brown crown and hindneck. Black belly. White greater wing-coverts form white line on folded wing. In flight, shows blackish upperwing with white wing stripe and grey or brown forewing. **Juvenile:** Like washed-out version of adult with grey bill and legs. Note white greater-coverts. **Voice** Call is a mellow, nasal *Tip-REE ra-REE-RE-REE.* [Sirirí Vientre Negro]

White-faced Whistling Duck *Dendrocygna viduata* 45 cm

Widespread in the N lowlands and foothills. **ID** Adult: White foreface contrasting with black rear crown and hindneck. Chestnut foreneck. In flight, shows blackish-brown wings and uniform black rump, tail and belly. **Juvenile:** Greyish or buff foreface contrasting with blackish rear crown and hindneck. Dull chestnut foreneck. **Voice** Call is a fast shrill whistled *SWee Ri-Ree.* [Sirirí Pampa]

Crested Duck *Lophonetta specularioides* 59 cm

NW Andes, Patagonia and the Falklands. Often on the sea. **ID** Large dabbling duck with a shaggy nuchal crest and long pointed tail. **Adult** *specularioides* (Patagonia and the Falklands): Red iris. Overall buffy with a dusky crown and blackish mask. Pale scapular spots and marbled flanks. **Adult** *alticola* (NW Andes): Differs from *specularioides* by orange-yellow iris and plainer underparts. In flight, appears notably long-necked. Shows dark brown upperwing with a pink speculum, bordered by a black subterminal bar and broad white trailing edge to secondaries. Brown underwing with white axillaries, median-covert bar and trailing edge. [Pato Crestón]

Spectacled Duck *Speculanas specularis* 57 cm

S Andes. Scarce, spectacular duck of Patagonian forest lakes and rivers. **ID** In pairs, but gregarious in winter. **Adult** ♂: Chocolate-brown head and neck, with large white oval pre-ocular patch, throat and spur on sides of neck. Bright orange legs. In flight, shows metallic pink speculum, and narrow white trailing edge. White axillaries. **Adult** ♀: Duller head. [Pato de Anteojos]

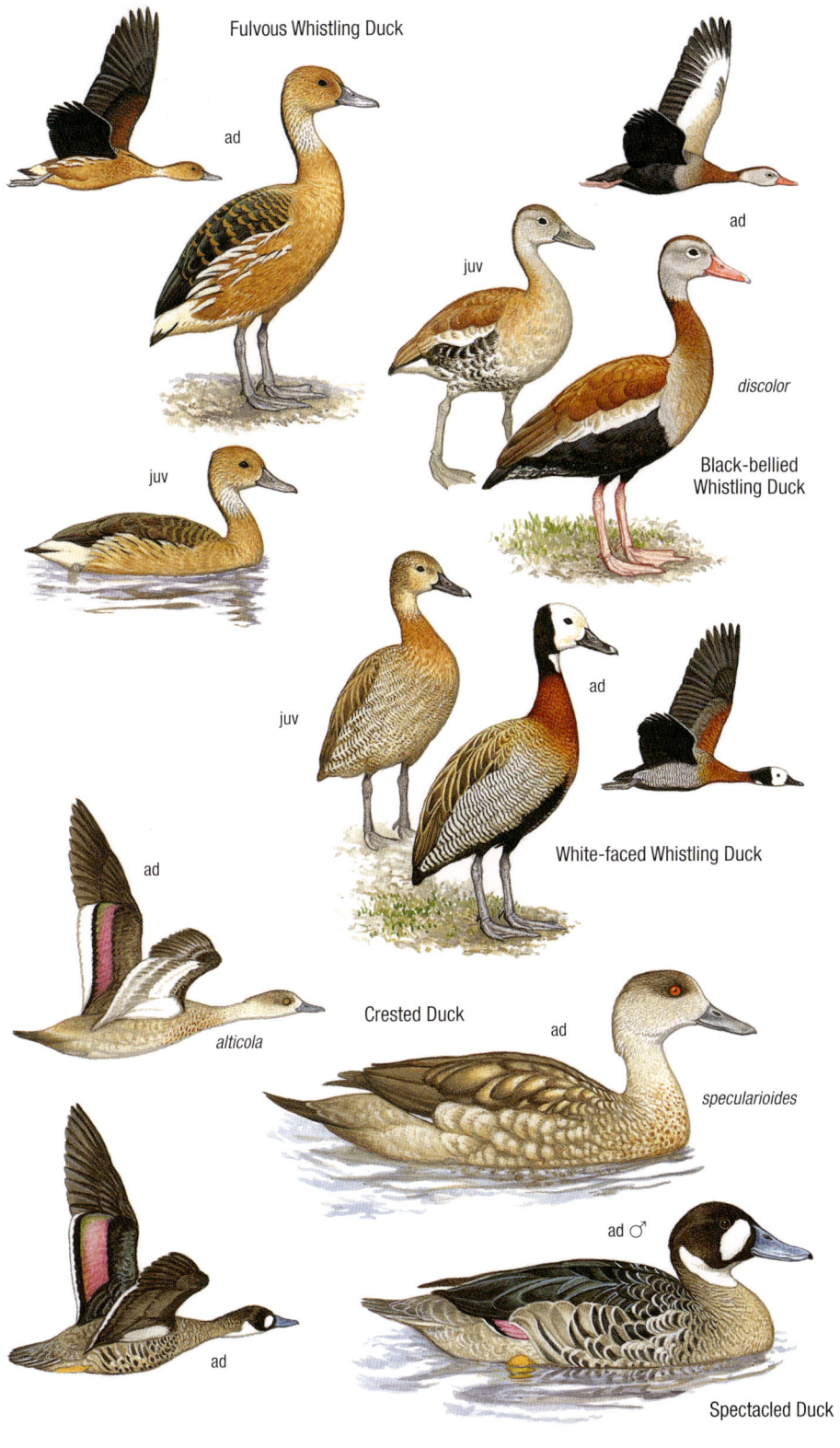

Fulvous Whistling Duck

ad

juv

juv

juv

Black-bellied
Whistling Duck

ad

discolor

ad

ad

White-faced Whistling Duck

ad

alticola

Crested Duck

ad

specularioides

ad

ad ♂

Spectacled Duck

Spatula Compact ducks, which up-end or 'dabble' but rarely dive. Flaps bordering the upper mandible are more prominent in larger species.

Blue-winged Teal *Spatula discors*   39 cm

Rare boreal vagrant. **ID Adult** ♂: Dull blue head with large white crescent on foreface. White patch on rear flanks and black vent. In fast and agile flight, upperwing shows grey-blue coverts, tipped white, and metallic green secondaries. Underwing with dusky leading edge, whitish coverts and grey flight feathers. **Adult** ♀: Dull yellow or brown tarsus, dark brown crown and nape, fine facial streaking, dark eye-stripe, narrow white eye-ring, and distinct white oval loral patch. Cold greyish-buff breast and flanks. In flight, ♀ is similar to ♂ but coverts duller grey, reduced white greater-covert bar and no green speculum. Body plumage colder than ♀ Cinnamon Teal. [Pato Media Luna]

Cinnamon Teal *Spatula cyanoptera*   42–47 cm

Widespread in the lowlands and Falklands; a larger ssp. *orinomus* (47 cm) occurs in the altiplano. **ID Adult** ♂: Head and underparts bright glossy chestnut with a black vent. In fast and direct flight, wing pattern like Blue-winged Teal. **Adult** ♀: Small flaps on sides of bill at close range. Dull yellowish tarsus, blackish crown and nape, coarser facial streaking than Blue-winged Teal, indistinct eye-stripe and, on some, an ill-defined whitish loral patch. Breast and flanks washed warm buff. In flight, wings of ♀ like Blue-winged Teal, but white greater-covert bar slightly longer. Warmer buff body plumage. [Pato Colorado]

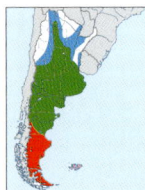

Red Shoveler *Spatula platalea*   47 cm

Widespread in lowlands; rare in NW foothills. **ID Adult** ♂: Heavy black spatulate bill. Light grey-brown head contrasts with rusty-brown to cinnamon-rufous body, heavily spotted black throughout. Black rump and pointed tail, edged white. Black vent with large white spot on rear flanks. In flight, upperwing with grey-blue coverts and broad white greater-covert stripe. Blackish remiges with small metallic green speculum. Whitish underwings with grey-brown flight feathers and whitish central stripe. Note pale head and pointed black tail, edged white. **Adult** ♀: Pale brown, lightly speckled face and coarsely spotted black on underparts with black chevrons on flanks. Blackish tail, edged buff. In flight, grey upperwing-coverts, narrowly tipped white. Brown primaries and darker secondaries. Underwing as ♂ but duller. Note paler head. [Pato Cuchara]

Northern Shoveler *Spatula clypeata*   48 cm

Not illustrated. Vagrant to Patagonia. **ID Adult** ♂: Outsized heavy bill. Bottle-green head. White breast with contrasting chestnut flanks and belly. Eclipse male shows white crescent on foreface. In flight, similar to Red Shoveler but with white underwing-coverts. **Adult** ♀: Like female Red Shoveler but with a dark crown and eye-stripe. [Pato Cuchara Boreal]

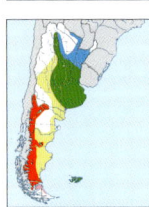

Silver Teal *Spatula versicolor*   37–42 cm

Widespread. **ID Adult** *versicolor* (37 cm; N lowlands): Blue bill with yellowish base. Sooty-brown crown and nape; creamy buff face and throat. Black and white barring of equal width on flanks. Belly and vent grey, vermiculated black. In fast and low flight, shows metallic green secondaries with a white trailing edge. White axillaries and underwing-coverts. **Adult** *fretensis* (42 cm; S Andes and the Falklands): Larger. From *versicolor* by narrow white and broad black bars on flanks, and strongly barred belly. [Pato Capuchino]

Puna Teal *Spatula puna*   39 cm

Puna lakes, mostly in Salta and Jujuy; lower in winter. **ID Adult** ♂: Large blue bill. Crown and centre of nape black, contrasting with creamy white face and throat. Flanks finely barred black and white. Flight is fast and low. **Adult** ♀: Wings as Silver Teal, but generally greyer. Flanks broadly barred brown with narrow cream bars, recalling Silver Teal more than the ♂. [Pato Puneño]

ad ♀

ad ♂   Blue-winged Teal

ad ♀

ad ♂

ad ♀

cyanoptera

Cinnamon Teal

ad ♂

ad ♂

ad ♂

Red Shoveler

ad ♀

ad ♀

versicolor   *fretensis*

Silver Teal   ad

ad

Puna Teal

ad ♂   ad ♀

Anas Compact ducks which up-end or 'dabble' but rarely dive. Most are partial austral migrants.

White-cheeked Pintail *Anas bahamensis* 49 cm

Chiefly in the N lowlands. **ID Adult:** Long grey bill with a red base. Warm brown crown and hindneck contrast with white cheeks, throat and upper foreneck. In rapid flight, upperwing brown with cinnamon bar on greater-coverts, metallic green speculum tipped black, and broad cinnamon trailing edge. Underwing shows white median-coverts and axillaries. [Pato Gargantilla]

Yellow-billed Pintail *Anas georgica* 49 cm

Widespread. **ID Adult:** Slender-necked with a long, pointed tail. Long yellow bill with a black culmen. Warm reddish-brown crown, finely streaked black. Large blackish spotting on flanks and vent. Flight is swift and agile. In flight, ♂ has dark green speculum with cinnamon trailing edge, ♀ has brown speculum and shorter tail than ♂. [Pato Maicero]

Speckled Teal *Anas flavirostris* 41 cm

Widespread; locally to 2000 m in the NW. **ID Adult:** Compact; rounded head and short, slightly pointed tail. Short yellow bill with a black culmen. Head densely speckled with dark brown. Chestnut tips to greater-coverts. Breast boldly spotted black. Plain flanks and vent. Flies low. In flight, black speculum with cinnamon trailing edge. White underwing-coverts and axillaries. Compare with Yellow-billed Pintail and Inca Teal. [Pato Barcino]

Inca Teal *Anas [flavirostris] oxyptera* 41 cm

Mostly on altiplano lakes; locally down to 1500 m on NW reservoirs. **ID Adult:** Recalls Speckled Teal but underparts white with only sparse breast spots. Darker, more contrasting blackish head. Overlaps with Speckled Teal in Tucumán without hybridising. **Tax note 3.** [Pato Inca]

Merganetta Slender-bodied duck with long stiff tail and red bill. Extremely agile in fast-flowing rivers.

Torrent Duck *Merganetta armata* 44 cm

Widespread in the Andes. **ID Adult** ♂ *armata* (S and C Andes): White head and neck, striped with black. Black bar below eye and black chin, breast and flanks. **Adult** ♂ *leucogenys* (N Andes): From *armata* by white throat, black restricted to upper breast and tawny underparts, striated black. In flight, grey upperwing with black distal primaries and green speculum. **Adult** ♀: Grey above with long, black-centred scapulars. Rufous below. Similar in both ssp. **Juvenile:** Greyish-brown crown and brown upperparts. White below, barred grey on sides of breast and flanks. [Pato de Torrente]

White-cheeked Pintail

rubrirostris

ad

Yellow-billed Pintail

spinicauda

ad

ad ♀

ad ♂

Speckled Teal

Inca Teal

ad

ad

ad ♀

Torrent Duck

ad ♂

armata

ad ♂

juv

♂

leucogenys

leucogenys

Callonetta and ***Amazonetta*** Two monotypic marshland dabbling ducks. Encountered in pairs or small groups. ***Mareca*** Sociable duck, freely mixing with others. Feeds mainly by dabbling and up-ending; rarely diving.

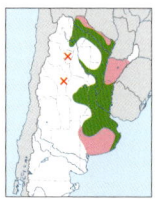

Ringed Teal *Callonetta leucophrys* 36 cm

N lowlands. Nests in tree holes or Monk Parakeet nests. **ID Adult** ♂: Black crown, nape and semi-collar contrasting with cream face. Chestnut scapulars. Cinnamon breast, spotted black. Black vent with two white lateral patches. In rapid flight, blackish upperwing with conspicuous white oval patch and green speculum. Blackish underwing. **Adult** ♀/ **juvenile**: Brown crown with long broken white supercilium and triangular brown patch on white cheeks. Whitish below, mottled with brown barring. Vent as adult ♂. In flight, diagnostic white oval on upperwing as ♂. Compare with Masked Duck (Plate 13). [**Pato de Collar**]

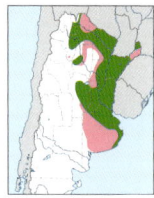

Brazilian Teal *Amazonetta brasiliensis* 44 cm

N lowlands and foothills. **ID Adult** ♂: Bright red bill and tarsus. Contrasting whitish patch on rear face and sides of neck, offset by dusky nape. Cinnamon-brown breast, mottled blackish. Flight typically low in pairs. In flight, blackish upperwing with broad central turquoise-green flash and contrasting white trailing edge. Black underwing-coverts and grey remiges with contrasting white trailing edge and axillaries. **Adult** ♀: Grey bill and red tarsus. Dark brown crown and foreface with white loral and pre-ocular spots. In flight, wings as ♂. Central abdomen whitish. [**Pato Cutirí**]

Chiloe Wigeon *Mareca sibilatrix* 48 cm

Widespread in the lowlands and Falklands, moving north in winter. **ID Adult** ♂: Sooty-brown head and neck with contrasting white foreface and cheek spot. Green sheen on side of head. Barred breast and chestnut-orange flanks. **Adult** ♀: Reduced white on face. In flight (both sexes), upperwing has large white patch on coverts and dark secondaries. Contrasting white rump and black tail. Flight is direct and measured. **Juvenile**: Lacks green on head and foreface is dingy grey. Flanks scalloped blackish, with little chestnut. **Voice** Highly vocal, giving audible nasal *weee-poorr*. [**Pato Overo**]

Netta Large, heavy, sociable dabbling ducks, which less commonly dive. Long-necked with a domed crown.

Rosy-billed Pochard *Netta peposaca* 56 cm

Resident in the Pampas, sporadic in Patagonia; wintering in N lowlands. **ID Adult** ♂: Red bill with swollen knob at base. Glossy black head and upperparts. Grey flanks, bordered by black ventral band and white undertail-coverts. In flight, black coverts with contrasting white remiges tipped black. Underwing mainly whitish. Strong flier; sometimes in V-formation. **Adult** ♀: Grey bill. Dark brown crown contrasts with paler brown face and neck. White chin and throat; sometimes with a white eye-ring. White vent. In flight, as ♂ but coverts brown. [**Pato Picazo**]

Southern Pochard *Netta erythrophthalma* 50 cm

Hypothetical, mainly in the NW Andes. **ID Adult** ♂: Red iris. Mostly blackish with maroon gloss in good light. Flanks chestnut. Brown vent. In fast and direct flight, black wing-coverts contrast with white remiges outlined in black (both upper- and underwing), unlike Rosy-billed Pochard. Compare with Cinnamon Teal, especially in the high Andes (Plate 10). **Adult** ♀: Sooty-brown head with white loral patch, throat and crescent arching around ear-coverts. In flight, wings as ♂. Central belly and vent white. **Juvenile**: Paler brown than ♀ with buffy-white throat and loral patch joining over bill. Caution is advised as perhaps not always safely distinguished from juvenile Rosy-billed Pochard, although underwing pattern is diagnostic. [**Pato Castaño**]

ad ♂

Ringed Teal

ad ♀

ad ♂

ad ♀ / juv

ipecutiri

ad ♂

ad ♂

Brazilian Teal

ad ♀

juv

ad ♀

ad ♂

Chiloe Wigeon

ad

ad ♀

ad ♀

ad ♂

ad ♂

Rosy-billed Pochard

juv

ad ♂

ad ♀

ad ♀

ad ♂

Southern Pochard

Heteronetta Long-bodied and short-tailed with flat crown and long sloping bill. Often swims with body partially submerged. Obligate nest parasite. *Oxyura* and *Nomonyx* Short-winged diving ducks with a swollen bill base. Tail cocked or lax. Wing pattern similar in both sexes. Regularly swim with body partially submerged. Juveniles of all species have narrow spiky tail feathers. *Nomonyx* has a smaller, more rounded head. *Mergus* Sleek forest diving duck with long, slender serrated bill for gripping fish. Deep pools are important for fishing.

Black-headed Duck *Heteronetta atricapilla*   42 cm

Widespread but local. **ID Adult** ♂: Long grey bill with a pink base. Black head and neck, sometimes with a white chin. Sooty-brown above. Finely vermiculated below; breast and flanks often tinged chestnut. In flight, head somewhat drooped, noticeably narrow wings; brown above with white trailing edge; axillaries and underwing-coverts mostly white. **Adult** ♀: Grey bill with a pinkish tinge at base in breeding season. Sooty crown with narrow whitish supercilium. Extensive white throat. **Immature:** All-dark bill and ill-defined supercilium. [Pato Cabeza Negra]

Ruddy Duck *Oxyura jamaicensis*   45 cm

Andes and Patagonia. **ID Adult** ♂ **breeding:** Long blue bill with a slightly swollen base. Black reaches further down hindneck than foreneck. Chestnut breast noticeably darker than back. Undertail-coverts tipped white. Erect tail never higher than top of crown. **Adult** ♀/ **eclipse** ♂: Mostly dark brown; cheeks slightly paler with an indistinct dusky cheek stripe. In flight, dark upperwing and grey below with contrasting dark leading edge and long white median stripe. **Juvenile:** As adult ♀ but paler below; more barred on body. Rectrices narrow and pointed. [Pato Zambullidor Grande]

Lake Duck *Oxyura vittata*   38 cm

Widespread in the lowlands. Overlaps with Ruddy Duck in the south. **ID Adult** ♂ **breeding:** Shorter blue bill than Ruddy Duck. Black of foreneck level with black of hindneck. Body paler, more cinnamon than Ruddy. Erect tail as high as top of crown. **Adult** ♀/ **eclipse** ♂: Grey bill. Brown head with whitish stripe below eye, and whitish stripe across lower cheeks. Rufous tones on breast of eclipse ♂. In flight, wings as Ruddy Duck but white ventral stripe indistinct. **Juvenile:** Like adult ♀ with duller facial pattern and paler brown underparts. [Pato Zambullidor Chico]

Masked Duck *Nomonyx dominicus*   33 cm

On floating vegetation in N lowlands. **ID Adult** ♂ **breeding:** Blue bill with large black nail. Black foreface and crown. Rest of plumage chestnut; mottled black on mantle and flanks. **Adult** ♀: Grey bill with bluish base and black nail. Dusky crown, eye-stripe and cheek bar across buffy-white face. Eclipse ♂ is similar to ♀ with broader, less distinct head stripes. **Juvenile** (not illustrated): Has whiter head than ♀ with narrower stripes. In flight, blackish upperwing with prominent white central square patch (larger in ♂); compare with Ringed Teal (Plate 12). Dark grey underwing with contrasting white axillaries. [Pato Fierro]

Brazilian Merganser *Mergus octosetaceus*   55 cm

Extinct? in n. Misiones. **ID** Sleek forest duck, with long serrated bill and wire-like nuchal crest. **Adult** ♂: Red tarsus and dark bottle-green head. **Adult** ♀: Shorter nuchal crest than ♂. In flight, both sexes show large white wing speculum and greater-coverts divided by a narrow black bar. [Pato Serrucho]

Black-headed Duck

imm

ad ♂

ad ♀

♂

juv

ad ♂ br

Ruddy Duck

ad ♀ /
eclipse ♂

♀

juv

ad ♂ br

ad ♀ /
eclipse ♂

Lake Duck

ad ♂ br

ad ♀

♀

Masked Duck

♀

ad ♂

Brazilian Merganser

ad

ad ♀

Tachybaptus, *Podilymbus*, *Rollandia*, *Podiceps* and *Podicephorus* Strictly aquatic birds with the legs situated towards the rear body, lobed toes and relictual tails. Sexes alike, but strong seasonal plumage variation in most species.

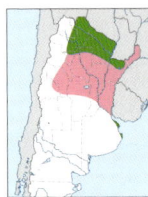

Least Grebe *Tachybaptus dominicus* 24 cm

Local in the N often on floating vegetation in forested areas. **ID** Tiny freshwater grebe with golden iris and short bill. **Adult breeding**: Grey head and neck with a black crown and throat patch. **Adult non-breeding**: From breeding adult by dull yellow lower mandible and white throat. **Voice** Long duetting trills recalling *Laterallus* crakes but more nasal. [Macá Gris]

Pied-billed Grebe *Podilymbus podiceps* 32 cm

Widespread in lowlands and foothills, but rare in the south. **ID** Robust freshwater grebe with a dark iris and notably stout pale bill. **Adult breeding**: Chalky white bill with a black transverse band. Sooty-grey head, white eye-ring and black throat. **Adult non-breeding**: Dull yellow bill, sometimes with an indistinct subterminal band. Dusky brown plumage with a whitish throat and warm brown foreneck. **Juvenile**: Sooty-brown above, with white face crossed by two narrow black stripes. White below with a brown breast band. **Voice** Call of male is a far-carrying long series which begins slowly and increases in volume and speed, comprising mainly disyllabic phrases. [Macá Pico Grueso]

White-tufted Grebe *Rollandia rolland* 26–37 cm

Widespread. **ID** A freshwater and coastal grebe with a peaked rear crown, slender pointed bill and red iris. Solitary or in small groups. **Adult breeding** *rolland* (37 cm; endemic to the Falklands): Note large size. Black head, neck and back. Shaggy white triangular cheek patch. Black eye-stripe or facial streaks. Chestnut abdomen and flanks. **Adult breeding** *chilensis* (26 cm; NE lowlands and Patagonia, moving north in autumn): As *rolland* but with a proportionately shorter and narrower bill. **Adult non-breeding** *chilensis*: Yellowish bill with a black culmen. White face and throat, streaked black. Buffy foreneck. **Juvenile** *chilensis*: Resembles non-breeding adult, but with a rufous supercilium, turning white behind eye, and two black facial stripes. Compare with juvenile Pied-billed Grebe. **Voice** Generally silent. [Macá Cara Blanca]

Silvery Grebe *Podiceps occipitalis* 29–31 cm

Widespread. **ID** Small compact, highly gregarious grebe with a short slender chisel shaped bill and red iris. **Adult breeding** *juninensis* (31 cm; resident above 3000 m in the NW Andes): Plumbeous-grey above with a pure white throat, underparts and flanks. Dull yellow facial plumes. Black line down nape and hindneck. **Adult breeding** *occipitalis* (29 cm; mainly C and S lowlands, wintering through the north and resident on the Falklands): From *juninensis* by paler grey upperparts and grey throat, broader and rounder black nuchal patch, and much brighter golden facial plumes. **Adult non-breeding/juvenile**: Lacks nuptial head plumes (both ssp). **Voice** Calls include high-pitched *chit* and *witchuwit*, usually delivered by several birds at once. [Macá Plateado]

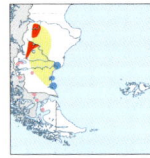

Hooded Grebe *Podiceps gallardoi* 35 cm

Endemic breeder; locally only in w. Santa Cruz; winters in coastal se. Santa Cruz. **ID** Small compact, highly gregarious freshwater and coastal grebe with a short slender chisel shaped bill and red iris. **Adult**: Black head and back, with white forehead bordering an erectile orange crest. White foreneck, flanks and underparts. **Juvenile**: Very similar to juvenile Silvery Grebe but with a blacker (not grey) back and blackish crown, contrasting only with the greyish forehead. **Voice** Loud, yet mellow, whistled *wut..CHRriiiii-óh* or *wot..KRREE-o*, with final syllable descending drastically. [Macá Tobiano]

Great Grebe *Podicephorus major* 62 cm

Widespread resident. **ID** Large, very long-necked, freshwater and coastal grebe with a peaked rear crown and very long dagger-like bill of variable colour. **Adult breeding** *major*: Black crown and line down hindneck, grey face and throat, and chestnut neck. **Adult breeding** *navasi* (Patagonian forest lakes): Slightly larger with a notably broader bill. Glossy green-black head and throat. Deeper, richer chestnut neck. **Adult non-breeding** *major*: Recalls a very faded breeding adult with a white face and throat. **Juvenile** *major*: From non-breeding adult by striped face and buffy flanks. **Voice** Far-carrying, mournful whistled cry *waaaaa…* fading towards the end. [Huala]

Least Grebe

speciosus

ad br

ad non-br

Pied-billed Grebe

ad br

antarcticus

ad non-br

juv

White-tufted Grebe

rolland

ad br

ad br

chilensis

ad non-br

juv

juninensis

ad br

occipitalis

ad br

Silvery Grebe

ad non-br

ad

Hooded Grebe

juv

major

ad br

ad non-br

Great Grebe

major

major

juv

navasi

ad br

Aptenodytes Large penguins with long slender bills and a neck patch. Sexes alike.

King Penguin *Aptenodytes patagonicus* 93 cm

Breeds on the Falklands, w. Tierra del Fuego and occasionally Isla de los Estados; ranging sporadically to continental waters. **ID Adult:** Straight bill with an orange or pink stripe on the lower mandible. Black head and throat with an orange oval on the side of the upper neck, joining a burnt orange patch on the side of the upper breast. Remainder of upperparts pearly-grey, bordered black at sides. **Immature:** Paler pink bill stripe, and narrower pale yellow 'spoon' on side of neck. Note semi-collar and broader black tip to underside of flipper compared to immature Emperor. [Pingüino Rey]

Emperor Penguin *Aptenodytes forsteri* 116 cm

Rare vagrant to the Falklands and Tierra del Fuego. **ID Adult:** Shorter bill than King Penguin and slightly decurved. More open, rounded neck patch than King with orange restricted to upper edge. Dark grey above, usually appearing black. **Immature:** Large open whitish neck patch. Note shorter, slightly decurved bill and reduced black tip to underside of flipper compared to immature King. [Pingüino Emperador]

Eudyptula and *Spheniscus* Small penguins with blunt-tipped bills. One is widespread in the region, the other a vagrant.

Little Penguin *Eudyptula minor* 42 cm

Vagrant, recorded once in the Magellan Straits. **ID** Tiny penguin with relatively slender bill and pale iris. **Adult:** Pink tarsus and bill base. Metallic blue-grey above and white below. White tail tip and trailing edge to dusky flippers. [Pingüino Azul]

Magellanic Penguin *Spheniscus magellanicus* 63 cm

Breeds on Patagonian and Falklands coasts, ranging north to Buenos Aires in winter. **ID** Small penguin with a very stout, blunt bill and two breast bands. **Adult:** Black bill crossed by white subterminal band. Black head and throat with pink supraloral and eye-ring. White supercilium encircling face. Two black breast bands, lower band forming a hoop around breast and flanks. **Juvenile:** Considerably greyer with smoky cheek and throat. No bill band or bare pink facial skin. [Pingüino Patagónico]

Pygoscelis Medium to fairly large penguins with stout, blunt-tipped bills and long pointed tails.

Gentoo Penguin *Pygoscelis papua* 74–94 cm

Breeds on the Falklands and Tierra del Fuego. **ID Adult** *ellsworthi* (74 cm; Antarctic vagrant to Tierra del Fuego in winter): Orange bill with black culmen and tip. Black above including throat with transverse white coronal patch. **Adult** *papua* (94 cm; Falklands and Tierra del Fuego): Up to 40% larger with longer, stouter and redder bill. **Juvenile** (not illustrated): Greyish throat, shorter bill and reduced coronal patch. [Pingüino de Vincha]

Chinstrap Penguin *Pygoscelis antarcticus* 74 cm

Vagrant to the Falklands and Patagonia. **ID Adult:** Dark red iris, and black crown and upperparts. Narrow black line on white face extends across lower throat. **Juvenile** (not illustrated): Similar but face and chinstrap may be less clean. [Pingüino de Barbijo]

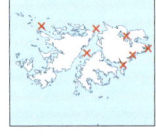

Adelie Penguin *Pygoscelis adeliae* 76 cm

Vagrant to the Falklands. **ID Adult:** Short, stout reddish-black bill, black head, throat and upperparts, and striking white eye-ring. **Juvenile:** White throat. Eye-ring is reduced or lacking. [Pingüino Ojo Blanco]

imm

ad

Emperor Penguin

imm

ad

King Penguin

novaehollandiae

ad

Little Penguin

ad

juv

Magellanic
Penguin

porpoising

Gentoo Penguin

ad

ad

ad

Adelie Penguin

juv

ad

papua

ellsworthi

Chinstrap Penguin

Eudyptes Small to medium-sized penguins with yellow crests. Two common species in region and five vagrants, some of which have bred. Extent and shape of supercilium and crest, and presence of fleshy gape are the most important features. Wet crests distort shape at sea, making identification often impossible.

Royal Penguin *Eudyptes schlegeli*   74 cm

Vagrant to the Falklands. **ID Adult typical:** Huge reddish bill with large pink fleshy gape. Crest joined on forehead. Entirely white face and throat. **Adult dark morph:** Face and throat blackish but chin and lower portion of ear-coverts are mainly white. Compare with Macaroni Penguin. [Pingüino Cara Blanca]

Macaroni Penguin *Eudyptes chrysolophus*   70 cm

Breeds on the Falklands and on Diego Ramírez; vagrant to Patagonia. **ID Adult:** Large reddish bill with contrasting pink fleshy gape. Crest joined on forehead. Face and throat black. **Juvenile:** Reduced head plumes but still joined on forehead. Greyish throat. [Pingüino Frente Dorada]

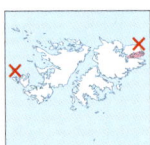

Tristan Penguin *Eudyptes moselyi*   60 cm

Vagrant to the Falklands; always in Rockhopper or Macaroni Penguin colonies. **ID Adult:** Unjoined supercilium and long, splayed and hanging yellow head plumes. [Alt: Northern Rockhopper Penguin] [Pingüino de Tristán]

Rockhopper Penguin *Eudyptes chrysocome*   60 cm

Breeds on the Falklands, Isla de los Estados, Diego Ramírez and off Santa Cruz, ranging to Buenos Aires in winter. **ID Adult:** Reddish bill without a visible gape line (pink in ssp. *filholi*; one Patagonian record). Spiky black crest and narrow yellow supercilium developing into crest behind the eye. **Juvenile:** Narrow yellow supercilium. Lacks head plumes. Throat mixed grey and white, finely speckled black. [Alt: Southern Rockhopper Penguin] [Pingüino Penacho Amarillo]

Erect-crested Penguin *Eudyptes sclateri*   65 cm

Vagrant to the Falklands and Patagonia. **ID Adult:** Broad reddish-pink bill with contrasting pale pink gape. Supercilium starts beside gape and develops into upstanding brush-like crest. **Juvenile:** Reduced supercilium from gape, though still slightly crested. White throat. [Pingüino Crestudo]

Snares Penguin *Eudyptes robustus*   64 cm

Vagrant to the Falklands and Tierra del Fuego. **ID Adult:** Broad reddish-pink bill with contrasting pale pink gape. Narrow yellow supercilium, broadening into tuft behind eye and spreading mainly downwards. **Juvenile/immature:** Bill brownish with pinkish gape. Supercilium very narrow and without tufts at rear. Throat white; ear-coverts washed grey. [Pingüino de Snares]

Fiordland Penguin *Eudyptes pachyrhynchus*   63 cm

Hypothetical vagrant to the Falklands. **ID Adult:** Broad reddish-orange bill lacking pink gape. Broad yellow supercilium, spreading into tuft at rear. Three white lines on cheeks though often concealed. **Juvenile:** Brownish bill. Supercilium narrower without tuft at rear. Throat white; ear-coverts washed grey. [Pingüino Pico Grueso]

Royal Penguin

ad

ad
dark morph

ad

Macaroni
Penguin

ad

ad

Tristan
Penguin

Rockhopper
Penguin

ad

chrysocome

Erect-crested
Penguin

ad

ad

ad

Fiordland Penguin

Rockhopper Penguin

Snares Penguin

juv

Macaroni Penguin

Fiordland Penguin

juv

juv

erect-crested Penguin

juv

Snares Penguin

juv/imm

Diomedea The largest of the albatrosses, characterised by white backs, a short tail cloaked by long uppertail-coverts, lack of a groove on the bill base, and nasal tubes situated near base of bill. Taxonomy uncertain, with Wandering sometimes considered to be four species (three in our region) and the Royal Albatrosses often separated as two species. Adult ventral illustrations are depicted at a different scale. Caution advised with many similar plumages. **Tax note 4.**

Wandering Albatross *Diomedea exulans* *c.*122 cm WS: *c.*320 cm

S Patagonia; deep offshore waters. **ID** Confusing and often depends on correct ageing. **Juvenile:** White face and chin contrasts with grey-brown to chocolate brown head and body. Black tail and upperwing. Large black thumbmarks at leading edge of white underwing with broad black tips. **Immature (stage 4):** Whitens from the rump, back and belly, sometimes with a scaly appearance. White blotches develop from the central wing outwards. Retains black tail tip. **Subadult:** Innerwing mostly white with black squares on inner secondaries. White or black tail tip. **Adult** ♂ 'Snowy': Wings mostly white above with black primaries, narrow trailing edge and white tail. **Adult** ♀ (not illustrated) tends to show a blotchy white innerwing, variable brown coronal patch, finely vermiculated mantle and scapulars, and black or partially black tail. Both sexes show white underwing with small black tip and very narrow trailing edge and leading edge to primaries. Nearly all lack the black cutting edge on the bill which is shown by both royal albatrosses (at close range), and differ by huge pink bill with yellowish or orangey tip. Mature birds show a pinkish or gingery patch on the rear ear-coverts, and have upward pointing nostrils. [Albatros Errante]

Antipodean Albatross *Diomedea [exulans] antipodensis* *c.*114 cm WS: *c.*290 cm

Vagrant in the Drake Passage. **ID** Adult ♂: Extremely similar to female Tristan though foreneck is at best vermiculated and back at most speckled. **Adult** ♀ (not illustrated): Very like juvenile Wandering but with an extensive white belly. [Albatros de las Antípodas]

Tristan Albatross *Diomedea [exulans] dabbenena* *c.*114 cm WS: *c.*290 cm

Vagrant recorded off Buenos Aires. **ID** Shorter bill than Wandering Albatross. **Adult** ♂: Clean white head and breast, but white spots in secondaries, perhaps inseparable from Wandering. **Adult** ♀: Black upperwing and tail with white restricted to scapulars and inner arm. Brown cap and mottled vermiculations on foreneck and breast suggest a pectoral band. **Juvenile/immature:** Inseparable from immature Wandering. [Albatros de Tristán]

Amsterdam Albatross *Diomedea amsterdamensis* *c.*115 cm WS: *c.*300 cm

Not illustrated. Very rare vagrant recorded off Tierra del Fuego. **ID** Adult (both sexes): Recalls juvenile Wandering. Brown above with variable amounts of white mottling on back, variable amounts of white on face and sides of neck, and white underparts with pale to dark brown collar and variable brown on flanks. At close range, pink bill with combination of black cutting edge and greener/yellower tip at all ages (but a very few dark-plumaged Antipodean Albatrosses may show these features). Juvenile/dark adult as Wandering, but note bill differences. [Albatros de Amsterdam]

Southern Royal Albatross *Diomedea epomophora* *c.*114 cm WS: *c.*320 cm

Widespread. Occurs closer inshore than Wandering Albatross. **ID** Juvenile: Unmarked black upperwing, distinguished from Northern Royal by very narrow white leading edge. Often shows black tail tip. **Immature (stage 2):** Similar narrow white leading edge on wing as on juvenile, but mottled white on inner wing like Wandering and Tristan. Tail usually white or barred black on outer rectrices. **Subadult (stage 4):** Broader white leading edge of wing and more heavily white mottled innerwing, unlike congeners. Tail usually white. **Adult:** Obvious white leading edge on wing and variable amount of pure white on innerwing. White tail. From Wandering by pale bill with black cutting edge, and from Northern Royal by narrow dark line (not wedge) on leading edge of primaries from below. Lacks any pink or ginger cheek patch, found on Wandering. [Albatros Real del Sur]

Northern Royal Albatross *Diomedea [epomophora] sanfordi* *c.*105 cm WS: *c.*300 cm

Uncommon in continental waters. Slighter build and shorter-billed than Southern Royal. **ID** Juvenile: Unmarked black upperwing, lacking white marginal coverts of Southern. More heavily streaked black on rump. **Adult:** Unmarked black upperwing and tail always white. Black cutting edge on bill visible at close range. From Southern by broader black wedge on leading edge of primaries from below. [Albatros Real del Norte]

exulans

Wandering Albatross

ad

ad ♂

subad

imm
(stage 4)

juv

Antipodean
Albatross

ad ♂

Tristan Albatross

ad ♂

ad ♀

Southern Royal
Albatross

ad

ad

subad
(stage 4)

imm
(stage 2)

juv

Northern Royal
Albatross

ad

ad

juv

Thalassarche From larger Great albatrosses by dark back; tail is dark in all species. Underwing and head patterns in combination with bill colours are the best features for identification. All species have nasal tubes situated far from the base of the bill. These smaller species are also known as mollymawks. See also Plate 19. **Tax note 4**.

Black-browed Albatross *Thalassarche melanophris*   85 cm WS: 240 cm

Breeds on the Falklands, Ildefonso and Diego Ramírez islands. Ranges throughout. **ID Adult:** Bill yellow with a reddish tip. White head and black brow. White central stripe on underwing bordered by thick black leading and trailing edges; shows less white on underwing than any other albatross. **Juvenile:** Olive-grey bill with black tip and small black area on base of upper mandible. White head with black brow as adult, and contrasting grey collar, often joining on breast. Underwing black at first before developing ghost pattern of adult underwing. **Subadult** (not illustrated) shows orangey bill with a black tip and may not have neat white stripe of adult underwing. [**Albatros Ceja Negra**]

Grey-headed Albatross *Thalassarche chrysostoma*   81 cm WS: 220 cm

Breeds on Ildefonso and Diego Ramírez. Ranges throughout, but scarce in continental waters. **ID Adult:** Black bill with yellow lines along culmen and base of lower mandible. Bluish-grey head with short dark brow, and greyish-brown mantle. Underwing shows more extensive white than Black-browed with a very narrow black trailing edge, but less extensive white than Yellow-nosed. **Immature:** Bill entirely black. Head variable, often with white on forehead or foreface, and a pronounced grey collar continuing onto hindcrown. Underwing entirely dusky but develops white with maturity. [**Albatros Cabeza Gris**]

Yellow-nosed Albatross *Thalassarche chlororhynchos*   78 cm WS: 198 cm

N continental waters; scarcer southwards. **ID Adult:** Black bill with bright yellow culmen and nail. Variable grey wash on head and dark brow, smaller than in Black-browed. Black triangle below eye. Tail darker than Black-browed. Ventral view shows much more white than Black-browed and Grey-headed, with a narrower black leading edge and very narrow trailing edge. **Juvenile:** Bill entirely black and head white. Underwing pattern as adult. [**Albatros Pico Fino**]

Black-browed Albatross

ad

melanophris

juv

ad

ad

Grey-headed Albatross

imm

ad

ad

ad

juv

ad

Yellow-nosed Albatross

chlororhynchos

Thalassarche From larger Great albatrosses by dark back; tail is dark in all species. Underwing and head patterns in combination with bill colours are the best features for identification. All species have nasal tubes situated far from the base of the bill. These smaller species are also known as mollymawks. See also Plate 18. **Tax note 4**.

Buller's Albatross *Thalassarche bulleri*   78 cm WS: 207 cm

Vagrant to Tierra del Fuego and the Falklands. **ID Adult**: Slender black bill with broad yellow-orange lines along culmen and base of lower mandible. Grey hood and throat with contrasting white forehead; some show a darker bluish-grey hood, then greatly enhancing the white forecrown. Blackish mantle concolorous with upperwing unlike congeners. Broad leading edge to underwing at all ages, broader than Yellow-nosed (Plate 18) and less bulging and more clean-cut than Grey-headed (Plate 18). **Juvenile**: Greyish bill, tipped black. Pale grey head and semi-collar arching down sides of neck and white forecrown. Separable from congeners by underwing pattern. [**Albatros de Buller**]

Shy Albatross *Thalassarche cauta*   95 cm WS: 235 cm

Continental and Falklands waters. **ID Adult** *steadi* 'Auckland' (common off Buenos Aires): Broad-based pale greyish-white bill, tipped yellowish. White crown contrasts with pale grey wash over ear-coverts and nape. Short blackish eyebrow. Slaty-black upperwing and somewhat greyer mantle. White distal primary shafts. **Adult** *cauta* 'Tasmanian' (rare in the region): Mostly white underwing with black thumbmark at the base of forewing, narrow black leading and trailing edges and a small dusky tip. From *steadi* only by bill colour; looks mostly yellow from a distance with a brighter tip. **Juvenile**: Resembles adult except for fairly uniform grey bill, tipped black, slightly larger black tips to underwing, and greyish semi-collar. [**Albatros Corona Blanca**]

Salvin's Albatross *Thalassarche salvini*   95 cm WS: 235 cm

Extreme vagrant to region. **ID Adult**: Considerably darker grey hood and throat than Shy Albatross with indistinct white forecrown. Dull olive bill with a yellowish culmen and dusky smudge near tip of lower mandible. Underwing similar to Shy but more extensive black tip. **Juvenile**: Virtually identical to Shy but tends to show darker and more extensive grey on the head, throat and semi-collar. [**Albatros de Salvin**]

Buller's Albatross

ad

ad

juv

Shy Albatross

ad

steadi

ad

cauta

juv

Salvin's Albatross

ad

ad

juv

Phoebetria Medium-sized brown albatrosses, distinguished from other species by their well-developed sulcus groove along the lower mandible and a strongly graduated, pointed tail.

Sooty Albatross *Phoebetria fusca*   86 cm WS: 203 cm

Very rare with sparse records, mainly north of the Falklands. **ID Adult:** Entirely sooty-brown with a slightly darker foreface. Note rounded crown, high position of eye, and broad orange-yellow sulcus groove. Yellowish or white primary shafts visible from a distance. **Juvenile:** Greyish nape and upper mantle; never as pale or extending to the rump as in Light-mantled. [**Albatros Oscuro**]

Light-mantled Albatross *Phoebetria palpebrata*   84 cm WS: 203 cm

Mostly in deep waters, in the Drake passage and south of the Falklands. **ID Adult:** Grey body with contrasting sooty hood. Grey saddle contrasts with dark head and wings. Note peaked crown, centralised eye position, and narrow pale blue sulcus groove. Fewer, and less distinct, pale primary shafts than Sooty Albatross. **Juvenile:** Resembles adult, but with crescentic markings on the mid and lower back. [**Albatros Manto Claro**]

Fregata Large seabird with notably long, pointed wings, prominently angled at the carpal, a scissor-shaped tail and a long slender bill with a hooked tip. Soars effortlessly, taking food from the surface while hovering or kleptoparasitises other seabirds. Males have an inflatable gular sac used in display.

Magnificent Frigatebird *Fregata magnificens*   101 cm WS: 230 cm

Sporadic visitor to Buenos Aires coasts (Oct–Feb); mostly juveniles; rarely inland. **ID** Considerable variation depending on age and sex. **Adult ♂:** Entirely black with scarlet gular sac. In favourable light shows purple and green gloss on head and mantle. **Adult ♀:** Black head and V-shaped throat stripe over white breast. Black belly, flanks and vent, and white lines on the axillaries. **Subadult ♀:** Black head and brownish pectoral band and variable black spotting on white breast and belly. **Young Juvenile (Stage 1):** White head, nape and underparts with a large wedge-shaped blackish spur over the breast sides. Black flanks and vent. **Older juvenile (Stage 2):** As younger bird but lacks the breast spurs. [**Ave Fragata**]

Sooty Albatross

ad

ad

juv

Light-mantled Albatross

juv

ad

ad

ad ♂

ad ♀

young juv
(stage 1)

older juv
(stage 2)

Magnificent Frigatebird

subad ♀

Macronectes Large petrels with dimensions similar to smaller albatrosses. Nasal tube extends more than half the length of the culmen. Wings do not extend beyond tail when folded. Tail is graduated with 16 rectrices.

Northern Giant Petrel *Macronectes halli* 87 cm WS: 190 cm

Scarce throughout the region. **ID** Dull or pale reddish tip to the bill at all ages. There are no adult dark or white morphs. **Adult:** Very similar to Southern but a high proportion shows a dusky cap and more extensive white continuing onto the breast; less demarcated. A much higher proportion shows white irides compared to Southern. **Juvenile:** Entirely dark brown. Note reddish bill tip. Fresh birds can appear darker-faced than Southern. **Immature:** Very similar to Southern but tends to show a darker cap. Some show a dark pectoral band. [Petrel Gigante Subantártico]

Southern Giant Petrel *Macronectes giganteus* 90 cm WS: 195 cm

Breeds in Patagonia and the Falklands, ranging throughout. **ID** Pale yellow bill, tipped pale olive or pale green at all ages; looks uniform at a distance. Highly variable plumage with several identifiable morphs. **Adult grey morph:** Rather variable. Usually shows a whitish head, throat and upper breast. Abdomen densely speckled with grey. **Adult white morph:** Entirely white with variable scattering of brown spots on the mantle, breast and wings. Usually only at S latitudes and much less common than grey morph; this plumage is unknown in Northern Giant Petrel. **Adult dark morph/Juvenile:** Entirely dark brown. **Immature:** Brownish cap, white forehead, throat and upper breast. Rest of body grey-brown. [Petrel Gigante Antártico]

Chionis A pigeon-like coastal scavenger. Sexes alike. Gregarious, especially in winter, often associates with pinniped colonies and hauling grounds.

Snowy Sheathbill *Chionis albus* 40 cm

Non-breeder, regular on the Falklands, and Patagonian and Buenos Aires coasts, mainly in winter with small numbers in summer. **ID Adult:** Olive or yellowish bill with a black tip and pale green sheath at the base. Pink skin around eye and caruncles at base of bill. In flight, rather dumpy with short broad wings. [Paloma Antártica]

Northern Giant Petrel

ad

juv

imm

ad / juv
dark morph

ad
grey morph

Southern Giant Petrel

imm

ad white morph

ads

Southern Giant Petrel

ad
white morph

ad
dark morph

Snowy
Sheathbill

Snowy Sheathbill

ads

Fulmarus Gull-like plumage. Long nasal tubes. *Thalassoica* Medium-sized, noticeably long-winged petrel. *Daption* Medium-sized pied petrel. Regularly follows ships. *Pagodroma* Entirely white medium-sized petrel. All species fly on stiff wings (except *Daption*).

Southern Fulmar *Fulmarus glacialoides*   45 cm WS: 117 cm

Throughout; reaches northern latitudes in Apr–Nov. **ID** Adult: Slender pink bill with blackish tip. Whitish head and underparts. Pale grey above with blackish trailing edge and white primary flashes. [Petrel Plateado]

Antarctic Petrel *Thalassoica antarctica*   43 cm WS: 105 cm

Vagrant to the Falklands and Drake Passage. **ID** Adult: Black bill and sooty hood, back and inner wing-coverts. Upperwing and tail mostly white with narrow black trailing edge and terminal band. [Petrel Antártico]

Cape Petrel *Daption capense*   39 cm WS: 86 cm

Throughout, reaches northern latitudes in Apr–Nov. **ID** Adult: Small black bill, black hood and wings with two separate white patches. White back and rump, checkered with black spots. Black tail band. [Petrel Damero]

Snow Petrel *Pagodroma nivea*   37 cm WS: 85 cm

Vagrant to the Falklands (Mar–Oct). **ID** Adult: Short stubby black bill. Entirely white with wedge-shaped tail. Compare with Snowy Sheathbill in flight (Plate 21). [Petrel Blanco]

Pterodroma Medium-sized petrels of deep offshore waters with short deep-based black bills; best identified by ventral patterns. Several species show a dusky M-marking across the upperwings. Flies in fast wheeling arcs. Another two species are illustrated on Plate 23 and one more on Plate 24.

Trindade Petrel *Pterodroma arminjoniana*   37 cm WS: 96 cm

Vagrant with sparse offshore records. **ID** Adult: Brown hood, often with a paler throat, contrasting with white breast and abdomen. Extensive white flash across underwing. Note fairly long, narrow tail. Brown above with vague M-marking across upperwings. An entirely dark morph has the same pale wing flashes of the pale morph. Intermediates occur. [Petrel Brasilero]

White-headed Petrel *Pterodroma lessonii*   43 cm WS: 109 cm

Mostly deep waters east of the Falklands in summer; scarcer southwards. **ID** Adult: White head, body and tail with dusky smudge through eye and mainly black underwing. M-marking across upperwings is usually browner than Mottled and Soft-plumaged Petrels (Plate 23). Partial or complete grey pectoral band. [Petrel Cabeza Blanca]

Atlantic Petrel *Pterodroma incerta*   43 cm WS: 104 cm

Deep offshore waters. **ID** Adult: Uniform brown hood and upperparts with contrasting white lower breast and belly, but brown undertail-coverts and undertail. Wings longer than congeners but 'hand' is proportionately short. Worn birds show a mask. [Petrel Cabeza Parda]

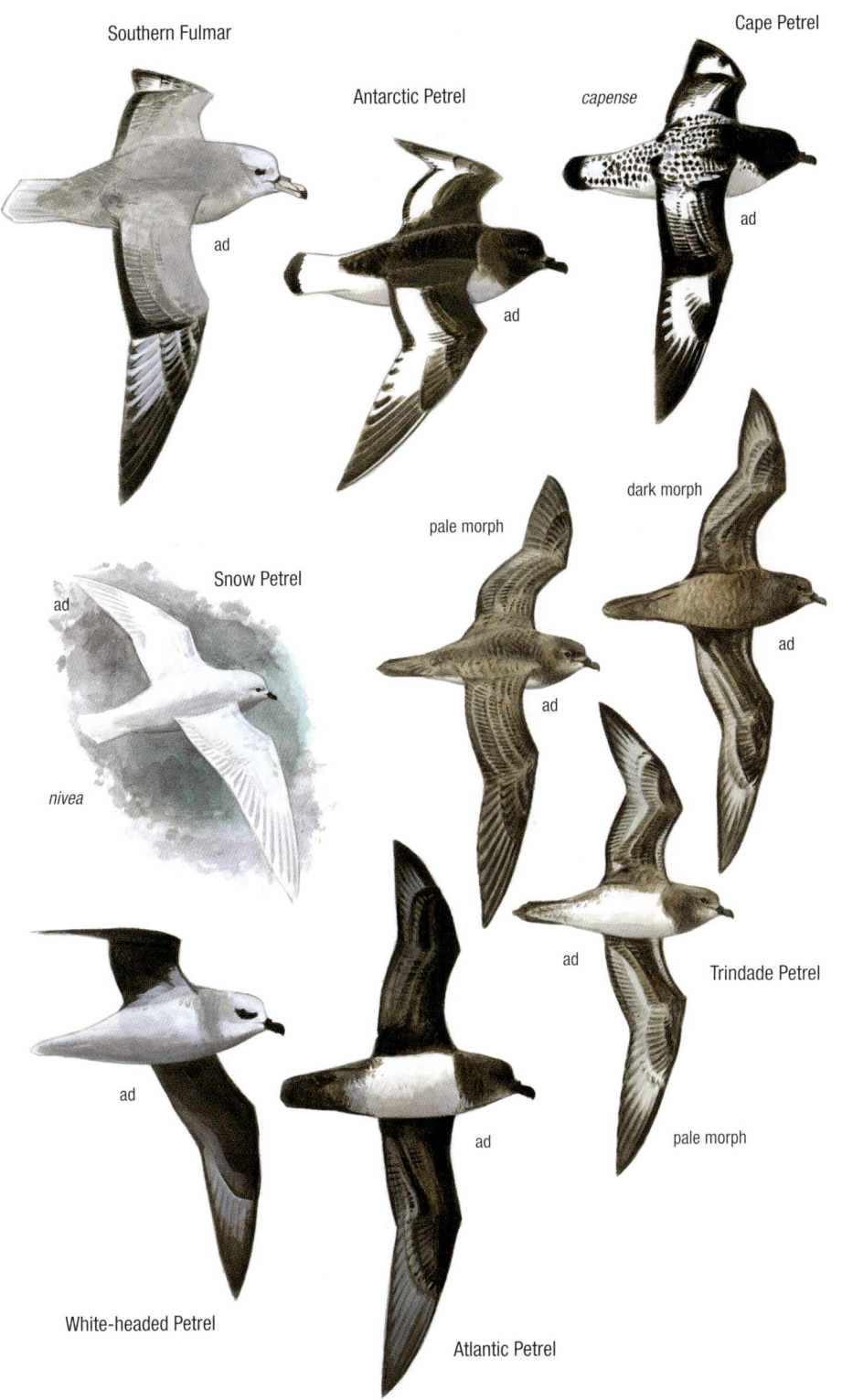

Southern Fulmar

ad

Antarctic Petrel

ad

Cape Petrel

capense

ad

Snow Petrel

ad

nivea

pale morph

dark morph

ad

ad

ad

Trindade Petrel

White-headed Petrel

ad

Atlantic Petrel

ad

pale morph

Pterodroma Medium-sized petrels of deep offshore waters with short, deep-based black bills; best identified by ventral patterns. Both species on this plate show a dusky M-marking across the upperwings. They fly in fast wheeling arcs. See also Plates 22 and 24.

Mottled Petrel *Pterodroma inexpectata*
34 cm WS: 85 cm

Very rare vagrant to the Falklands and Drake Passage in Dec. **ID Adult**: Grey breast and belly contrasting with white throat and vent. Conspicuous transverse black carpal bars on white underwings do not reach body. [Petrel Moteado]

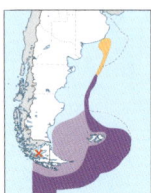

Soft-plumaged Petrel *Pterodroma mollis*
34 cm WS: 89 cm

Mainly deep offshore waters. **ID Adult light morph**: Grey crown and pectoral band. Dusky eye smudge. Underwing varies from being entirely dark to showing pale primary flashes and axillaries. From above, shows classic M-marking of many congeners. **Adult dark morph**: Sooty or dusky grey underparts with trace of pectoral band and mask. [Petrel Collar Gris]

Pachyptila Small blue-grey petrels with rounded black tail tip and blackish M-marking across upperwings. Ventral pattern similar in all species. Low weaving flight. See ID Keys in Appendix 2 (p. 443). *Halobaena* Prion-like petrel but with square-ended tail.

Fairy Prion *Pachyptila turtur*
26 cm

Breeds on Beauchêne Island, Falklands and winters nearby. **ID Adult**: Relatively short bill and rounded head. Pale whitish face without a contrasting post-ocular streak. Prominent black M-marking across upperwings and proportionately larger black tail tip than congeners. Indistinct grey extensions on sides of breast. [Prión Pico Corto]

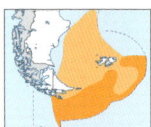

Antarctic Prion *Pachyptila desolata*
29 cm

Common in the Drake Passage in Feb–Mar; scarcer off S Patagonia and the Falklands; vagrant to s. Buenos Aires (not mapped). **ID Adult**: Perhaps not safely distinguished from Broad-billed in the field. Crown generally darker than mantle. Deep shovel-like pale bill. Narrow white supercilium, and dusky eye-stripe. Prominent black M-marking across upperwings. [Prión Pico Grande]

Broad-billed Prion *Pachyptila vittata*
27.5 cm

Vagrant to the Falklands. **ID Adult**: Perhaps not safely distinguished from Antarctic in the field. Tends to show a darker and more capped appearance with a steeper forehead, and at close range bill is much deeper, often appears blacker and the culmen is concave. Inconspicuous facial pattern. [Prión Pico Ancho]

Slender-billed Prion *Pachyptila belcheri*
29 cm

Breeds on the Falklands, ranging to continental waters. **ID Adult**: Fairly long, narrow bill and steep forehead. Broad white supercilium, often highlighted by narrow dark post-ocular streak. Appears rather washed out with indistinct M-marking across upperwings and proportionately small black or brown tail tip. In flight, with or without blue-grey extensions on sides of breast. [Prión Pico Fino]

Blue Petrel *Halobaena caerulea*
30 cm

Breeds at Ildefonso, Diego Ramírez and Cape Horn; ranges in the Drake Passage, and Fuegian and Falklands waters. **ID Adult**: Striking black cap and obvious white tail tip readily distinguishes this species from all prions. Large dusky spur on sides of breast. Flight is swooping with rapid changes of direction. [Petrel Azulado]

Mottled Petrel

ad

light morph

ad

Soft-plumaged Petrel

ad

dark morph

Fairy Prion

ad

Antarctic Prion

ad

ad

Broad-billed Prion

ad

Slender-billed Prion

Blue Petrel

PLATE 24: *PROCELLARIA* PETRELS AND ALLIES

Pterodroma Medium-sized, all-dark petrel of deep offshore waters with short, deep-based, black bill. Flies in fast wheeling arcs. *Aphrodroma* Smaller and differs by proportionately narrower and more pointed wings, steeper forehead and different flight action.

Great-winged Petrel *Pterodroma macroptera*　　　　40 cm WS: 99 cm

Vagrant in deep offshore waters. **ID Adult:** Stout black bill; deeper than congeners. Entirely dark brown with whitish bill surround at close range, and silvery-grey primary bases on underwing. Higher flight than congeners with pendulum motion in rough seas. [**Petrel Alas Grandes**]

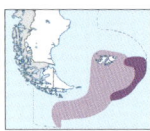

Kerguelen Petrel *Aphrodroma brevirostris*　　　　34 cm WS: 81 cm

Scarce in deep waters, mostly east and south of the Falklands. **ID** Fast weaving flight with swoops and high kestrel-like hovers. **Adult:** Brownish-grey or slate. In favourable light, reflective silvery underside of primaries, marginal and lesser wing-coverts, and sides of tail. Hooded or capped appearance is also diagnostic. Compare with Sooty Shearwater (Plate 25). [**Petrel Pizarra**]

Procellaria Stocky petrels with thickset cylindrical bodies, long powerful wings, wedge-shaped tails and robust pale bills with prominent nasal tubes. Regularly follow ships.

Spectacled Petrel *Procellaria conspicillata*　　　　55 cm WS: *c.*140 cm

Rare, mostly in deep waters from Buenos Aires to Patagonia and north of the Falklands, mainly Feb–Mar. **ID Adult:** From congeners by conspicuous white spectacles; variable in extent. Bill usually tipped black. [**Petrel de Anteojos**]

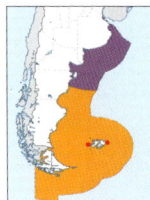

White-chinned Petrel *Procellaria aequinoctialis*　　　　53 cm WS: 140 cm

Breeds on the Falklands; common in continental waters. **ID Adult:** Sooty black with tiny white chin patch; variable in size and usually invisible. Greenish-yellow bill with black cutting edge and pale tip. From below, paler area in base of primaries. [**Petrel Barba Blanca**]

Parkinson's Petrel *Procellaria parkinsoni*　　　　*c.*46 cm WS: *c.*115 cm

Vagrant off Buenos Aires (Feb, Apr). **ID Adult:** Very similar to Westland Petrel, but with reduced black tip to shorter, narrower bill. Head more rounded and neck more slender. Folded primaries extend well beyond the tail. [**Petrel de Parkinson**]

Westland Petrel *Procellaria westlandica*　　　　51 cm WS: 137 cm

Rare in Fuegian waters; vagrant around the Falklands. **ID Adult:** Identical to White-chinned Petrel except for black bill tip. Folded primaries extend a little beyond the tail. [**Petrel Negro**]

Grey Petrel *Procellaria cinerea*　　　　50 cm WS: 120 cm

Deep offshore waters, mainly near the Falklands. **ID Adult:** Bluish bill with yellow stripe at close range. Blackish cap contrasts with greyish back and rump. Narrower, more pointed wings than congeners. Flight usually higher. Strong contrast between white body and dusky underwings (compare with Trindade Petrel, Plate 22) showing indistinct silvery-grey primary bases. Grey extensions on the breast sides. [**Petrel Ceniciento**]

Great-winged Petrel

ad

Kerguelen Petrel

ad

ad

Spectacled Petrel

ad

White-chinned
Petrel

Parkinson's Petrel

ad

ad

Westland Petrel

ad

Grey Petrel

Calonectris Sturdy shearwaters with proportionately long, bowed wings. ***Ardenna*** Slender wings held stiffly. Faster than *Calonectris*, slower than *Puffinus*. ***Puffinus*** Smaller-headed and slimmer-bodied than *Ardenna* with neat pied plumage. Flight action is considerably faster than in *Ardenna*, especially in Little which does not follow ships (unlike most other species).

Cory's Shearwater *Calonectris diomedea*   49 cm WS: 117 cm

Buenos Aires waters south to N Patagonia (Dec–May). **ID** Adult: Black-tipped yellow bill. Greyish-brown crown and back grading below ear-coverts into pure white underparts. Uniform brown upperwing. Indistinct white horseshoe over the uppertail-coverts, and fairly long rounded blackish tail. From below, white wing-linings with black trailing edge and dusky primaries. [Pardela Grande]

Cape Verde Shearwater *Calonectris edwardsii*   c.45 cm WS: c.106 cm

Buenos Aires waters (Nov–Apr, mainly Feb–Mar). **ID** Adult: Proportionately smaller-headed and longer-tailed than Cory's. Slender pinkish or grey bill with a black subterminal band and grey tip. Dusky M-marking across upperwings. More white across uppertail-coverts than Cory's. Clear-cut white throat. [Pardela de Cabo Verde]

Pink-footed Shearwater *Ardenna creatopus*   46 cm WS: 109 cm

Vagrant recorded once in May in N Patagonia. **ID** Adult: Recalls Cory's but bill pink, tipped black; more extensive brownish over sides of neck and leading edge of underwing; flanks and undertail-coverts brown; lacks white on uppertail-coverts. Flight more laboured than congeners. [Pardela Patas Rosas]

Flesh-footed Shearwater *Ardenna carneipes*   c.47 cm WS: 115 cm

Vagrant recorded once in Oct, N of the Falklands (not mapped). **ID** Adult: Large, stocky blackish-brown shearwater with fairly broad wings. Large rounded head. Slender pale yellowish or grey bill, tipped black. Pale fleshy tarsus diagnostic. Compare with Sooty Shearwater, and Westland and Parkinson's Petrels (Plate 24). [Pardela Patas Claras]

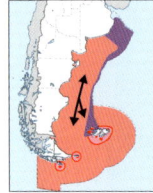

Sooty Shearwater *Ardenna grisea*   44 cm WS: 99 cm

Breeds on Fuegian islands and the Falklands, ranging throughout continental waters, often in large numbers. **ID** Adult: Entirely blackish-brown with silvery flashes on underwing-coverts. Slender body and wings; very agile compared to congeners. Long and very fine drab bill; dull grey tarsus. [Pardela Oscura]

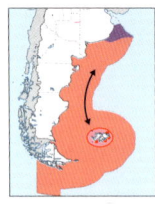

Great Shearwater *Ardenna gravis*   50 cm WS: 115 cm

Local Falklands breeder. Common from Buenos Aires to N Patagonia. **ID** Adult: Slender black bill. Sooty cap and white nuchal collar. White horseshoe over uppertail-coverts contrasts with black tail. Mostly white underwings, tipped dusky and with a prominent dusky trailing edge and narrow leading edge; blackish diagonal bar across inner wing-coverts. Brown belly patch. [Pardela Cabeza Negra]

Little Shearwater *Puffinus assimilis*   28.5 cm WS: 61 cm

Scarce around the Falklands; vagrant N to Buenos Aires. **ID** Adult: More compact than Manx Shearwater with less pointed wing tips, narrower trailing edge and smaller black wing tip. Faster wingbeats with less gliding, closer to the water. Upperwing often shows a pale stripe along the greater-coverts, lacking in Manx. When swimming, note fine delicate black bill with blue-grey base. Smaller breast spur than Manx, rounder head, and folded primaries just reach the tail tip. Diagnostic blue tarsus is hard to see. [Pardela Chica]

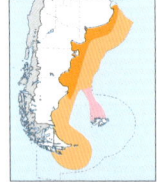

Manx Shearwater *Puffinus puffinus*   34 cm WS: 82 cm

Common in continental waters. **ID** Adult: White underparts and underwing with broad black wing tips, trailing edge and narrow leading edge. Upperparts black. When swimming, note slender black bill, broad blackish breast spurs and folded primaries extending beyond the tail tip. [Pardela Boreal]

Cory's Shearwater

ad

borealis

Cape Verde Shearwater

ad

Pink-footed
Shearwater

ad

Flesh-footed
Shearwater

ad

ad

Sooty
Shearwater

Great Shearwater

ad

elegans

Little Shearwater

Manx Shearwater

ad

ad

Garrodia, ***Fregetta*** and ***Oceanites*** Small storm petrels with rounded wings and white rumps (grey in *Garrodia*). Three breed in the region, one is a summer visitor and two are vagrants. All images are adults; immatures are similar. See also Plate 27.

Grey-backed Storm Petrel *Garrodia nereis* 17 cm

Breeds on the Falklands, ranging into surrounding, and Fuegian, waters. Often on floating kelp. **ID** Adult: Small boldly patterned storm petrel with short rounded wings held straight out from body. Tarsus reaches or extends beyond the tail tip. Grey rump and most of tail with black terminal band. Grey greater wing-coverts and centre of mantle. White abdomen and underwing-coverts with broad black leading edge. [Paíño Gris]

Black-bellied Storm Petrel *Fregetta tropica* 20.5 cm

Common in deep offshore waters of the Drake Passage and east of the Falklands. **ID** Adult: Robust storm petrel with a white rump, underwing-coverts and much of abdomen. Erratic zig-zagging flight. Indistinct upperwing stripe, usually lacking altogether. Blackish stripe through centre of abdomen reaching the vent, variable in width and sometimes broken. Tarsus extends slightly beyond tail. [Paíño Vientre Negro]

White-bellied Storm Petrel *Fregetta grallaria* 19.5 cm

Very rare vagrant, mostly to the NE of the Falklands. **ID** Adult: Robust storm petrel with a white rump, underwing-coverts and much of abdomen. Erratic zig-zagging flight. Greyish-white upperwing stripe, and slight greyish cast to upperparts. Slightly shorter tarsus than Black-bellied, and unmarked white abdomen with black undertail-coverts. [Paíño Vientre Blanco]

Wilson's Storm Petrel *Oceanites oceanicus* 17–19 cm

Common throughout; breeds on the Falklands. **ID** Adult *oceanicus* (17 cm): White-rumped storm petrel with comparatively short, rounded wings, held straight out from body, and squarish tail. Tarsus has yellow webs and extends beyond tail in fast travelling flight. White rump wraps onto lateral tail-coverts and sometimes gives false impression of a white vent. Upperwing stripe usually prominent. Foot-pattering foraging habit is distinctive. Adult *exasperatus* (19 cm; not illustrated): Breeds in Antarctica and the Scotia sea; ranges rarely north to Buenos Aires and differs only by larger size. [Paíño de Wilson]

Fuegian Storm Petrel *Oceanites* [*oceanicus*] *chilensis* 17 cm

Breeds on Fuegian islands, ranging into SW Atlantic. **ID** Adult: Differs from Wilson's by showing a second less distinct and short white underwing bar on some (fresh) individuals. **Tax note 5**. [Paíño Fueguino]

Pincoya Storm Petrel *Oceanites pincoyae* 17 cm

Occasional at Lago Puelo, Chubut (from the Pacific). **ID** Adult: White belly, often smudged brown usually connects with white rump. Upperwing stripe variable but white in fresh plumage. Striking white underwing stripe. Throat and breast paler than mantle. **Tax note 6**. [Paíño Pincoya]

Grey-backed Storm Petrel

ad

chubbi

Black-bellied Storm Petrel

tropica

ad

leucogaster

ad

White-bellied Storm Petrel

oceanicus

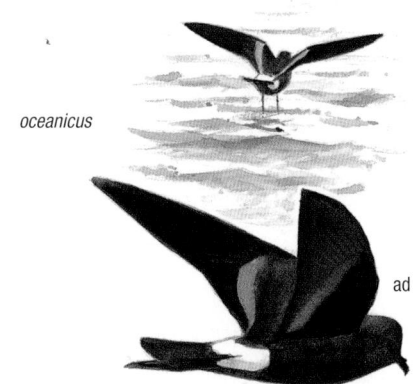

ad

Wilson's Storm Petrel

ad

Fuegian Storm Petrel

ad

Pincoya Storm Petrel

Pelagodroma and *Cymochorea* Long-winged vagrants. All images are adults; immatures are similar. See also Plate 26.

White-faced Storm Petrel *Pelagodroma marina* 20.5 cm

Extremely rare vagrant. **ID Adult:** Large, boldly patterned storm petrel with proportionately long wings and tarsus that projects far beyond the tail. Rapid jerky travelling flight mixed with bounding glides. Dark cap and post-ocular stripe with contrasting white supercilium. Grey-brown mantle and upperwing-coverts, often with paler wing stripe. Pale grey rump and white underparts. [Paiño Cara Blanca]

Leach's Storm Petrel *Cymochorea leucorhoa* 20 cm

Extremely rare vagrant. **ID Adult:** Fairly large storm petrel with proportionately long angled wings and long forked tail. Striking pale upperwing stripe. White rump with dusky central line. [Paiño Boreal]

Hornby's Storm Petrel *Cymochorea hornbyi* 22 cm

Large, vagrant storm petrel recorded once in sw. Neuquén (Apr). **ID Adult:** Grey above with black cap and dark wings. Pale wing-stripe. Forked tail. White forehead and underparts with broad grey breast band. [Paiño de Collar]

Pelecanoides Small compact pied petrels with short wings and tail. Mainly around subantarctic islands to 60°S. Flight is straight and rapid, less than 1 m above surface, followed by a sudden dive into the sea, surfacing elsewhere.

Common Diving Petrel *Pelecanoides urinatrix* 19–21 cm

Widespread offshore. **ID Adult** *berard* (21 cm; breeds on the Falklands; winters north to continental waters): Greyish ear-coverts and extensive mottling on sides of breast suggesting a pectoral band. Pale scapular line sometimes present. **Adult** *coppingeri* (19cm; Santa Cruz and Fuegian region): As *berard*, but smaller and without breast mottling. [Yunco Común]

Magellanic Diving Petrel *Pelecanoides magellani* 21 cm

Breeds on Fuegian islands; ranges to the Falklands, and to Chubut in winter. **ID Adult:** White crescent on sides of neck and blackish extension on sides of breast. Scapulars fringed white when fresh. [Yunco Magallánico]

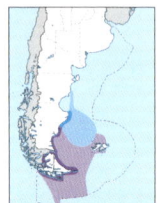

South Georgia Diving Petrel *Pelecanoides georgicus* 20.5 cm

Extremely rare vagrant to the Drake Passage and the Falklands. **ID Adult:** Very similar to Common Diving Petrel but pale grey swathe extends over the eye and curves around darker ear-coverts. White scapular line often notable. Usually less grey on sides of breast than ssp. *coppingeri* of Common. [Yunco Geórgico]

Leach's Storm Petrel

ad

ad

marina

White-faced Storm Petrel

ad

Hornby's Storm Petrel

ad

Common Diving Petrel

Magellanic Diving Petrel

ad

berard

ad

ad

coppingeri

Common Diving Petrel

ad

South Georgia Diving Petrel

Anhinga Large, extremely long-necked aquatic bird with webbed toes, fine pointed bill and proportionately long tail. Swims low in the water; soars high in the sky. Sexually dimorphic.

Anhinga *Anhinga anhinga* 83 cm

N lowlands. **ID Adult ♂**: Black with long white streaks on scapulars, mostly white wing-coverts and white tail band. **Adult ♀**: Brown head, hindneck and mantle. Tawny throat and upper breast contrasts with black belly. Scapulars as ♂, but wing-coverts mostly cream, and tail band brown. In flight, note slender wings and long neck and tail. **Juvenile** (not illustrated): Differs by whiter foreneck and upper breast, and browner belly. [Aninga]

Phalacrocorax Large, long-necked aquatic, mostly marine, birds with webbed toes, slender bill with a hooked tip, and proportionately short rigid tail. Breeding birds show filoplumes, tufts or brightly coloured caruncles. Sexes alike.

Neotropic Cormorant *Phalacrocorax brasilianus* 72 cm

Throughout the lowlands, Andean foothills and coasts south to Tierra del Fuego. **ID Adult breeding**: Essentially black with a white line surrounding the gular pouch, and small white plumes on the side of the head. **Adult non-breeding**: Browner throughout with olive tones on the back and wings. Lacks white head markings. Often holds wings out to dry. In flight, large all-black cormorant with distinctively kinked neck. **Juvenile** (not illustrated): Similar but very pale brown on throat and breast. [Biguá]

Red-legged Cormorant *Phalacrocorax gaimardi* 63 cm

Coastal Santa Cruz, occasionally wintering to ne. Chubut. **ID Adult**: Stunning pale blue-grey cormorant with red sheath around bill base, bright coral-red tarsus, white oval neck patch and white wing-spotting. In flight, entirely grey; white neck patch usually visible from a distance. [Cormorán Gris]

Guanay Cormorant *Phalacrocorax bougainvillii* 70 cm

Restricted to one headland in e. Chubut where possibly hybridised out with Imperial Shag. **ID Adult**: Slender yellowish bill and bare red skin around eye. Black with narrow white line extending from throat down foreneck, expanding in an inverted U over breast. Larger and more slender-necked, but shorter-tailed, than Rock Shag. In flight, tarsus extends just beyond rather short tail. [Guanay]

Rock Shag *Phalacrocorax magellanicus* 67 cm

Patagonia and the Falklands. **ID Adult breeding**: Black bill, red facial skin, white patch on ear-coverts, small frontal tuft and solid black foreneck separate from Guanay Cormorant, but beware of aberrant birds with white throats or neck stripes. In flight, black foreneck and contrasting white abdomen. Fairly long tail and tarsus not visible. **Juvenile**: Brown head, foreneck and upperparts; white belly with brown mottling and blotches. [Cormorán Cuello Negro]

Imperial Shag *Phalacrocorax atriceps* 73–82 cm

Widespread in Patagonia and SW Atlantic islands. **ID** Prominent recurved erectile crest in breeding birds of all ssp. **Adult non-breeding** *atriceps* (73 cm; Patagonian coasts and lakes from sw. Neuquén to w. Chubut): Yellow caruncles and blue eye-ring. Bottle-green above with white carpal bar and usually a small white interscapular patch. White ear-coverts and foreneck. **Adult breeding** *albiventer* (73 cm; Falklands): From *atriceps* by black ear-coverts and less likely to show an interscapular patch. **Adult non-breeding** *albiventer*: From continental form by yellow or pink caruncles. **Adult non-breeding** *bransfieldensis* (82 cm; Antarctica): Historical Fuegian records. Larger and proportionately longer-tailed than other ssp. Extent of white on face similar to *atriceps*. Usually has a large white interscapular patch. Yellow-orange caruncles. In flight, all forms show white foreneck and contrasting white underparts. **Tax note 7**. [Cormorán Imperial]

Neotropic Cormorant

brasilianus

ad

ad non-br

ad br

Anhinga

ad ♀

ad ♀

ad ♂

anhinga

Neotropic

ad

Red-legged Cormorant

ad

Red-legged

ad

Guanay

ad

Rock

ad

Imperial

ad

cirriger

ad

Guanay Cormorant

ad

Rock Shag

ad br

juv

ad non-br

ad br

albiventer

ad non-br

atriceps

ad non-br

bransfieldensis

Imperial Shag

Phoenicoparrus Deep, keel-billed flamingos which feed on unicellular organisms, and lack a hind toe.

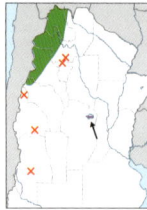

James's Flamingo *Phoenicoparrus jamesi*

90 cm

NW Andes above 3000 m south to La Rioja, and Mar Chiquita, ne. Cordoba. **ID Adult breeding**: Relatively short red legs. Pink restricted to head and upper neck, chest, and lower scapulars. Short black distal portion of yellow bill imparts blunt appearance. Dark red lores and border to bill. In flight, note rich pink rump and axilliaries. Lacks pink on upperwing, unlike Andean and Chilean. **Adult non-breeding**: Mostly white. Remiges mostly concealed. **Juvenile**: Note bill shape. Head and neck cleaner than congeners. Streaks restricted to lower back. [Parina Chica]

Andean Flamingo *Phoenicoparrus andinus*

110 cm

Range as James's with more sporadic lowland records. **ID Adult breeding**: Long yellow legs. Head and neck pink, becoming bright magenta on chest. Pink wing-coverts. Long hooked black distal portion of bill with yellow base; complex orange and green pattern at brief height of breeding season. In flight, pink wing-coverts with a whitish leading edge. **Adult non-breeding**: Mostly white with, at most, pink chest and wing-coverts. Remiges mostly exposed unlike James's and Chilean. **Juvenile**: Bill can be shorter than adult. Head and neck duskier than congeners. Streaked back. [Parina Grande]

Phoenicopterus From *Phoenicoparrus* by less specialised feeding habits due to distinct bill structure, and presence of a small hind toe.

Chilean Flamingo *Phoenicopterus chilensis*

100 cm

Throughout. **ID Adult breeding**: Grey legs with pink 'knees' and feet. Pink chest. Reddish wing-coverts and lower scapulars. Long hooked black distal portion of bill with whitish base. Pale iris. In flight, entirely red wing-coverts from above and below. **Adult non-breeding**: Similar to Andean except for completely concealed remiges. **Juvenile**: Note bill shape. Distinctly browner (less grey) than congeners. [Flamenco Austral]

James's

ad br

Andean

ad br

Chilean

ad br

James's Flamingo

juv

ad non-br

ad br

James's

ad

Chilean

ad

Andean

ad

juv

ad non-br

ad br

Andean Flamingo

ad br

ad non-br

juv

Chilean Flamingo

Mycteria, *Jabiru* and *Ciconia* Wood Stork and Jabiru are both terrestrial and arboreal; Maguari Stork is terrestrial. All three species are often with other storks and herons.

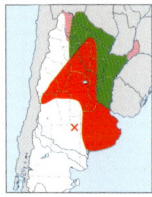

Wood Stork *Mycteria americana*   86 cm

N lowlands (Oct–Apr). **ID** Small stork with a drooping bill and bare head. **Adult:** Pink feet and greyish head and neck with a pale forehead. Appears mostly white when standing. In flight, from Maguari Stork by black remiges contrasting with reduced white wing linings, and mostly black tail visible from below. **Juvenile:** Brown head and pale yellow bill. [**Tuyuyú**]

Jabiru *Jabiru mycteria*   136 cm

Resident in N chaco wetlands and NE esteros. **ID** A giant stork with bare head and bulbous neck, and huge bill with upturned lower mandible. **Adult:** Blackish legs, bill, head and neck with red base. Rest of plumage white. In flight, all white with contrasting dark neck. **Juvenile:** Browner head, often with pink on base of neck. Plumage washed with brown, especially on wing-coverts. [**Yabirú**]

Maguari Stork *Ciconia maguari*   119 cm

Resident in N and C lowlands. **ID** Large stork with straight, thick-based bill. **Adult:** Red legs and loral skin. Mostly white with exposed black remiges. In flight, from Wood Stork by very broad white wing linings, almost reaching trailing edge, and only black tail borders visible from below. **Juvenile:** Dark bill. Mostly black with a white face. [**Cigüeña Americana**]

Pelecanus Well-known, fish-eating, aquatic birds with huge bills, broad wings and short tail. Flies with neck folded.

Peruvian Pelican *Pelecanus thagus*   *c.*140 cm

Vagrant from Chile. **ID Adult breeding:** Long, deep, yellow and red bill with blue gular pouch. Pale yellow head and foreneck. Brown to blackish hindneck. Mostly grey above with pale wing-coverts contrasting with blackish remiges and dark underwings. **Adult non-breeding:** Has drab bill colours and whitish head. **Juvenile:** Brownish head, neck and upperparts; white breast and belly. [**Pelícano Pardo**]

Peruvian Pelican

Wood Stork

ad

juv

Jabiru

juv

ad

Maguari Stork

ad

juv

ad

ad

Wood Stork

Maguari Stork

ad

Jabiru

Botaurus and *Ixobrychus* One large species and two smaller marsh bitterns with a residual tail and long, broad wings. All are reclusive and secretive.

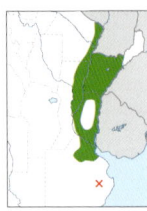

Pinnated Bittern *Botaurus pinnatus*   71 cm

Local resident in the NE, south to ne. Buenos Aires. **ID** A very secretive, mostly solitary, large bittern of reed- and rushbeds with a short tail and very large, broad wings. Often detected by its crepuscular booming. **Adult:** Buff upperparts with a black-barred hindneck and dorsal tyre-track markings. Creamy-white below with long, thick cinnamon streaks on the foreneck. Compare juvenile Rufescent Tiger Heron (Plate 33). In flight, contrasting black flight feathers and primary wing-coverts. Tarsus extends well beyond the tail. **Voice** Extremely low-pitched booming. *BHMM…m-m…BHMM…m-m*, recalling blowing over an empty bottle. Flushes with a loud croak and dangling legs. [Mirasol Grande]

Azara's Bittern *Ixobrychus* [*exilis*] *erythromelas*   31 cm

Local in e. Formosa, e. Chaco, n. Corrientes and n. Misiones (Sep–Apr). **ID Adult** ♂: Black crown and back with contrasting buff wing-coverts. Chestnut face and sides of neck. Whitish throat, foreneck and belly. **Adult** ♀: Similar to ♂ but cap and back brown; some streaking on the foreneck. In flight, slaty flight feathers contrast with brown wing-coverts. **Juvenile:** Rusty upperparts with black checkered pattern. White below finely streaked on the breast. **Voice** Mostly crepuscular, in spring is a series of 3–6 well-spaced, deep resonant gurgling *grraw* or *grrahh* notes, more drawn-out than Stripe-backed Bittern. **Tax note 8.** [Mirasol Caoba]

Stripe-backed Bittern *Ixobrychus involucris*   36 cm

N and C lowlands. Prone to wander. **ID Typical adult:** Mainly buff with black and white streaks on the back, and variable fine breast streaking. In flight, buff wing-coverts and mainly black flight feathers with a broad chestnut trailing edge. **Juvenile:** Warmer throughout than adult. Thick rusty brown streaks on the foreneck. **Voice** Call is a series of 4–9 resonant, gulping *KuOkk* notes, usually preceded, or followed, by several well-spaced, low-pitched stuttered croaks. [Mirasol Estriado]

Butorides Diurnal, found alone or in pairs. *Cochlearius* Strictly nocturnal, gregarious at colonies otherwise solitary and reclusive in gallery and riverine forest. *Nycticorax* and *Nyctanassa* Crepuscular and nocturnal, and usually gregarious.

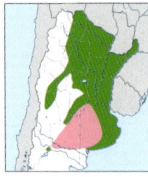

Striated Heron *Butorides striata*   42 cm

N and C lowlands. **ID Adult:** Black cap contrasting with grey face and neck. Dark green wing-coverts, scaled with white. Bright orange tarsus. Shows a white trailing edge to the wings in flight. **Juvenile:** Crown streaked cinnamon. Brown back and wings with white triangular flecks. White below, streaked brown on the breast. **Voice** On take-off and during flight, a short, raspy *keow!* [Garcita Azulada]

Boat-billed Heron *Cochlearius cochlearius*   59 cm

Very local resident in e. Formosa, e. Chaco and n. Misiones forest. **ID** A chunky nocturnal heron recalling Black-crowned Night Heron, but with a broad flattened bill, notably large iris, and long broad nuptial plumes on the sides of the nape. **Adult:** White forehead and creamy ear-coverts. Black crown and long nuptial plumes. Pale grey back and wings. White below with a chestnut central abdomen and vent, and black flanks. In flight, pale grey back and wings, and white rump and tail. Note chestnut belly and black flanks. **Juvenile:** Brown crown and chestnut-mauve back and wings. White below with a brown wash on the foreneck. [Garza Cucharona]

Black-crowned Night Heron *Nycticorax nycticorax*   63 cm

Widespread. **ID** Stocky crepuscular and nocturnal heron with a stout pointed bill and very fine long white nuptial plumes. **Adult** *hoactli* (N lowlands): Small white forehead and supraloral streak. Black cap and back, and grey wings. White face and underparts; tinged grey or brown on breast. **Adult** *obscurus* (Patagonia): From *hoactli* by brown ear-coverts and underparts. Most high Andean birds are similar to *obscurus*. **Juvenile:** Orange iris. Head and neck streaked brown and buff. Brown back and wings with white drop-shaped spots. **Voice** Characteristic hollow *kwok* call at dusk and dawn in flight. [Garza Bruja]

Yellow-crowned Night Heron *Nyctanassa violacea*   67 cm

Not illustrated. Vagrant recorded once at 2500 m in Andes of Jujuy. **ID Adult:** Mostly grey with black head, creamy crown and white cheek stripe. **Juvenile:** Very similar to Black-crowned but longer-legged, darker with reduced wing-covert markings and stouter, blunter bill. [Garza Bruja Coronada]

Pinnated Bittern

Azara's Bittern

ad ♂

♀

juv

ad

ad

pinnatus

ad

juv

ad

juv

Stripe-backed
Bittern

juv

ad

striata

Striated Heron

juv

Boat-billed Heron

ad

juv

ad

hoactii

juv

Black-crowned
Night Heron

ad

ad

obscurus

Ardea Large herons with a long S-shaped neck. Slower wingbeats and broader wings than smaller and slender-necked *Egretta*. *Pilherodius* and *Bubulcus* are distinctive monotypic genera. All species develop elongated nuptial plumes on nape and scapulars in breeding plumage.

Capped Heron *Pilherodius pileatus* — 57 cm

Sporadic in n. Misiones, e. Formosa and e. Chaco. **ID** A thick-necked, solitary marsh heron with long wire-like nuptial plumes, fairly short rounded wings and a slender bill. **Adult breeding:** Bluish bill and bright blue loral skin extending around eye. White forehead and black cap. Otherwise white with a yellowish wash over the neck and breast; lacking in juveniles and non-breeders. [Garcita Real]

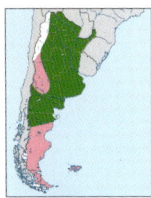

Cocoi Heron *Ardea cocoi* — 116 cm

Widespread in the N lowlands and Andes to 3500 m; scarce in S lowlands. **ID Adult breeding:** Black cap and scapular tuft over shoulder. White neck and grey back and wings. Pink legs. Non-breeding birds show more uniform yellow-orange bill and less contrasting lores. **Juvenile:** Greyish neck and greenish dusky bill. [Garza Mora]

Cattle Egret *Bubulcus ibis* — 50 cm

Widespread in the lowlands; occasional in the Andes. **ID** Thickset, gregarious egret with a relatively short neck and legs. **Adult breeding:** Yellow or reddish bill and legs. White with ginger crown, breast and plumes. **Adult non-breeding:** Entirely white. Tarsus extends beyond tail. [Alt: Western Cattle Egret] [Garcita Bueyera]

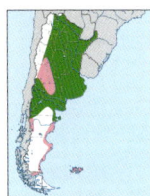

Snowy Egret *Egretta thula* — 54 cm

Widespread in the lowlands and Andes. **ID Adult breeding:** Entirely white. Black bill with a yellow base and loral skin. Black legs with yellow toes. **Juvenile:** Recalls juvenile Little Blue Heron but olive tarsus often shows either yellow at rear or black at front. Bill is longer with a straighter culmen. Lacks dusky primary tips in flight. [Garcita Blanca]

Little Blue Heron *Egretta caerulea* — 52 cm

Rare in the north; once in Chubut. **ID Adult breeding:** Blue-grey with a dark purplish head and neck. **Juvenile:** Recalls adult and juvenile of Snowy Egret but lores tend to be grey (never bright yellow), and legs tend to be olive (never with yellow at rear, or black on front). Bill shorter with very slightly curved culmen, and usually grey with distal black tip. Dusky primary tips in flight diagnostic. [Garza Azul]

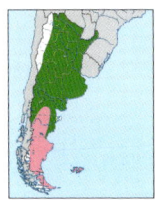

Great Egret *Ardea alba* — 94 cm

Widespread in lowlands and Andes to 3000 m. **ID Adult non-breeding:** Entirely white. Yellow bill and blackish legs, extending well beyond tail in flight. Broader wings and slower flight than smaller Snowy Egret. **Adult breeding** (not illustrated): Bright green lores contrasting with yellow bill and elaborate scapular plumes. [Garza Blanca]

Capped Heron

ad br

Cocoi Heron

ad br

ad non-br

ad br

ibis

Cattle Egret

ad br

thula

Snowy Egret

ad

ad

juv

ad br

ad non-br

Great Egret

egretta

ad

Little Blue Heron

Tigrisoma Two large, solitary, long-necked resident herons with a long, pointed bill, bare yellow loral skin and gular sac, and a loose ruff on the hindneck. The neck is folded in flight.

Rufescent Tiger Heron *Tigrisoma lineatum* 80 cm

N lowlands and foothills. **ID Adult:** Rufous crown and hindneck. White throat and foreneck with dark brown 'neck-tie'. Grey-brown back and wing-coverts. **Juvenile:** Head and neck barred buff and black; sometimes with a black crown. Back and wing-coverts spotted or barred buff and black. Buff thighs feathered to the intertarsal ('knee') joint and barred black. **Voice** Loud, deep and resounding series *ho-KÓ...ho-KÓ...*; also far-carrying cow-like and drawn-out moans. [**Hocó Colorado**]

Fasciated Tiger Heron *Tigrisoma fasciatum* 54 cm

Yungas forest rivers up to 1700 m (*pallescens*). Nominate ssp. extinct in Misiones? **ID Adult:** Black crown. Brown to greyish face and dark olive sides of neck, vermiculated throughout and bordered by a black line. Brown 'neck-tie' and pinkish belly. **Juvenile:** Distinguished from Rufescent Tiger Heron by thighs not strongly barred black, proportionately shorter tarsus, back and wing-coverts vermiculated and never blotched with black. [**Hocó Oscuro**]

Syrigma A conspicuous heron of open country and light woodland with a short slender bill and filamentous nuptial plumes. Gregarious in winter.

Whistling Heron *Syrigma sibilatrix* 57 cm

N lowlands and NW Andes to 2000 m. **ID Adult:** Pink bill, tipped black. Sky-blue facial skin. Dark cap and buff neck. Blue-grey back and wings with black-streaked cinnamon wing-coverts. In flight, mostly grey above and creamy-buff below with a white rump and tail. Note bowed shallow wingbeats. **Juvenile:** Pale grey above with buff on the upper back. Cinnamon wing-coverts, finely streaked black. **Voice** 2–4 distinctive far-carrying flute-like whistles. [**Chiflón**]

Roseate Spoonbill *Platalea ajaja* 80 cm

N lowlands, moving north in winter; vagrant to the high Andes and Falklands. **ID** Long, flattened bill with distinctive spatulate tip. Often gregarious or found with egrets. **Adult breeding:** Yellowish to black bill. Orange lores. Naked, sulphur crown. Long, reddish legs. Pink pectoral tuft, lower back, wings and belly. Magenta scapulars, rump and vent. **Immature:** Lacks magenta in plumage, and head is all white. Note small black primary tips. [**Espátula Rosada**]

Green Ibis *Mesembrinibis cayennensis* 62 cm

Paraná forest and gallery forest in the NE. **ID** Large ibis tied to riverine forest, with short legs, broad-based bill and a nuchal ruff. Solitary or in pairs. Often reclusive. **Adult breeding:** Olive bill with a yellowish tip. Dusky legs. Dark brown head and underparts. Metallic green nuchal ruff and bronze-green back and wing-coverts with bluish-black flight feathers. In flight, tarsus does not extend beyond tail. **Voice** Delivers an excitable trilling and low-pitched gurgled cacophony while perched or in flight, recalling Greater Ani. [**Tapicurú**]

Limpkin *Aramus guarauna* 68 cm

N lowlands and foothills to 1400 m. **ID** Large marsh bird with a long straight bill, notably long slender neck and slender tarsus. Solitary or in pairs but highly gregarious in winter. **Adult:** Yellowish bill, tipped dusky. Dark brown, streaked white on the hindneck and upper back. In flight, high stiff mechanical flaps with long downward angled legs. **Voice** Very vocal, loud *krrraaawww!* and growling noises, frequently heard also at night. [**Carau**]

ad

Rufescent
Tiger Heron

juv

juv

ad

pallescens

Fasciated Tiger Heron

Whistling Heron

juv

imm

Roseate Spoonbill

ad

sibilatrix

ad

ad br

Green Ibis

ad

ad br

ad

Limpkin

carau

ad

Bare-faced Ibis *Phimosus infuscatus*   54 cm

N lowlands. **ID** Short-legged marsh ibis with bare facial skin and a thick bill base. Overlaps with White-faced and rarely with Puna. **Adult breeding:** Red face. Pink bill and legs. Mostly black, glossed green. **Adult non-breeding** (not illustrated): Yellowish bill, and plumage browner with less gloss. In flight, wings fairly broad and rounded, feet do not extend beyond the tail. Note loose formation of group in flight. [Cuervillo Cara Pelada]

Plegadis Gregarious marsh ibises with a slender bill and long legs.

White-faced Ibis *Plegadis chihi*   49 cm

Widespread in the lowlands; NW Andes to 2000 m. Overlaps with Bare-faced and with Puna in winter. **ID Adult breeding:** Grey or pink bill with a red base and lores. Narrow white spectacles. Reddish tarsus. Coppery-chestnut head, neck and underparts. Black wings, glossed green. **Adult non-breeding:** Dusky head and neck with fine white streaking. Lacks the spectacles and chestnut gloss. Grey legs. May show pink on the bill base. In flight, fairly pointed slender wings with chestnut lesser wing-coverts. Legs extend far beyond the tail. Note V-formation of group in flight. [Cuervillo de Cañada]

Puna Ibis *Plegadis ridgwayi*   56 cm

Puna lakes in Jujuy and Salta, rarely south to Catamarca. Descends as low as 1300 m in winter when it overlaps with Bare-faced and White-faced Ibises. **ID** Shorter and thicker-necked than White-faced. **Adult breeding:** Pink lores and reddish bill with a broader base than White-faced. Blackish legs with rarely visible reddish 'knees'. Deep chestnut head, neck and breast. Dark purplish back and bronze-lilac wing-coverts. Green gloss over rest of wings. **Adult non-breeding:** Dull pink bill. Blackish without chestnut tones. Head and upper neck finely speckled white. Bottle-green wings. In flight, only the feet project beyond the tail. Note broad wings and short neck. [Yanavico]

Theristicus Large ibises with a shaggy nuchal ruff and salmon-pink legs. Loud trumpeting calls. Gregarious; especially in winter. No known overlap between resident Buff-necked and migratory Black-faced.

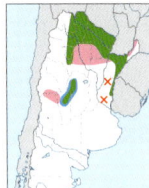

Buff-necked Ibis *Theristicus caudatus*   75 cm

N lowlands, Andean foothills to 1700 m, C sierras to 1800 m. **ID Adult:** Two pendulant black throat sacs. Cinnamon-buff head and neck with a burnt chestnut crown and upper breast, extending around the lower hindneck. Wing-coverts mostly white. Lower breast to vent blackish. In flight, whiter upperwing-coverts than Black-faced Ibis. **Voice** Flight call is a two-note and trumpeting *Tas-Tás!* Longer voice is a slow, cacophonous descending laughter with a seriema-like quality. [Bandurria Tastás]

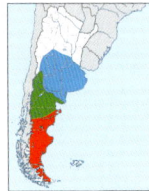

Black-faced Ibis *Theristicus melanopis*   71 cm

Patagonian forest, reaching the C lowlands in winter; vagrant to the Falklands. Overlap with Buff-necked is possible in San Luis and Catamarca in winter. **ID Adult:** From Buff-necked by large single gular pouch, grey pectoral band, lack of strong chestnut on the crown, chest or nape, less black on the abdomen, and greyer wing-coverts. In flight, grey upperwing-coverts contrast with black flight feathers. **Voice** Very much like Buff-necked Ibis but with a deeper, more nasal quality. [Bandurria Austral]

Plumbeous Ibis *Theristicus caerulescens*   74 cm

NE savanna; scarcer through the N chaco. **ID Adult:** Mainly grey with a white forehead and thick nuchal ruff. In flight, toes extend just beyond the tail. Broad-winged. **Voice** A rapid series of stuttered, piping *ko-ko-ko-ko-ko-ko* notes…, faster and more ringing than Buff-necked Ibis. [Bandurria Mora]

Bare-faced Ibis

infuscatus

ad br

ad

White-faced Ibis

ad br

ad non-br

ad

Puna Ibis

ad br

ad

ad non-br

hyperorius

Buff-necked Ibis

ad

ad

melanopis

Black-faced Ibis

ad

Plumbeous Ibis

Coragyps Short-tailed, dumpy vulture; detects carrion visually. ***Cathartes*** Long-tailed vultures with long wings and bare coloured heads; detect carrion by smell and often gregarious. ***Sarcoramphus*** Very broad wings; usually alone in forest, but joins other vultures at large carcasses. ***Vultur*** A huge sexually dimorphic vulture; often alone, but highly gregarious at carcasses.

Black Vulture *Coragyps atratus*   60 cm WS: 155 cm

Widespread reaching 1750 m; rare in S Patagonia. **ID** Short-tailed, dumpy vulture. Flies with stiff, fast and deep wingbeats interspersed with glides. **Adult:** Slender grey bill, tipped yellowish. Bare wrinkled grey head and throat. Black with a slight green gloss on the wings. Whitish legs. **Flight:** Bulging secondaries form a wedge-shaped wing. Large silvery-white patch on underside of primaries. Notably short tail. Soars on flat or slightly elevated wings. [Jote Cabeza Negra]

Turkey Vulture *Cathartes aura*   65 cm WS: 175 cm

Widespread, reaching 3450 m. **ID Adult** *ruficollis* (from N Patagonia northwards): Creamy bill. Bare red head with striking white hindcrown and nape. Rest of plumage entirely black. Blacker overall than *jota*. Tarsus pale pink. **Adult** *jota* (S Patagonia; migrates through the range of *ruficollis*): Dark red nape, sometimes with black feathering. Browner than *ruficollis*, with broad white fringes to secondaries and greater coverts, also visible in flight. **Adult** *falklandica* (resident on Falklands, S. Patagonia?; not illustrated). As *jota*, but smaller with shorter wings and more contrasting secondary fringes. **Juvenile:** Grey head and black bill. **Flight:** Black underwing-coverts contrast with silvery-grey underside of flight feathers. Fairly long tail with a slightly rounded tip. Head-on, wings are held in a deep V. **Tax note 9.** [Jote Cabeza Colorada]

Lesser Yellow-headed Vulture *Cathartes burrovianus*   60 cm WS: 160 cm

NE savanna and dry chaco to n. Córdoba. **ID Adult:** Yellow or orange-yellow cheeks and pale blue crown, pink band around base of hindcrown and broad pink subterminal bill band. Usually has a black spot in front of the eye, and a black line behind and below the eye. Rest of plumage entirely black. Folded primaries extend beyond the tail. **Flight:** Very similar to Turkey Vulture but wings proportionately narrower. From above, white primary quills form a distinct patch, unlike Turkey Vulture. Head-on, wings are held in a shallow V. Often undertakes low quartering flights unlike Turkey and Greater Yellow-headed Vultures. [Jote Cabeza Amarilla Chico]

Greater Yellow-headed Vulture *Cathartes melambrotus*   77 cm WS: 180 cm

Yungas forest to 1400 m (also Paraná forest?); vagrant to San Juan. **ID Adult:** Similar to Lesser Yellow-headed, but paler yellow cheeks extend over the forehead and onto the throat, thus any pink is much reduced. Rest of plumage entirely black. Folded primaries reach, but do not extend beyond, tail tip. More bulky than Lesser Yellow-headed. **Flight:** Wings considerably broader than Lesser with thick primaries; indistinct white primary quills on upperside. Diagnostic pale grey 'window' on innermost secondaries, sometimes extending to some or most of tertials, contrasts with darker outermost secondaries and dark primaries. Long, fairly narrow tail. Appears very small-headed. Head-on, wings are almost flat. [Jote Cabeza Amarilla Grande]

King Vulture *Sarcoramphus papa*   76 cm WS: 190 cm

NW Andes to 1900 m, C Sierras, chaco and Misiones. **ID** Very broad wings. **Adult:** Multicoloured head with large cere and white iris. White back, body and wing-coverts contrast with grey collar, and black wings and tail. **Flight: Adult:** White body and underwing-coverts contrast with black flight feathers and tail. **Immature:** Contrasting white axillaries. Underparts vary from black, streaked white to white blotched black. [Jote Real]

Andean Condor *Vultur gryphus*   123 cm WS: 295 cm

Andes to 4500 m; C sierras to 2200 m, coastal s. Santa Cruz and Tierra del Fuego; occasional in the C lowlands. **ID** Soars effortlessly. **Adult ♀:** Red iris and bare blackish-grey head lacking appendages. Fluffy white collar. Black body and tail; most of folded wings white. **Adult ♂:** Grey-brown iris and dull reddish head, with fleshy comb and caruncles. **Flight: Adult:** Mostly black with splayed primaries and white collar. From above, has mostly white greater-coverts and secondaries. Note large pinkish comb on crown of ♂. **Juvenile:** Creamy underparts and underwing-coverts with contrasting dark brown tail and flight feathers becoming blacker on the primaries. From above, all-dark. [Cóndor Andino]

Black

ad

Black Vulture

foetens

ad

ruficollis

Turkey
Vulture

ad

ad

Turkey

jota

ad

ruficollis

ad

juv

ad

jota

ruficollis

Lesser Y-h

ad

Lesser
Yellow-headed
Vulture

ad

ad

Greater
Yellow-headed
Vulture

ad

Greater Y-h

ad

ad

imm

King
Vulture

juv

ad ♀

Andean Condor

ad ♂

ad ♂

Ictinia and *Elanoides* Migratory grey or pied forest kites with relatively small heads and short tarsus. All perch in the canopy or bare branches, feed aerially, and are often gregarious. *Elanus* Widespread open country kite that hovers to detect prey. *Gampsonyx* Tiny, unobtrusive kite of chaco woodlands.

Plumbeous Kite *Ictinia plumbea*   36 cm WS: 80 cm

N forests in Sep–Mar; abundant in Paraná forest. **ID** Adult: Ash-grey with contrasting slaty-black wings; primaries extending beyond the tail tip with rufous fringes visible at close range. Bright orange toes. Juvenile: Brown mantle and wings. Tail proportionately longer than adult, but primaries still protrude. Buffy to white below, streaked black on breast, becoming more rounded spots on belly and vent. Overlaps with Mississippi Kite. **Flight:** Adult is grey below with rufous flash in primaries. Two narrow white tail bands. Less distinct rufous primaries from above; tail bands not visible. Juvenile is white or buffy below, streaked brown. Underwing grey with narrow white trailing edge; sometimes with a hint of rufous on distal primaries. Three narrow white tail bands. [Milano Plomizo]

Mississippi Kite *Ictinia mississippiensis*   35 cm WS: 79 cm

Boreal migrant to chaco woodlands (Oct–Feb); scarce in N Mesopotamia. **ID** Adult: Greyish-white head contrasts with grey back, breast and belly. Slaty wing-coverts with whitish folded secondaries. Primaries extend minimally beyond the tail tip, but never show rufous when folded. Chrome-yellow toes. Juvenile: Tail tip equal to, or longer than primary tips. Variable broad rufous or warm brown ventral streaks; sometimes wholly rufous on undertail-coverts. Overlaps with Plumbeous Kite. **Flight:** Adult is grey below with unmarked black tail. Blackish primaries with indistinct rufous webs. Contrasting white head. From above, white head and white secondaries form striking contrast. Juvenile has blotchy rufous ventral streaking and flecked underwing-coverts. No rufous in primaries. Tail bands ill-defined. [Milano Boreal]

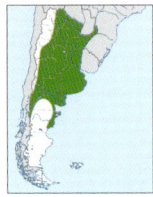

White-tailed Kite *Elanus leucurus*   37 cm WS: 95 cm

N lowlands south to N Patagonia. **ID** Medium-sized open-country kite. Frequently hovers. Sexes alike. Adult: Yellow cere and tarsus. White head, underparts and tail. Soft grey above with contrasting black shoulder. Juvenile: Rusty breast contrasts with white face. Streaked crown and spotted back. Black shoulders and flight feathers. **Flight:** Adult is mainly white with blackish primaries; contrasting black carpal spot visible at close range. Splays tail when hovering. [Milano Blanco]

Pearl Kite *Gampsonyx swainsonii*   24 cm WS: 50 cm

Light woodlands in the N lowlands. **ID** Diminutive open-country kite. Perches high and takes most prey from the ground. Adult: Buff forehead and cheeks with black skullcap and white nuchal collar. Black above, white below with cinnamon thighs and black breast spurs. **Flight:** Adult is essentially white from below with cinnamon wash on underwing-coverts, and neat black breast spurs. Short squarish tail, becoming distally grey. Silhouette recalls a *Progne* martin. [Milano Chico]

Swallow-tailed Kite *Elanoides forficatus*   55 cm WS: 127 cm

Chiefly in Yungas and Paraná forest (Aug–Apr). **ID** Large, mostly gregarious, black-and-white kite with extremely long scissor-shaped tail. Sexes alike. Adult: White head, nape and underparts with contrasting black mantle, wings and tail. **Flight:** White head, underparts and underwing-coverts contrast with black flight feathers and tail. [Milano Tijereta]

juv

ad

ad

juv

ad

Plumbeous
Kite

Mississippi
Kite

juv

ad

ad

juv

ad

ad

juv

White-tailed Kite

leucurus

ad

swainsonii

Pearl Kite

yetapa

ad

Swallow-tailed Kite

Circus Long-tailed, small-headed, open-country raptors with owl-like facial discs and small white rump visible in flight. Quarters low on V-shaped wings in open country. Sexes differ.

Cinereous Harrier *Circus cinereus* ♂ 42 cm; ♀ 48 cm WS: 102 cm

Widespread; some move north in winter. **ID Adult** ♂: Ashy-grey head, upper breast and upperparts. Contrasting rust and white ventral barring. **Adult** ♀: Brown head and mantle. Whitish supercilium and nuchal streaking. Grey tail with narrow blackish bands and broader subterminal band. Ventral barring as ♂. **Juvenile**: Prominent cinnamon-fringed wing-coverts unlike Long-winged Harrier. Buff underparts with dense, blurry brown streaking. Primaries do not reach tail tip. **Flight**: Wings held in a shallow V. **Adult** ♂ is pale grey above with black distal primaries, trailing edge to secondaries and subterminal tail band. Underparts rusty-barred but underwing pure white with black wing tips and trailing edge. **Adult** ♀ has rusty-barred abdomen and underwing-coverts. Flight feathers finely barred. **Juvenile** lacks dark head of juvenile Long-winged, and shows dense breast streaking, sometimes suggesting a pectoral band. Whitish flash in base of primaries and underwing-coverts flecked with brown. [Gavilán Ceniciento]

Long-winged Harrier *Circus buffoni* ♂ 50 cm; ♀ 52 cm WS: 137 cm

Widespread but common only in NE marshes. **ID Adult** ♂ **pale morph**: Black-and-white head with black collar and contrasting white underparts. Black above with grey wings and up to 5 broad black tail bands. **Adult** ♀ **pale morph**: Resembles pale morph ♂, but black plumage replaced by brown, and underparts washed buff or with fine brown streaks. **Juvenile pale morph**: Resembles pale morph ♀ but richer cinnamon below with thicker streaking, wing-coverts more prominently fringed. **Juvenile dark morph**: Rust below with thick blurry brown streaking. Some variant juveniles have white face and underparts with heavy ventral markings. **Flight**: Notably longer-winged than Cinereous; held in a deeper V. **Adult** ♂ **pale morph**: Black upperwing-coverts with grey flight feathers, finely barred black. 3–5 black tail bands. Mainly white below with finely barred underwing. **Adult** ♀ **pale morph**: As with other pale morphs, has a dark head unlike Cinereous. Densely marked underwing-coverts and grey secondaries with black trailing edge. Creamy below, with fine streaking on breast at close range. **Adult dark morph**: Dorsally like pale morph ♂. Underparts and underwing-coverts sooty-brown with a chestnut vent; appears black from a distance. **Juvenile** ♀ **pale morph**: Buff underparts and underwing-coverts; stronger breast streaking than adult ♀ and solid grey secondaries. [Gavilán Planeador]

Cinereous Harrier

juv

ad ♂

ad ♂

ad ♂

juv

ad ♀

ad ♀

Long-winged Harrier

ad
dark morph

ad ♂
pale morph

ad ♀
pale morph

juv ♀
pale morph

ad ♀
pale morph

juv
pale morph

juv
variant

ad ♂
pale morph

juv
dark morph

Buteo Robust, compact hawks, all with relatively short tails and broad wings. ♀♀ are somewhat larger. Sexes alike except for tail pattern in Zone-tailed Hawk. All readily soar.

Grey-lined Hawk *Buteo nitidus*
42 cm WS: 84 cm

Rare vagrant in the NE. **ID Adult:** Yellow cere, iris and tarsus. Ash-grey above with white barring on wing-coverts. Black tail with narrow white tips and uppertail-coverts, and broad white median band. White below with dense grey barring. Compare with Crane Hawk (Plate 39) and ♂ Hook-billed Kite (Plate 46). **Juvenile:** Creamy-white head with black post-ocular streak. Brown back and wings, often barred white on tertials. 5 black tail bands, evenly barred with cream. Creamy-white below with brown tear-shaped breast spots. **Flight: Adult:** Mostly white below with fine grey breast barring, and grey-tipped primaries. White undertail with 3 black bands; only the broadest distal band likely to be seen at distance. **Juvenile:** Creamy-white below with obvious brown breast spotting. Similar to adult but without a dusky trailing edge, and at least 5 more prominent narrow tail bands. [Aguilucho Gris]

Broad-winged Hawk *Buteo platypterus*
38 cm WS: 86 cm

Rare boreal migrant to Yungas foothill forest (Oct–Feb); vagrant elsewhere. **ID Adult:** Brown above, black tail with one broad and one narrower white band. White throat, finely streaked brown. Rufescent breast, spotted white, grading to irregular rufous- and white-barred belly, and white vent. **Adult dark morph** (rare; not illustrated): Much as dark morph Short-tailed Hawk (Plate 47) but with obvious clear-cut median white tail band. **Juvenile:** Brown head with indistinct buff supercilium, and dusky malar. Brown uppertail with indistinct narrow blackish bands. Cream below, thickly streaked brown; often more prominent at the sides. **Flight: Adult:** From congeners by black trailing edge extending around primary tips. Note black tail with obvious white band; narrow white inner band sometimes visible. **Juvenile:** Highly variable spotted breast. Best identified by blackish primary tips, dusky trailing edge to wing, and very indistinct tail bands. [Aguilucho Alas Anchas]

Swainson's Hawk *Buteo swainsoni*
♂ 45 cm; ♀ 53 cm WS: ♂ 124 cm; ♀ 131 cm

Boreal migrant to the N lowlands; mainly the Pampas in Oct–Apr. **ID Adult barred morph:** Dark brown above, often with a white forehead. Primaries reach tip of dull grey tail with indistinct narrow black bands. White throat contrasts with broad dark brown pectoral band, densely barred dark rufous below, except for creamy undertail-coverts. **Adult typical pale morph:** Warm brown pectoral band, and largely unmarked creamy abdomen; some with barring on sides of belly. **Juvenile pale morph:** White or buffy head, streaked brown on the crown; often with a narrow brown post-ocular streak, buff supercilium and broader brown malar streak. Scapulars usually mottled white. Less distinct tail bars than adult. Round brown breast spots, often forming patches at the sides. **Flight: Adult dark morph:** Dark brown body and contrasting white throat. Underwing-coverts mixed dark brown and rufous. Dark grey underside of flight feathers, paler only at the base of the outer two primaries, with black-tipped primaries and trailing edge. **Adult typical pale morph:** White throat and warm brown pectoral band contrasts with creamy to white body and underwing-coverts. Undertail and underwing otherwise as dark adult but paler. **Juvenile:** Plain or spotted cream underwing-coverts and body contrast with grey flight feathers, blackish trailing edge and primary tips. Greyish undertail indistinctly barred black with narrow white tips. [Aguilucho Langostero]

Zone-tailed Hawk *Buteo albonotatus*
52 cm WS: 128 cm

Scarce in the N chaco and Yungas foothills; very rare in Misiones. **ID Adult ♀:** Entirely black or dark slate except for small white forehead. Primaries reach or exceed tail tip. Uppertail with broad greyish subterminal band and two narrower basal bands (only two bands in ♂). Compare with dark morphs of White-tailed and Variable Hawks (Plate 40), Hook-billed Kite (Plate 46) and Short-tailed Hawk (Plate 47). **Flight: Adult ♂:** Black body and underwing-coverts. Note straight trailing edge to wing. Contrasting silvery-white underside of flight feathers, densely barred black with a black trailing edge. Broad white subterminal tail band and narrow white inner band (two or more inner bands in ♀; some ♂♂ rarely show two inner bands). From above, note typical Turkey Vulture-like flight on elevated dihedral wings, and grey bars on uppertail. **Juvenile** (not illustrated) can show white ventral speckling, a less distinct dusky trailing edge to the wing, and whitish undertail with numerous fine grey bands and often an indistinct broader subterminal band. [Aguilucho Jote]

Grey-lined Hawk

juv

ad

pallidus

Broad-winged Hawk

juv

ad

juv

ad

juv

ad
pale
morph

ad
dark
morph

ad

Zone-tailed Hawk

juv
pale
morph

ad
pale
morph

ad
barred
morph

ad ♂

ad ♀

Swainson's Hawk

Roadside Hawk *Rupornis magnirostris*   38–42 cm WS: 81 cm

Widespread in the N and C lowlands, Yungas and Paraná forest. **ID** Adult *pucherani* (N and C lowlands; 42 cm): White iris. Blackish-brown head and throat forming a hood. Dark brown above with a rufous tail crossed by 4 black bands. Warm brown to dull rufous streaks on upper breast and barred the same colour on the rest of the underparts. **Adult** *magniplumis* (Paraná forest; 38 cm): Brownish-grey tail, and less contrast between hood and back. **Juvenile**: Brown crown and nape, streaked buff. Notable buff supercilium. Buff below, thickly streaked brown on breast and with wavy brown barring on the belly and thighs. Cinnamon tail with 6–7 narrow dark bands. **Flight**: **Adult**: Black hood. Rufescent breast streaking, and finely barred abdomen and underwing-coverts. Creamy underwings, indistinctly barred grey. Dusky tipped primaries. Greyish-buff undertail with indistinct black bars. From above, large round rufous patch on base of primaries. Tail barred with rufous. **Juvenile**: Brown-streaked breast, and barred abdomen and underwing-coverts. Pale buff primary window. Indistinct tail barring. **Tax note 10.** [Taguató]

Crane Hawk *Geranospiza caerulescens*   ♂ 50 cm; ♀ 54 cm WS: 98 cm

Humid N woodlands and forest islands. **ID** Small-headed, long-tailed hawk with noticeably long bare orange tarsus that it uses to reach prey in cavities. Sexes alike. **Adult**: Ash-grey, vermiculated white on wing-coverts, and barred white ventrally with white undertail-coverts. Creamy tail with 3 broad black bands; upper band usually concealed. **Juvenile**: From adult by whitish forehead and coronal streaks, whiter ear-coverts and throat, and buff undertail-coverts. **Flight**: Oval wingtip with bulging secondaries and long tail. Finely barred body and wings with diagnostic white crescent bordering black primary tips. White undertail with black subterminal band. [Gavilán Patas Largas]

Harris's Hawk *Parabuteo unicinctus*   50 cm WS: 105 cm

Andes to 2000 m and lowlands southwards to N Patagonia; expanding south. **ID** *Buteo*-like hawk with relatively small head and long tail. Sexes alike. **Adult**: Dark brown with contrasting chestnut shoulders and thighs, and white undertail-coverts. **Juvenile**: Cream supercilium and underparts; spotted black on breast and barred on flanks. Compare with juvenile Variable Hawk (Plate 40). **Flight**: Long slender tail with rounded tip, but *Buteo*-like wings. White rump band. **Adult**: Chestnut underwing-coverts offset by whitish bases to black flight feathers. White vent and black tail, tipped white. Sooty-brown above with chestnut shoulders, white rump patch and long brown tail tipped white. **Juvenile**: Heavily streaked below. Little contrast on undertail and wings except notable white flash in base of primaries. Compare with juvenile Variable Hawk (Plate 41). [Gavilán Mixto]

Snail Kite *Rostrhamus sociabilis*   43 cm WS: 107 cm

N marshes, mainly Sep–Feb. **ID** Marsh hawk with fine, sickle-shaped bill adapted for feeding on snails. Sexes differ. **Adult** ♂: Reddish cere and lores. Orange legs. Slaty overall with white undertail-coverts and tail base. **Adult** ♀: Red iris like ♂. Yellow-orange legs. Mostly dusky brown with whitish foreface; thickly blotched on belly. Clean white vent and tail base. **Juvenile**: Resembles juvenile Harris's Hawk, but note short squarish tail with black distal band, bill shape, habitat and hanging flight. Flanks never barred. **Flight**: Broad squarish wings and wedge-shaped tail; frequently hovers. **Adult** ♂: Slaty with contrasting white rump and black tail, narrowly tipped white. **Adult** ♀: Tail as ♂, but underwing-coverts brown and remiges barred. Brown upper breast, densely blotched below and white vent. **Juvenile**: Resembles ♀ with paler coverts, whiter base of primaries, duller tail base and finer ventral streaking. Flight behaviour facilitates identification. [Caracolero]

juv

pucherani

ad

magniplumis

ad

ad

pucherani

juv

ad

Roadside Hawk

juv

Crane Hawk

ad

ad

ad

Harris's Hawk

ad

unicinctus

juv

ad

flexipes

sociabilis

ad ♂

Snail Kite

ad ♀

juv

ad ♀

juv

Geranoaetus Recalls *Buteo* with slightly broader, more bulging wings. Gregarious in winter. At least four age-classes are discernible by plumage. Highly polymorphic (especially Variable Hawk). ♀♀ noticeably larger with red backs in adult pale morphs of Variable. **See flight images on Plate 41.**

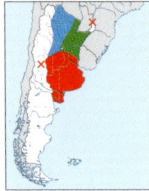

White-tailed Hawk *Geranoaetus albicaudatus* ♂ 51 cm; ♀ 55 cm

Widespread in N and C lowlands, C sierras and in winter to the NW Andes (to 1800 m). **ID** Primary tips always exceed tail tip. Sexes alike. **Adult grey morph:** Entirely grey except for white rump and tail with broad black subterminal band. **Adult pale morph (dark-throated form):** Dark brown head, throat and upperparts with contrasting shoulders. Tail as grey morph. White below with fine rufescent barring on the abdomen and thighs. **Adult pale morph (white-throated form):** Like dark-throated form but throat white, and lacks the ventral barring. Can show a black or rufous patch on the sides of the breast. **Juvenile pale morph:** Brown above with white patch on ear-coverts. White rump with a brown centre. Greyish tail with fine dusky bars. Brown or white throat. Solid brown or streaked sides of the breast highlight unmarked white or cream patch in centre of breast. Belly blotched with brown. [Aguilucho Alas Largas]

Variable Hawk *Geranoaetus polyosoma* ♂ 47–51 cm; ♀ 51–55 cm

Widespread. **ID** Very variable. Highland forms are larger. **Adult ♀ typical highland barred morph** (NW Andes above 3000 m): Barred and dark morphs dominate in the high NW. Usually shows whitish throat streaks. Note long rectangular-shaped folded secondaries and tertials, while primaries exceed the tail tip. **Adult ♀ dark morph:** Dark, rufous-bellied and pale morphs are widespread but absent from Mesopotamia. In all, note tapered folded secondaries and primaries reaching or barely exceeding tail tip, unlike barred morph. Sooty-grey head, wings and underparts with a white barred vent. White tail with a black subterminal band. **Adult ♀ rufous-bellied morph:** Like dark morph with a rufous abdomen and unbarred vent. **Adult ♀ typical pale morph:** Rufous nape and back. Grey cap and wings. White cheeks and underparts. **Adult ♂ typical pale morph:** Essentially grey above and white below. Pale morph is the most common morph. **Juvenile ♀ pale morph:** Dark brown above, large chestnut facial patch and broad blackish moustachial. Buff spots on wing-coverts and tertials. Grey tail with fine indistinct bars. Buff below, streaked dark brown on breast, and checkered on belly and flanks. **Second-year ♀ pale morph:** Fine moustachial. White throat and breast, finely barred brown or rufous below. [Aguilucho Ñanco]

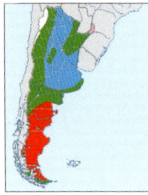

Black-chested Buzzard-Eagle *Geranoaetus melanoleucus* ♂ 61 cm; ♀ 75 cm

Widespread. Andes, C sierras and Patagonia; also in the NE lowlands. **ID** Stocky with robust bill, short tail and short tarsus. Sexes alike. **Adult *australis*** (Andes and Patagonia): Grey head, becoming blackish on mantle and breast. Contrasting pale grey wing-coverts. White or greyish-white below, finely vermiculated black. **Juvenile:** Tawny or buff breast, thickly spotted dark brown; contrasting black vest. Unlike adult, primaries do not reach tail tip. **Third-year:** Essentially black, flecked white on nape. Thighs barred chestnut. [Águila Mora]

ad ♂
grey morph

ad ♀
pale morph

White-tailed Hawk

ad ♂
pale morph

white-throated
form

albicaudatus

dark-throated
form

juv ♂
pale morph

Variable Hawk

ad ♀
barred morph

ad ♀
dark morph

2nd-yr ♀
pale morph

ad ♂

ad ♀

ad ♀

juv ♀
pale morph

typical
pale morph

rufous-bellied
morph

typical
pale morph

ad

juv

australis

3rd-year
imm

Black-chested
Buzzard-Eagle

Geranoaetus Recalls *Buteo* with slightly broader, more bulging wings. Gregarious in winter. At least four age-classes are discernible by plumage. Highly polymorphic (especially Variable Hawk). ♀♀ noticeably larger with red backs in adult pale morphs of Variable. **See perched images on Plate 40.**

White-tailed Hawk *Geranoaetus albicaudatus* WS: ♂ 126 cm; ♀ 135 cm

Widespread in N and C lowlands, C sierras and in winter to the NW Andes (to 1800 m). **ID** Flight: **Adult pale morph (white-throated form)**: White underwing-coverts with dark wrist bar and white patch at base of outer primaries. Barred secondaries with black-tipped primaries and trailing edge. White tail with black subterminal band. **Adult dark morph**: Black to dark brown body, rump and underwing-coverts. Note pale patch in base of primaries, and paler secondaries. **Juvenile**: Pale patch across central breast. Wings finely barred with darker underwing-coverts, black-tipped primaries and pale bases to outer primaries. Finely barred tail appears unmarked at a distance. [Aguilucho Alas Largas]

Variable Hawk *Geranoaetus polyosoma* WS: ♂ 122–139 cm; ♀ 132–145 cm

Widespread. **ID** Flight: **Adult ♀ typical highland barred morph**: Tends to show whitish throat streaks. Note very broad secondaries. **Adult ♀ typical pale morph**: White tail with black subterminal band in all adult morphs. White underwing-coverts and paler underwings than barred morph with more rounded wing tips. Dusky trailing edge, narrowly fringed white. **Adult ♂ rufous-bellied morph**: Grey back, head and underwing-coverts. Contrasting rufous belly. **Adult ♂ dark morph**: Black upperparts, body and underwing-coverts. Rather pale underwings with black-tipped primaries. Compare with rufous-bellied morph and dark morph Short-tailed Hawk (Plate 47). **Adult ♂ black-chested rufous morph**: One of several barred morphs. Note that flexed primaries alter wing shape. **Adult ♂ grey morph**: As black morph but body and underwing-coverts grey. Similar morph in White-tailed Hawk would have longer pointed wing tips, darker remiges and no fine white trailing edge. **Juvenile ♂ pale morph**: Cinnamon breast and underwing-coverts, streaked brown. Whitish primary window. Pale grey tail, finely barred at close range. **Juvenile dark morph** (not illustrated): Resembles adult black morph but with tail pattern of juvenile pale morph; best distinguished from juvenile Zone-tailed Hawk by silhouette. Note also slightly paler cheek patch. [Aguilucho Ñanco]

Black-chested Buzzard-Eagle *Geranoaetus melanoleucus* WS: 160 cm

Widespread. Andes, C sierras and Patagonia; also in the NE lowlands. **ID** Flight: Extremely broad-winged with tapering primaries. Adults appear almost tailless. **Adult *melanoleucus* (NE lowlands)**: Black breast, contrasting white abdomen and underwing-coverts. **Juvenile**: Chestnut underwing-coverts contrast with pale brown remiges and darker primary tips and secondaries; less bulging than adult. Wedge-shaped tail longer than adult. Dark belly contrasts with pale breast in youngest birds. [Águila Mora]

ad ♂
dark morph

juv ♀

white-throated
form

White-tailed Hawk

ad ♀
pale morph

ad ♂
rufous-bellied
morph

ad ♂
black morph

Variable Hawk

juv ♂
pale morph

ad ♂
grey morph

ad ♂
black-chested
rufous morph

ad ♀
barred morph

ad ♀
pale morph

melanoleucus

juv

ad

Black-chested Buzzard-Eagle

Savanna Hawk *Buteogallus meridionalis*　　48–52 cm WS: 120 cm

N lowlands. **ID** Large, long-legged raptor which delivers far-carrying shrill whistles. **Adult** *australis* (52 cm, Widespread in N lowlands): Mostly pale cinnamon including shoulder. Finely barred abdomen. Primaries almost reach tail tip; black with white central band. **Juvenile**: Creamy head with black post-ocular streak and crown. Solid brown patches on sides of breast. Cinnamon secondaries. Finely streaked abdomen and cinnamon thighs. Primaries reach or extend beyond tail tip, unlike Black-collared and Great Black Hawks (Plate 43). **Flight**: **Adult** *australis*: Cinnamon with black-tipped primaries, trailing edge and tail band with white base. **Adult** *meridionalis* (48 cm; N Misiones): Brighter and more rufous on underparts and underwing-coverts than *australis*, but tail band poorly defined. **Juvenile**: Creamy underwings with dusky-tipped primaries and trailing edge. White flash in base of primaries. Indistinct dark tail band. Dark patches on sides of breast and streaked abdomen. [Aguilucho Colorado]

Black-collared Hawk *Busarellus nigricollis*　　55 cm WS: 129 cm

N marshes; mainly in the NE. **ID** Large, chunky hawk of rivers and marshes with short tail. Soles of feet with spiny scales to help grip fish. Sexes alike. **Adult**: Rich chestnut with contrasting white head, black pectoral band and distal tail bar. **Juvenile**: White of head continues onto upper back and breast, streaked with black. Brown above. Narrow black pectoral band; sometimes reduced. Belly mottled with brown or rufous. **Flight**: Small-headed with very broad, squarish wings and short fanned tail. **Adult**: Rich chestnut with white head and black pectoral band. Contrasting black primaries, trailing edge and tail band. **Juvenile**: Pale brown wings with darker primary tips and blackish smudge on carpal. White head and breast, usually with a narrow black pectoral band. [Aguilucho Pampa]

Pandion Non-breeding boreal migrant; feeds exclusively on fish. Sexes differ slightly.

Osprey *Pandion haliaetus*　　60 cm WS: 164 cm

Scarce but annual boreal migrant to the lowlands, south to N Patagonia; mainly Sep–Feb. **ID Adult** ♂: White head and underparts. Brown upperparts, crown-stripe and arc over eye. **Adult** ♀: White body and underwing-coverts with striking black carpal patch, barred undertail and remiges with black fingers. Differs from ♂ by finely streaked breast. **Flight**: Long, relatively narrow, wings with notably angled carpals. Short tail. [Águila Pescadora]

australis

ad

juv

australis

ad

juv

meridionalis

ad

Savanna Hawk

leucocephalus

ad

juv

juv

ad

Black-collared Hawk

ad ♀

carolinensis

ad ♂

Osprey

Buteogallus Large grey to black forest hawks and eagles of Yungas, riverine and dry forest. All soar with fanned tails, have broad wings, and deliver long, shrill, far-carrying cries. Juveniles show large dark breast patches.

Great Black Hawk *Buteogallus urubitinga*   63 cm WS: 130 cm

N lowlands and foothills, usually close to water. **ID** Stocky, long-legged with fairly long neck and small head. Sexes alike. **Adult:** Entirely sooty-black with broad white basal tail band. **Juvenile:** Creamy head and underparts; blackish crown, eye-stripe and thickly streaked nape and breast, especially on the sides. Primaries do not reach tail tip. **Flight:** Long, broad wings with fingered primaries. Tail held in wedge. Readily soars. **Adult:** Entirely black with broader white tail base than Solitary Eagle. **Juvenile:** Slightly bulging secondaries. Cream or tawny underwing, finely barred blackish with black-tipped primaries. Black streaks along greater coverts; suggestion of a dark trailing edge and tail band. Black-streaked breast, especially at the sides. Overlaps with Savanna Hawk (Plate 42) and Chaco Eagle. [Águila Negra]

Chaco Eagle *Buteogallus coronatus*   ♂ 75 cm; ♀ 79 cm WS: 176 cm

Scarce in open or light woodland of the N lowlands and foothill forest. **ID** Large, long-legged raptor with fingered primaries and slightly bulging secondaries. Sexes alike. **Adult:** Ash-grey to brownish-grey with a distinct crest. Black tail with white median band. **Juvenile:** Brown above and white below. Crest longer than adult. Distinctive brown wedges on sides of breast. Overlaps with Great Black Hawk. Compare with darker Solitary Eagle. **Flight: Adult:** Grey with black trailing edge; primaries tipped black with a white flash at base. Black tail with white median band and narrow tips. **Juvenile:** Dusky primary tips with white flash in base. Black spots on carpal. Brown thighs and wedges on sides of breast. **Voice** Delivers far-carrying shrill whistles. [Águila Coronada]

Solitary Eagle *Buteogallus solitarius*   ♂ 75 cm; ♀ 79 cm WS: 168 cm

Rare in Yungas forest. **ID** Stocky, long-legged with fairly long neck and small head. Sexes alike. **Adult ♀:** Yellow iris and broad cere. Dark slate-grey unlike Great Black Hawk, but appears black at a distance. White median tail band. **Juvenile ♂:** Buff head, streaked black with black smudge behind eye, extending into broad blackish patches on sides of breast. Dark brown above without distinct tail bands. Abdomen thickly spotted brown. **Flight: Adult ♀:** Extremely broad wings with an almost tailless appearance. Dark slate, often appears black, with distinctly paler flash on base of primaries unlike Great Black Hawk. White median tail band. Compare with paler Chaco Eagle which overlaps in foothill forest. **Immature ♂:** Black breast patches and thighs. Heavily marked underwing-coverts; black-tipped primaries and trailing edge with white flash in base of primaries. No tail bands. Compare with juvenile Chaco Eagle and Great Black Hawk. [Águila Solitaria]

Great Black Hawk

juv

urubitinga

ad

juv

ad

Chaco Eagle

juv

ad

juv

ad

solitarius

Solitary Eagle

juv ♂

ad ♀

ad ♀

imm ♂

Rufous-thighed Kite *Harpagus diodon*   31 cm

Yungas and Paraná forest; mainly Oct–Mar. **ID** Smaller, rounder head than *Accipiter* hawks, with a comparatively short, square-ended tail. Ill-defined superciliary arc and large rounded eye imparts innocent look. **Adult:** From Bicoloured Hawk by red iris, although sometimes orange. Narrow dark throat streak diagnostic. Blackish hood and 2–3 narrow pale bars on uppertail. Grey below with rufous thighs and a white vent. **Juvenile:** Red or orange iris. Dark brown ventral streaking. Juvenile accipiters never have rufous thighs. Compare with juvenile Rufous-thighed Hawk. **Voice** Gives an unusual high-pitched tyrannid-like *SWEEee-Swit* call. [Milano de Corbata]

Accipiter and *Hieraspiza* Sleek, narrow-tailed hawks of forest and open country. All have yellow legs, cere and bare loral skin; latter intensifies in colour when breeding. Well-developed superciliary arc imparts fierce look. Most species are polymorphic. ♀♀ are considerably larger than ♂♂. In flight, identification aided by comparative tail length and shape of tail tip, shape of leading edge of wing, and extent that head protrudes from wing edge. All readily soar. **See flight images on Plate 45.**

Bicoloured Hawk *Accipiter bicolor*   ♂ 36 cm; ♀ 42 cm

Widespread in the N. **ID** Adult ♂ '*pileatus*' (Paraná forest): From Rufous-thighed Kite by yellow or pale orange iris. Grey ear-coverts with a contrasting blackish cap, and longer, narrower tail with broader pale bands. Adult ♀ '*guttifer*' (chiefly Yungas forest and humid chaco): From '*pileatus*' by a dull rufous wash over lower breast and belly, often barred or spotted greyish-white. Both ssp. can show diffuse throat streak recalling Rufous-thighed Kite. **Juvenile** ♀ '*guttifer*': Pale yellow or white iris. Dark brown above with white and rufous nuchal streaking suggesting a collar. Underparts white, cream, buff or cinnamon-rufous with dark brown tear-drops. **Voice** A measured nasal squawking *kut-kut-kut-kut-kut* series at 4–5 notes per sec; lower-pitched than Chilean Hawk. [Esparvero Variado]

Chilean Hawk *Accipiter* [*bicolor*] *chilensis*   ♂ 36 cm; ♀ 43 cm

Patagonian forest. **ID** Adult ♀: Resembles Bicoloured Hawk (no overlap) but grey breast and brown abdomen with white checkered barring. Deep orange to golden-yellow iris. **Juvenile** ♂: Probably not always separable from Bicoloured Hawk but tends to show more extensive ventral spotting and more striated (less barred) sides of breast and flanks. **Voice** A fast cackling *kekekekekekekeke* series at 10–12 notes per sec; higher-pitched than Bicoloured Hawk. **Tax note 11.** [Peuquito]

Tiny Hawk *Hieraspiza superciliosa*   ♂ 24 cm; ♀ 28 cm

Scarce in Paraná forest. **ID** Adult ♂: Red iris. Black cap, greyish ear-coverts and brown upperparts, often tinged grey. White throat and underparts, finely barred grey except on the throat. **Juvenile** ♀: Orange iris. Brown above with rufous fringes and a dark cap. Pale buff below, finely barred brown except on the throat. **Voice** An accelerating-decelerating whistled liquid *pwi-pwi-pwi-pwi* series of upward-inflected notes at 5–6 notes per sec; a little higher-pitched than Rufous-thighed Hawk. [Esparvero Chico]

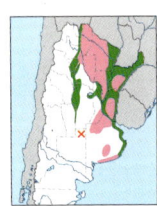

Rufous-thighed Hawk *Accipiter* [*striatus*] *erythronemius*   ♂ 27 cm; ♀ 31 cm

Widespread in N and C lowlands and foothills. **ID** All ages and plumages have rufous thighs. Adult ♀ **rufous morph:** Yellow iris. Dark brown above with rusty ear-coverts. Whitish throat and mainly rufous underparts, mottled white on the breast. Adult ♀ **typical morph:** Resembles rufous morph but breast and belly white with variable brown or rufous barring. Contrasting rufous thighs. **Juvenile** ♂: Brown above, finely fringed rufous and with whitish nuchal streaking. White or cream below, streaked pale brown, with rufous flanks and thighs and a white vent. Compare with juvenile Rufous-thighed Kite in Yungas and Paraná forest. **Voice** A pacey whistled, becard-like *tew-tew-tew-tew-tew* series at 5–6 notes per sec; much higher-pitched than Bicoloured Hawk. **Tax note 12.** [Esparvero Estriado]

Grey-bellied Goshawk *Accipiter poliogaster*   ♂ c.40 cm; ♀ 45 cm

Rare in Paraná forest. **ID** Adult **hooded morph:** Yellow iris. Black head and nape. Dark grey to sooty-brown back and wings. 2–3 indistinct pale tail bands. White below with a white spur on the sides of the neck; some are washed pale buffy-grey except on the throat. Adult **capped morph:** Grey ear-coverts accentuate a black cap. Compare with Collared Forest Falcon (Plate 50) and Short-tailed Hawk (Plate 47). **Juvenile:** Resembles adult Ornate Hawk-Eagle but is notably smaller, thighs unfeathered and lacks a crest. **Voice** A slow, high-pitched, rich liquid *pwi pwi pwi pwi pwi* series of upward-inflected notes at 3–4 notes per sec; surprisingly similar to Tiny Hawk although slower and with fewer notes. [Esparvero Grande]

ad

juv

Rufous-thighed Kite

'guttifer'

'guttifer'

'pileatus'

juv ♀

ad ♀

ad ♂

Bicoloured Hawk

ad ♀

juv ♂

juv ♀

ad ♂

Tiny Hawk

Chilean Hawk

ad capped morph

ad hooded morph

juv ♂

ad ♀ rufous morph

ad ♀ typical morph

juv

Rufous-thighed Hawk

Grey-bellied Goshawk

Rufous-thighed Kite *Harpagus diodon*   WS: 65 cm

Yungas and Paraná forest; mainly Oct–Mar. **ID** Flight: Resembles an *Accipiter* but note compact head, straighter, less bulging secondaries, much more pointed wings, and relatively short, narrow, square-ended tail compared to Bicoloured Hawk, and larger head, longer wings and shorter tail than Tiny Hawk. Readily soars like *Accipiter*. Adult: Best distinguished from Bicoloured by overall proportions (see above), but secondary barring and tail bands also better defined. Dark gular streak sometimes visible (at most faint in Bicoloured). Juvenile: Dark breast streaking. Rufous underwing-coverts and thighs. **Voice** Gives an unusual high-pitched tyrannid-like *SWEEee-Swit* call. [Milano de Corbata]

Accipiter and *Hieraspiza* Sleek, narrow-tailed hawks of forest and open country. All have yellow legs, cere and bare loral skin; latter intensifies in colour when breeding. Well-developed superciliary arc imparts fierce look. Most species are polymorphic. ♀♀ are considerably larger than ♂♂. In flight, identification aided by comparative tail length and shape of tail tip, shape of leading edge of wing, and extent that head protrudes from wing edge. All readily soar. **See perched images on Plate 44.**

Bicoloured Hawk *Accipiter bicolor*   WS: ♂ 68 cm; ♀ 92 cm

Widespread in the N. **ID** Flight: Straight leading edge of wing with large protruding head. Proportionally long-tailed with a rounded tip, and poorly defined secondary barring on adult, unlike congeners and Rufous-thighed Kite. Adult ♂ '*pileatus*': Rufous underwing-coverts and thighs; uniform grey breast. Adult '*guttifer*': Rufous lower breast and belly separate from Rufous-thighed Kite. Juvenile: Spotted underwing-coverts and sides of breast. **Voice** A measured nasal squawking *kut-kut-kut-kut-kut* series at 4–5 notes per sec; lower-pitched than Chilean Hawk. [Esparvero Variado]

Chilean Hawk *Accipiter [bicolor] chilensis*   WS: ♂ 65 cm; ♀ 80 cm

Patagonian forest. **ID** Flight: Adult ♀: Grey breast and browner belly, checkered with wavy white barring. Juvenile: As Bicoloured Hawk but usually shows a streaked breast. **Voice** A fast cackling *kekekekekekeke* series at 10–12 notes per sec; higher-pitched than Bicoloured Hawk. **Tax note 11.** [Peuquito]

Tiny Hawk *Hieraspiza superciliosa*   WS: ♂ 40 cm; ♀ 46 cm

Scarce in Paraná forest. **ID** Flight: Adult ♂: Diminutive hawk, with finely barred abdomen lacking rufous, and notably short notched tail with ill-defined bands. Wings straighter and less bulging than Rufous-thighed Hawk. Juvenile ♀: Fine brown barring on abdomen and underwing-coverts. Tail longer than adult, and more square-ended. **Voice** An accelerating-decelerating whistled liquid *pwi-pwi-pwi-pwi* series at 5–6 notes per sec; a little higher-pitched than Rufous-thighed Hawk. [Esparvero Chico]

Rufous-thighed Hawk *Accipiter [striatus] erythronemius*   WS: ♂ 52 cm; ♀ 60 cm

Widespread in N and C lowlands and foothills. **ID** Flight: Note more rounded leading edge of wing than congeners and Rufous-thighed Kite, shorter and squarer tail tip than Bicoloured Hawk with a narrower base than Rufous-thighed Kite, and narrower bands than both, but more prominent than Tiny Hawk. Head barely protrudes beyond wings. Typical adult ♀: Finely barred underwing-coverts and abdomen. Secondaries prominently barred, and tail bands narrow. Adult ♂ rufous morph: Rufous underwing-coverts and abdomen; white-speckled breast. Juvenile ♂: Tail proportionally longer than adult. Breast streaked brown or rusty, with coarse wavy barring on belly, unlike straighter-winged Tiny Hawk. **Voice** A pacey whistled, becard-like *tew-tew-tew-tew-tew* series at 5–6 notes per sec; much higher-pitched than Bicoloured Hawk. **Tax note 12.** [Esparvero Estriado]

Grey-bellied Goshawk *Accipiter poliogaster*   WS: 77 cm

Rare in Paraná forest. **ID** Flight: Adult hooded morph: Usually the plainest *Accipiter* from below, but some can show barred remiges. Juvenile: Very similar to much larger adult Ornate Hawk-Eagle (see Plate 48) with straight leading wing edge, less bulging secondaries and broader bands on narrower tail. Appears very small-headed in flight. Compare also juvenile Grey-headed Kite typical streaked morph (Plate 46). **Voice** A slow, high-pitched, rich liquid *pwi pwi pwi pwi pwi* series of upward-inflected notes at 3–4 notes per sec; surprisingly similar to Tiny Hawk although slower and with fewer notes. [Esparvero Grande]

Bicoloured Hawk

ad

juv

'guttifer'

'pileatus'

ad ♂

ad ♀

Rufous-thighed
Kite

juv

superciliosa

juv ♀

ad ♂

Tiny Hawk

ad ♀

juv

Chilean Hawk

juv

ad ♀
hooded
morph

ad ♀
typical
morph

juv

Grey-bellied Goshawk

ad ♂
rufous
morph

juv ♂

Rufous-thighed
Hawk

Hook-billed Kite *Chondrohierax uncinatus* 43 cm WS: 91 cm

N lowlands and Yungas foothills forest. **ID** Small-headed, polymorphic, forest raptor with bare loral skin, white iris and bare yellow tarsus. Long decurved nail on the tip of the upper mandible, used for feeding chiefly on snails. **Typical adult** ♂: Grey head, and grey below mostly barred white. **Typical adult** ♀: Narrow tawny nuchal collar. Chestnut below with irregular cream barring. **Adult black morph:** Entirely sooty-brown. All morphs show 1–2 grey or white tail bands. **Juvenile white morph:** Black cap, brown back and wings. Mostly white below. Broad white tail bands. **Flight:** Outsized, blunt paddle-like wings and a fairly long wedge-shaped tail. **Adult** ♀: Cinnamon body and undertail-coverts with dusky bars. Two broad black tail bands. **Adult black morph:** Blackish body and underwing-coverts. Blackish remiges with, or without, narrow white bars. Compare with Zone-tailed Hawk (Plate 38). **Juvenile white morph:** Pure white body and whitish underwing-coverts; heavily barred wings and three black tail bars. Compare with Grey-headed Kite and Black-and-white Hawk-Eagle. [Milano Pico Garfio]

Grey-headed Kite *Leptodon cayanensis* 55 cm WS: 100 cm

Humid chaco and Paraná forest. **ID** Small-headed, long-tailed, sleek forest raptor with unfeathered tarsus. **Adult:** Pale grey head. Black lower nape, back, wings and tail with two grey or whitish bands. White below. **Juvenile typical streaked morph:** Yellow cere and legs. Black cap and tawny ear-coverts. Dark brown above and white below, with variable blackish ventral streaking. Greyish undertail and grey-brown uppertail with 2–4 black bars. **Juvenile pale morph:** Rarer than streaked morph. Strongly recalls adult Black-and-white Hawk-Eagle, but more obvious pale bands on longer tail, small bill with yellow cere and lores, dark eyes and bare tarsus. **Flight:** Huge wings with bulging secondaries. **Adult:** White body contrasts with black underwing-coverts and two broad black tail bands. **Juvenile pale morph:** White head, body and underwing-coverts. Tail narrower than Black-and-white Hawk-Eagle and with more prominent barring. Compare with juvenile white morph Hook-billed Kite. [Milano Cabeza Gris]

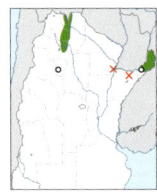

Black-and-white Hawk-Eagle *Spizaetus melanoleucus* 53 cm WS: 122 cm

Paraná and Yungas forest; local in the humid chaco. **ID** Stocky forest raptor with an orange-yellow cere, feathered tarsus and relictual crest. **Adult:** At close range, piercing yellow eyes, black loral mask and erectile black skullcap. White head and body. Black back and wings. 3 pale and 3 black bars on uppertail. **Juvenile** (not illustrated): Reduced skullcap and a browner back with 4–5 brown tail bars. **Flight:** Broad wings with bulging secondaries and a squarish tail. **Adult:** White body and underwing, barred blackish on the distal primaries and with a dusky trailing edge. Three narrow black tail bars. Compare with Grey-headed Kite (juvenile pale morph) and Hook-billed Kite (juvenile white morph). [Águila Viuda]

Mantled Hawk *Pseudastur polionotus* 49 cm WS: 123 cm

Very rare in Paraná forest of n. Misiones. **ID** Large, thickset and thick-necked, short-tailed forest raptor with unfeathered tarsus. **Adult:** White head and body, slaty back and wings. Primaries reach tip of broad white tail. Black bill and grey cere; eye appears very dark. **Juvenile** (not illustrated): Similar but with some streaking on the head. **Flight:** Extremely broad-winged with a short wedge-shaped tail, splayed open in flight. **Adult:** Appears mainly white with blackish primary tips, and greyer secondaries bordering a white trailing edge. [Aguilucho Blanco]

Hook-billed Kite

uncinatus

ad black morph

ad ♀ typical morph

ad ♂ typical morph

juv white morph

ad black morph

ad ♀

juv white morph

ad black morph

Grey-headed Kite

ad

juv pale morph

monachus

juv pale morph

ad

juv typical streaked morph

Black-and-white Hawk-Eagle

ad

Mantled Hawk

ad

Parabuteo A small stocky hawk, recalling a *Buteo* but with a proportionately large squarish head, larger iris and shorter tail. *Buteo* Robust, broad-winged hawks with relatively short, broad tails The tarsus is usually feathered to the intertarsal joint.

White-rumped Hawk *Parabuteo leucorrhous* 36 cm WS: 73 cm

Yungas forest; very rare in Paraná forest. **ID Adult:** Yellow cere and iris. Orange tarsus. Mostly black with narrow white uppertail-coverts, short narrow median greyish tail band, rufous thighs and a white vent. **Juvenile:** Yellowish cere and tarsus. Dark brown above, heavily streaked buff on rear crown and nape. White uppertail-coverts and 1–2 narrow, grey-brown tail bars. Buff or white below, blotched brown with barred thighs; rufous at the rear. Clean white vent. **Flight: Adult:** Unmistakable. Black flight feathers and body contrast with creamy underwing-coverts and white undertail-coverts. **Immature:** Streaked body, especially dark on flanks, contrasting creamy underwing-coverts and white undertail-coverts. Compare with juvenile Roadside (Plate 39) and Broad-winged Hawks (Plate 38). [Taguató Negro]

Short-tailed Hawk *Buteo brachyurus* 41 cm WS: 93 cm

Resident, mainly in Yungas, Paraná and N chaco forest. **ID Adult:** Buffy or whitish forehead and mainly dark brown above. Folded primaries almost reach tail tip. Four narrow pale greyish tail bands. White below with a small dark brown extension on the sides of the breast. **Juvenile:** Resembles adult but with whitish facial streaking. Brown tail with narrow blackish bars. Buff below; sides of breast with black spots/streaks and flanks streaked blackish. **Flight: Adult pale morph:** Appears mainly white from below with blackish-tipped primaries, black hood and extensions on sides of breast. Dark carpal crescents. From a distance, hood can be difficult to see (compare with Black-and-white Hawk-Eagle on Plate 46). **Adult dark morph:** Black body and underwing-coverts. Compare with dark morph Variable Hawk (Plate 41) and Zone-tailed Hawk (Plate 38). **Immature:** As adult pale morph but sides of breast and flanks streaked dusky. Tail proportionately longer. [Aguilucho Cola Corta]

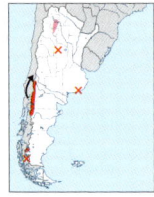

White-throated Hawk *Buteo albigula* 45 cm WS: 93 cm

Patagonian Forest (Sep–Apr); recorded on passage in the NW Andes. **ID Adult:** White underparts with tawny sides of breast spotted black, and brown sides of abdomen, streaked black. Flanks densely barred chestnut. **Juvenile:** Similar to adult but ear-coverts streaked buff, underparts creamier (except throat) and more spotted at sides. Compare with juvenile Rufous-tailed Hawk in Patagonia and juvenile Short-tailed Hawk in the NW, where the species can occur on passage. **Flight: Adult:** Broken pectoral band and vest. Diagnostic white window in distal primaries. **Immature:** Flanks more streaked and underparts creamier than adult. [Aguilucho Andino]

Rufous-tailed Hawk *Buteo ventralis* 56 cm WS: 126 cm

Rare in Patagonian forest to 2000 m. **ID Adult:** Blackish above with dull whitish lores. Rufous tail with 8–9 narrow blackish bands and broad subterminal band, tipped white. White throat, bordered by a blackish gorget and thick black malar streak. Tawny breast, streaked black and white. Blackish spots on abdomen. Thighs barred rufous and white, with a whiter vent. **Juvenile:** Olive-yellow cere. Dark brown above, densely speckled white on the nape, back, scapulars and wing-coverts with boldest markings on the greater coverts. Folded primaries reach halfway along the brown tail with 6–8 narrow blackish bands and a slightly broader subterminal band, tipped whitish. Pure white below with blackish spots on the sides of the breast and across the central abdomen. **Flight: Adult pale morph:** Dusky pectoral band and vest. Conspicuous brown patagial bar. **Adult dark morph** (scarce): As pale morph but body and underwing-coverts black. Note dark subterminal band. Resembles adult dark morph Variable Hawk (Plate 41) and especially juvenile dark morph Variable Hawk (not illustrated) which has a paler cheek patch, greyer tail, lacks an obvious subterminal tail band and has a much paler, less well-defined, trailing edge to the wing. **2nd-year immature:** Generally whiter below than adult pale morph but lacks a pectoral band, and shows a dark patagial bar from 1st year. Indistinctly barred primaries and tail which is tipped white (prominent from above). [Aguilucho Cola Rojiza]

White-rumped
Hawk

ad

juv

ad

imm

Short-tailed
Hawk

brachyurus

imm

ad

juv

ad
pale
morph

ad
dark morph

ad

juv

White-throated
Hawk

ad

imm

ad
pale morph

imm
2nd-yr

ad

juv

ad
dark
morph

Rufous-tailed
Hawk

Grey-bellied Goshawk *Accipiter poliogaster*   ♂ 40 cm; ♀ 45 cm WS: 77 cm

Included here for comparison – see main entry on Plates 44–45. Rare in Paraná forest. **Juvenile:** Resembles adult Ornate Hawk-Eagle but notably smaller, thighs unfeathered and lacks a crest. **Flight:** Juvenile: Less bulging secondaries than hawk-eagles and narrower tail. Appears very small-headed in flight.

Spizaetus Erectile crest at all ages. Feathered tarsus. Broad oval wings, pinched-in at the base with hand frequently angled forwards. Readily soars, although only Black-and-chestnut fans the tail. Sexes alike. ♀♀ slightly larger. Ornate and Black Hawk-Eagles give multi-noted shrill whistles, often while soaring, greatly aiding their detection.

Ornate Hawk-Eagle *Spizaetus ornatus*   63 cm WS: 117 cm

Yungas and Paraná forest; rare in e. Formosa. **ID Adult** ♀: Black crown and long crest; often upstanding. Large rufous nuchal patch reaching sides of breast. Black mantle and wings barred warm brown. Lower underparts barred black. **Juvenile** ♂: White head, neck and underparts. Short black crest. Narrow black moustachial. Upperparts like adult but with numerous narrow black tail bands. Sides of breast, belly and thighs barred with black. **Flight: Adult** ♀: Mostly white underwing, heavily spotted black on underwing-coverts. Chestnut sides of neck and breast. 3–4 narrow black bars on squarish undertail. Compare with Grey-bellied Hawk. **Juvenile** ♂: Poorly marked underwing; white patch in base of primaries visible from distance. 3–6 very fine tail bars are barely visible. Compare with much darker juvenile Black Hawk-Eagle. **Voice** 4–6 mellow whistles that rise and fall, with last note clearly distinct and ending abruptly, *we WEE we wo*, with emphasis on the 2nd or 3rd syllable. [Águila Crestuda Real]

Black Hawk-Eagle *Spizaetus tyrannus*   65 cm WS: 131 cm

Rare in Paraná and humid chaco forest. **ID Adult** ♀: Mostly slaty-black with short erectile bushy crest with white feather bases. Tarsus fully feathered and narrowly barred white. **Juvenile** ♂: Crown, neck and sides of breast densely streaked brown and white. Narrow white supercilium and dusky ear-coverts. Contrasting white throat and upper breast. Belly and thighs densely barred black and white. **Flight: Adult** ♀: Black body and underwing-coverts; boldly barred black-and-white flight feathers. Four white tail bands. **Juvenile** ♂: Contrasting white throat and upper breast. Prominently barred flight feathers and 4 whitish-and-black tail bands. Compare with much paler juvenile Ornate Hawk-Eagle. **Voice** While perched or in flight, a far-carrying, 5–6 note distinctive whistled *wup woop wu-wu-WHEEER*, often repeated frantically. [Águila Crestuda Negra]

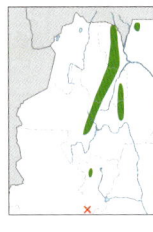

Black-and-chestnut Eagle *Spizaetus isidori*   69 cm WS: 146 cm

Scarce in Yungas forest. **ID Adult** ♀: Glossy black head, throat and upperparts, and deep chestnut below. White tail base. **Juvenile** ♂: White underparts and head, washed brown on crown and nape. Brown wing-coverts, streaked white. Three narrow black bands on greyish-brown tail. **Flight: Adult** ♀: Black head and deep chestnut body and underwing-coverts; can appear black at distance. Flight feathers mostly grey, barred black distally and with a striking white flash in base of primaries and broad black trailing edge. White tail with a black terminal band. **Juvenile** ♂: White body and creamy underwing-coverts. Flight feathers mostly grey, barred black distally and with a less obvious white flash in base of primaries. 2–3 narrow black tail bands. [Águila Poma]

juv

Grey-bellied
Goshawk

juv

Ornate Hawk-Eagle

ornatus

ad ♀

juv ♂

juv ♂

ad ♀

juv ♂

ad ♀

ad ♀

juv ♂

tyrannus

Black Hawk-Eagle

ad ♀

juv ♂

juv ♂

ad ♀

Black-and-chestnut Eagle

Crested Eagle *Morphnus guianensis* ♂ 81 cm; ♀ 86 cm WS: 146 cm

Very rare in Paraná forest. **ID** Very large with erectile crest. Unfeathered tarsus. Polymorphic. Sexes alike. **Adult ♀ dark morph:** Dusky grey head, upper breast and upperparts; thickly barred black and white below. Compare with Black Hawk-Eagle, which has feathered tarsus. **Adult ♀ pale morph:** Soft grey head and upper breast contrasts with creamy-white underparts. Blackish crest. Dusky-brown above; wing-coverts finely fringed white. Long black tail with 3 greyish bands. **Juvenile ♂:** White head and underparts. Fairly long white crest on centre of hindcrown (compare with juvenile Harpy Eagle). Soft greyish-white mantle and wing-coverts contrast with dusky-banded secondaries. **Flight:** Very broad wings with bulging secondaries, relatively long, broad tail with rounded tip. Proportionately large-headed. Readily soars. **Adult ♀ pale morph:** Soft grey throat and upper breast contrast with white underparts. Creamy underwing-coverts; flight feathers prominently barred black, especially on primaries. 3 broad white and 3 narrower black tail bands. **Second-year ♂:** Buffish-white body and underwing-coverts contrast with greyish flight feathers. 7–8 indistinct dusky tail bands. [Águila Monera]

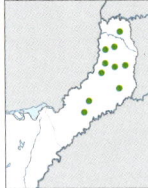

Harpy Eagle *Harpia harpyja* ♂ 93 cm; ♀ 98 cm WS: 188 cm

Rare in Paraná forest. **ID** Huge with broad tail, bifurcated crest at all ages and extremely thick tarsus. Sexes alike. **Adult ♀:** Pale grey head. Slaty-black back, wings and broad pectoral band. Black tail with 2–3 visible greyish bands. **Juvenile ♂:** White head and underparts; marbled grey on back and wings; numerous but distinct tail bands. Note bifurcated crest. **Flight:** Extremely long, broad wings, with relatively short, broad square-ended tail. Head protrudes well, but is dwarfed by overall size. Travels low in or over the canopy, but never soars. Sexes alike. **Adult ♀:** Bold black pectoral band and median-covert bar. Striking black-barred flight feathers and 3 black tail bands. **Second-year ♂:** Creamy underparts with contrasting grey pectoral band. Creamy underwing with some grey markings on coverts and barred remiges; especially primaries. Indistinct blackish tail bands unlike Crested Eagle. [Harpía]

ad ♀
pale morph

ad ♀
dark morph

juv ♂

Crested Eagle

♂
2nd-yr

ad ♀
pale morph

ad ♀

juv ♂

Harpy Eagle

ad ♀

♂
2nd-yr

Laughing Falcon *Herpetotheres cachinnans*
47 cm WS: 83 cm

Mainly humid chaco woodlands, but locally in Paraná forest. **ID** Chunky, noticeably full-headed falcon with upright stance and broad rounded wings in flight. Sexes alike. **Adult:** Creamy head and underparts with large black mask joining across the nape. Squarish black tail with 3–5 cream or white bands. Compare with Yellow-headed Caracara (Plate 51). **Flight:** Cream underparts and underwing-coverts. Narrow black tail bands with broader subterminal band. Black mask. Flaps and glides at canopy height. **Voice** Delivers long, far-carrying duets. [Guaicurú]

Micrastur Sleek polymorphic forest raptors with long graduated tails and very long tarsus. Reminiscent of *Accipiter*. Reclusive, and only likely to be seen flying inside forest and not soaring. Highly vocal at dawn and dusk giving characteristic monotonous calls and songs. Little sexual size difference.

Barred Forest Falcon *Micrastur ruficollis*
34–38 cm

Humid chaco and Paraná forest (*ruficollis*; 34 cm) and Yungas forest to 2600 m (*olrogi*; 38 cm). **ID Adult typical morph:** Brown above with grey cap. 3–5 narrow white tail bars. Rufous ear-coverts, breast and semi-collar. White below with fine or coarse black barring. **Adult rufous morph:** As typical morph but with rufous crown and upper back. Intermediate morphs occur. **Juvenile:** Dark brown above. White necklace, broken on sides of neck which is largely rufous. 3–6 narrow white tail bars. White or buff below with sparse narrow dark bars. **Flight: Adult grey morph:** Underwing-coverts and abdomen finely barred grey. Barred flight feathers and tail, although tail bands not well defined. **Voice** A long series of barking *kiów* notes every 1–2 secs (e.g. for up to 20 mins), which often then runs into cacophonous phrases of ratchety sputtered notes e.g. *ko-ko-ko-ko-Ko-KO KARRAra*, repeated over and over. [Halcón Montés Chico]

Collared Forest Falcon *Micrastur semitorquatus*
56 cm

Humid chaco, Paraná and Yungas forest to 1700 m. **ID Adult white morph:** Black crown and triangular 'sideburns' on white ear-coverts. Dark brown to blackish upperparts with 3–4 white tail bands, white nuchal collar and underparts. **Juvenile white morph:** Dark brown above with dark brown 'sideburns' and a rusty or white nuchal collar; barred cinnamon on back and wings. 4–5 narrow buff tail bands. White below, coarsely barred black. **Flight: Adult white morph:** Black 'sideburns' and breast spur. White below with conspicuous white barring on black tail and flight feathers. **Adult buff morph:** Underwing-coverts and abdomen tawny. Tail bands either white or tawny. Uppertail has 3 narrow bands in both morphs. **Voice** A series of well-spaced deep, resonant *Áow* notes every 5–7 secs, lasting 1–4 mins. Cackling voice speeds up, rises and then falls in pitch e.g. *ko ko ko-ko-kok-ok-ok oH.. aow... Áow...* finishing with progressively more spaced and longer notes. [Halcón Montés Grande]

cachinnans

ad

Laughing Falcon

ad
grey
morph

ruficollis

ad
rufous
morph

ad
typical
morph

juv

Barred Forest Falcon

ad
white
morph

juv
white
morph

ad
white
morph

semitorquatus

ad
buff
morph

Collared Forest Falcon

Spot-winged Falconet *Spiziapteryx circumcincta*

29.5 cm WS: 54 cm

W and C lowlands and foothills. **ID** Small falcon of light woodlands with broad rounded wings and graduated tail. Sexes alike. **Adult:** Bright yellow tarsus, cere and eye-ring. White supercilium and throat with broad black malar. Brown wash over breast with black shaft streaks. In flight, wings finely spotted white. White rump patch. Outertail barred white. **Voice** Highly vocal, including a loud cackling duet and a piercing descending-ascending whistle. [Halconcito Gris]

Milvago Smallest caracaras with narrow rectangular wings, pale flash in base of primaries and pale rump. Often gregarious, foraging terrestrially for carrion. Sexes alike.

Chimango Caracara *Milvago chimango*

37–41 cm WS: 81 cm

Widespread. **ID** Adult *chimango* (37 cm): Bill and tarsus yellow, peach/orange cere (♂); duller with a pink cere (♀). Uniform brown with a dusky eye-stripe. In flight, whitish flash in primaries and large creamy rump. Adult *temucoensis* (41 cm; Patagonian forest): Larger and considerably darker than *chimango*. Lacks a dark eye-stripe. **Juvenile:** Resembles adult but cere dull grey, mantle and scapulars fringed buff, and lacks dusky tail band. [Chimango]

Yellow-headed Caracara *Milvago chimachima*

40 cm WS: 88 cm

Forest and savanna in the NE. **ID** Adult: Creamy head and underparts with narrow black post-ocular streak. Contrasting blackish-brown mantle and wings. In flight, blackish upperwing with white primary flash. Contrasting creamy head, rump and tail base with black terminal band. **Juvenile:** Black crown and ear-coverts with cinnamon patch on sides of neck. Throat and breast densely streaked dark brown. Upperparts browner than adult. In flight, browner above than adult with duller primary flash. Dark cap with contrasting creamy nuchal band and creamy rump. **Immature:** Head approaching adult, but heavily streaked brown. Back, wings and breast similar to juvenile. [Chimachima]

Caracara Largest caracara with huge bill and flattened crest. Usually alone or in pairs. Sexes alike. Feeds on carrion but also attacks and kills fairly large prey.

Crested Caracara *Caracara plancus*

60 cm WS: 120 cm

Widespread. **ID** Adult: Large pink or orange cere. Black cap and creamy nape. Upper mantle and breast finely barred black and white. Dark brown wings and belly. In flight, long-necked. Mostly white primaries visible at great distance. Tail mostly white with black terminal band. **Juvenile:** Brown cap. Breast and upper mantle striated buff on brown. Paler brown wings and belly than adult. **Tax note 13.** [Carancho]

Spot-winged
Falconet

ad

temucoensis

ad

chimango

ad

juv

ad

Chimango Caracara

imm

ad

ad

juv

ad

juv

Yellow-headed Caracara

juv

ad

ad

Crested Caracara

Phalcoboenus Large caracaras with white terminal tail bands in adults. Proportionately longer-necked than other caracaras in flight, when tail is often fanned. Sometimes gregarious. Sexes alike.

Mountain Caracara *Phalcoboenus megalopterus*   54 cm WS: 117 cm

N Andes south to Neuquén. **ID** Adult: Large orange-red cere. Bushy peak on rear crown. Glossy black upperparts, throat and breast; white belly and rump. In flight, black throat and breast contrast with white belly and underwing-coverts. Black tail with a white distal band. Juvenile: Pink cere. Fairly uniform pale biscuit-brown with light spotting on wing-coverts and tertials. Creamy rump. Immature: Darker brown than juvenile; otherwise similar with small cinnamon primary flash and creamy rump. Very narrow white trailing wing edge at close range. [Matamico Andino]

White-throated Caracara *Phalcoboenus albogularis*   56 cm WS: 114 cm

Scarce in the S Andes from Neuquén to Tierra del Fuego. **ID** Adult: Chrome-yellow to orange cere. Lacks black throat and breast of Mountain Caracara. Mature adults have a black spur on side of upper breast. In flight, white underparts and underwing-coverts contrast with black flight feathers. Black tail tipped white. Juvenile: Pale yellow cere. Contrasting dark brown forehead and ear-coverts. Darker brown below with buff streaking on breast and finely spotted belly. Immature: Darker than juvenile and somewhat darker than Mountain, with larger cinnamon primary flash. Dusky mask and breast streaking usually evident. [Matamico Blanco]

Striated Caracara *Phalcoboenus australis*   61 cm WS: 120 cm

Rare in coastal s. Tierra del Fuego; common on Hornean islands and the Falklands. **ID** Adult: Small pale orange or pink cere. Mostly black with chestnut thighs and pointed white striations on hindneck and breast. In flight, small white handprint in base of primaries. Rusty bar along greater coverts. Chestnut thighs, white breast streaking and terminal tail band. Juvenile: Blackish bill and pale pink cere. Dark brown with chestnut streaking on upper back. In flight, dark brown with large cinnamon flash in primaries and bronzy tail. Immature: Grey bill and yellow cere. More extensive chestnut streaking than juvenile, extending onto breast. Bronzy-chestnut tail as juvenile. [Matamico Marino]

ad

juv

imm

ad

Mountain Caracara

juv

Striated Caracara

imm

ad

ad

White-throated Caracara

juv

juv

imm

ad

ad

Falco Sleek raptors with long, pointed wings and long tails. Very fast travelling flight in some species. Sexes alike except in American Kestrel. Mainly ♀♀ illustrated; ♀♀ are proportionately larger than ♂♂, with a greater difference in larger species.

American Kestrel *Falco sparverius*   27.5 cm WS: 53 cm

Widespread. **ID** Adult ♂: Blue-grey cap and wings. Black moustachial and stripe bordering white ear-coverts. Bright chestnut back and tail with black subterminal band. White below, washed cinnamon on upper breast and sparsely spotted black. **Adult** ♀: Grey crown, often with a brown centre. Face as ♂ but duller. Chestnut-brown back and wings, densely barred black. Tail as ♂ but finely barred black. White below with blurry brown streaking. **Juvenile** (not illustrated): Similar to adult but with more heavily barred back and heavier breast streaking. In flight, proportionately slender wings and tail compared to congeners. Black subterminal tail band. Hovers for long periods. [Halconcito Colorado]

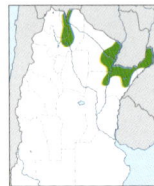

Bat Falcon *Falco rufigularis*   ♂ 26 cm; ♀ 28 cm WS: 67 cm

Yungas forest, humid chaco and Paraná forest. **ID** Adult ♀: Small bill. Black hood and upperparts. Primaries do not reach tail tip. Throat, upper breast and neck crescent varies from white to buff or orange-buff. High black vest, finely barred white or orange; sometimes only at sides. Lower belly and vent rufous. Unmarked undertail-coverts. In flight, blackish underwing, finely spotted white. Note narrow wing base compared to Orange-breasted Falcon. **Juvenile** ♂: Pale orange throat, neck crescent and upper breast; streaked at lower edge. Unmarked black vest. Rufous belly and vent; undertail-coverts spotted with black. [Halcón Negro Chico]

Orange-breasted Falcon *Falco deiroleucus*   ♂ 32 cm; ♀ 37 cm WS: 77 cm

Rare in Yungas (and Paraná?) forest. **ID** Adult ♀: Proportionately larger-headed than Bat Falcon with a much larger bill and tarsus with thick and proportionately longer toes; shape recalls Peregrine. Clean white throat only turns orange at rear of neck crescent and at lower border where small black streaks or spots are diagnostic. Low black vest scalloped with wavy barring; not linear as in Bat Falcon. Black bars on rufous undertail-coverts unlike adult Bat Falcon, but compare with juvenile Bat Falcon. In flight, black underwing and vest with rufous belly and vent distinguishes from congeners except Bat Falcon, but note broader wing base creating more triangular, less narrow, wing silhouette. White spots on remiges and narrow white tail bands more visible than in Bat Falcon. **Immature** ♀: White throat, becoming buffy or orange on upper breast with diagnostic streaks or spots. Dark brown vest (never black as in Bat Falcon) and spotted with buff and scalloped cinnamon, unlike Bat Falcon. Belly and vent white or buff with black chevrons on thighs. Note large feet and bill. [Halcón Negro Grande]

Aplomado Falcon *Falco femoralis*   ♂ 36 cm; ♀ 41 cm WS: 89 cm

Widespread. **ID** Adult *femoralis*: Grey cap, thick eye-stripe and upperparts. White or buff supercilium joins across nape. White or buff throat and breast with long pointed black moustachial. Black band across cinnamon belly and vent. In flight, wings more slender than Peregrine with black underwing-coverts spotted white. Black tail narrower than Peregrine with up to 7 white bars. **Adult** *pinchinae* (High NW Andes): Differs from *femoralis* by broken band across belly, darker belly and back, and larger size. **Juvenile**: Resembles adult but cap and upperparts dark brown. Cinnamon supercilium. Buff throat and breast, thickly streaked black at lower edge. Indistinct creamy tail bars. [Halcón Plomizo]

Peregrine Falcon *Falco peregrinus*   ♂ 40 cm; ♀ 47 cm WS: 105 cm

Widespread. **ID** Adult ♀ *cassini* (Resident): Dark hood with broad triangular lower border (lacking moustachial of migrant *tundrius*). Underparts variable, from white to buff or rich ochre, strongly barred black on belly, thighs and flanks. In flight, black helmet identifies ssp. Throat and breast variable but averages rich buff. **Adult** ♂ *cassini* pale morph (very rare in Patagonia): Lacks a hood, but brownish crown and spiky moustachial with white ear-coverts suggest *tundrius*. Black mantle and wing-coverts fringed creating barred effect. White below, sparsely flecked black on belly. **Juvenile** ♂ *cassini*. Whitish forehead and throat. Helmet-shaped hood as adult. White or buff nuchal patch. Buff below thickly streaked black. **Adult** ♀ *tundrius* (Nov–Apr, mainly in the N): From *cassini* by white ear-coverts and spiky black moustachial. Tends to show a greyer mantle and wings than *cassini*, and averages paler below; white to buff. Lower underparts finely barred black. In flight, wings and tail broader than congeners. Spiked moustachial and pale throat and breast identify ssp. [Halcón Peregrino]

hovering

Bat Falcon

rufigularis

ad ♀

ad ♂

juv ♂

ad ♂

ad ♀

ad ♀

ad ♀

ad ♂

cinnamominus

American Kestrel

ad ♂

femoralis

Aplomado Falcon

ad

ad ♀

imm ♀

ad ♀

ad ♀

ad

ad

Orange-breasted
Falcon

juv

femoralis

pichinchae

ad ♂
pale
morph

ad ♀

ad ♀

ad ♀

tundrius

cassini

cassini

tundrius

Peregrine Falcon

cassini

juv ♂

ad ♀

cassini

Speckled Crake *Coturnicops notatus*   14.5 cm

Rare in NE marshes, chiefly in e. Buenos Aires; wintering very locally in humid chaco marshes. **ID** Tiny, enigmatic (nocturnal?) crake. Flushes with a low weak flight of up to 5 m; rarely twice. **Adult:** Large red iris. Blackish with white spots, flecks, chevrons and streaks; most prominent on the foreneck. In flight, striking white speculum. Legs not visible. **Voice** A syncopated (sometimes duetted), crepuscular series of 3–4 (rarely 2) nasal, medium-pitched notes, e.g. *kít-kít-kít* or *düt-düt-düt* in less than 1 sec, sometimes with a slightly different final note. [Burrito Enano]

Dot-winged Crake *Laterallus spilopterus*   15 cm

Local in saltmarshes from C Argentina to N Patagonia. **ID Adult:** Red iris. Essentially dark, with unmarked slate-grey throat and breast, and white barring on wing-coverts, flanks, belly and undertail-coverts. Legs brown. In flight, notable white flecking on the wing-coverts and dangling legs. **Voice** Song is a disyllabic *kee-krrrr* repeated over and over. Also mellow whistles and harsh rolling chatters. [Burrito Negruzco]

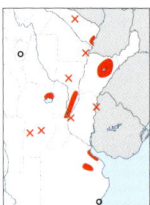

Yellow-breasted Crake *Hapalocrex flaviventer*   14 cm

Oct–Apr in floating marsh vegetation in the NE. **ID** Small nocturnal crake, with a very extensible neck and unusually long toes. **Adult:** Long pink or orange tarsus which dangle in flight. White supercilium broken above the eye and bordered by black eye-stripe. Brown back with white-streaked scapulars. Extensive black and white barring from sides of breast to vent. **Voice** High-pitched *dit* calls. [Burrito Amarillo]

Grey-breasted Crake *Laterallus exilis*   15 cm

Tall marsh grass in e. Formosa, e. Chaco, Santa Fe and Corrientes. **ID Adult:** Red iris. Grey head and breast, black-and-white barring on flanks and vent. Bill greenish-yellow, legs yellowish-brown. No other crake in range has a contrasting rufous nape. **Voice** Trill is fairly slow and metallic, sometimes preceded by ticking calls, or harsh ratchet-like trill, starting slowly. Common call is a fast monotone *dit-dit-dit-dit-dit* series (recalling some calls of Chestnut-capped Blackbird), less often a high-pitched *chep*. Sonogram on p. 445. [Burrito Pecho Gris]

Ocellated Crake *Micropygia schomburgkii*   14 cm

Local in tall, dry rolling campos grasslands of sw. Misiones. **ID** Generally found by voice. **Adult:** Turquoise bill with black culmen. Bright pink tarsus. Mostly ochraceous-buff with a contrasting rusty forehead and white centre to the belly. Nape, mantle and wing-coverts sprayed with pure white ocellations, outlined in black. **Voice** Common call is an unmistakable incessant series of bizarre raspy nasal brays *kreearh, kreearh, kreearh,…* even during the heat of the day, but mostly at dusk. Also gives an explosive descending trill, ending in piping notes. [Burrito Ocelado]

Red-and-white Crake *Laterallus leucopyrrhus*   18 cm

Local in sedges, especially *Scirpus giganteus* and lilac beds, often inside light woodlands, chiefly in ne. Buenos Aires and e. Entre Ríos; locally in Misiones. **ID Adult:** Rather rich rufous head, sides of neck and breast. Upperparts generally warmer and more rufescent than sympatric Rufous-sided Crake. Note diagnostic white undertail-coverts and red tarsus. Bill base usually much brighter yellow than Rufous-sided. **Voice** Trill (often in duet) has short, soft, mellow introductory notes followed by slower, less harsh and lower-pitched trill than Rufous-sided Crake, but very similar. Frequent call is a brief, resonant, liquid whistle. Slow chatter often given after playback and fast rattle (alarm?) diagnostic. Sonogram on p. 445. [Burrito Colorado]

Rufous-sided Crake *Laterallus melanophaius*   17 cm

Widespread in reed and rush marshes throughout the N lowlands. Easy to observe compared to congeners. **ID Adult:** Brown above with contrasting rufous sides to the neck and breast only. Unlike Red-and-white Crake shows chestnut undertail-coverts, greyish face and dull olive-yellow tarsus. **Voice** Trill (often in duet) is a long series of soft introductory notes followed by a faster, harsher and higher-pitched trill than Red-and-white Crake, but very similar. Frequent call is a dry *chep* or *tip*, sometimes in long series with harsh notes. Alarm is a piercing *TEEW!* Rapid trill call is medium-pitched, liquid and descending. Sonogram on p. 445. [Burrito Canela]

Speckled Crake
ad
ad

Dot-winged Crake
ad
ad

Yellow-breasted Crake
flaviventer
ad

ad
Grey-breasted Crake

ad
chapmani
Ocellated Crake

Red-and-white Crake
ad

ad
melanophaius
Rufous-sided Crake

Paint-billed Crake *Mustelirallus erythrops*   19 cm

Poorly known (Oct–Apr) in N marshes, lagoon edges and flooded woodlands; chiefly in the NW. **ID** Medium to large crake with a stout, broad-based bill. **Adult:** Conspicuous bright red tarsus and bill base. Drab olive-brown above. Slate-grey below; contrasting whitish throat, and black and white-barred vent. **Juvenile:** Dark iris and olive tarsus. Lacks the red bill base (entire bill can appear pinkish), but (unlike Ash-throated) bill is noticeably short and upperparts are unmarked. **Voice** Loud, mournful, resonant somewhat descending clucking series of notes, and a hollow purring. [Burrito Pico Rojo]

Ash-throated Crake *Mustelirallus albicollis*   23 cm

Widespread in N lowland marshes. **ID Adult:** Medium to large crake with a stout, broad-based bill. **Adult:** Olive-brown above with prominent blackish feather centres. White throat contrasts with ash-grey underparts. Blackish flanks and vent narrowly barred white. **Juvenile** (not illustrated): Differs by dark eye, drab tarsus and bill. **Voice** Raucous, wailing, antiphonal voice at dawn and dusk is unmistakable. [Burrito Grande]

Spot-flanked Gallinule *Porphyriops melanops*   26.5–32 cm

Widespread in the N lowlands and S Andes. **ID** A small gallinule which once out of water can appear quite crake-like but note the frontal shield and lobed toes. **Adult** *melanops* (26.5 cm; N lowlands): Stout pea-green bill. Blackish lores and forecrown. Slaty neck and breast; contrasting chestnut wing-coverts; white spots on flanks and pure white undertail-coverts. **Adult** *crassirostris* (32 cm, S Andes; not illustrated): Larger and with a thicker bill. **Juvenile:** Brown head and hindneck with a slight rufous wash on the wing-coverts. White throat, pale grey below with indistinct white flank spots and some streaks. [Pollona Pintada]

Purple Gallinule *Porphyrio martinicus*   29.5 cm

Oct–Apr in the N lowlands. **ID** Long-legged, reclusive, migratory marsh gallinule with a small frontal shield. **Adult:** Red bill, tipped yellow. Pale blue frontal shield. Purplish-blue head and underparts with white undertail-coverts. Bronzy-green upperparts. **Juvenile:** Bluish frontal shield bordered by black lores. Drab olive upperparts with bluish wing-coverts. White belly and undertail-coverts. [Pollona Azul]

Azure Gallinule *Porphyrio flavirostris*   26 cm

Local in e. Formosa and n. Corrientes in Oct–Feb. **ID** Long-legged, reclusive, migratory marsh gallinule with a small frontal shield. **Adult:** Mainly turquoise and white, with short yellow bill and long yellow tarsus. **Juvenile:** Brown upperparts with buff wash on face and sides of breast, and white central underparts. [Pollona Celeste]

Heliornis Small long-necked river bird with black-striped head and neck, and lobbed toes. Reversed sexual roles.

Sungrebe *Heliornis fulica*   29 cm

Local along NE forest rivers. **ID** Breeding ♀: Red bill and eye-ring. Cinnamon ear-coverts. Breeding ♂: Yellowish bill and white ear-coverts. Non-breeding ♀ resembles ♂ but with a dull red bill. **Voice** Resonating, low-pitched *kwoot-kwoot* carries well. [Ipequí]

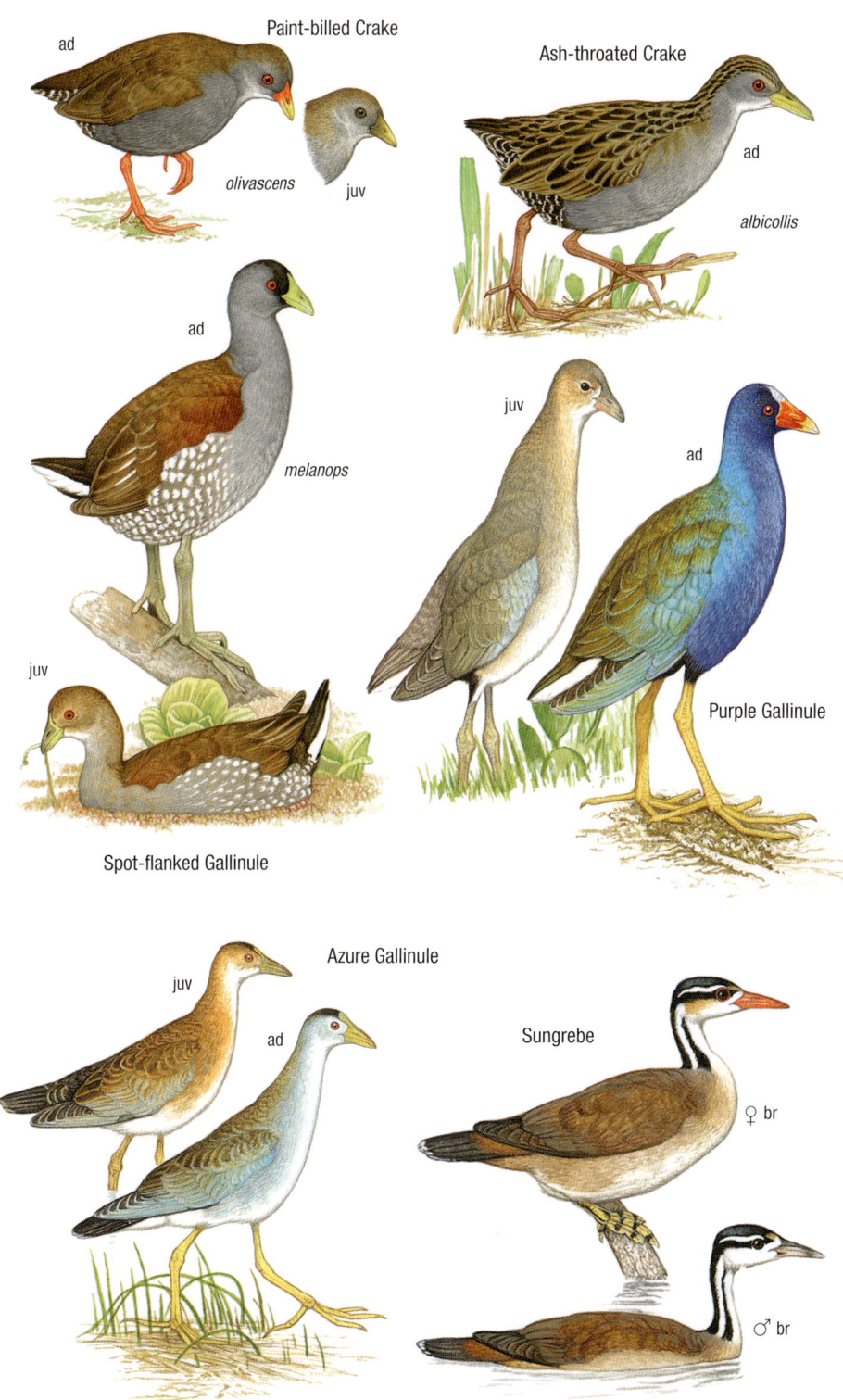

Paint-billed Crake

ad

olivascens

juv

Ash-throated Crake

ad

albicollis

ad

melanops

juv

juv

ad

Purple Gallinule

Spot-flanked Gallinule

Azure Gallinule

juv

ad

Sungrebe

♀ br

♂ br

Austral Rail *Rallus antarcticus*   19 cm

Rushbeds in Santa Cruz and Chubut; some moving north in winter? **ID** Small and compact, reclusive rail with a slender bill. **Adult:** Red bill, pink tarsus. Grey face and breast. Buff supraloral turning whitish behind eye. Flanks barred black and white. **Voice** Detected by fast, monotonous metallic couplets. In duet, female delivers simultaneous deep hollow booming. [Gallineta Chica]

Spotted Rail *Pardirallus maculatus*   29.5 cm

Dense marsh vegetation in the N Lowlands (Sep–Mar). **ID** Long-billed and long-legged, medium-sized marsh rail. **Adult:** Densely streaked and barred white throughout. Sky-blue base to upper mandible and red spot on base of lower mandible. Pink tarsus. **Juvenile** (not illustrated): Paler throughout, ill-defined supercilium and dusky bill with paler lower mandible. **Voice** Song is an accelerating, then decelerating, dry raspy emphatic *kwek… kwek…kwek… kwek kwek kwek kwek-kwek-kwek….* Low 5–6 sec drumming begins very slowly and accelerates toward the end (much slower than somewhat similar drumming of Paint-billed Crake). Call is a dry scolding *krek* alternating with *krek-teek*. [Gallineta Overa]

Blackish Rail *Pardirallus nigricans*   32.5 cm

Flooded forest and dense marshes in Misiones to w. Corrientes. **ID** Long-billed and long-legged, medium-sized marsh rail. **Adult:** Green bill. Olive upperparts and grey underparts with contrasting whitish throat. Overlaps with Plumbeous Rail and Slaty-breasted Wood Rail. **Voice** Song switches between delicate trills, sweet upswept notes and final descending laughter *teerrrr sweeeEP sweeeEP kee-kee-kee-kee…* Calls include a high-pitched *ti-dik* and isolated *sweeeEP* notes. [Gallineta Negruzca]

Plumbeous Rail *Pardirallus sanguinolentus*   29–38 cm

Widespread. **ID** Long-billed and long-legged, medium-sized marsh rail. **Adult** *sanguinolentus* (29 cm): Green bill with sky-blue base of upper mandible, and red spot on base of lower mandible (duller in non-breeders). Olive above with blackish centres to the secondaries and tertials. Grey face and underparts. **Adult** *luridus* (38 cm; Tierra del Fuego): Reddish brown above without any blackish wing markings. **Juvenile:** Dusky bill. Entirely brown with a whitish throat. **Voice** Song is a shrill, penetrating *creeaw-tree creeaw-tree peeerrrrr* or *creeaw-tree creeaw-tree poowoorrr poowoorrr* by male, while female delivers a low-pitched booming during duet. Also gives a low buzzy *züm…züm…züm…züm zümzümzümzüm* which rises and falls, accelerating toward the end. Calls include high-pitched *weeP, kleeP* or a two-note *kill-it*! Some vocalisations sound similar to calls of Austral Rail. [Gallineta Pico Pintado]

Giant Wood Rail *Aramides ypecaha*   48 cm

Marshes in the N lowlands, often in the open. **ID Adult:** Chestnut hindneck and pink abdomen. **Voice** Often in duet, a loud hysterical too *WA-KAA …. too….. WA-KAA waaAH waaAH waaAH*, with low guttural sounds (female?) at close range. [Ipacaá]

Slaty-breasted Wood Rail *Aramides saracura*   42 cm

Paraná forest. **ID Adult:** Bronzy hindneck and grey abdomen. From confusingly similar Blackish Rail by broader-based, straighter bill and bronzy hindneck, but voices very different. **Voice** Song, often in duet, mostly at dusk and dawn, is a strident *WAK WAO… KA POW* series, often lasting a few minutes. Also *pow, wow* or *wok* calls. [Saracura]

Grey-necked Wood Rail *Aramides cajaneus*   39 cm

Woodlands in the N lowlands and foothills, and nearby watercourses. **ID Adult:** Grey hindneck and chestnut abdomen. **Voice** Mostly crepuscular duet is a long, loud series: *chiri cot chiri cot cu-cu-cu-cu* or *KIK-kok KIK-kok cu cu cu CU CU CU*, with low guttural sounds audible at close range, repeated for several minutes. Call is a muted *chaok*. [Chiricote]

Austral Rail

ad

Blackish Rail

ad

nigricans

Spotted Rail

maculatus

ad

juv

Plumbeous
Rail

ad

luridus

ad

sanguinolentus

ad

Slaty-breasted
Wood Rail

ad

ad

cajaneus

Giant Wood Rail

Grey-necked
Wood Rail

Common Gallinule *Gallinula galeata*   35–40 cm

Widespread in lowlands and foothills; also Puna lakes. **ID** Aquatic and terrestrial marsh bird with large frontal shield and elongated slender toes. **Adult** *galeata* (35 cm; lowlands and foothills): Red frontal shield and bill with yellow tip. Yellowish tarsus. Blackish head and plumbeous body plumage with prominent white flank stripe. Brownish wings. **Adult** *garmani* (40 cm; Puna lakes; not illustrated): Larger. Black back concolorous with wings. **Juvenile:** Dull yellow bill and reduced frontal shield. Faded plumage with whitish throat. [**Pollona Negra**]

Fulica Gregarious black waterbirds with frontal shields and lobed toes and loud multi-syllabic vocalisations.

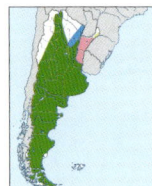

White-winged Coot *Fulica leucoptera*   35 cm

Widespread. **ID Adult:** White, yellow or orange-yellow rounded and slightly bulbous frontal shield. Whitish bill. In flight, white trailing edge to secondaries. **Juvenile:** Generally sooty, sometimes with whitish throat. [**Gallareta Chica**]

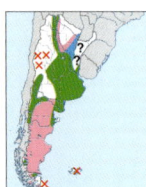

Red-fronted Coot *Fulica rufifrons*   39 cm

Widespread; usually in reedbeds and seldom on open water. **ID** Swims with erect tail. More reclusive than other lowland coots. **Adult:** Steep and pointed dark red frontal shield, and yellow bill. In flight, note large white sides to the vent. **Juvenile:** Noticeably steep forehead. Whitish throat. [**Gallareta Escudete Rojo**]

Red-gartered Coot *Fulica armillata*   42 cm

Widespread. **ID Adult:** Pointed yellowish or white frontal shield with reddish transverse line across base. White or yellowish bill. In flight, shows less white on sides of vent than congeners. Note position of red garters. **Juvenile:** Palest of the lowland coots. [**Gallareta Ligas Rojas**]

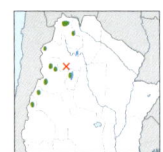

Horned Coot *Fulica cornuta*   54 cm

Rare and very local in high NW Andes. **ID Adult:** Feathered black proboscis droops along culmen. Bright orange bill base when breeding; otherwise drab yellow. Legs greenish. Head and neck blacker than body. Whitish trailing edge to wing in flight. **Juvenile** (not illustrated): Somewhat paler without the contrasting black head and with a white patch over chin and throat. [**Gallareta Cornuda**]

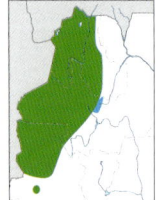

Giant Coot *Fulica gigantea*   56 cm

Widespread on Puna and high NW Andean lakes. **ID Adult:** Conspicuous knobs over eye sockets and furrowed crown. Large orange-yellow frontal shield. Bright red tarsus. **Juvenile:** Cranial structure as adult. Blackish area around eye and crown, contrasting with white lores and ear-coverts. [**Gallareta Gigante**]

Andean Coot *Fulica ardesiaca*   45 cm

NW Andean and Puna lakes. **ID Adult:** Bulbous dark red frontal shield and yellowish and white bill. Head and neck black, contrasting with grey body plumage. **Juvenile:** White area around eye. [**Gallareta Andina**]

Common Gallinule

ad

galeata

juv

ad

White-winged
Coot

juv

ad

ad

juv

ad

Red-fronted
Coot

ad

juv

ad

juv

ad

Horned Coot

ad

Red-gartered Coot

ad

ad

Giant Coot

juv

ardesiaca

juv

Andean Coot

Haematopus Large, robust shorebirds with combination of black, brown and white plumage, and a long, straight red bill, often with a yellow-orange tip, and relatively short, stout legs. Often gregarious, forming mixed oystercatcher flocks.

Magellanic Oystercatcher *Haematopus leucopodus* 43 cm

Patagonia and the Falklands; many breed inland. **ID** Adult: Bill more slender than American Oystercatcher. Uniform black head, breast, mantle and wing-coverts. Yellow eye-ring. In flight, white secondaries form a wedge, unlike American Oystercatcher. Thick black leading edge to underwing. [Ostrero Austral]

American Oystercatcher *Haematopus palliatus* 43 cm

Throughout continental coasts; scarcer in S Patagonia. **ID** Adult: Broader chisel-shaped bill than Magellanic Oystercatcher. Contrasting brown back and wing-coverts. Red eye-ring. In flight, enclosed white wing-stripe; dusky trailing edge to underwing. [Ostrero Pardo]

Blackish Oystercatcher *Haematopus ater* 44 cm

Patagonian coasts and the Falklands; usually on rocky shores. **ID** Adult: Black with a brown back and wing-coverts; lacking white of congeners. Notably deep, blunt bill with a narrower base. In flight, completely black underparts and no white on wings. [Ostrero Negro]

Black-necked Stilt *Himantopus mexicanus* 37 cm

Widespread, except Tierra del Fuego. **ID** Slender, long-necked pied wader with extremely long pink legs, and a needle-like bill. **Adult** ♂: Black nape, hindneck and breast spurs. White band across upper back. Black mantle and wings. In flight, long pointed wings contrast with white tail and V-shaped rump. **Adult** ♀ (not illustrated): Browner mantle and wings. **Juvenile** differs by greyish-brown crown and hindneck with contrasting white forehead. **Voice** Noisy groups even at night. Standing and flight calls are a quick series of high-pitched barks. Take-off call is remarkably similar to that of Greater Yellowlegs, *tew tew tew!* Tax note 14. [Tero Real]

Andean Avocet *Recurvirostra andina* 46 cm

High NW Andes. **ID** Sturdy wader with a slender, upcurved black bill, and long blue tarsus. **Adult:** White with black back and wings. In flight, note contrasting white underwing-coverts and rump; tarsus extends well beyond tail. **Voice** Very nasal, metallic and emphatic *weENK* calls, sometimes as a series. [Avoceta Andina]

Jacana A long-legged marsh wader with extremely long toes for walking on aquatic vegetation. Males incubate eggs and care for young.

Wattled Jacana *Jacana jacana* 24 cm

Widespread in the N lowlands and foothills. **ID** Adult: Glossy black with chestnut back and wing-coverts. Bright yellow flight feathers. **Juvenile:** Narrow white supercilium and underparts. Brown back. [Jacana]

Magellanic Oystercatcher

ad

ad

typical
display
posture

durnfordi

American Oystercatcher

ad

juv

melanurus

ad ♂

Blackish Oystercatcher

Black-necked Stilt

ad

ad

juv

Andean Avocet

Wattled Jacana

Magellanic Plover *Pluvianellus socialis* — 19.5 cm

Breeds in S Patagonia; winters north to se. Buenos Aires. **ID** An enigmatic dove-like wader of pebble-lined lake shores. Unique among waders in possessing a crop and regurgitating food to its young. Sexes alike. **Adult**: Red iris. Bright pink legs. Soft grey upperparts and breast. White forehead, throat and belly. In flight, long white wing-stripe. Grey centre of rump and black central tail bordered white. **Juvenile**: Red or pink iris. Orange tarsus. Paler grey and spangled white above. Pectoral band streaked whitish. [Chorlito Ceniciento]

Tawny-throated Dotterel *Oreopholus ruficollis* — 26.5 cm

Breeds in Patagonia; winters in the Pampas. Resident populations in the Puna, NW Andes and C sierras. **ID** Elegant, long-necked, pot-bellied wader with long tarsus and very fine bill. **Adult**: Pale grey head and neck with prominent creamy supercilium. Pale orange throat. Black spot on creamy belly. Black-streaked wing-coverts. **Juvenile**: Faded version of adult with little orange on throat and belly spot reduced. Prominent brown cap. Wing-coverts scalloped rather than streaked. In flight, long-winged and long-tailed; low to high, fast, sometimes twisting flight. Gleaming white underwings, and prominent wing-stripe on upperwing. Black belly patch. **Voice** Flight calls include a whistled and mellow *Pi-chew-chew-chew* and *PU-Pi-u*. [Chorlo Cabezón]

Rufous-chested Dotterel *Charadrius modestus* — 21.5 cm

Breeds in S Patagonia; winters in the N lowlands near the coast. Resident on the Falklands. **ID Adult breeding**: Bold white supercilium. Rufous breast bordered by black band. **Adult non-breeding**: Brown above with buffy eye-ring and post-ocular streak. White chin, brown breast, clearly demarcated from white abdomen. In flight, no wing-stripe but rump bordered white and outer tail-feathers white. **Juvenile**: Resembles non-breeding adult but tarsus yellowish, darker cap and blackish back, scalloped with white. Breast mottled brown. [Chorlito Pecho Canela]

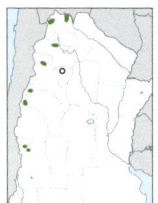

Diademed Sandpiper-Plover *Phegornis mitchellii* — 17.5 cm

Rare and local in high C and N Andes. **ID** Delightful wader of cushion-plant bogs and pebble-lined rivers. Slender drooping black bill. Flies little and easily overlooked, but loud melancholic whistle draws attention. Sexes alike. **Adult**: Fine black bill. Bright orange tarsus. Black head with white diadem and rufous nape. White gorget; finely vermiculated black below. In flight, broad pointed wings with a short white trailing edge. Short white-tipped tail. **Juvenile**: Yellow-orange tarsus. Brown crown and upperparts, barred rusty. Diadem lacking or indistinct. Pale rufous nape. Indistinct and more widely spaced ventral vermiculations. **Voice** A drawn-out, rich, whistled *weeeo*, which carries well. Also short *klip* or *pip* calls. [Chorlito de Vincha]

South American Painted-snipe *Nycticryphes semicollaris* — 21 cm

N lowlands south to NE Patagonia. **ID** Large-headed, secretive wader of rushbeds with long drooping bill. Sexes alike. Influxes into areas with suitable higher water levels. **Adult**: Green or yellow bill, tipped orange when breeding. Chocolate-brown or maroon head. Buffy white coronal stripe and narrow white supercilium. White breast spurs extend into golden scapular braces. White belly and large white spots on wing-coverts. **Juvenile** (not illustrated): Paler than adult with buff-fringed wing-coverts and whitish throat. Legs extend beyond the tail in flight. Flushes with mechanical bat-like turns, often with dangling legs. [Aguatero]

ad

Magellanic Plover

Tawny-throated
Dotterel

ad

ruficollis

juv

ad

juv

ad

ad non-br

ad non-br

ad br

juv

Rufous-chested Dotterel

ad

juv

ad

ad

Diademed
Sandpiper-Plover

ad

South American
Painted-snipe

Pluvialis Robust, non-breeding boreal migrant waders with a short straight bill. ***Vanellus*** Robust large-headed waders with a short, stubby, black-tipped pink bill. Often gregarious.

American Golden Plover *Pluvialis dominica*   26 cm

N grasslands (Aug–Apr), scarcer through E Patagonia. **ID Adult non-breeding:** Slighter build and smaller bill than Grey Plover. Dark brown cap and obvious white supercilium. Upperparts, especially the wing-coverts, notched buffy-yellow to greyish-white. Folded primaries extend beyond tail tip. Whitish throat and belly, otherwise drab greyish breast, with a dusky streaked effect. In flight, dark cap and dusky smudge behind eye. Indistinct whitish wing-stripe. Grey underwings. Finely barred tail. **Adult breeding:** Spangled black-and-white above, notched with gold. White supercilium, extending onto sides of breast. Black face and underparts (looks blotched when moutling into non-breeding plumage). [Chorlo Pampa]

Grey Plover *Pluvialis squatarola*   29 cm

Mainly coastal (Nov–Apr); on passage elsewhere. **ID Adult non-breeding:** Grey-brown above, fringed whitish. Whitish supercilium. White below with variable brown or grey streaking over breast; less prominent than American Golden Plover. In flight, easily distinguished from American Golden Plover by contrasting black axillaries. Also, stouter-billed, lacks a dark cap, has an obvious white wing-stripe, white rump and more conspicuously barred tail. **Adult breeding:** Spangled black-and-white above (giving grey effect), lacking gold. Face and underparts solid black. White supercilium, extending onto sides of breast. Folded primaries reach tail tip. [Alt: Black-bellied Plover] [Chorlo Ártico]

Pied Plover *Hoploxypterus cayanus*   21 cm

Rare vagrant on N rivers. **ID** Small riverine plover recalling a *Charadrius* plover in behaviour, or a lapwing in flight. Lacks a hind toe, but has a carpal spur. **Adult:** Broad black mask with large sandy skullcap, ringed white. Black breast and mantle with white V-shaped scapular braces. Contrasting white throat and belly. Conspicuous white inner wings, rump and black tail band in flight. [Chorlo de Espolón]

Andean Lapwing *Vanellus resplendens*   32 cm

High NW Andes; lower in winter. **ID Adult:** Shorter, brighter legs than Southern Lapwing. Soft grey head and breast, demarcated from white belly. Brown above, glossed green on mantle and lilac on wing-coverts. Local overlap with Southern Lapwing (ssp. *lampronotus*) in winter. In flight, proportionately broader wings and narrower black tail band than Southern Lapwing. **Voice** Calls given in flight and on the ground are metallic and tern-like *kerr kerr kerr…*, faster and higher-pitched than Southern Lapwing. [Tero Serrano]

Southern Lapwing *Vanellus chilensis*   34–39 cm

Widespread. **ID Adult *fretensis*** (39 cm; Patagonia): Grey head with vestigial crest. Black foreface and large black breast patch. **Adult *lampronotus*** (34 cm; N lowlands and Andes to 2000 m, rarely higher). Smaller and slighter build than *fretensis* with less bulbous, pale greyish-brown head; prominent crest; less black on foreneck and breast; and relatively longer tarsus. In flight, white wing-stripe and rump contrasts with black flight feathers and tail and broad black subterminal tail band. **Voice** Loud metallic *Teú* and higher-pitched, louder and raucous *TE-RU, TE-RU* repeated over and over. Patagonian birds have a very different, higher-pitched and harsher, parrot-like *KE-RREU, KE-RREU*. **Tax note 15.** [Tero]

ad non-br

ad non-br

ad br (moulting)

American
Golden Plover

ad non-br

ad non-br

ad br

Grey Plover

ad

Pied Plover

ad

fretensis

ad

ad

ad

lampronotus

Andean Lapwing

Southern Lapwing

Charadrius Small, migratory and resident plovers of marsh, beach and grassland, with relatively large head, straight bill and compact build. Distinguished by head and abdomen pattern. Most have a distinct non-breeding plumage.

Semipalmated Plover *Charadrius semipalmatus* 18 cm

Boreal migrant, chiefly to coasts (Sep–May). **ID** Narrow white nuchal collar in all plumages. **Adult breeding**: Stubby orange bill, tipped black. Dull orange tarsus. Narrow yellowish eye-ring. White forehead enclosed by black. Broad black pectoral band. **Adult non-breeding**: Little or no black on head, but supercilium prominent; usually joins with white forehead. Complete or broken blackish-brown pectoral band. White nuchal collar. In flight, notable white wing-stripe. Narrow white outer tail-feathers. **Juvenile**: Olive-brown tarsus. Indistinct eye-ring, but supercilium well defined. Pale-fringed wing-coverts. Pectoral band broken or complete. [Chorlito Palmado]

Collared Plover *Charadrius collaris* 15.5 cm

Resident in the N lowlands and Andes. **ID Adult**: Pink tarsus. Black lores, coronal bar and narrow pectoral band. Rufous hindcrown, nape and sides of neck, with brown centre. Non-breeders similar. In flight, rusty fringes to back and scapulars. Notable white wing-stripe. More white in the outer tail-feathers than Puna, Semipalmated and Two-banded Plovers. **Juvenile**: White lores, forehead and short post-ocular streak; black breast spurs. Note pale legs. [Chorlito de Collar]

Puna Plover *Charadrius alticola* 17.5 cm

Resident in the Puna and high NW Andes. **ID Adult breeding**: Black tarsus. Chestnut hindcrown with brown skullcap. Black spur on sides of breast, and complete grey band across lower breast; sometimes mixed with chestnut. **Adult non-breeding**: Usually lacks rufous on crown but retains black coronal band. Brown breast spurs and faded or broken lower band. In flight, wing-stripe only prominent on the inner primaries. Narrow white outer tail-feathers. **Juvenile**: Tiny white forehead and no coronal band. Buff-fringed scapulars and wing-coverts. Small brown breast spur. Black tarsus unlike juvenile Collared. [Chorlito Puneño]

Two-banded Plover *Charadrius falklandicus* 19.5 cm

Breeds in Patagonia and very locally in the N; winters in the N lowlands. Resident on the Falklands. **ID Adult breeding**: Only *Charadrius* with two complete pectoral bands (although most breeding adult Falklands birds usually lack the upper pectoral band which is restricted to spurs). **Adult non-breeding**: Lacks black and chestnut head pattern. Complete brown pectoral band and faded lower band. In flight, indistinct wing-stripe, mostly restricted to the inner primaries. Narrow white outer tail-feathers. **Juvenile**: As non-breeding adult but lower band usually lacking altogether, and wing-coverts fringed buff. [Chorlito Doble Collar]

Lesser Sand Plover *Charadrius mongolus* 17 cm

Vagrant, recorded once from coast of Buenos Aires. **Adult ♂ breeding**: Black mask and frontal bar highlighting white forehead and throat. Orange breast. Pattern reduced and faded in ♀. **Adult non-breeding**: Lacks black and orange head pattern. Rather plain grey-brown above, white below with prominent breast spurs. [Chorlito Mongol]

ad non-br

ad non-br

ad br

Semipalmated Plover

juv

ad non-br

juv

ad non-br

juv

ad

ad non-br

Collared Plover

Puna Plover

ad br

ad non-br

juv

ad non-br

ad non-br

stegmanni

ad br

Two-banded Plover

ad ♂ br

Lesser Sand Plover

Medium to large, long-billed (except Upland Sandpiper) and long-legged shorebirds. Predominantly grey or brown, with brighter breeding plumages in *Limosa* and *Limnodromus*.

Upland Sandpiper *Bartramia longicauda*   32 cm

Grasslands and crops in the N lowlands (Aug–Apr). **ID** Peculiar long-necked, small-headed and pot-bellied wader with a notably long tail. Non-breeding boreal migrant. **Adult:** Straight yellow bill with black culmen and tip. Yellow tarsus. In flight, long narrow wings with paler scaling on the inner wing and narrow white trailing edge, but no wing-stripe. Long tail, barred black. Flanks and underwing-coverts barred black. **Voice** Mellow twittering phrases given in flight; also at night. [Batitú]

Eskimo Curlew *Numenius borealis*   34 cm

Possibly extinct; recorded in the E, principally in Pampas grasslands (Sep–Apr). **ID** Non-breeding boreal migrant with long decurved bill and black lateral crown stripes. **Adult:** Slender, blackish bill 1.5 x length of head, slightly decurved on distal half. Buffy supercilium, bordered by distinct black eye-stripe. Folded primaries extend beyond tail tip. Y-shaped chevrons on lower breast and flanks. In flight, toes do not extend beyond tail. Uniform, unmarked blackish flight feathers; grey-brown from below. Buff undewing-coverts, finely barred brown. **Juvenile** (not illustrated): Differs by buff-streaked crown, profuse cinnamon notches on upperparts, wing-coverts almost wholly buff and brighter, more cinnamon, underparts with similar markings. [Playero Esquimal]

Whimbrel *Numenius phaeopus*   43 cm

Scarce throughout coastal region (Sep–Apr); rare inland. **ID** Non-breeding boreal migrant with long decurved bill and black lateral crown stripes. **Adult:** Bill up to 3 x length of head, decurved on distal half. Heavily streaked breast and barred flanks. Primaries fall short of tail tip when folded. In flight, toes extend slightly beyond tail. Flight feathers notched with buff; only the primary wing-coverts are unmarked. **Juvenile** (not illustrated): Smaller, shorter-billed but still with a blunt tip, and shows brighter cinnamon-buff fringing on upperparts. [Playero Trinador]

Eurasian Curlew *Numenius arquata*   55 cm

Vagrant recorded once in ce. Buenos Aires. **ID Adult:** Extremely long bill. Lacks a black eye-stripe or coronal stripe. White rump and lower back. Streaked breast and mostly white abdomen. [Playero Gigante]

Hudsonian Godwit *Limosa haemastica*   39 cm

Mainly coastal (Sep–May); inland on passage. **ID** Large wader with distinctive thick-based, slightly upswept bill. Non-breeding boreal migrant. **Adult ♂, coming into breeding plumage:** Fleshy-pink bill, tipped dusky. Deep chestnut breast and belly mixed with broad buffy-white bars. Barred undertail-coverts. ♀ shows more washed-out underparts with more white, less chestnut and thicker black bars. **Adult non-breeding:** Pale grey-brown upperparts with white supercilium, most prominent in front of eye and accentuated by black loral line. Pale brown wash across breast. In flight, black wings with narrow white wing-stripe. White rump contrasts with black tail. **Juvenile** (not illustrated): As adult non-breeding but darker and with buff fringing on scapulars and wings. [Becasa de Mar]

Short-billed Dowitcher *Limnodromus griseus*   28 cm

Vagrant recorded mostly from Buenos Aires coast. **ID** Medium-sized non-breeding boreal migrant with long straight bill. Prominent white supraloral, accentuated by black lores in all plumages. **Adult** *hendersoni* breeding: The only long-billed wader with cinnamon-rufous underparts. White bellied ssp. could occur. **Adult non-breeding:** Grey-brown upperparts, and grey breast becoming speckled at lower edge. In flight, striking white V-shaped wedge on lower back. Indistinct white trailing edge. Finely barred tail. **Juvenile:** Scapulars and wing-coverts crisply fringed cinnamon or rufous. Cinnamon wash over throat and breast becoming white below; finely spotted on sides of breast and undertail-coverts. **Voice** Typical flight call is a whistled *tu tu tu*. Compare with Willet (Plate 66). [Becasa Gris]

Upland Sandpiper

ad

Eskimo Curlew

ad

Whimbrel

ad

hudsonicus

Eurasian Curlew

ad

orientalis

ad non-br

ad non-br

ad br

ad ♂ br

ad non-br

hendersoni

ad non-br

Hudsonian Godwit

juv

Short-billed Dowitcher

Calidris Small to fairly large gregarious sandpipers, the smaller species are generally known as peeps. All are boreal migrants, and arrive in non-breeding plumage. Many achieve breeding plumage before departing. Intermediate plumages are common by Feb–Mar. Juveniles are closest to breeding adults with crisper fringing and brighter colours. Many species overwinter in small numbers. See also Plate 64.

Stilt Sandpiper *Calidris himantopus* — 21 cm

Aug–Apr, mainly in the north. **ID** Long, straight, tube-like black bill. Yellowish legs; longer than congeners. **Adult non-breeding:** Notable supercilium. Streaked upper foreneck. In flight, ill-defined wing-stripe and contrasting white rump. Yellow legs extend well beyond tail. **Adult breeding:** Rusty cheeks. Marked with black dorsally, streaked on foreneck and barred on belly. [Playero Zancudo]

Sanderling *Calidris alba* — 20 cm

Sep–Apr, throughout the coast; especially Patagonia and sparsely inland. **ID** Straight black bill. **Adult non-breeding:** White forehead, supercilium and underparts. Grey above and on sides of breast. Contrasting black shoulder usually visible. In flight, broader and more striking wing-stripe than congeners, highlighted by black flight feathers and primary wing-coverts. Dark centre to rump. **Adult breeding** (not illustrated): Checkered pattern above with rufous head and breast. **Juvenile:** More striking black shoulder than adult. Black eye-stripe, streaked crown and sides of breast, and spotted back (recalling pattern of adult breeding). [Playerito Blanco]

Dunlin *Calidris alpina* — 20 cm

Rare vagrant. **ID** Sturdy peep with slender, slightly decurved bill. **Adult non-breeding:** Brownish-grey above and wash over breast with a paler throat and white belly. In flight, bold white wing-stripe and dark centre to rump. **Adult breeding** (not illustrated): Rufous-brown cap and upperparts, streaked breast and diagnostic black belly patch. [Playerito Vientre Negro]

Curlew Sandpiper *Calidris ferruginea* — 21 cm

Not illustrated. Known from one historical specimen from eastern Patagonia and one recent record at Mar Chiquita (Córdoba). **ID Adult non-breeding/Juvenile:** resembles a slim, more attenuated Dunlin with a bolder supercilium, finer-tipped drooping bill and longer black legs. [Playero Zarapitín]

Baird's Sandpiper *Calidris bairdii* — 17.5 cm

Aug–May, in the Andes to 4325 m and Patagonia; on passage elsewhere. **ID** Shorter, more slender black bill than White-rumped with a finer tip. Primaries extend far beyond the tertials and tail tip. **Adult non-breeding:** From White-rumped by indistinct supercilium, neater browner pectoral band restricted to upper breast and no flank streaks. In flight, brown centre of rump. White wing-stripe weaker than in White-rumped. **Adult breeding** (until Oct): Thicker black coronal streaking than White-rumped. Flanks are unmarked unlike White-rumped. [Playerito Unicolor]

White-rumped Sandpiper *Calidris fuscicollis* — 17.5 cm

Aug–May, throughout the N lowlands; abundant in Patagonia. **ID** Fine black bill; very slightly decurved towards the tip. Primaries extend just beyond tail tip. Often with diagnostic pinkish base to lower mandible. **Adult non-breeding:** Grey-brown above (usually greyer than Baird's) with a streaky head and breast. Indistinct whitish supercilium. Blackish dorsal shaft streaks and centres to scapulars. In flight, the only small *Calidris* with a white rump. Fairly prominent white wing-stripe. **Adult breeding** (from Mar, but occasionally in Aug on arrival): Warm brown crown, spotted black. Whitish supercilium. Back mixed with brown and black, fringed chestnut. Lines of blackish spots over breast, becoming larger on the flanks. [Playerito Rabadilla Blanca]

Red Knot *Calidris canutus* — 26 cm

Aug–May, throughout the coast; especially Patagonia. **ID** Broad-based, straight black bill. **Adult non-breeding:** Grey to grey-brown above, with whitish-fringed wing-coverts. Whitish supercilium and underparts with a grey wash on the breast and fine brown spotting. In flight, white wing-stripe enhanced by blackish primary wing-coverts. Finely barred rump and tail. **Adult breeding:** Rich rusty-orange supercilium, face, throat and breast. [Playero Rojizo]

ad non-br

ad br

ad non-br

Stilt Sandpiper

ad non-br

juv

ad non-br

ad non-br

Sanderling

Dunlin

ad non-br

ad non-br

Baird's
Sandpiper

ad br

ad non-br

ad non-br

non-br

ad br

White-rumped
Sandpiper

ad br

ad non-br

rufa

Red Knot

ad non-br

Calidris Small to fairly large gregarious sandpipers, also known as peeps. All are boreal migrants, and arrive in non-breeding plumage. Many achieve breeding plumage before departing. Intermediate plumages are common by Feb–Mar. Juveniles are closest to breeding adults with crisper fringing and brighter colours. Many species overwinter in small numbers. See also Plate 63.

Least Sandpiper *Calidris minutilla* — 14 cm

ID Yellow or greenish legs unlike other small peeps. No primary projection. **Adult non-breeding**: Indistinct darker dorsal feather centres. High pectoral band. In flight, weak white wing-stripe and dark centre to rump. **Adult breeding**: Prominent white fore-supercilium. Extensive breast streaking. [Playerito Menor]

Semipalmated Sandpiper *Calidris pusilla* — 14 cm

Vagrant. **ID** Relatively short blunt black bill. Short primary projection. **Adult non-breeding**: Indistinct supercilium. Grey breast sides. In flight, grey above and white below. Blackish wings with white wing-stripe. Dark centre to rump. **Adult breeding**: Dark crown. Black centres to scapulars. Neat brown breast streaking. Compare with larger White-rumped and Baird's Sandpipers (Plate 63). [Playerito Enano]

Pectoral Sandpiper *Calidris melanotos* — 21.5 cm

Aug–Apr, almost throughout, scarcer southwards. **ID** Sturdy peep with yellowish legs. **Adult non-breeding**: Extensively streaked breast, clean-cut from white belly. In flight, narrow white wing-stripe. Dark centre to rump. Brown breast sharply demarcated from white belly. **Adult breeding** (not illustrated): Similar, but brighter, with chestnut tones. [Playerito Pectoral]

Buff-breasted Sandpiper *Tryngites subruficollis* — 19 cm

Aug–Mar in the Pampas; on passage in the north. **ID** Unmistakable, non-breeding boreal migrant to grasslands and grazing land. **Adult non-breeding**: Yellow legs. Mostly buff. In flight, no wing-stripe. White underwing with a dusky crescent on the primary coverts, and dusky trailing edge. **Adult breeding** (not illustrated): Almost identical. [Playerito Canela]

Phalaropus Delicate, non-breeding boreal waders which spend much foraging time swimming on open water, and spinning in circles. Females are brighter than males.

Wilson's Phalarope *Phalaropus tricolor* — 22 cm

Common throughout (Aug–Apr). **ID** **Adult non-breeding**: Long, slender black bill. Grey crown and hindneck. Contrasting white supercilium and face. Paler grey dorsally than congeners. **Adult ♀ breeding**: White throat contrasts with chestnut and black neck markings. Black legs. No wing stripe in flight. Only phalarope with a clean white rump. Adult ♂ is like duller version of ♀. **Juvenile**: Scaled scapulars and wing-coverts. Only phalarope with yellow legs. [Falaropo Común]

Red-necked Phalarope *Phalaropus lobatus* — 18 cm

Very rare vagrant. **ID** **Adult non-breeding**: Needle-like black bill, finer and longer than Red Phalarope. Head pattern similar to non-breeding Red Phalarope but more subdued. Pale scapular fringing suggests streaks. **Adult ♀ breeding** (not illustrated): White throat contrasting with rufous on sides of neck and grey breast and flanks. Upperparts dark grey with bright buff scapular stripes. Adult ♂ is like duller version of ♀. [Falaropo Pico Fino]

Red Phalarope *Phalaropus fulicarius* — 22 cm

Widespread vagrant (Sep–Apr). **ID** **Adult non-breeding**: Relatively thick straight bill with yellow base. Black smudge behind eye and patch on hindcrown. Upperparts plain grey. **Adult ♀ breeding**: Yellow bill, tipped black. Golden scapular stripes. White face and rufous underparts. Adult ♂ is like duller version of ♀. [Alt: Grey Phalarope] [Falaropo Pico Grueso]

ad non-br

ad br

Least Sandpiper

ad non-br

ad non-br

ad br

Semipalmated Sandpiper

ad non-br

ad non-br

ad non-br

ad non-br

Buff-breasted
Sandpiper

ad non-br

Pectoral Sandpiper

ad ♀ br

ad non-br

ad non-br

Red-necked Phalarope

Wilson's Phalarope

ad ♀ br

Red Phalarope

ad non-br

juv

ad non-br

Ruddy Turnstone *Arenaria interpres*   22.5 cm

Boreal migrant (Sep–May) on coasts mainly south to Chubut; on passage elsewhere. **ID** Coastal wader of rocky shores, sand and pebble beaches. Unusual strongly pointed bill. Orange legs. **Adult breeding**: Chestnut back. Intricate black face pattern and broad pectoral band. **Adult non-breeding**: Dark grey-brown back and reduced facial markings. In flight, white dorsal stripe, parallel scapular bands and wing-stripe. White rump and a mostly black tail. Broad blackish pectoral band. [Vuelvepiedras]

Surfbird *Calidris virgata*   25 cm

Very rare in Tierra del Fuego (Nov–Mar). **ID** Sturdy plump wader with stout blunt bill. Favours rocky shorelines or floating kelp. Yellow legs and bill base. **Adult breeding**: Chestnut scapulars. Finely streaked head and breast. Black chevrons on flanks. **Adult non-breeding**: Grey upperparts, head and breast contrast with white belly and vent. In flight, mostly pale brownish-grey, except white belly. Narrow white wing-stripe. Broad white uppertail, and broad black distal band. [Playero de Rompiente]

Gallinago Cryptically patterned waders with long straight bills. Crepuscular display flights involve mechanical, diagnostically distinct, drumming/winnowing produced by air rushing through modified tail feathers which are opened and closed during diving flights. Small species flush with loud *chek*. Inhabit marshes and bogs.

South American Snipe *Gallinago paraguaiae*   26 cm

Mainly resident in the N lowlands, Andes to 1800 m and C Sierras. **ID** Adult: Pinkish bill base. Blackish back and densely marked foreneck and breast, strongly demarcated from the white belly. **Voice** Mechanical winnow is a series of whooshing accelerating, then decelerating harsh notes, that in the final dive sound like a car skidding to a halt, *höhöhöhö hÖ hÖÖ HÖÖÖ HÖÖÖÖ höö*. Ground call includes a steady series of slow (5 per sec) or fast (11 per sec) monotone *chep* notes, repeated over [Becasina de Bañado]

Magellanic Snipe *Gallinago magellanica*   28 cm

Breeds in Patagonia and the Falklands; some moving north in winter. **ID** Adult: Yellow-olive legs. Long black-tipped straw bill. Paler and browner-backed than South American Snipe with sparser breast markings; secondaries barred rufous. Long tail with rufous subterminal patch. Unmarked buffy-white belly. **Voice** Mechanical winnow is a series of scratchy couplets of notes like a rusty toy *pheSHE… phe-SHE… PHE-SHE… phe-she*, slower than South American Snipe. Ground call is a strident series of *chek* notes at 3 per sec, or alternating with high/low-pitched notes. **Tax note 16**. [Becasina Patagónica]

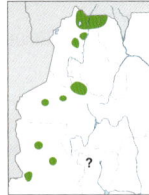

Puna Snipe *Gallinago andina*   24 cm

NW Andes above 3000 m, south to Catamarca. **ID** Adult. Yellow legs diagnostic in range. Yellowish bill base. Rather black-backed. In flight, legs do not extend beyond the tail. Note white-spotted upperwing-coverts, barred underwing with an unmarked buff leading edge, relatively short and rather rounded wings, and extensive rufous in tail. No overlap with congeners. **Voice** Mechanical winnow recalls that of South American Snipe, but with monotone and shorter harsh notes throughout. Ground call includes a series of *chip* notes. [Becasina Andina]

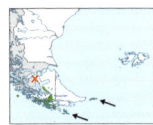

Fuegian Snipe *Gallinago stricklandii*   34 cm

Tussock grass and stunted woods on Isla Navarino, Hornean islands and Isla de los Estados; very rare on Isla Grande. **ID** Adult: Stout pale-based bill, often slightly kinked at the tip. Short, thick tarsus. Short tail, barred black and rufous. Buff belly. In flight, note heavy bill, broad wings and lack of rufous in tail. Clipped, woodcock-like flight. [Becasina Fueguina]

Giant Snipe *Gallinago undulata*   42 cm

Rare in ne. Corrientes. **ID** Adult: Extremely broad-based and long pale bill, tipped dusky. Broad black stripe across buff cheeks. Rather black-backed. Cinnamon rump and uppertail. Breast and flanks heavily barred brown. **Voice** A loud, guttural and slow *wag-ag-gAHh,.. wag-ag-gAHh,…… wag-ag-gAHh* with notes repeated 3–6 times. Simultaneously or separately, delivers a loud air-rushing 'jet plane' sound. Ground call is a nasal, springy *gaH-a, gaH-a, gaH-a, gaH-a, gaH-a…* [Becasina Gigante]

Ruddy Turnstone

ad non-br

ad br

ad non-br

ad br

ad non-br

Surfbird

ad non-br

ad

South American
Snipe

ad

Magellanic
Snipe

ad

Puna Snipe

ad

ad

Fuegian Snipe

Giant Snipe

Xenus*, *Actitis* and *Tringa Rather long-billed and long-legged (especially *Tringa*) shorebirds. All are non-breeding boreal migrants. Differences between breeding and non-breeding plumages are less pronounced (except in Spotted Sandpiper).

Terek Sandpiper *Xenus cinereus* 23 cm

Very rare vagrant to ce. Buenos Aires. **ID** Chunky, short-legged wader with long, slightly upcurved bill. Assumes horizontal stance while dashing along shore with continual changes in direction. **Adult non-breeding**: Dull orange bill base. Bright orange-yellow tarsus. Brownish-grey with white face, supercilium and underparts. Grey patches on sides of breast suggest non-breeding Solitary Sandpiper. Blackish bar on lesser wing-coverts sometimes visible. In flight, grey rump and notable white trailing edge. [Playerito Pico Curvo]

Spotted Sandpiper *Actitis macularius* 18 cm

Aug–Mar in the N lowlands, south to N Patagonia; often near coast. **ID** Continuously wags hind body. Whirring flight on level wings. **Adult breeding** (from Jan): Long orange bill, tipped black. Speckled with dusky notches above. Spotted black below. In flight, rapid quivering flaps, interspersed with glides on down-bowed wings showing white wing-stripe. **Adult non-breeding**: Narrow white supercilium. Broken brown pectoral band. Compare with Solitary Sandpiper. [Playerito Manchado]

Solitary Sandpiper *Tringa solitaria* 21–22 cm

N lowlands on freshwater (Aug–Apr). **ID** Adult *solitaria* breeding (21 cm): Olive tarsus. Dark brown mantle and tertials spangled with white dots. White supraloral streak and eye-ring. Fine black streaking on foreneck and upper breast. Tail barred black and white. In flight, dark unmarked upperparts with long slender wings. Barred tail sides. Underwing also uniformly dark. **Adult *cinnamomea* non-breeding** (22 cm): Prominent eye-ring and supraloral. Brown wash on sides of breast. Differs from *solitaria* by larger size, less distinct dorsal spotting and by broader tail barring. Compare with Spotted Sandpiper. [Pitotoy Solitario]

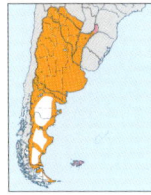

Lesser Yellowlegs *Tringa flavipes* 26 cm

Throughout (Aug–Apr). **ID** From Greater Yellowlegs by having straight bill only as long as head and without an obvious grey or olive base. **Adult breeding**: Long yellow or orange-yellow tarsus. Not as black above as breeding Greater and with an unmarked white belly; sometimes a few brown chevrons on the flanks. In flight, a smaller and slighter version of Greater. **Adult non-breeding**: Apart from size and bill distinctions, rarely shows indistinct white spotting on the upperparts unlike Greater. **Voice** Single or double notes when taking flight; typically higher-pitched than Greater. [Pitotoy Chico]

Greater Yellowlegs *Tringa melanoleuca* 34 cm

Throughout (Sep–Apr). **ID** From Lesser Yellowlegs by having bill longer than head with slight upswept impression and greyish base (beware of shorter-billed birds which can be more easily confused with Lesser). **Adult breeding**: Long yellow or orange-yellow tarsus. Blackish above mixed with white. Heavy black streaking on foreneck and breast. Black chevrons on sides of breast and flanks. In flight, no wing-stripe, contrasting white rump and yellow legs extending beyond tail. **Adult non-breeding**: Plainer, but note presence of whitish spotting on mantle. **Juvenile** (not illustrated): Like non-breeding adult but browner, more thickly spotted white on mantle, and stronger breast streaking. **Voice** Gives 3, rarely 4, loud ringing *tew* notes when taking flight. [Pitotoy Grande]

Willet *Tringa semipalmata* 37–40 cm

Very rare in the NE but regular at Punta Rasa. **ID** Large, plain sandpiper with long stout bill. **Adult breeding** (not illustrated): Heavily streaked and notched black with contrasting white supraloral. From Greater Yellowlegs by longer blue-grey tarsus. 'Eastern Willet' **Adult non-breeding** *semipalmata* (37 cm): Blue-grey basal half of bill. Soft brownish-grey above with prominent white supraloral. Juvenile slightly darker than non-breeding adult. In flight, black underwing with broad white stripe, and black distal upperwing with broad white stripe and pale innerwing. The black primary coverts enhance this pattern. White rump and greyish distal tail. So-called 'Western Willet' **Adult non-breeding** *inornata* (40 cm) has also been recorded at Punta Rasa. Differs by its average larger build, longer legs, longer more slender bill, and paler and greyer (less brown) plumage. **Voice** Three-note flight call recalls that of Greater Yellowlegs. **Tax note 17**. [Playero Ala Blanca]

ad non-br

Spotted Sandpiper

ad br

ad non-br

ad non-br

Terek Sandpiper

ad non-br

ad non-br

ad br

solitaria

Solitary
Sandpiper

Lesser
Yellowlegs

cinnamomea

ad br

ad non-br

ad non-br

ad non-br

ad br

ad non-br

ad non-br

Greater Yellowlegs

ad non-br

ad non-br

ad non-br

inornata

semipalmata

ad non-br

Willet

Thinocorus Small, sexually dimorphic, pot-bellied and almost dove-like. Stout yellow bill with a black tip. Songs are monotonous series recalling a pygmy-owl. Both give a snipe-like *chek* flush call. ***Attagis*** Large robust birds recalling grouse, with cryptic dorsal patterns, relatively small heads and stout bills used for browsing vegetation. Sexes alike. Gregarious when not breeding. Loud unusual shrill calls often create a cacophony.

Grey-breasted Seedsnipe *Thinocorus orbignyianus* 21 cm

Throughout the Andes; above 2300 m in north or 1000 m in south. **ID** Adult ♂: Orange tarsus. Grey face, neck and upper breast, bordered by a narrow black pectoral band. Black stirrup around white chin. In flight, black underwing-coverts, bordered white, and narrow white trailing edge. Adult ♀: From ♀ Least by orange tarsus, and thickly streaked foreneck and upper breast. Black chin stirrup and pectoral band diagnostic when present (rare). [**Agachona de Collar**]

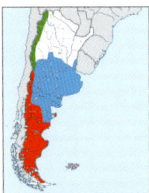

Least Seedsnipe *Thinocorus rumicivorus* 18.5 cm

N Andes above 3000 m (*bolivianus*); Patagonian steppe to 1750 m, migrating to C lowlands in winter (*rumicivorus*). **ID** Adult ♂ *rumicivorus*: From Grey-breasted by black necktie. Flight is very much faster than Grey-breasted with diagnostic sudden rolling from one side to the other and changes of direction. Adult ♀: From Grey-breasted by yellow tarsus, and less coarsely streaked, more spotted, foreneck and breast. Some show a black necktie. Adult *bolivianus* (not illustrated): Lacks black in the dorsal pattern and has a slight buff wash on the belly. [**Agachona Chica**]

Rufous-bellied Seedsnipe *Attagis gayi* 28.5 cm

Throughout the Andes to Santa Cruz above 3000 m; usually much higher. **ID** Adult: Complex dorsal and breast pattern of brown, black, buff and pale rufous vermiculations. Pale pinkish-cinnamon belly and vent. In flight, belly pattern reflected on underwing-coverts. Overlaps with White-bellied in W Patagonia. [**Agachona Grande**]

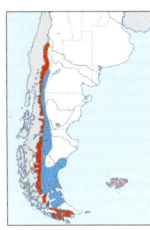

White-bellied Seedsnipe *Attagis malouinus* 28.5 cm

Cushion heath and scree in C and S Andes to Tierra del Fuego; in winter descends to lowlands and north to w. Chubut. **ID** Adult: Intricate dorsal and breast pattern with black feather centres, vermiculated buff, cinnamon and white. Pure white belly. In flight, white belly and underwing-coverts. Narrow white tail tip. Overlaps with Rufous-bellied in W Patagonia. [**Agachona Patagónica**]

Grey-breasted Seedsnipe

♂

ad ♀

ad ♂

orbignyianus

♂

ad ♀

ad ♂

Least Seedsnipe

rumicivorus

ad

ad

gayi

Rufous-bellied
Seedsnipe

ad

ad

White-bellied
Seedsnipe

Morus Large white coastal bird with feathered throat, contrasting black remiges and pointed tail. Catches fish by plunge-diving from considerable heights. ***Sula*** Habits and shape similar to *Morus* but with a bare gular sac.

Cape Gannet *Morus capensis*   89 cm WS: 171 cm

Rare summer vagrant off Patagonia. **ID Adult**: White with black flight feathers and solid black tail. Sulphur cast over head and nape. **Second-year immature**: Upperwing-coverts mostly white. Recalls adult but with brown-spotted back. **Juvenile**: Dark bill. Dusky head, neck and breast, grading to whitish on belly. Narrow white horseshoe on uppertail-coverts and white underwing stripe. **Note** Some records may refer to Australian Gannet *Morus serrator* (or Cape X Australian hybrids?) with extensive white outer tail-feathers, shorter gular stripe and burnt orange crown. [Piquero del Cabo]

Brown Booby *Sula leucogaster*   69 cm WS: 141 cm

Rare summer vagrant off Buenos Aires. **ID Adult**: Yellow bill. Sooty-brown above with head and neck neatly demarcated from white abdomen, and white underwing-coverts. **Juvenile**: Greyish bill. Belly washed brown but whitish underwing-coverts usually evident. [Piquero Pardo]

Peruvian Booby *Sula variegata*   73 cm WS: 150 cm

Not illustrated. Two historical records: specimen from Falklands and unconfirmed sighting in Tierra del Fuego. **ID Adult**: White head, neck and underparts contrasting with brownish-black back, rump and upperwing-coverts, finely barred white, and white horseshoe on uppertail. **Juvenile**: Resembles adult, but finely speckled brown over head, neck and underparts. [Piquero Peruano]

Stercorarius Chunky, gull-like, square-tailed seabirds which are fiercely aggressive predators and opportunistic scavengers. Only South Polar is polymorphic, but Brown Skua is variable. See smaller species on Plate 69.

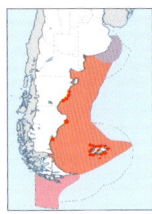

Brown Skua *Stercorarius antarcticus*   56 cm

Widespread. **ID Adult** *antarcticus* (Breeds in Patagonia and the Falklands; ranges north to Buenos Aires): Huge blackish bill with hooked tip. Brown to greyish-brown with contrasting blackish brown back, wings and tail. Foreface or forehead is often darker. Extensive white primary flash. Hindneck and breast streaked or flecked with cream. **Juvenile**: Black-tipped bill. Dark hood, unstreaked back and plumage warmer than adult. Reduced primary flash. **Adult** *lonnbergi* (Breeds in Antarctica and Antarctic islands; vagrant to Tierra del Fuego): Proportionately smaller head and bill than *antarcticus*, somewhat warmer overall without a contrasting dark forehead and with less pronounced dorsal streaking. Beware of hybrids with South Polar Skua. [Escúa Antártica]

Chilean Skua *Stercorarius chilensis*   57 cm

Breeds in Patagonia and Tierra del Fuego; ranges throughout. **ID Adult**: Fairly stout blackish or dark-tipped bill. Brown above with a contrasting dusky cap. Creamy nuchal streaking. Rusty ear-coverts, underparts and underwing-coverts. Narrow white primary flash. **Juvenile**: Grey bill, tipped black. Cleaner rusty face and underparts. Reduced primary flash. [Escúa Canela]

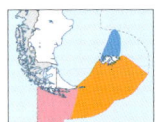

South Polar Skua *Stercorarius maccormicki*   53 cm

Antarctic breeder. Adults occasional in Tierra del Fuego in summer. Juveniles migrate north in Feb, returning in Oct. **ID Adult light morph**: Creamy head and underparts. Black underwing-coverts and notable white primary flash. **Adult intermediate morph**: Creamy nuchal collar, and variable amount of cream to pale brown below. **Adult dark morph**: Dark brown. Recalls juvenile Brown Skua, but bill slighter, unhooked and entirely dark. **Juvenile**: Dusky-tipped bill. Reduced primary flash. More uniformly dark than juvenile Brown, slighter build, smaller-headed and shorter, slenderer unhooked bill. [Escúa Polar]

Cape Gannet

ad

imm
2nd-yr

juv

Brown Booby

leucogaster

ad

juv

juv

antarcticus

ad

lonnbergi

Brown Skua

ad

Chilean Skua

juv

ad

ad
light morph

juv

ad
dark morph

South Polar Skua

ad
intermediate
morph

Stercorarius Non-breeding migratory seabirds which feed mainly by kleptoparasitism. All three species (except adult Long-tailed) are polymorphic, although there is immense individual plumage variation. Key features include: proportional length and width of the hand compared to the arm of the wing; general bulk and centre of gravity on the abdomen; strength of tail-covert barring; width and pattern of the primary flash; extent of white primary shafts on the upperwing; shape of the central tail feathers in juveniles and immatures (these are worn on adults); underwing-covert pattern; and bill size and extent of black tip. Bicoloured bills correspond to most plumages, except adult or well-advanced Arctic and Long-tailed.

Arctic Skua *Stercorarius parasiticus* 44 cm

Sep–Apr throughout the region. **ID** Centre of gravity at mid-belly. Arm and hand of equal length. **Adult light morph**: Blackish cap or hood (like Long-tailed and Pomarine), grey pectoral band and creamy barring on the upper- and undertail-coverts. Broad white primary flash. **Adult dark morph**: Dark brown, usually with a blacker cap. Note numerous white primary shafts. Primary flash like light morph. **Second-year light morph**: Resembles adult light morph but with barred juvenile underwing-coverts and bill usually has a dark tip. Pointed tail projections. These birds can also be hooded like non-breeding adult Long-tailed. **Juvenile dark morph**: Resembles adult dark morph but darker throughout with indistinct barring on the undertail-coverts. Broad primary flash. **Juvenile light morph**: Scarce. Creamy head contrasts with warm brown upperparts. Pale brown below sometimes with a whitish belly. Single, or suggestion of a double, white flash in the base of the primaries. Short pointed tail projections. **Juvenile intermediate morph**: Mainly brown with barred underwing, white half-moon in primaries and indistinctly barred undertail-coverts. [Alt: Parasitic Jaeger] [Salteador Chico]

Long-tailed Skua *Stercorarius longicaudus* 34 cm

Nov–Apr in continental and offshore waters, south to the Falklands. **ID** Some overlap in size with small Arctic but slighter build with the centre of gravity at the breast, although bill thicker and shorter. Hand is slightly longer than the arm. Usually shows only 2–3 white primary shafts whereas Arctic and Pomarine show 3–8 pale shafts. Flight is buoyant with shearwater-like gliding. **Adult non-breeding**: Capped or hooded effect and pectoral band like Arctic and Pomarine but underwing more uniformly dark with little or no primary flash. **Adult breeding** (not illustrated and unlikely to be seen in region): Pale grey-brown upperparts with black cap, pure white underparts and very long tail streamers. Graceful flight. **First-year**: In this and all juvenile morphs, the bill is more broadly tipped black than in Arctic and Pomarine. Grey hood with contrasting creamy nuchal collar. Barred underwing like second-year Arctic but little or no primary flash like adult non-breeding Long-tailed. Prominent pointed tail projections. **Juvenile dark morph**: The only dark morph skua with prominent barring on the upper- and undertail-coverts. The underwing can be dark or barred. Round-tipped tail projections in all juvenile morphs aids separation from Arctic. **Juvenile light morph**: Creamy-yellow head, and greyish mantle and wing-coverts with contrasting black flight feathers form a unique pattern. Underwing like intermediate morph. **Juvenile intermediate morph**: Note hooded effect, grey nuchal band and dark pectoral band; tail-coverts more prominently barred than intermediate morph Arctic, but primary flash always narrow. [Alt: Long-tailed Jaeger] [Salteador Coludo]

Pomarine Skua *Stercorarius pomarinus* 56 cm

Rare in continental waters. **ID** Heavy-bodied with centre of gravity between breast and belly. Flies like Brown and Chilean Skuas, showing a very broad arm, and shorter, disproportionately narrow hand. Bill is always bicoloured and appreciably hooked at close range. **Adult light morph non-breeding**: Stout, bicoloured bill unlike Arctic and Long-tailed. Quite variable, sometimes resembling Arctic but sides of abdomen always barred. Tail extensions usually lacking, but if present are long, spoon-shaped and twisted towards the tip (as in breeding plumage). **Adult dark morph**: Entirely blackish. Note rounded tail extensions unlike dark morph Arctic, and bicolored bill. **Immature**: Highly variable but primary flash is split by a dark bar. **Juvenile intermediate morph**: Barred underwing like intermediate morph Arctic and Long-tailed but primary flash split by a dark bar. Head and body vary from blackish to warm brown. [Alt: Pomarine Jaeger] [Salteador Grande]

Arctic Skua

juv
light morph

juv
dark morph

2nd-yr
light morph

ad non-br
light morph

juv
intermediate
morph

ad
dark morph

juv
intermediate
morph

Long-tailed Skua

juv
light morph

ad non-br

juv
dark morph

1st-yr

imm

Pomarine Skua

juv
intermediate
morph

ad
dark morph

ad non-br
light morph

Leucophaeus Medium-sized gulls with hooded non-breeding and juvenile plumages. *Larus* Large black-backed gulls with stout bills.

Dolphin Gull *Leucophaeus scoresbii*   47 cm

Resident on Patagonian coasts and the Falklands. **ID** Adult breeding: Stout, blunt, orange-red bill. Red tarsus. Contrasting pale iris. Greyish-white head and somewhat darker grey below. Blackish mantle and wings with white primary windows. White tail. **Adult non-breeding**: Dark grey hood and prominent white eye-ring. Note blackish primaries on underwing. **First-winter**: Fleshy bill, tipped black. Dark brown hood and brown wing-coverts. Note crisp white trailing edge in flight unlike other first-year gulls. **Second-winter**: Yellow bill, tipped dark red. Dull pink tarsus. Contrasting blackish hood. Black subterminal tail band. [Gaviota Gris]

Grey Gull *Leucophaeus modestus*   41 cm

Rare vagrant recorded in the Andes of Salta and on the Falklands. **ID** Adult breeding: Slender black bill and tarsus. White head and throat contrast with grey underparts. Slate-grey back and wings. **Adult non-breeding**: Brown hood. Black tail with a very narrow white terminal band. Note neat white trailing edge to wing in flight but lack of primary windows. **Juvenile**: Pale brown, often whitish on the crown. In flight, blackish flight feathers with contrasting pale brown wing-coverts, fringed white. Black tail with narrow whitish terminal band. **Second-winter**: Brown hood and brownish-grey mantle and wing-coverts. [Gaviota Garuma]

Olrog's Gull *Larus atlanticus*   53 cm

Endemic breeder on islands off s. Buenos Aires and Chubut; ranges sparsely through continental seaboard. **ID** Adult breeding: Orange-yellow bill with a black subterminal band and blood red tip; many (older or peak breeding?) birds lack black on the lower mandible. Black mantle and wings (compare with Kelp Gull) but no white primary mirrors and broad black subterminal tail band, also lacking on Kelp Gull. **Adult non-breeding**: Variable grey hood. Note lack of white primary mirrors in flight, dark bill tip and clear-cut tail band edged white at the sides. **First-winter**: Straw-coloured bill with black tip. Creamy tarsus. Brown hood and breast with contrasting white foreface, flanks and belly. Strongly scaled mantle and wing-coverts. Black tail. In flight, note pale innerwing with dark secondary stripe. **Second-winter**: From breeding adult and first-winter by grey-brown hood extending onto central breast, usually only retained from Jan to Apr. Whitish forehead and lores like first-winter. [Gaviota Cangrejera]

Belcher's Gull *Larus belcheri*   51 cm

Not illustrated. Vagrant from the Pacific, known from a sight record in w. Rio Negro. **ID** Adult breeding: Similar to Olrog's Gull but lacks black on the bill. **First-year**: Much darker than first-year Olrog's Gull and lacks the pale forehead. [Gaviota Peruana]

Kelp Gull *Larus dominicanus*   55 cm

Widespread throughout the region. **ID** Adult breeding: Yellow bill with red spot near tip of lower mandible. Slaty-black mantle and wings (compare with Olrog's Gull) with 3–4 white wing mirrors. White tail. In flight, note white wing mirrors, narrow white leading wing edge and unmarked tail. **First-winter**: Stout black bill. Pink tarsus. Usually shows dusky ear-coverts and blackish centres to scapulars and wing-coverts. Compare with very different Olrog's. In flight, broad black tail band usually mixed with a few whitish shafts or barring at the sides suggesting Olrog's, but darker throughout without contrasting pale foreface, and bill entirely black. **Third-winter**: Yellow bill with black subterminal band and narrow yellow tip suggests adult Olrog's, but tail band broken into a series of spots. Lacks primary mirrors. [Gaviota Cocinera]

Lesser Black-backed Gull *Larus fuscus*   52 cm

Not illustrated. One historical specimen from s. Buenos Aires. **ID** Smaller than Kelp Gull with a more slender bill. **Adult**: From Kelp Gull by size, proportions and chrome yellow legs. Some ssp. have grey back and wings. **First-winter**: From first-winter Kelp Gull by more slender bill, overall build and blackish patch around eye. [Gaviota Sombría]

Dolphin Gull

1st-winter

ad non-br

2nd-winter

ad br

Grey Gull

juv

ad non-br

2nd-winter

ad br

Olrog's Gull

1st-winter

ad non-br

1st-winter

2nd-winter

ad br

Kelp Gull

3rd-winter

1st-winter

1st-winter

ad br

1st-winter

ad br

dominicanus

Chroicocephalus and *Leucophaeus* Small to medium-sized gulls with slender bills. Breeding birds have a black or grey hood (Franklin's has a hood or half-hood at all ages). Adults have white tails while juveniles show a black tail band.

Andean Gull *Chroicocephalus serranus*　　45 cm

Resident in the N Andes, south to NW Patagonia. **ID Adult breeding** (Aug–Dec): Black hood, more extensive than Brown-hooded Gull. Dusky red bill and tarsus. **Adult non-breeding**: Black bill. Black spot or crescent on ear-coverts. In flight, black distal primaries enclosing large white patch. **First-winter**: Black-tipped yellowish to dull orange bill. Blackish smudge in front of eye, and cheek spot. Some brown on wing-coverts. In flight, dark trailing edge to wing. White patch in distal primaries. [Gaviota Andina]

Brown-hooded Gull *Chroicocephalus maculipennis*　　38 cm

Throughout seaboard, N lowlands and S Andes, moving north in winter; resident population on the Falklands. **ID Adult breeding**: Reddish bill and tarsus. Brown hood cut squarely from hindcrown. May show pink wash on breast. **Adult non-breeding**: Black crescent on ear-coverts. In flight, white wedge on distal primaries. Underwing-coverts whiter than congeners. **First-winter**: Red bill tipped black. Contrasting black bar on wing-coverts. In flight, mostly white primaries. Broad black trailing edge and dusky bar across median wing-coverts. [Gaviota Capucho Café]

Grey-hooded Gull *Chroicocephalus cirrocephalus*　　40 cm

Resident mainly in the NE, wandering to the NW and south to N Patagonia. **ID Adult breeding**: Red bill with black tip. Pale iris. Orange-red tarsus. Grey hood, outlined in black. **Adult non-breeding**: Dusky ear-spot. In flight, underwing mostly dark with small white window in outermost primaries. **First-winter**: Yellowish or pale orange bill, tipped black. Pinkish tarsus. Indistinct ear-spot. Brown markings on lesser and median wing-coverts. In flight, broad white wedge on distal primaries with black trailing edge. [Gaviota Capucho Gris]

Franklin's Gull *Leucophaeus pipixcan*　　38 cm

Scarce but widespread (locally common in Mendoza) non-breeding boreal migrant. **ID Adult breeding**: Redder bill than Andean and Brown-hooded Gulls. Well-developed black hood with very prominent white eyelids. Dark grey mantle and wings. **Adult non-breeding**: In flight, distinctive white trailing edge to wing, highlighting black distal primaries with white mirrors. Mostly white underwing. **First-winter**: Black bill. White forehead and loral patch with contrasting black hindcrown and ear-coverts with prominent white eyelids. In flight, white trailing edge to most of wing. Neat back tail band. [Gaviota Chica]

Sabine's Gull *Xema sabini*　　35 cm

Rare boreal vagrant to Tierra del Fuego. **ID** Striking wing pattern with black outer primaries contrasting with white triangle formed by inner primaries and secondaries. White rump. **Adult non-breeding**: Dark bill tipped yellow. Brownish-black patch on upper hindneck and nape separated from grey mantle by white lower hindneck. Moderately forked white tail. **Adult breeding** (not illustrated): Slate grey hood edged black and blacker bill tipped yellow. **Juvenile**: Black bill. Brownish mantle (scaled white), upperwing-coverts and breast spur. Black terminal tail band. [Gaviota de Sabine]

ad non-br

juv

Sabine's Gull

Andean Gull

ad non-br

1st-winter

1st-winter

ad br

Brown-hooded Gull

ad non-br

1st-winter

1st-winter

ad br

Grey-hooded Gull

ad non-br

1st-winter

1st-winter

cirrocephalus

ad br

Franklin's Gull

ad non-br

1st-winter

1st-winter

ad br

Sterna Medium-sized terns which either undergo a short migration or the longest migrations known among birds. All show a black cap in breeding plumage, which is otherwise reduced to a nuchal band, although none are crested. Bill colours change through the breeding cycle.

Common Tern *Sterna hirundo*   33 cm

Common boreal migrant throughout continental coasts (Oct–Mar) and sporadically inland. **ID** Adult non-breeding: Black bill. White forehead and solid black skullcap extending onto ear-coverts. Opaque primaries with a suggestion of a dark wedge. Blackish carpal bar of varying strength is diagnostic. Juvenile: Darker above than adult with a diagnostic blackish carpal bar. [Gaviotín Golondrina]

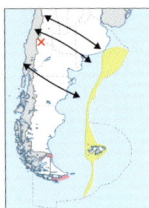

Arctic Tern *Sterna paradisaea*   34 cm

Boreal migrant, rarely on passage (Oct). **ID** Adult non-breeding: Primaries more attenuated than Common Tern and translucent, with much neater and more clear-cut blackish trailing edge. Black legs considerably shorter than in Common Tern. Juvenile: Carpal bar weaker than Common, secondaries and underwing whiter and tail proportionately longer. [Gaviotín Artico]

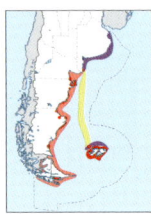

South American Tern *Sterna hirundinacea*   41 cm

The commonest *Sterna* tern in continental waters; mainly breeds in the south and winters in the north. **ID** Adult breeding: The only black-capped, red-billed tern likely in continental waters. Grey wash on the underparts. Adult non-breeding: Bill varies from black to red. White forehead, black eye-stripe and hindcrown. Litlle or no dusky border to primaries unlike Common and Arctic Terns. Underparts cleaner and whiter than breeding adult. Juvenile: Black bill. Reduced white forehead. Scapulars and back mixed with dusky barring. [Gaviotín Sudamericano]

Antarctic Tern *Sterna vittata*   37 cm

Antarctic breeder; no confirmed records in the region. **ID** Adult *gaini* breeding: Scarlet bill. Black crown bordered by a white streak across the cheeks. Soft grey underparts. Adult *gaini* non-breeding: Black bill, often red at the base. Black eye-stripe and nape band. White below. Juvenile *gaini*: Similar to juvenile South American but with a dusky spur on the sides of the breast. Adult *georgiae* breeding (Breeds on South Georgia): Smaller and generally darker than *gaini* and with a short, more orangey-red bill. [Gaviotín Antártico]

Sooty Tern *Onychoprion fuscatus*   44 cm

Known from an historical Falklands specimen. **ID** Large sturdy sea tern with a forked tail. Adult: Mostly black above and white below. White forehead extends to eye. [Gaviotín Sombrío]

Bridled Tern *Onychoprion anaethetus*   37 cm

Known from a specimen taken off Diego Ramírez. **ID** Slender sea tern with a forked tail. Adult: Black crown and nape contrasts with sooty-brown upperparts. Narrow white forehead extends just beyond the eye. [Gaviotín Embozado]

Black Noddy *Anous minutus*   35 cm

Known from an historical Falklands specimen. **ID** Slender sea tern with a long forked tail. Adult: Blackish-brown with a white cap extending to the nape; reduced in juvenile [Gaviotín Corona Blanca]

Sooty Tern

Bridled Tern

Black Noddy

juv

ad non-br

hirundo

ad non-br

Common Tern

juv

ad non-br

ad non-br

Arctic Tern

ad non-br

ad non-br

ad br

juv

South American Tern

ad br

ad br

gaini

ad non-br

ad br

juv

georgiae

Antarctic Tern

Snowy-crowned Tern *Sterna trudeaui*   36 cm

Resident in the N and C lowlands and coast. **ID** Adult breeding (late Jul–Jan): Orange bill with a black subterminal band and yellow tip. White head with a contrasting black mask. **Adult non-breeding** (Feb–Jul): Bill black with a yellow tip, otherwise similar to breeding adult but primaries darker. **Juvenile:** Black bill. Black mask as adult but crown and nape grey speckled black. [Alt: Trudeau's Tern] [Gaviotín Lagunero]

Gull-billed Tern *Gelochelidon nilotica*   37 cm

Breeds in the Pampas region and Buenos Aires coast; winters in N lowlands. **ID** Large, short-tailed estuarine tern with an unusually thick gull-like bill. **Adult breeding:** Short, thick, black bill. Black crown contrasts with soft grey plumage. **Adult non-breeding:** Black restricted to a post-ocular mask. Note blackish trailing edge to primaries. **Juvenile:** As non-breeding adult but with brown shaft streaks on crown and nape, and ginger mottling on mantle. [Gaviotín Pico Grueso]

Thalasseus Large, stout billed, marine terns with a nuchal crest, which breed in the region and undertake short migrations.

Royal Tern *Thalasseus maxima*   48 cm

Resident on coasts from Buenos Aires to ne. Santa Cruz. **ID** Large, stout-billed, marine tern with a nuchal crest. **Adult breeding:** Thick cigar-shaped body and broad wings. Stout orange bill and black cap; crest visible when perched. **Adult non-breeding:** Bill paler and yellower than breeding adult, black nuchal band thicker than non-breeding Sandwich Tern. **Juvenile:** Head as non-breeding adult, wings show a dark grey carpal bar and secondaries show a dark trailing edge. [Gaviotín Real]

Sandwich Tern *Thalasseus sandvicensis*   39 cm

Widespread. **ID** Slim-bodied with angled wings. **Adult *eurygnathus* breeding** (Argentine form): The commonest form; breeds on Patagonian coasts; winters north to Buenos Aires. Slender, slightly drooped orange to orange-yellow bill. Black cap; crest visible when perched. **Adult *eurygnathus* non-breeding:** Orange base and yellow distal portion of bill. White forehead and some streaking/spotting on central crown, and black from eyes around nape. **Juvenile *eurygnathus*:** Bill black, sometimes with yellow cutting edge. Head like non-breeding adult, and back and scapulars mottled with dark centres. **Adult *eurygnathus* non-breeding** (Brazilian form): Regular on Buenos Aires coasts. Bill smaller and straighter than Argentine form and entirely yellow. Black restricted to hindcrown. **Adult '*acuflavidus*' non-breeding** (Scarce, south to Patagonia): Straight black bill, tipped yellow. Plumage like Brazilian *eurygnathus*. Tax note 18. [Gaviotín Pico Amarillo]

Black Tern *Chlidonias niger*   24 cm

Vagrant to N and C lowlands and coast (Sep-Apr). **ID** Tiny, non-breeding, boreal migrant, saltmarsh tern. **Adult breeding** (not illustrated): Mainly black with grey wings and rump, white vent and black legs. **Adult moulting:** Underparts mixed with black. Reduced black on crown. **Juvenile:** Slender black bill. Black cap with contrasting white forehead and nuchal band. Dark grey upperwing with blackish carpal bar. Dark smudge on sides of breast. [Gaviotín Negro]

White-winged Tern *Chlidonias leucopterus*   22 cm

Not illustrated. Recorded twice in Córdoba and Chaco. **ID Adult breeding:** Recalls Black Tern with darker back, pure white upperwings with white lesser and median coverts, black underwing-coverts, white rump and tail, and red legs. **Adult non-breeding:** Similar to non-breeding Black Tern but paler, with dark ear spot and lacking dark smudge on sides of breast. **Juvenile:** Similar helmeted effect as juvenile Black Tern with more white on sides of hindneck, redder legs and darker back. [Gaviotín Ala Blanca]

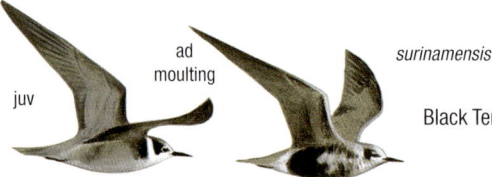

juv

ad
moulting

surinamensis

Black Tern

ad non-br

Snowy-crowned Tern

juv

ad br

ad br

Gull-billed Tern

juv

ad non-br

ad br

gronvoldi

ad non-br

ad non-br

juv

Royal Tern

ad br

ad non-br

ad br

maxima

ad non-br

eurygnathus

eurygnathus

eurygnathus
'Brazilian'

ad br

'acuflavidus'

rygnathus
rgentine'

ad non-br

juv

ad non-br

ad non-br

eurygnathus
'Argentine'

eurygnathus
'Brazilian'

ad non-br

ad br

'acuflavidus'

ad non-br

Sandwich Tern

Yellow-billed Tern *Sternula superciliaris* 23 cm

Resident in N lowlands; occasionally on the coast. **ID** Tiny estuarine tern with a relatively short forked tail. **Adult breeding:** Yellow bill and legs. Black crown with contrasting white supraloral. Outer primaries blackish. **Adult non-breeding:** Bill tipped black. Greenish-yellow legs. Crown mostly white. **Juvenile:** Duller yellow bill, often tipped dusky. Duller throughout with some coronal and dorsal speckling and dusky carpal bar (weaker than in juvenile Least Tern). [Gaviotín de Río]

Least Tern *Sternula antillarum* 23 cm

Rare vagrant, recorded Nov–Feb on Buenos Aires coasts. **ID** Tiny estuarine tern with a relatively short forked tail, and yellow bill and legs when breeding. **Adult non-breeding:** Black bill (shorter than Yellow-billed) and extensive white crown with black eye-stripe and nuchal band. Black restricted to outermost two primaries. Legs dull black to blackish-orange. **Juvenile:** Compared to adult shows a dusky carpal bar (darker than in juvenile Yellow-billed) and considerably darker upperwing except for contrasting white secondaries. [Gaviotín Chico Boreal]

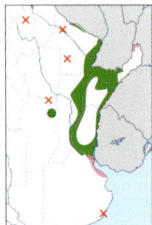

Large-billed Tern *Phaetusa simplex* 37 cm

Partial austral migrant in N lowlands. **ID** Largest species of tern found at inland marshes and rivers, with a proportionately outsized bill and relatively short tail. **Adult breeding:** Huge orange-yellow bill. Black cap. Wings mostly white with contrasting grey shoulders and broad black primary tips. **Adult non-breeding:** Yellower bill than breeding adult and grey cap with blackish smudge through eye. **Juvenile:** White forehead and blackish eye-stripe joining across hindcrown. Grey-brown mantle and blackish carpal bar. Underparts washed grey. [Atí]

Rynchops Gregarious tern-like bird with pied plumage and unique laterally compressed razor-like bill with the lower mandible longer than the upper. Flies in V-formation.

intercedens

cinerascens

Black Skimmer *Rynchops niger* 43 cm

Throughout the lowlands and coast. **ID** Adult *intercedens* breeding: Large orange bill with black distal half. Black above with broad white trailing edge to wing. White forehead, underparts, underwing-coverts and outer tail feathers. **Juvenile** *intercedens*: Pink bill with black tip. Blackish smudge through eye and brown crown with white forehead. Brown above with buff flecking on wing-coverts. Creamy-buff below. **Adult** *cinerascens* breeding: Rare migrant. Differs by ash-grey underwing, narrower white trailing edge and an entirely black tail. Non-breeders of both ssp. are browner above with a white nuchal band. **Tax note 19.** [Rayador]

juv

ad non-br

ad br

Yellow-billed Tern

ad non-br

juv

ad non-br

Least Tern

ad non-br

juv

Large-billed Tern

ad non-br

ad br

ad non-br

juv

ad br

ad br

cinerascens

intercedens

ad br

Black Skimmer

Patagioenas Large arboreal and terrestrial pigeons. Often gregarious. Sexual dimorphism more marked in some species. See also Plate 76.

Spot-winged Pigeon *Patagioenas maculosa*   35 cm

Widespread. **ID** Adult *maculosa*: Whitish iris. Mainly grey with brown mantle and wing-coverts conspicuously spotted white. In flight, from Picazuro Pigeon by heavily spotted wing-coverts. **Adult** *albipennis* (above 2800 m in villages of extreme n. Jujuy and n. Salta): White greater wing-coverts; more notable than in Picazuro Pigeon. **Voice** In pattern essentially identical to Picazuro Pigeon, but deep, throaty *wrraaÓh… wroh-wro wrooh… wroh-wro wrooh…* [Paloma Manchada]

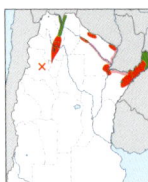

Pale-vented Pigeon *Patagioenas cayennensis*   32 cm

Yungas, humid chaco and Paraná forest (mainly Sep–Apr). **ID** Adult: Red tarsus and iris. Maroon with a rich ruddy back, contrasting grey ear-coverts and green gloss on hindcrown. In flight, ruddy upperwing-coverts contrast with brown remiges. Underwing and rump grey. Broad pale grey distal tail band. **Voice** Recalls Picazuro Pigeon but faster, more obviously rising and falling notes in both pitch and volume *uhwOOo… wu cucu wOOo… wu cucu wOOo…* [Paloma Colorada]

Picazuro Pigeon *Patagioenas picazuro*   38 cm

Widespread. **ID** Adult: Orange iris. Mainly vinaceous-pink with browner mantle and wing-coverts. Heavily scaled hindneck. In flight, whitish fringes on greater-coverts form a wing-stripe. Extensive grey rump and dark tail. **Voice** Spirited cooing is a clean, medium-pitched *wooaa… woo-wo woou… woo-wo woou…* [Paloma Picazuro]

Plumbeous Pigeon *Patagioenas plumbea*   35 cm

Very rare in ne. Misiones. **ID** Adult: Red eye-ring. Slender build with long neck and tail. Grey crown and greyish neck and abdomen. Brown back, wings and tail, sometimes with some purplish gloss. In flight, mostly drab grey-brown with a notably long tail. [Paloma Plomiza]

Scaled Pigeon *Patagioenas speciosa*   32 cm

Sporadic and rare in the NE. **ID** Adult ♂: Red tarsus and bill with a pale yellow tip. Head, mantle and wing-coverts rich reddish-brown. Heavily scaled green ruff on nape and breast with whitish centres. Adult ♀: Ruddy coloration of ♂ replaced by brown. White belly and vent conspicuous in flight. [Paloma Trocal]

ad

maculosa

Spot-winged Pigeon

ad

ad

maculosa

albipennis

ad

sylvestris

Pale-vented Pigeon

ad

picazuro

Picazuro Pigeon

ad

plumbea

Plumbeous Pigeon

ad ♀

ad ♂

Scaled Pigeon

Patagioenas Large arboreal and terrestrial pigeons. Often gregarious. Sexual dimorphism is more marked in some species. See also Plate 75.

Band-tailed Pigeon *Patagioenas fasciata*   36 cm

Yungas forest; lower in winter. **ID Adult** ♂: Yellow bill and tarsus. Lilac crown and breast. White nuchal collar and green gloss on hindneck. **Adult** ♀: Differs by brown crown and underparts. Pale grey tail band conspicuous in flight. **Voice** Very low-pitched monotonous series of c. 8–15 *woÓ* or doubled *ug-WÓ* notes at 1 per sec. **Tax note 20**. [Paloma Nuca Blanca]

Chilean Pigeon *Patagioenas araucana*   37 cm

N Patagonian forest. **ID Adult** ♂: Red tarsus and iris. Mostly rich maroon with white nuchal collar and glossy green hindneck. **Adult** ♀: Ruddy coloration restricted to crown and upper mantle. Underparts washed grey. Black basal tail band conspicuous in flight. **Voice** Begins with an introductory drawn-out, hollow *URrrrruuH*, followed by 4–5 low-pitched *cu-CUU* phrases. [Paloma Araucana]

Leptotila Mainly rufous underwing clearly visible in flight. Alone or in pairs. Easily detected by distinctive voices.

White-tipped Dove *Leptotila verreauxi*   29 cm

Widespread. **ID Adult**: Blue-green hindneck is diagnostic. Grey skin around eye. Pinkish-red tarsus. In flight, tail more graduated than congeners. **Voice** Typical call consists of double cooing notes *woo whoooo*, with the second more drawn-out, and sometimes a little higher-pitched. [Yerutí Gris]

Grey-fronted Dove *Leptotila rufaxilla*   29.5 cm

Paraná forest. **ID Adult**: Clearly defined grey crown and red skin around eye. Purple gloss on hindneck. Pinkish-red tarsus. **Voice** Typical call consists of single cooing notes *whOOu*, not as deep as calls of quail-doves and recalling last note of White-tipped, but somewhat throaty. [Yerutí Colorada]

Yungas Dove *Leptotila megalura*   29.5 cm

Yungas forest. **ID Adult**: Purple gloss on hindneck and upper back. Warm brown tail. Grey skin around eye. Pinkish-red tarsus. **Voice** Typical call consists of a fast series of 4–5 cooing notes *boou boo-boo-bu bOOU*, with longer initial and final notes and especially louder towards the end, which carries well. [Yerutí Yungueña]

albilinea

ad ♂

ad ♀

Band-tailed Pigeon

ad ♂

ad ♀

Chilean Pigeon

ad

chalcauchenia

White-tipped Dove

ad

Grey-fronted Dove

ad

saturata

Yungas Dove

Claravis and *Paraclaravis* Sexually dimorphic forest ground doves with barred wing-coverts. Underwing and tail pattern greatly aid identification.

Blue Ground Dove *Claravis pretiosa* 20.5 cm

Yungas and Paraná forest; local in the humid chaco. Somewhat nomadic; more common Sep–Mar. **ID** Adult ♂: Tarsus straw or pink. Greenish bill. Blue-grey with contrasting white forehead, face and underparts. Wings heavily spotted black; two obvious bars across wing-coverts. In flight, greyish-white underwing and blackish outer tail feathers. Adult ♀: Tarsus flesh. Brown upperparts and breast with contrasting rufous rump and most of tail. Reddish-brown wing spots and bars as male. In flight, underwing and outer tail as ♂. **Voice** Short, evenly-spaced, monotone couplets *pút...pút...* [Palomita Azulada]

Purple-winged Ground Dove *Paraclaravis geoffroyi* 22.5 cm

Very rare, critically endangered, nomadic bamboo specialist of Misiones. **ID** Adult ♂: Red tarsus. Black bill. Bluish upperparts and breast, with contrasting white forehead, face and belly. One small and two broad black bars (purple in favourable light), edged white, across wing-coverts. In flight, blackish underwing and white outer tail feathers. Adult ♀: Brown upperparts and breast with contrasting buffy face and undertail-coverts. Centre of belly white. Wing pattern as ♂. In flight, underwing as ♂, but outer tail feathers buff. **Voice** A deep, hollow and upward-inflected *kwooo-ÚP* [Palomita Morada]

Geotrygon and *Zentrygon* Secretive, sexually dimorphic (except White-throated), chunky terrestrial forest doves; they typically retire to tree cover to rest or when alarmed. All deliver a very low-pitched single resonating note.

Ruddy Quail-Dove *Geotrygon montana* 25 cm

Paraná forest. **ID** Adult ♂: Rich ruddy upperparts with violet gloss and broad reddish malar streak. Adult ♀: Bronzy olive-brown upperparts and thick malar. Chestnut forehead. **Voice** Low-pitched, sad *wuuuuuu* trailing off at the end, deeper than Grey-fronted Dove. [**Paloma Montera Castaña**]

Violaceous Quail-Dove *Geotrygon violacea* 24 cm

Notably rare in Paraná forest. **ID** Adult ♂: Whitish face and underparts contrast strongly with iridescent violet hindneck and upper mantle. Rump and tail reddish-chestnut. Adult ♀: Like a duller washed-out version of ♂ with green gloss on the crown and a brown breast, indistinctly tinged violet on the upper back. [**Paloma Montera Violácea**]

White-throated Quail-Dove *Zentrygon frenata* 32 cm

N Yungas forest. **ID** Adult: Pale orange iris. Black eye-stripe and malar streak. Contrasting purplish gloss on mantle, white throat and dark scaling on sides of neck and breast. **Voice** Very low-pitched resounding *whuuuoo*, occasionally disyllabic with second part lower-pitched *wu-oooh*. [**Paloma Montera Grande**]

Blue Ground Dove
ad ♂
♂
ad ♀
♀

Purple-winged Ground Dove
ad ♂
♂
♀
ad ♀

Ruddy Quail-Dove
ad ♀
montana
ad ♂

Violaceous Quail-Dove
ad ♀
violacea
ad ♂

White-throated Quail-Dove
ad
margaritae

Columbina Small compact, sexually dimorphic (except Scaled) lowland doves with rhythmic voices.

Plain-breasted Ground Dove *Columbina minuta* 15.5 cm

Very rare in Misiones and ne. Corrientes. **ID** Adult ♂: Grey-brown upperparts with blue-grey forecrown. Pinkish ear-coverts and breast. Three blackish spots on lesser wing-coverts and series of streaks on scapulars and tertials. Adult ♀: Upperparts lack grey tones, duller pinkish-brown below. In flight, underwing entirely chestnut. Narrow white tips to all but central rectrices. [Torcacita Enana]

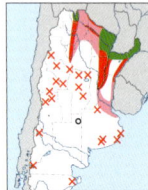

Ruddy Ground Dove *Columbina talpacoti* 18 cm

N lowlands and foothills from Aug–Apr. **ID** Adult ♂: Rich chestnut with blue-grey crown and pinkish face. In flight, contrasting black flight feathers and outertail. Adult ♀: Mainly brown with contrasting chestnut-brown rump and vent. Wings more heavily marked than Plain-breasted. **Voice** Fast cooing with slightly ascending last note *cucu-wOOh... cucu-wOOh...* repeatedly (6 notes every 5 secs), sometimes with short introductory notes, or fast briefer cooing series in duet. [Torcacita Colorada]

Picui Ground Dove *Columbina picui* 19 cm

Widespread in lowlands and Andes (to 2950 m) south to N Patagonia. **ID** Adult ♂: Pale brown above with a blue-grey crown. Violet-black bar across lesser wing-coverts. Whitish below. In flight, white edge of wing-coverts contrasts with blackish flight feathers. White outer tail feathers. Adult ♀: Brown crown. **Voice** Fast clearly upward inflected cooing, *cuwOOP... cuwOOP...* (8 notes every 5 secs). [Torcacita Picuí]

Scaled Dove *Columbina squammata* 21 cm

Misiones, n. Corrientes, e. Chaco and e. Formosa; expanding. **ID** Adult: Brown with prominent white area on wing-coverts, and scaled blackish over much of upperparts and belly. Note long tail. In flight, conspicuous chestnut patch in primaries and white outer tail feathers. **Voice** Monotonous and loud *POOW... pa-POW...* repeated over and over; much higher-pitched than other *Columbina* doves. [Torcacita Escamada]

ad

Plain-breasted
Ground Dove

ad ♂

ad ♀

Ruddy Ground Dove

ad ♂

♂

talpacoti

ad ♀

♂

ad ♂

Picui Ground Dove

picui

ad ♀

Scaled Dove

ad

squammata

ad

Metriopelia Gregarious, terrestrial Andean doves of open terrain. Coloured ocular skin, except Golden-spotted. Sexes alike; moderate dimorphism in Bare-faced. Soft vocalisations (except Black-winged); notable wing rattle in flight.

Golden-spotted Ground Dove *Metriopelia aymara* 18.5 cm

Puna. **ID Adult:** Chunky, short-tailed pale brown dove with whiter throat. At close range, one or two black spots on tertials and row of iridescent bronze-gold spots on lesser wing-coverts. In flight, chestnut patch in base of primaries visible on underwing. Short, square, unmarked blackish tail. **Voice** A deep, vibrating, purring *pwoorh… pwoorh… pwoorh.* [Palomita Dorada]

Moreno's Ground Dove *Metriopelia morenoi* 19 cm

Endemic. NW Andes. **ID Adult:** Bare reddish-orange skin around eye. Grey bill, tipped black. Cold brown above; grey below becoming pinkish-brown on belly and vent. In flight, small white tail corners. **Voice** Most common is a series of fast, nasal *weh… weh, weh, weh, weh* notes. [Palomita Ojo Desnudo]

Bare-faced Ground Dove *Metriopelia ceciliae* 19 cm

Extreme n. Jujuy and n. Salta. **ID Adult** ♂: Bare reddish-orange skin around eye. Scapulars and wing-coverts heavily mottled with whitish spots. Vinaceous breast. **Adult ♀:** Differs from adult ♂ by pinkish-brown underparts. Extensive white tail corners visible in flight. **Voice** Fast series of wheezy notes *whoh whoh whoh whoh whoh* and harsh, deep, springy call *bawr!* [Palomita Moteada]

Black-winged Ground Dove *Metriopelia melanoptera* 23.5 cm

Andes south to Chubut. **ID Adult:** Bare yellow or orange skin below eye. Brown with blackish wings and white carpal patch. In flight, conspicuous white carpal; long unmarked black tail. Juveniles lack bare skin and are browner throughout. **Voice** Loud, high-pitched and ringing, rising and falling *prruuee treew*, recalling Little Nightjar. Call is a passerine-like *chip*. Both voices are highly unusual for a dove. [Palomita Cordillerana]

Zenaida Medium-sized doves of open areas, with broad white tail tips.

West Peruvian Dove *Zenaida meloda* 30 cm

Expanding through towns and villages in low monte desert (650–1450 m) from NW to N Patagonia. **ID Adult:** Bare violet-blue skin around eye. Red tarsus. White stripe on closed wing. In flight, striking white transverse wing band and tail band. **Voice** Comprises three melodic phrases with an introduction, *who… wu-LO-ro, OHh,… wu-LO-roh, OHh,…wu-LO-roh…* [Torcaza Ala Blanca]

Eared Dove *Zenaida auriculata* 25.5 cm

Widespread. **ID** Gregarious, nondescript, fast-flying, medium-sized brown dove. **Adult** ♂: Pinkish or brown forecrown and face; notable black line on upper and lower edge of ear-coverts. Rest of crown blue-grey. Iridescent gold on sides of neck. Black spots on tertials. Vinaceous underparts. **Adult ♀:** Crown and underparts browner. Little or no iridescence on sides of neck. In flight, underwing greyish; white tips to tail. **Voice** Low-pitched, mournful and throaty *huuuh…huuuuh… huuh…* [Torcaza]

Golden-spotted Ground Dove

ad

Moreno's Ground Dove

ad

ad ♀

zimmeri

ad ♂

Bare-faced Ground Dove

melanoptera

ad

Black-winged
Ground Dove

ad

West Peruvian Dove

ad ♀

auriculata

ad ♂

Eared Dove

Primolius Small forest macaws with bare facial skin. Far-carrying raucous cries.

Blue-winged Macaw *Primolius maracana*   40 cm

Misiones; seemingly extinct. **ID Adult:** Bluish head with small red patch on forehead. Red patch on belly. Tail red, tipped blue. Pairs or small groups. In flight, note undulating flight, pale face and red patch on rump. [Maracaná Lomo Rojo]

Golden-collared Macaw *Primolius auricollis*   40 cm

N Yungas forest. **ID Adult:** Similar to Blue-winged Macaw but has blackish forecrown; easily distinguished by yellow nuchal collar. Tail red, tipped blue. In flight, similar to Blue-winged but lacks red belly and red rump (no overlap). **Voice** Raucous *wrreegh… wrreegh…* in flight and when perched, notably higher-pitched than Military Macaw. [Maracaná Cuello Dorado]

Anodorhynchus and *Ara* Unmistakable, large, long-tailed parrots. The two species in *Ara* have a large bare facial patch, finely crossed by feathered lines. Far-carrying voices. Crepuscular flight to and from feeding grounds.

Glaucous Macaw *Anodorhynchus glaucus*   71 cm

Presumed extinct. Formerly known from palm savanna in Corrientes and Entre Ríos. **ID Adult:** Unmistakable large sky blue macaw. Bare yellow skin around eye and on malar region. [Guacamayo Azul]

Red-and-green Macaw *Ara chloroptera*   82 cm

Formerly (erratic?) in e. Formosa and w. Misiones. Forest, gallery woodlands and forest islands in savanna. **ID Adult:** Unmistakable, the only red macaw in range. **Voice** Far-carrying voice. [Guacamayo Rojo]

Military Macaw *Ara militaris*   68 cm

Rare in foothill Yungas forest of n. Salta. **ID. Adult:** Head and upperparts green with red forehead, pale blue rump and vent, and red uppertail tipped blue. **Voice** Far-carrying, dry *WRRAAAGH* in flight and when perched; deeper and more rolling than Golden-collared Macaw. [Guacamayo Verde]

ad

Blue-winged
Macaw

ad

Golden-collared
Macaw

ad

ad

ad

Glaucous Macaw

Red-and-green Macaw

boliviana

Military Macaw

Psittacara, *Thectocercus* and *Enicognathus* Gregarious, green, long-tailed, medium-sized parakeets found in a variety of forest types. Head, wing and tail patterns aid identification. *Cyanoliseus* Spectacular, gregarious, open country, long-tailed parrot of macaw-sized proportions.

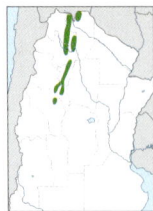

Mitred Parakeet *Psittacara mitratus* · 37 cm

NW Andes. **ID** Long, pointed tail and pale bill. Small to large flocks. **Adult:** Red foreface, restricted to forehead in immature. In flight, rather uniform; note flight features of White-eyed and Blue-crowned which can overlap. **Voice** More nasal and cleaner than White-eyed and Blue-crowned Parakeets, but very similar. [Calancante Cara Roja]

White-eyed Parakeet *Psittacara leucophthalmus* · 33 cm

Widespread in the north. **ID** Long, pointed tail and pale bill. Small to large flocks. **Adult:** Sparse red spotting on head. Red bend of wing. In flight, scarlet and yellow on bend of underwing. **Voice** Less nasal than calls of Mitred Parakeet, not as harsh as those of Blue-crowned, although very similar. [Calancate Ala Roja]

Blue-crowned Parakeet *Thectocercus acuticaudatus* · 33 cm

Widespread in N & C areas. **ID** Long, pointed tail and bicoloured bill. Small to large flocks. **Adult:** Bluish crown and face. Pink upper mandible and feet. In flight, reddish underside of tail. **Voice** Harsher than White-eyed and Mitred Parakeets, but very similar. [Calancate Cabeza Azul]

Austral Parakeet *Enicognathus ferrugineus* · 32 cm

Patagonian forest; usually the only parrot species in range. **ID** Sturdy, long-tailed gregarious forest parakeet with a swift direct flight. **Adult:** Green with dusky scaling throughout. Dull red lores, forehead and patch on belly. Compare with Slender-billed Parakeet in extreme NW Patagonia. **Voice** Gargled churring notes in series [Cachaña]

Slender-billed Parakeet *Enicognathus leptorhynchus* · 39 cm

Sporadic in NW Patagonia? **ID Adult:** Differs from Austral Parakeet by its long, curved upper mandible, more extensive bright red forehead and eye-ring, considerably brighter green upperparts and yellower underparts. Proportionately longer, brighter red tail. **Voice** Similar to Austral Parakeet but shriller and much more strident. [Choroy]

Burrowing Parrot *Cyanoliseus patagonus* · 45–47 cm

Widespread in the N and C Andes, C sierras and Patagonia. **ID** Large, long-tailed, gregarious macaw-like parrot of arid open terrain. Frequently flies low over ground. Nests communally in cliff burrows. **Adult** *patagonus* (47 cm; Patagonia and C Andes): Yellow rump and belly with red centre. Whitish sides to breast. In flight, note yellow belly and laboured flight. **Adult** *andinus* (45 cm; NW Andes): Smaller with olive rump and belly with dull red centre. **Adult** *conlara* (C sierras; not illustrated): intermediate between *andinus* and *patagonus*. **Voice** Nasal, grating and often drawn-out; very vocal for much of the time. [Loro Barranquero]

mitratus

leucophthalmus

acuticaudatus

ad

ad

ad

Mitred Parakeet

White-eyed Parakeet

Blue-crowned Parakeet

ad

Austral Parakeet

ad

ferrugineus

Slender-billed Parakeet

andinus

ad

patagonus

ad

ad

Burrowing Parrot

Pyrrhura Slender, gregarious, long-tailed, forest-canopy parakeets. Rapid twisting flight. *Myiopsitta* Gregarious open-country parakeets with a long, pointed tail. Swift and direct flight, while rolling from side to side.

Maroon-bellied Parakeet *Pyrrhura frontalis* 26 cm

NE forests. **ID Adult**: Green crown and small dark red forehead. Reddish undertail and bronze-olive uppertail. Red patch on centre of belly. **Black-headed variant**: Found in mixed flocks with normal adults in the humid chaco. Black crown, nape, lores and often the entire throat. Recalls Green-cheeked Parakeet, but note tail pattern. **Voice** High-pitched screeched calls. [Chiripepé Cabeza Verde]

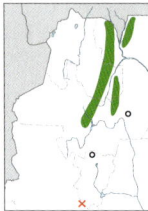

Green-cheeked Parakeet *Pyrrhura molinae* 27 cm

Yungas forest. **ID Adult**: Sooty-brown crown and lores, contrasting green cheeks. Red uppertail and undertail. Throat and breast heavily fringed yellowish-olive. Extensive red patch on belly. **Voice** High-pitched screeched calls. [Chiripepé Cabeza Parda]

Monk Parakeet *Myiopsitta monachus* 28.5 cm

Widespread in the lowlands south to N Patagonia. **ID** Builds conspicuous huge communal thorny nests. **Adult**: Bluish-grey forecrown; bright green rear crown and nape. Grey throat, cheeks and breast with variable buffy-white scaling across breast. In flight, compare with Grey-hooded Parakeet (Plate 83) in the C sierras and NW Andes, and Cliff Parakeet in n. Salta. [Cotorra]

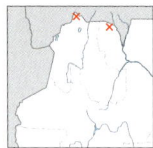

Cliff Parakeet *Myiopsitta [monachus] luchsi* 28.5 cm

Very rare (unconfirmed) in inter-Andean dry valleys. **ID Adult**: Brighter than Monk Parakeet. Dusky mark on base of upper mandible, unscaled grey breast and yellow belly. In flight, paler underwing than Monk Parakeet (no overlap). Compare also with Grey-hooded Parakeet (Plate 83). **Tax note 21**. [Cotorra de Acantilados]

Peach-fronted Parakeet *Eupsittula aurea* 26 cm

Rare in e. Formosa and e. Chaco; very rare in n. Misiones. **ID** Small parakeet with pointed tail, feathered orange eye-ring and black bill. **Adult**: Orange or yellow-orange forehead. Greenish-yellow lower underparts. In flight, uniform with yellower belly. [Calancate Frente Dorada]

Nanday Parakeet *Aratinga nenday* 34 cm

NE savanna; feral populations in Buenos Aires. **ID** Stocky, long-tailed green parakeet with a large hooked black bill. **Adult**: Crown and foreface blackish. Blue wash across breast. In flight, contrasting blackish remiges from below. Yellowish underwing-coverts and rump. [Ñanday]

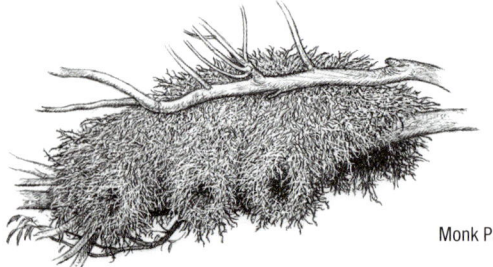

Monk Parakeet nest

Maroon-bellied Parakeet

ad

black-headed
variant

chiripepe

Green-cheeked Parakeet

ad

australis

ad

Monk Parakeet

ad

ad

Cliff Parakeet

ad

Nanday Parakeet

Peach-fronted
Parakeet

ad

Andean Parakeet *Bolborhynchus orbygnesius* 17.5 cm

Rare or local above 2500 m in Jujuy and Salta. **ID** More robust and broader-tailed than Grey-hooded and Mountain Parakeets. Direct flight. **Adult**: Grey or whitish bill. From Mountain Parakeet mainly by shorter and broader tail. In flight, less blue in the primaries than Mountain and distinct tail shape. **Voice** Fast, metallic twittering extremely like Mountain Parakeet, although less shrill, more chattery and slightly lower-pitched. [Catita Andina]

Mountain Parakeet *Psilopsiagon aurifrons* 18.5 cm

N and C Andes, and local in C sierras. **ID** Small gregarious parakeet of open mountainous terrain. Bill pink in ♂, grey in ♀. Pointed tail. High, slightly undulatory flight. **Adult** *rubrirostris* (CW Andes and C sierras): Crown, face, underparts and tail washed powder blue. **Adult** *margaritae* (NW Andes): Entirely green, brighter on rump and underparts. Pointed olive tail. In flight, note bluish primaries and fairly long pointed tail. **Voice** Fast, metallic twittering, unlike Grey-hooded Parakeet. [Catita Serrana Chica]

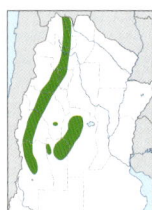

Grey-hooded Parakeet *Psilopsiagon aymara* 20 cm

N and C Andes, and C sierras. **ID** Small gregarious parakeet of open mountainous terrain. Pointed tail. High, slightly undulatory flight. **Adult**: Brownish cap and sometimes a dusky mask. Contrasting pale grey throat and breast, washed yellow at the sides. In flight, can recall Monk and Cliff Parakeets (Plate 82), but much smaller and faster. Note whitish central underparts. **Voice** Dry stuttered calls. [Catita Serrana Grande]

Yellow-chevroned Parakeet *Brotogeris chiriri* 20.5 cm

Scarce in the chaco; feral populations in Misiones and Buenos Aires. **ID** Small, slender parrot with a graduated, pointed tail. **Adult**: Fairly bright green with contrasting sulphur-yellow greater wing-coverts. In flight, prominent yellow wing bands. Note pointed wings and tail, and fast, twisting flight. **Voice** Sharp chattering calls. [Catita Chiriri]

Pileated Parrot *Pionopsitta pileata* 21.5 cm

Paraná forest. **ID** Compact, fairly short-tailed parrot. Pairs or small to large groups. **Adult** ♂: Mostly green with a scarlet foreface. **Adult** ♀: Lacks red of ♂. In flight, green with violet-blue primary-coverts, remiges, tips and edges to tail. Direct flight; often high over canopy. **Voice** High-pitched jingling flight call; disyllabic calls when perched. [Catita Cabeza Roja]

Blue-winged Parrotlet *Forpus xanthopterygius* 13 cm

NE forests. **ID** Small, compact, sexually dimorphic parrot with short tail. Found in pairs or small groups. **Adult** ♂: Pink or grey bill. Green with brighter face and underparts. Violet-blue rump, greater coverts, secondaries, shoulder and underwing-coverts. In flight, contrasting violet-blue wings and rump. **Adult** ♀: Lacks violet-blue and has a bright green rump. **Voice** Gives high-pitched metallic calls. [Catita Enana]

Andean Parakeet

ad

Mountain Parakeet

ad ♂

ad ♀

ad ♂

rubrirostris

margaritae

ad

Grey-hooded Parakeet

ad

Yellow-chevroned Parakeet

ad ♂

ad ♀

ad ♂

ad ♀

validus

ad ♀

Pileated Parrot

ad ♂

Blue-winged Parrotlet

Pionus Medium-sized, robust and short-tailed parrot. More agile and faster-flying than *Amazona* and with downcurved paddle-like wings. Pairs and small flocks. *Amazona* Large robust parrots which commute at dawn and dusk to feeding grounds on shallow wingbeats. Calls generally raucous and far-carrying.

Scaly-headed Parrot *Pionus maximiliani*   28–30 cm

Widespread in the N lowlands and Yungas forest. **ID** Adult *melanoblepharus* (28 cm; NE): Dull green above, scaled dusky. Throat, sides of neck and breast washed violet-blue. Undertail-coverts scarlet. Adult *siy* (28 cm; chaco): Considerably brighter throughout than *melanoblepharus* with strong bronze tones above and conspicuous, broad, broken white eye-ring. Adult *lacerus* (30 cm; Yungas; not illustrated): Resembles *siy* but larger. **Voice**. Dry, scratchy flight calls, usually delivered in couplets. Remarkably loud pure whistle. [Loro Maitaca]

Red-spectacled Amazon *Amazona pretrei*   32 cm

Extremely rare wanderer to Misiones. **ID** Adult ♂: Scarlet forecrown and prominent eye-ring. Blackish-fringed back, scapulars and abdomen. Extensive scarlet shoulders, carpal and primary wing-coverts. In flight, extensive red on forewing. Adult ♀: Red restricted to smaller patch on primary coverts. All other amazons in range have a red speculum. [Charao]

Vinaceous-breasted Amazon *Amazona vinacea*   37 cm

Rare and local; mainly *Araucaria* forest in e. Misiones. **ID** Adult: Reddish bill. Narrow red forehead and lores. Pale blue nape, fringed black. Breast mixed mauve and pale blue, fringed blackish. In flight, notable red speculum. Carpal edged yellow and/or red. Tail green, tipped yellowish and with red in the base. [Loro Vinoso]

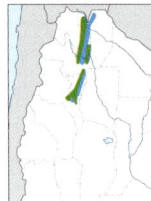

Tucumán Amazon *Amazona tucumana*   32 cm

Yungas forest; lower in winter. **ID** Adult ♂: Red forehead. Black scaling throughout, especially on nape and underparts. Red primary wing-coverts. Tail green, tipped yellowish. In flight, large red patch on forewing. Adult ♀: Reduced red on primary wing-coverts. Compare with Scaly-headed Parrot and Turquoise-fronted Amazon. No overlap with Red-spectacled Amazon. **Voice** More metallic, strident and ringing than Turquoise-fronted. [Loro Alisero]

Turquoise-fronted Amazon *Amazona aestiva*   38 cm

Widespread in the N lowlands and Yungas foothills. **ID** Adult: Sky-blue forecrown and lores (also sometimes upper breast) contrasting with yellow mid-crown, face and throat. Head pattern highly variable. In flight, large red speculum. Yellow shoulders and red carpal. Tail broadly tipped yellowish, except on central pair. Yungas foothills birds may lack yellow on the head and exhibit green shoulders and carpal. **Voice** Pleasant and leisurely, gurgling calls *grraowl… krreow…* are slower and less metallic than Tucumán Amazon. [Loro Hablador]

Scaly-naped Amazon *Amazona mercenarius*   31 cm

Historical vagrant to the NW Andes. **ID** Adult: Whitish spot on base of upper mandible. Forehead (sometimes tinged red), cheeks and underparts bright yellowish-green. Black fringing on back, upper breast and especially nape. Tail tipped yellowish with a red subterminal band. In flight, note small red speculum, yellow carpal mixed with red, and tail pattern. [Loro Nuca Escamada]

melanoblepharus

ad

ad

siy

ad

ad ♂

ad ♂

Scaly-headed Parrot

ad ♀

Red-spectacled Amazon

ad

Vinaceous-breasted Amazon

ad ♂

ad ♂

Tucuman Amazon

ad ♀

xanthopteryx

ad

ad

mercenarius

Turquoise-fronted Amazon

Scaly-naped Amazon

Coccyzus and *Coccycua* Slim-bodied migratory cuckoos with long, graduated tails and white spots visible on the undertail which are indistinct on juveniles. Sexes are alike. Voices greatly aid location and identification.

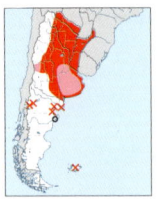

Dark-billed Cuckoo   *Coccyzus melacoryphus*   27 cm

Oct–May in the N lowlands and foothills. **ID Adult:** Stout black bill. Blackish mask and yellow eye-ring. Grey crown and sides of neck and rich cinnamon-buff underparts. Tail is largely black with brown central rectrices, and striking white spots. In flight, brown flight feathers. **Juvenile:** Usually lacks the adult's mask and has a grey eye-ring. Grey wash on the sides of the neck and buff wash below. Rufescent flight feathers. **Voice** Song is a slow throaty *kwuh...kwuh...kwuh...kwuh...kwuh... kwuh...wuhh* with final notes harsher and lower-pitched. Calls include a fast *kwo-kwo-kwo-kow-kokorr* which fades in volume, and a faster and accelerating *kekekerrr*. [Cuclillo Canela]

Black-billed Cuckoo   *Coccyzus erythropthalmus*   28 cm

Very rare boreal vagrant recorded in Jujuy and Misiones. **ID Adult:** Slender black bill and red eye-ring. Little to no rufous in wings. Small tail spots with dusky subterminal tips. White underparts. **Juvenile:** Black bill with a grey lower mandible. Yellow or grey eye-ring. Wing-coverts fringed whitish, and may show some rufous in primaries. Distinctive olive-buff wash on the throat. Tail spots very small and indistinct. **Voice** Generally silent. [Cuclillo Ojo Colorado]

Yellow-billed Cuckoo   *Coccyzus americanus*   28.5 cm

Boreal migrant; Sep–Apr in the N and C lowlands. **ID Adult:** Stout yellow bill with a black culmen and tip. Yellow eye-ring. Rufous flight feathers. Black undertail with large prominent white spots. In flight, striking rufous flight feathers are brighter than any juvenile congener. **Juvenile:** Grey eye-ring. Youngest birds lack yellow on bill. Otherwise similar to adult with brown undertail and indistinct spots. **Voice** Generally silent; some voices recall Dark-billed Cuckoo. [Cuclillo Pico Amarillo]

Pearly-breasted Cuckoo   *Coccyzus euleri*   27.5 cm

Oct–Jan in Paraná forest; sparse records from the humid chaco. Some overlap with Yellow-billed Cuckoo at forest edge. **ID Adult:** Very similar to Yellow-billed but lacks rufous in the wings. Very pale grey wash on the lower throat and breast, and sometimes a very pale yellow wash on the belly. In flight, brown flight feathers. **Juvenile:** Very similar to juvenile Yellow-billed, but lower mandible yellow and usually with a grey wash on the belly. Flight feathers rufescent in flight. **Voice** Song is a mournful trogon-like series of 5–20 well-spaced notes *kúo...* or *kóu....* Excited call is a soft and quick muffled ticking, sometimes followed by short rattles *tictictictictictictictictictictictbrreu-brreu-brreu...* [Cuclillo Ceniciento]

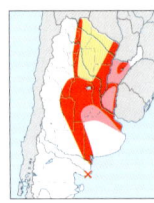

Ash-coloured Cuckoo   *Coccycua cinerea*   22.5 cm

Sep–Apr in the N and C lowlands. **ID** Smaller than the four *Coccyzus* species. **Adult:** Short black bill. Red iris and eye-ring. Grey-brown above with a relatively short, square-ended tail. White tail tips best seen from below. Buffy-brown wash on throat and breast. In flight, greyish flight feathers. Cinnamon underwing-coverts rarely seen. Note white tail tip. **Juvenile:** Iris varies from brown to reddish. Eye-ring grey at first, rapidly becoming red. Flight feathers fringed rufous. Warm brown tail lacks white tips. Ashy throat and breast. **Voice** Song is a series of 6–18 mournful low-pitched pure notes *úow...úow...úow...úow...úow...*, descending slightly in pitch at c. 2 per sec. Grating churring call (alarm?) *brreew... brreew...* [Cuclillo Chico]

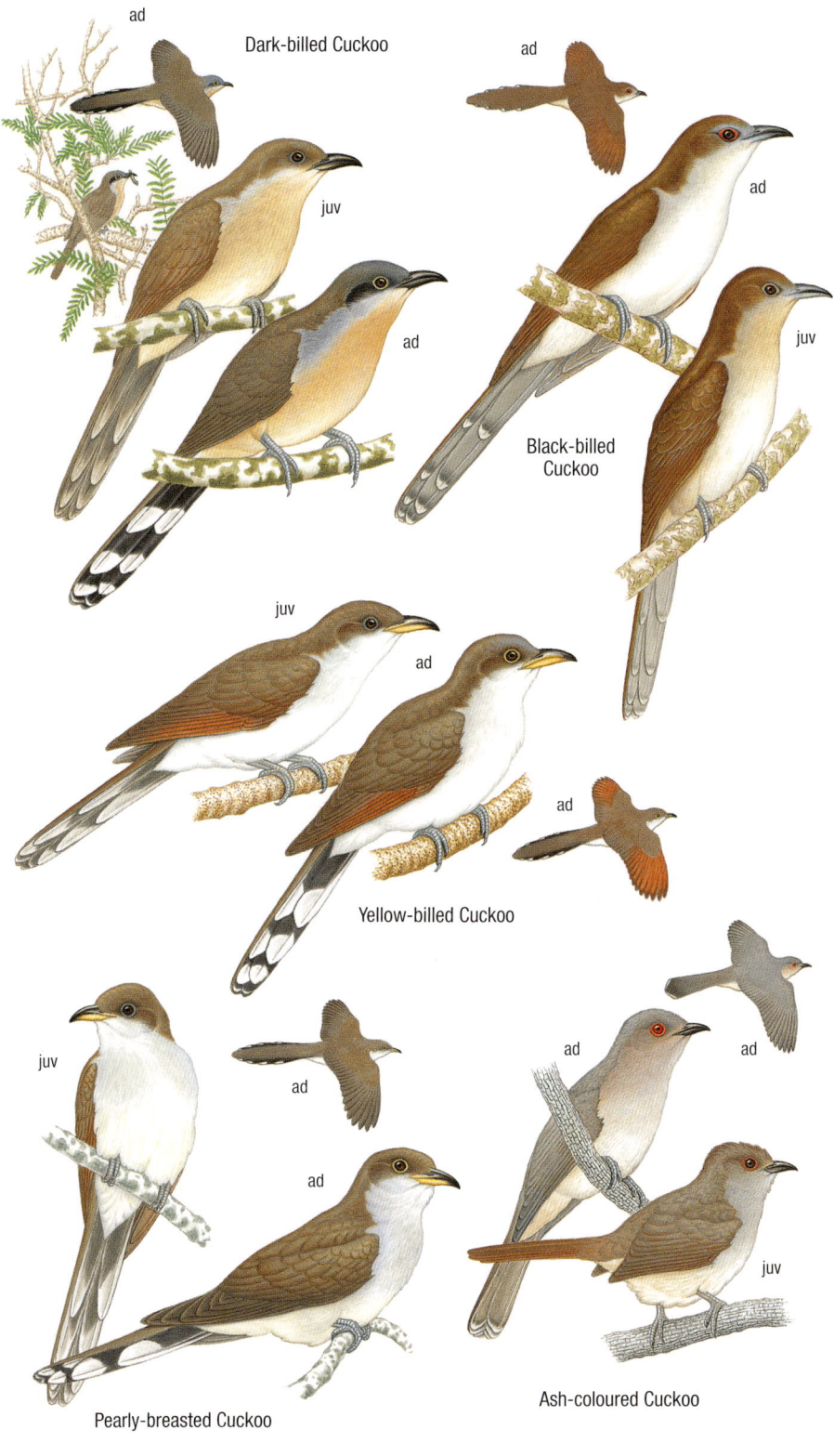

ad

Dark-billed Cuckoo

ad

juv

ad

ad

Black-billed
Cuckoo

juv

juv

ad

ad

Yellow-billed Cuckoo

juv

ad

ad

ad

ad

ad

juv

Pearly-breasted Cuckoo

Ash-coloured Cuckoo

Tapera Reclusive cuckoo of open country and open woodland with long narrow tail and very long uppertail-coverts.

Striped Cuckoo *Tapera naevia* 28 cm

N lowlands; partial austral migrant (mainly Sep–Mar). **ID** Adult: Erectile rufous crown. Upperparts streaked black. Juvenile: Crown, mantle and wing-coverts thickly spotted buff. More rufescent tail than adult. **Voice** Best detected by repetitive two-note whistle with higher second note *Ya..Sí*. Long song is a series of ascending whistles with final quavering note *fee...fee...fee...fee... feeoe*. Also, 2–4 long, upswept notes, each successively higher-pitched *feee-a... feee-a...* [Crespín]

Piaya Large forest cuckoo with extremely long graduated tail.

Squirrel Cuckoo *Piaya cayana* 51 cm

Widespread in the north. **ID** Adult *macroura* (NE; illustrated): Unmistakable. Rufous head, breast and upperparts. Grey belly. Bare red skin around eye. Adult *mogenseni* (Yungas forest): Brighter rufous above and paler below. **Voice** Song is a series of 8–15 emphatic *uíp* notes at c. 1.5 per sec (much slower than Rusty-breasted Nunlet). Calls include a harsh *wrrreeea*, a loud *meeah k-krreough*, a two-note *pík!...kwaaa*, and an endless rattling chatter (slower than Olivaceous Woodcreeper). [Tingazú]

Guira Large open country cuckoo with shaggy crest and long graduated tail. Often gregarious.

Guira Cuckoo *Guira guira* 40 cm

Widespread in lowlands and Andean foothills south to N Patagonia. **ID** Adult: Orange or yellow bill. Thick white dorsal streaking. White rump. Black tail with white base and broad white tips. **Voice** Series of rising whistles, each successively slightly lower-pitched and drawn-out, finishing in a gurgled trilling series *ooeee... ooeee... ooeee... wurrr... wurrr... wurrrrrrr*. Long, higher-pitched oscillating trill eventually followed by shorter trills *prrrrrrrrrr... chrr chrr chrr*. [Pirincho]

Crotophaga Gregarious black cuckoos with laterally compressed bills.

Groove-billed Ani *Crotophaga sulcirostris* 34 cm

Sparse records from n. Salta. **ID** Adult: Very similar to Smooth-billed Ani but upper mandible is less bulging and with three grooves. **Voice** Calls include a nasal *tea klee* like a hoarse *Synallaxis* spinetail, and a metallic *Campephilus* woodpecker-like *ps-cléeu* call. [Anó Pico Surcado]

Smooth-billed Ani *Crotophaga ani* 34 cm

Widespread in the N lowlands and Andean foothills. **ID** Adult: Distinctive bulge at base of upper mandible. Both mandibles smooth. **Voice** A whistled *weeoo-klíp* alone or in series. [Anó Chico]

Greater Ani *Crotophaga major* 48 cm

Oct–May in gallery and riverine forest in the NE; sparse records from the NW. **ID** Adult: Greenish-white iris. Razor-like ridge at base of culmen. Glossy blue-black with very long purple tail. Juvenile (not illustrated): Dark eye and reduced bill ridge. **Voice** Variable series of querulous and raucous notes that accelerate *ukrreau krrau krreorreorreo...* (very similar to Green Ibis), frequently in a chorus and sometimes morphing into an endless gurgled 'boat engine' chatter. [Anó Grande]

Striped Cuckoo

naevia

juv

ad

ad

macroura

Squirrel Cuckoo

Guira Cuckoo

ad

Groove-billed Ani

sulcirostris

Smooth-billed Ani

ad

Greater Ani

ad

ad

Dromococcyx Reclusive understorey cuckoos with small heads, nuchal crests and broad tails cloaked by extremely long uppertail-coverts. Mostly crepuscular and nocturnal, often in or near bamboo and usually only detected by voice or bill-snapping.

Pavonine Cuckoo *Dromococcyx pavoninus* 28.5 cm

Paraná forest. **ID** Adult: Yellow lower mandible. Rufescent crown and nuchal crest. Yellow eye-ring and long buff post-ocular streak. Blackish back, fringed white. Dull rufous throat and breast, becoming white on centre of throat. **Voice** Song is a ventriloquial whistled *Ya..Si...ya-te-ré*, flatter-sounding and faster than that of Pheasant Cuckoo. [Yasiyateré Chico]

Pheasant Cuckoo *Dromococcyx phasianellus* 40 cm

Scarce in Paraná forest, very rare in humid chaco. **ID** Adult: Dark chestnut crown and nuchal crest. White post-ocular streak, bordered by thick blackish eye-stripe. Fine black malar streak and blackish spots on buffy breast, unlike Pavonine Cuckoo. **Voice** Song recalls that of Pavonine Cuckoo but is more liquid, with the final notes slurred together. [Yasiyateré Grande]

Nyctibius Strictly nocturnal and arboreal; position and plumage mimics tree stump during day. Large-headed, extremely wide gape and large-eyed with elongated body.

Common Potoo *Nyctibius griseus* 36 cm

N lowland forest (Sep–Apr); resident in Misiones. **ID** Adult grey morph: Primaries extend from half length of tail to tip. In flight, long-winged with rather long, broad tail. **Adult rufous morph**: When alert, compact, hunched look with noticeably large iris. **Voice** Eerie long series of rich descending notes *whoooh-k... whoooh-k... whoooh-k... whoooh-k....* [Urutaú]

Long-tailed Potoo *Nyctibius aethereus* 63 cm

Local in n. Misiones. **ID** Adult rufous morph: Much larger than Common Potoo. At most, primaries only extend halfway along tail. Whitish patch on lesser wing-coverts. **Voice** Low-pitched, hollow, rising *waHUU* note, echoes strangely from a distance; repeated occasionally. [Urutaú Coludo]

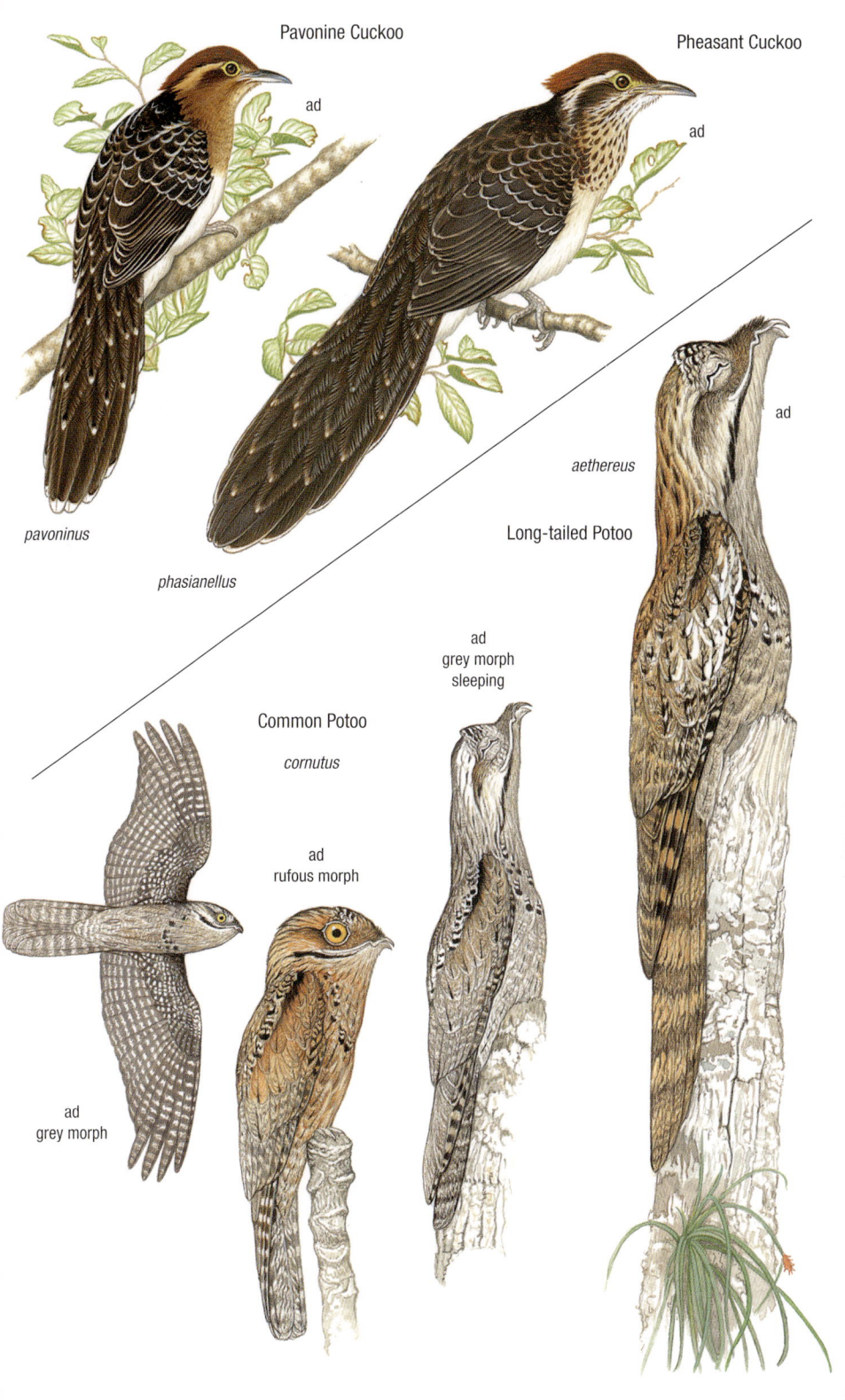

Pavonine Cuckoo

ad

pavoninus

phasianellus

Pheasant Cuckoo

ad

aethereus

Long-tailed Potoo

ad
grey morph
sleeping

Common Potoo

cornutus

ad
rufous morph

ad
grey morph

ad

Chordeiles Small to large compact nighthawks with proportionately large blade-like wings and white gorget. Sometimes diurnal and gregarious. ***Lurocalis*** Nocturnal canopy nighthawk which lacks a white primary band. Loud monosyllabic and somewhat metallic single notes reveal presence at dusk.

Common Nighthawk *Chordeiles minor* 23 cm

Sep–Mar in N lowlands. Arboreal and terrestrial; forages 15–30 m above ground. **ID Adult** ♂: Mainly grey, spangled black. Wing-band in closed wing looks diamond-shaped. White underparts, barred black. Wing tip extends beyond tail tip. In flight, white wing band near base of primaries. Forked tail with white subterminal band. **Adult** ♀ (not illustrated): Lacks subterminal tail band of ♂. [Añapero Boreal]

Lesser Nighthawk *Chordeiles acutipennis* 22 cm

Rare in the N chaco (Dec–Jan). Arboreal; forages within 10 m of ground. **ID** Proportionately smaller-headed than Common Nighthawk. **Adult** ♂: Rectangular white wing band is positioned halfway to the tip and beyond the longest tertial. Buff spots at base of primaries. Wing tip barely reaches tail tip. In flight, very similar to but paler on average than Common Nighthawk. Wing band closer to primary tips than Common. **Adult** ♀ (not illustrated): Has buff wing band. [Añapero Alas Cortas]

Least Nighthawk *Chordeiles pusillus* 15.5 cm

Local in humid grasslands of s. Misiones to ne. Corrientes with rocky outcrops (Oct–Mar). Terrestrial. **ID Adult**: Spangled and vermiculated rufous above. White gorget and barred underparts. In flight, undertakes high diurnal flight with mechanical turns. White band near primary tips. Notable white trailing edge to wing and white vent. **Voice** Flight song is a very fast, fairly high-pitched and slightly ascending *kokokokokokebréee* with distinctive longer final note. Flight call is a frog-like dry *peent crreeek!* or *prréent!* [Añapero Chico]

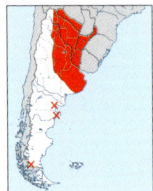

Nacunda Nighthawk *Chordeiles nacunda* 28 cm

Widespread in the N lowlands (Sep–Feb). Terrestrial. **ID Adult**: Fawn-brown, speckled black. White gorget. **Adult** ♂: In flight, white belly, underwing-coverts and broad white band across base of contrasting black primaries. Broad white terminal tail band (lacking in ♀). **Voice** Song in flight is a series of quickly delivered deep resounding notes *buuuu-bururu…* sometimes preceded by a chuckling *kokoro*. Contact call in flight is a soft *uhrr* or *brru puhrr*. [Ñacundá]

Short-tailed Nighthawk *Lurocalis semitorquatus* 24 cm

Chiefly Misiones, local in the N lowlands (Sep–Mar). Arboreal and aerial. **ID** Chunky and dark. **Adult**: Speckled black and tawny with frosted whitish area on tertials and inverted white gorget. In flight, usually appears all dark, with outsized blade-like wings. **Voice** Loud, rich whistle, inflected upwards towards the end while perched or in flight, *Doo-Wit*, sometimes followed by a harsh note. [Añapero Castaño]

Common Nighthawk

ad ♂

ad ♂

ad ♂

ad ♂

Lesser Nighthawk

saturatus

ad

Least Nighthawk

Nacunda Nighthawk

ad ♂

nacunda

ad

Short-tailed Nighthawk

ad

nattereri

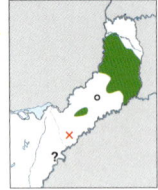

Long-trained Nightjar *Macropsalis forcipata*

♂ 80 cm; ♀ 29 cm

Local in rocky ce. Misiones clearings, often at the top of hills. **ID** Small-bodied, rufous-collared nightjar with long narrow elongated outer tail-feathers in ♂. Arboreal and terrestrial. **Adult ♂:** Tail held straight when perched and in travelling flight, or depressed vertically during display; fringed white on inner webs. **Adult ♀:** Dull rufous nuchal collar. Heavily spotted wing-coverts. In flight, tail longer and more deeply forked than ♀ Lyre-tailed and with more pointed outer tail feathers. Flight feathers narrowly barred cinnamon. **Voice** Generally silent. Territorial song with scissor movements of pendant tail while hovering is a soft *pt-sssssii*. Aggressive males deliver hissing *ptss ptss pts pts pts*. [Atajacaminos Coludo]

Lyre-tailed Nightjar *Uropsalis lyra*

♂ 89 cm; ♀ 26 cm

Local at Yungas forest cliffs in Jujuy. **ID** Small-bodied, rufous-collared nightjar with elongated, broad, incurved outer tail feathers in ♂. Arboreal and terrestrial. **Adult ♂:** Tail held in same manner as Long-trained. Broad outer tail feathers incurved and tipped white. **Adult ♀:** Greyish scapulars and tertials stand out. In flight, dark nightjar with no primary band and fairly long forked tail with rufous notches on outer webs. **Voice** Loud complex yodelling song is a series of 5–8 *wo wipple* or *wopo wipple* phrases which increase successively in pitch and volume, becoming emphatic and barmy at the end. [Atajacaminos Lira]

Scissor-tailed Nightjar *Hydropsalis torquata*

♂ 47 cm; ♀ 28 cm

Widespread in the N lowlands and foothills to N Patagonia; moves north in winter. **ID** Small-bodied, rufous-collared nightjar with elongated, deeply forked, black-and-white tail feathers in ♂, and closed elongated central rectrices in both sexes. Arboreal and terrestrial. **Adult ♂:** Narrow black outer rectrices, barred white, and fringed white on inner web. **Adult ♀:** Grey-brown with rufous nuchal collar extending onto sides of neck. Notably long-tailed. Unique trident-shaped tail in flight. **Voice** Common call is a series of sharp ticking *tchik* notes; flight display includes faster calls, low frog-like chattering and mechanical sounds. [Atajacaminos Tijera]

Sickle-winged Nightjar *Eleothreptus anomalus*

17.5 cm

Local in the NE. **ID** Tiny, large-headed nightjar with modified incurved outer primaries in ♂. Mainly terrestrial, sometimes arboreal. Foraging flight is very fast and low; quarters several times over same area. **Adult ♂:** Greyish with white supraloral and breast spots. Folded black primaries curve upwards. Tail tipped white. In flight, club-shaped wing with incurved primaries and cinnamon primary wing-coverts are very difficult to see in the field. Narrow white trailing edge and tail tips. **Adult ♀:** Resembles ♂ but primaries brown, barred cinnamon and not curved. Narrow buff tips to outer rectrices. Compare with Little Nightjar. In flight, proportionately large-headed. Unmodified flight feathers, narrowly barred cinnamon. **Voice** In flight high-pitched *pwick* calls and a loud metallic dry trill *zrrrrreee*. In high display flight, gives a series of 5–17 mechanical *tuk* wing-sounds, also on landing. [Atajacaminos Ala Negra]

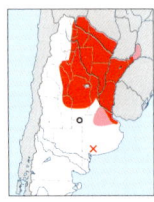

Little Nightjar *Setopagis parvula*

20 cm

N lowlands and foothills to 1300 m (mainly Sep–Apr; winters in Misiones). **ID** Small, large-headed, rufous-collared nightjar. Arboreal and terrestrial. **Adult ♂:** Striking large white gorget and rufous nuchal collar. In flight, white primary band and small white tips to all but central tail feathers. **Adult ♀:** Tawny nuchal collar and large white or buffy gorget. In flight, flight feathers and outer tail feathers finely barred cinnamon. Compare with Patagonian Nightjar in winter and Siku Nightjar above 1200 m. **Voice** A rich, liquid and fast *cluk vrEEE vri-vr-vr-vra* with a strong gurgled quality and a strange 'wood-knocking' background harmony. [Atajacaminos Chico]

Spot-tailed Nightjar *Antiurus maculicaudus*

20 cm

Rare and local in ne. Misiones understorey. **ID** Small, large-headed, rufous-collared nightjar. **Adult ♂:** Black malar wedge. Prominent buff scapular line and large spots on wing-coverts. Broad white tip to tail and 2 rows of spots on undertail. White terminal tail spots are broad on the underside. In flight, lacks a wing band. **Adult ♀** (not illustrated): Similar to ♂ but tail spots are buff. **Voice** Song is a high-pitched, fast disyllabic *ptt-touit* or trisyllabic *ptt-tsi-ii*, rising in pitch at the end, given from a low snag, log or termite mound; faster during flight display. [Atajacaminos Ceja Blanca]

Long-trained Nightjar

ad ♀

ad ♂

ad ♀

Lyre-tailed Nightjar

argentina

ad ♂

ad ♀

ad ♀

Scissor-tailed Nightjar

furcifera

ad ♀

ad ♂

ad ♀

Sickle-winged Nightjar

ad ♂

ad ♂

ad ♀

ad ♀

*

ad ♀

parvula

ad ♂

ad ♀

ad ♂

Little Nightjar

Spot-tailed
Nightjar

Systellura Sexually dimorphic, compact, open country nightjars that perch on the ground or in shrubs. *Antrostomus* Big, large-headed arboreal and terrestrial forest nightjars. No wing bands. Males have tail spots. *Nyctidromus* Notably long-tailed, sexually dimorphic forest nightjar. *Nyctiphrynus* Small-headed, arboreal nightjar with proportionally long, broad tail.

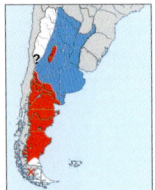

Patagonian Nightjar *Systellura [longirostris] bifasciata* 25 cm

Widespread Patagonian breeder; to N lowlands in winter. **ID** The only nightjar in most of its breeding range, but compare Little Nightjar. No overlap with Siku Nightjar. **Adult ♂ dark morph**: Blackish plumage. In flight, white primary band and large white spots on outer tail feathers. **Adult ♀ pale morph**: Pale greyish with heavily spotted wing-coverts. No tail spots. In flight, narrow cinnamon primary band. **Voice** Drawn-out whistled *TSeerruuii* rising at the end with a notable shrill quality in the middle. Series of soft metallic *pic* notes in flight. **Tax note 22**. [Atajacaminos Patagónico]

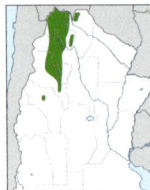

Siku Nightjar *Systellura [longirostris] atripunctata* 23.5 cm

Resident in NW Andes above 1200 m; typically in rocky grasslands at forest edge. Narrow overlap with Little Nightjar. **ID Adult ♂**: Smaller, paler and browner than Patagonian Nightjar (no overlap) with smaller white tail spots. **Adult ♀** (not illustrated): Lacks tail spots. Cinnamon primary band. **Voice** Drawn-out whistled *SHRREEuii* rising at the end; sounding more disyllabic than Patagonian and with a shrill quality at the start. Similar *pic* notes in flight. **Tax note 22**. [Atajacaminos Ñañarca]

Silky-tailed Nightjar *Antrostomus sericocaudatus* 29 cm

Resident in Paraná forest. **ID** Dark, large-headed arboreal and terrestrial forest nightjar. No wing bands. **Adult ♂**: White gorget, black breast with white spots at lower edge and on belly. Striking broad white tips to outer three tail feathers; easily viewed from below during upright perching stance or in flight. **Adult ♀**: Considerably darker than Rufous Nightjar without rufescent tones. Buffy or white gorget, white spots on abdomen and narrow buff tips to outer three tail feathers. Compare with Rufous Nightjar and Ocellated Poorwill. **Voice** Song, from a high perch, is a rich, resounding *poor-will-WEoh*. [Atajacaminos Oscuro]

Rufous Nightjar *Antrostomus rufus* 29 cm

N forested lowlands and foothills (late Aug–Apr). **ID** Big, large-headed arboreal and terrestrial forest nightjar. No wing bands. **Adult ♂**: Less rufescent than ♀ and with blackish breast. Large tawny spots on outer tail feathers from below. In flight, large white spots with tawny fringes on outer three tail feathers. **Adult ♀**: buff or white gorget, and usually buff spots on lower breast. Rows of large buff spots on wing-coverts. In flight, cinnamon tips to outer three tail feathers. Compare with Silky-tailed Nightjar and Ocellated Poorwill in Paraná forest (Misiones). A rare grey morph is also known. **Voice** Song, from a high perch, is a fast *chuck whip-wip-wrreeo*, sometimes followed by mechanical wing sounds. [Atajacaminos Colorado]

Pauraque *Nyctidromus albicollis* 30.5 cm

Common in Misiones; rare in n. Corrientes, humid chaco and Yungas forest. Terrestrial. **ID** Notably long-tailed, sexually dimorphic forest nightjar. **Typical adult ♂**: Grey crown contrasts with rufous ear-coverts. White in outer tail feathers sometimes visible at rest. In flight, white band across primaries and striking white penultimate outer tail feathers and inner web of outer rectrix. **Typical adult ♀**: Lacks white gorget and outer tail feathers of ♂. In flight, narrow cinnamon band across primaries, and small white tips to penultimate two outer tail feathers of long tail. Scarcer rufous morph has uniform rufous upperparts. **Voice** Song is a highly-inflected whistle repeated every 2–4 secs, *kwu-eAHh-o*, less often *bup bup bup bup weerrAo*. [Curiango]

Ocellated Poorwill *Nyctiphrynus ocellatus* 21 cm

Paraná forest. **ID** Small-headed, arboreal nightjar with proportionally long, broad tail. **Adult**: Dark rufous with white gorget and frosted scaling across breast suggesting second gorget. 2–3 small white scapular spots and scattering of white ocellated spots over central abdomen. White tips to outer four tail feathers. Usually seen from below flying over canopy; small and dark with relatively long, broad tail. Compare with Silky-tailed and Rufous Nightjars. **Voice** Song, from a high perch, is a quavering *whrrelll* or *brrrauu* with a vibrant quality. [Atajacaminos Ocelado]

ad ♀
pale morph

Patagonian Nightjar

ad ♀
pale morph

ad ♂
dark morph

ad ♂
dark morph

ad ♂

Siku Nightjar

ad ♂

Silky-tailed Nightjar

sericocaudatus

ad ♀

ad ♂

♀

♂

Rufous Nightjar

ad ♂

rutilus

ad ♀

♀

♂

♀

♂

derbyanus

Ocellated Poorwill

ocellatus

ad ♀
rufous
morph

ad ♂
typical

ad

Pauraque

Megascops Small nocturnal owls of forest and light woodland with ear-tufts. Identification usually depends on voice. Birds pose in alert posture with erect ear-tufts, compressed facial disk, and attenuated body, or in relaxed posture with puffed-out body and folded ear-tufts. Highly variable iris colour, especially in Hoy's, Long-tufted and Black-capped, does not aid identification, while rufous morphs tend to have dark irides.

Tropical Screech Owl *Megascops choliba* 23 cm

Widespread in N lowlands and foothills. **ID** Adult (rufous/grey morph): Prominent black facial rim of uniform width, highlighted by white. Black ventral shaft-streaks are larger but straight on upper breast and with fine herringbone pattern throughout. Overlaps with Long-tufted and Black-capped Screech Owls, and locally also with Hoy's around 1200–1400 m. **Voice** Song is a rolling trill followed by one or more (usually two) emphatic notes *burrrrrrrrrrrr CU CU*, ending abruptly with much variation. Also a more purring *bu-bu-bu-bu-bu-bu-bu*, sometimes in duet. [Alilicucu Común]

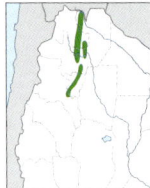

Hoy's Screech Owl *Megascops hoyi* 25 cm

Yungas forest above 1150 m; limited overlap with Tropical Screech Owl. **ID** Adult brown morph: Pale lemon-yellow eyes; dark in rare rufous morph. Narrow facial rim, broader at base but little or no white sub-border. White-spotted nuchal collar unlike Tropical. Ventral streaks thicker than Tropical but laddering ill-defined. **Voice** Long song is a fairly slow hollow trill gradually increasing in volume, lasting 8–14 secs; lower-pitched and slower than Buff-fronted Owl. Short song is a slow hooting series with a bouncing ball effect, lasting 2.5–4 secs. [Alilicucu Yungueño]

Long-tufted Screech Owl *Megascops sanctaecatarinae* 28 cm

Local in upland forest of ne. Misiones. **ID** Adult (rufous/grey morph): Larger than Tropical and Black-capped Screech Owls with larger white scapular spots. Very thick facial rim. At night underparts appear two-toned with darker base colour on breast and paler belly. Large sparse black triangles on upper breast also distinguish from Tropical and Black-capped. Notably long ear-tufts. Grey morphs also occur. **Voice** Harsh song is an almost snipe-like series of rasping notes *who who who who whowhowhowho who... who... who...who*, increasing in speed at the start, then slowing and lasting 4–6 secs. Trilled song is a uniform low-pitched series; recalls Black-capped, but shorter with a hoarse, rasping quality and lasting 4–5 secs without an obvious increase in volume. [Alilicucu Orejudo]

Black-capped Screech Owl *Megascops atricapilla* 25 cm

N. Misiones. **ID** Closely resembles Hoy's Screech Owl (no overlap) and Long-tufted. Adult grey morph: Darker than Long-tufted, usually with a blackish cap and shorter ear tufts though not diagnostic; safely identified only by voice. Facial rim tends to be narrower than Long-tufted. Smaller black triangles on upper breast than Long-tufted. Adult rufous morph: Note narrower facial rim, smaller black triangles on upper breast and shorter ear-tufts than Long-tufted. **Voice** Long song is a fast, hollow toad-like trill, lasting 7–10 secs or more, that increases gradually in volume. Compare with Long-tufted Screech Owl and Buff-fronted Owl. Short song is a soft *bub bub bub bub bub-bu-bu-bu-bu-bu-bu* in 2–3 secs, but is rarely heard. [Alilicucu Grande]

Aegolius Reclusive, earless, nocturnal owl of subtropical forest with high-pitched toad-like song.

Buff-fronted Owl *Aegolius harrisii* 21.5 cm

Yungas, chaco and Paraná forest. **ID** Adult: Yellowish to dull orange eyes. Black cap, eyebrows and narrow facial rim. Cinnamon face and underparts. Wings spotted white. Ssp. *dabbenei* (NW) and *iheringi* (NE) are similar and possibly not distinguishable. **Voice** Long song is a higher-pitched, softer and faster toad-like trill than long songs of Hoy's and Black-capped Screech Owls. Short song is a short trill and drawn-out whistle *bu-bu-bu WEEEE*, inflected upwards at the end. [Lechucita Canela]

Tropical Screech Owl

wetmorei

ad
rufous morph

ad
grey morph

ad
brown morph

Hoy's Screech Owl

ad
rufous
morph

ad
rufous
morph

Black-capped
Screech Owl

ad
brown
morph

ad
grey
morph

Long-tufted
Screech Owl

ad

dabbenei

Buff-fronted Owl

Glaucidium Small, polymorphic, nocturnal and diurnal earless owls of forest and light woodlands with streaked bellies. Best distinguishing features include pattern of crown, tail and sides of breast. All species show false eyespots on the hindcrown.

Austral Pygmy Owl *Glaucidium nana* 18 cm

Patagonian forest and sparsely in Patagonian steppe. No definite overlap with Ferruginous Pygmy Owl. **ID Adult rufous morph:** Crown streaked as is usual in rufous-morph congeners. White or buffy spots on sides of breast and flanks, usually more prominent than in Ferruginous. Rufous tail with 9–11 black bars. **Adult brown morph:** Far more numerous than rufous morph. Crown warmer brown than mantle and nearly always streaked; less commonly spotted at rear. Spots on breast and flanks usually more prominent than Ferruginous. Tail as rufous morph, distinguishing it from Ferruginous. Scarce grey morph resembles brown morph on plumage features. **Juvenile:** Poorly marked crown, reduced supercilium and lacks breast spotting. **Voice** A long series of tooting notes *wükwükwükwükwükwük*, higher-pitched, faster-paced (18–20 notes in 5 secs) and less resonant than Ferruginous Pygmy Owl. [Caburé Austral]

Ferruginous Pygmy Owl *Glaucidium brasilianum* 17 cm

Widespread in N and C lowlands and foothills. **ID Adult brown morph:** Crown finely spotted, streaked or with a combination of both. Sides of breast rarely with ill-defined spots. Blackish tail with 4–6 narrow white spot-bars. **Adult rufous morph:** Finely streaked crown. Tail either bright uniform rufous or rufous with 7–8 indistinct dusky or pale brown bars. **Adult variant:** Tail suggestive of Austral Pygmy Owl, but proportionately shorter with fewer brown bars; breast spotting restricted or absent; presence of spots on forecrown eliminates Austral. **Voice** Song is a series of quavering notes followed by a fast series of twangy or tooting whistles, *prrrrrlíu… fufufufufu….* Tooting resembles that of Austral Pygmy Owl but is lower-pitched, more resonant and slower-paced (13–15 notes in 5 secs). [Caburé Chico]

Yungas Pygmy Owl *Glaucidium bolivianum* 17 cm

Yungas cloud forest, usually above 1650 m. Overlap with Ferruginous Pygmy Owl uncertain. **ID Adult:** Fairly large white crown spots, ocellated in black. Poorly-defined supercilium. Mantle distinctly spotted white unlike Austral and Ferruginous Pygmy Owls. More prominent white spots on sides of breast than Ferruginous. **Voice** Song is a series of 2–3 burry notes falling in pitch towards the end followed by a series of mellow slow tooting whistles, *weeeurrrr… hu hu hu hu hu hu….* (8–9 notes in 5 secs). [Caburé Yungueño]

Athene Diurnal and nocturnal earless owl of open country. Nests in burrows.

Burrowing Owl *Athene cunicularia* 21–26 cm

Widespread. **ID Adult *partridgei*** (24 cm; widespread, illustrated): Prominent white supercilium and facial rim. Wings profusely spotted white and 4 white tail bands. **Adult *juninensis*** (*c.*26 cm; altiplano, not illustrated): much warmer brown. **Adult *grallaria*** (*c.*21 cm; n. Misiones, not illustrated): notably smaller, darker and shorter-tailed than *partridgei*, and more heavily spotted white on the sides of the breast. **Voice** A scolding note and a fast chatter *cheeshhh che-che-che-che-che-che-che-che*. Song at night is an inquisitive *h hWooo*. [Lechucita Vizcachera]

ad rufous morph

Austral Pygmy Owl

juv

ad brown morph

hindcrown

ad brown morph

ad brown morph

ad variant

ad rufous morph

brasilianum

Ferruginous Pygmy Owl

ad

ad

ad

partridgei

Yungas Pygmy Owl

Burrowing Owl

Asio Medium-sized with ear-tufts of varying length. Voices comprise hoots, whistles, screams and hisses. *Bubo* Very large owls with prominent ear-tufts and large white gorget.

Striped Owl *Asio clamator*   41 cm

N woods and forest, often marshy, to 1300 m in Yungas forest. **ID** Adult: Large erectile black ear-tufts. White face outlined with black rim. White below with dense black streaking on breast and long, fine streaks on the belly and flanks. **Juvenile:** Mainly buff to ochre with fine brown dorsal barring. Rufous-brown facial disk with striking white moustachial and forehead. Overlaps with Stygian and Great Horned Owls; also with Chaco, Rusty-barred, Mottled and Black-banded Owls (Plate 94). **Voice** Explosive hooting *WÚu*, much like Stygian Owl but longer and fading. Puppy-like series of barks and an eerie descending whistle. [Lechuzón Orejudo]

Stygian Owl *Asio stygius*   42 cm

Humid chaco, Paraná and Yungas forest to 1700 m. **ID** Adult: Erectile black ear-tufts. Sooty-brown with a creamy abdomen with blotchy dark brown streaking. Overlaps with Striped and Great Horned Owls; also with Rusty-barred, Mottled and Black-banded Owls (Plate 94). **Voice** Explosive hooting *WÚ*, similar to Striped Owl, but shorter and ending abruptly. [Lechuzón Negruzco]

Great Horned Owl *Bubo virginianus*   51 cm

Widespread in woodlands of N and C lowlands. **ID** Adult: Orange iris. Very large erectile eartufts. Coarser ventral barring than Magellanic Horned Owl. **Voice** Very deep hooting *bu...bu bu bu... búuuu...*, often in duet with second bird higher-pitched. [Ñacurutú]

Magellanic Horned Owl *Bubo [virginianus] magellanicus*   45 cm

Patagonian, Andean and Sierran shrub-steppe. **ID** Adult: Yellow iris. Small ear-tufts compared to Great Horned Owl. Dense ventral barring. **Voice** Deep staccato introduction followed by rapid quavering *tu-cú...cucurrr...* very different from Great Horned Owl. Tax note 23. [Tucúquere]

Short-eared Owl *Asio flammeus*   37 cm

Widespread in open country; irruptive. Diurnal and nocturnal. **ID** Medium-sized with very short ear-tufts. Adult *suinda* (mainland): Yellow iris. Short ear-tufts rarely visible. Buff facial disc. Black dorsal streaks and on upper breast. Adult *sanfordi* (Falklands): Brighter than *suinda* with a rich cinnamon plumage. **Voice** Common call is a nasal *wék...* or *wâk...* Song is a fast, huet-huet like series of 14–19 low-pitched notes *kwu-kwu-kwu-kwu...* rising slowly in pitch, ending abruptly. [Lechuzón de Campo]

Tyto Cosmopolitan nocturnal owl well adapted to open country and forested habitats.

American Barn Owl *Tyto furcata*   36 cm

Widespread. **ID** Adult: White heart-shaped facial disc. Greyish to cinnamon above and white or buff below. In flight, gives ghostly white impression. **Voice** Hisses, screeches and long clicking series. Tax note 24. [Lechuza de Campanario]

Striped Owl

ad

juv

midas

Great Horned Owl

ad

nacurutu

Stygian Owl

ad

stygius

Magellanic Horned Owl

ad

American
Barn Owl

ad

Short-eared Owl

ad

suinda

ad

tuidara

sanfordi

Strix Large, earless, nocturnal forest owls. Vocal repertoire includes simple stereotyped phrases, often in duet.

Chaco Owl *Strix chacoensis* 40 cm

Monte desert, dry and sierran chaco. **ID Adult:** Whitish facial disc with black rim. Barred underparts with rufous flanks and vent. 4–5 buff or creamy bars on uppertail and 3–4 bars on undertail. **Voice** Characteristic call is a deep rasping *currrrrr-currrrr* repeated every 3–10 secs. Less frequently heard song is an accelerating/decelerating dry *cucucrucrucrucruCRU currru CRÚ crru-crru.* [Lechuza Chaqueña]

Rufous-legged Owl *Strix rufipes* 38 cm

Patagonian forest. **ID Adult:** Resembles Chaco Owl (no overlap) but facial disc rufescent, and thighs and vent vaguely washed cinnamon. 6–8 very narrow buff bars on uppertail and 5–6 bars on undertail. **Voice** Song is a long series *Juou...Juou...jou...jol...col...col...jol...juou,* recalling an excited chimpanzee, especially when duetting. [Lechuza Austral]

Rusty-barred Owl *Strix hylophila* 36 cm

Paraná forest. **ID Adult:** Cinnamon facial disc with blackish rim. White throat and brown pectoral band distinctive at night. White belly scalloped with black bars. Overlaps with Mottled and Black-banded Owls; also with Stygian and Striped Owls (Plate 93). **Voice** Song is a deep, loud, resonant series of 4–5 rising *brr* notes, followed by 3–4 descending or more drawn-out monotone *brrr-brrer* couplets. Isolated single or double *brru* notes are heard commonly. [Lechuza Listada]

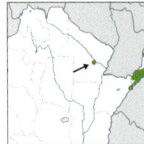

Mottled Owl *Strix virgata* 36 cm

Misiones; very local in ne. Corrientes and e. Chaco. **ID Adult:** Dark brown facial disk with long creamy or white eyebrows connecting with narrow creamy facial rim. Buff below with long black streaks. Breast mixed brown and white. Overlaps with Rusty-barred and Black-banded Owls; also with Stygian and Striped Owls (Plate 93). **Voice** Song is a monotone pumping series of 4–7 slightly disyllabic muffled notes, *kwo kWO kWO kWO kwo.* [Lechuza Estriada]

Black-banded Owl *Strix huhula* 40 cm

Yungas forest (to 1200 m) and Paraná forest in n. Misiones. **ID Adult** *albomarginata* (n. Misiones): Unmistakable finely-barred black owl with striking orange-yellow bill and bright yellow toes. **Adult** *huhula* (Yungas forest; not illustrated): Smaller with broader ventral barring. **Voice** Song is a rather weak *wu-wu-wu-w...WU,* with a clear pause before the final, louder note. [Lechuza Negra]

Pulsatrix Large earless forest owls with prominent eyebrows and pectoral band. Little eye-shine at night. Unique resonant tremulous voices are similar in both species.

Tawny-browed Owl *Pulsatrix koeniswaldiana* 42 cm

Paraná forest. **ID Adult:** Resembles Spectacled but eyebrows always cinnamon. 4–5 narrow white tail bands. Abdomen tends to be more ochraceous and flanks often barred brown. **Juvenile:** White head with black ocular region. **Voice** Song is very similar to that of Spectacled, but faster and less guttural *gku-gku-gku-gku-gku-gk*u. Local overlap with Spectacled in Misiones. [Lechuzón Mocho Chico]

Spectacled Owl *Pulsatrix perspicillata* 49 cm

Yungas foothill forest (illustrated) and humid chaco; local in n. Misiones. **ID Adult:** White eyebrows and lores form a distinctive X-mark over blackish facial disc. 8–9 pale tail bands. **Juvenile:** White head with black heart-shaped facial disc, turning pale cream on belly. Yellow iris. **Voice** Song is very similar to Tawny-browed, a slow guttural and muffled series of bouncing notes *gu-gu-gu-gu-gu-gu* becoming weaker and lower-pitched towards the end, but variable. Recalls a wobbling metal sheet. Duets frequently, with female singing at a slightly lower-pitch. Local overlap with Tawny-browed in Misiones. [Lechuzón Mocho Grande]

Chaco Owl

Rufous-legged Owl

rufipes

ad

ad

ad

Rusty-barred Owl

borelliana

ad

albomarginata

ad

Mottled Owl

Black-banded Owl

juv

ad

Spectacled Owl

ad

Tawny-browed Owl

juv

boliviana

Cypseloides Chunky swifts with long, fairly broad wings and a broad squarish tail; often associate with waterfalls for breeding and roosting, where they cling upright. Vocalisations are slow chatters; all species are very similar. ***Streptoprocne*** Giant swifts with a white collar. Breed and roost at waterfalls. Shrill and squeaky vocalisations.

Rothschild's Swift *Cypseloides rothschildi* 15 cm

Yungas forest and nearby fallow fields (mainly Oct–Mar); occasional in the N lowlands. **ID Adult:** Slightly paler and larger than Sooty Swift (no overlap), with pale primary shafts in optimum light. [Vencejo Pardo]

Sooty Swift *Cypseloides fumigatus* 14 cm

Local in Paraná forest (mainly Aug–Mar). **ID Adult:** Blackish-brown body, brown wings with blacker underwing-coverts. Size difficult to judge but considerably smaller than Great Dusky Swift. No overlap with Rothschild's Swift. [Vencejo Negruzco]

Great Dusky Swift *Cypseloides senex* 18 cm

Paraná forest. **ID Adult:** Brown with blackish underwing-coverts. Reflective whitish crown imparts a pale-headed appearance (unlike Sooty Swift). Throat varies from brown to buff. Pale primary shafts are sometimes visible. [Vencejo de Cascada]

White-collared Swift *Streptoprocne zonaris* 21.5 cm

Sep–Apr in the NW Andes (to 3000 m), C sierras and Paraná forest. **ID Adult:** Mostly brown with a blacker back and rump, and complete white collar (can appear broken on perched birds). Notched tail tip. [Vencejo de Collar]

Biscutate Swift *Streptoprocne biscutata* 21 cm

Very rare in Paraná forest. **ID Adult:** White collar, broken at the sides (can appear broken on perched White-collared Swift), is broader on the nape and much broader on the breast; pointing towards the chin and not downwards as in White-collared. Note the contrasting greyish forehead, sometimes extending to the chin unlike White-collared, and the less notched tail with rounded corners. [Vencejo Nuca Blanca]

Fork-tailed Palm Swift *Tachornis squamata* 12.5 cm

Vagrant (?) recorded in n. Misiones. Usually associates with palms. **ID** Sleek-bodied and narrow- and bow-winged swift with bifurcated tail, usually held in a point. **Adult:** Brown above, barred and mottled dusky below except on whitish throat and central abdomen. [Vencejo Palmero]

Rothschild's Swift

Sooty Swift

Great Dusky Swift

adults

zonaris

White-collared Swift

Biscutate Swift

Fork-tailed Palm Swift

biscutata

semota

Chaetura Small swifts with cigar-shaped bodies, and a blunt squarish or wedge-like tail when banking, with short protruding spines. Little or no visible carpal bend. Rapid wingbeats. ***Aeronautes*** Slender swifts of the Andes with proportionately very long narrow wings and a long deeply-forked tail (pointed when closed). Highly agile and acrobatic in small to large flocks.

Sick's Swift *Chaetura meridionalis* 12 cm

N forests from late Aug–Mar. **ID** Adult: Pale brown rump and tail contrast with dark brown upperparts. From below, whitish throat contrasts with brown body and underwing. Also differs from Chimney by vent being paler than belly (visible in good light) and penultimate primary slightly longer than outermost. **Voice** Flight calls are high-pitched metallic chatters *trripip trripip trrip trri trri*, similar in quality to those of Chimney Swift but generally slower. Often a succession of unevenly patterned rich *chit* notes but overall slower at the start, faster in the middle, and slower at the end. Lacks the shrill quality of Grey-rumped. [Vencejo de Tormenta]

Chimney Swift *Chaetura pelagica* 10 cm

Vagrant to the NW Andes; most likely over open ground. **ID** Adult: Rather uniform coloration above with a slightly warmer brown rump can distinguish this species from Sick's Swift in good light. From smaller Sick's Swift by concolorous or darker vent than belly. On spread wings, outer two primaries are of equal length, unlike Sick's. **Voice** Flight calls comprise a series of high-pitched metallic chatters *trripip trripip trrip trri trri*, similar in quality to those of Sick's Swift but generally faster. [Vencejo de Chimenea]

Grey-rumped Swift *Chaetura cinereiventris* 11.5 cm

Resident in Paraná forest. **ID** Adult: Blue-black crown, back, upperwing and tail; all with a green sheen in favourable light. Extensive pale grey rump. Pale grey throat and upper breast, becoming darker ashy-grey on abdomen and blackish on vent. **Voice** Flight calls include fairly high-pitched *chip* notes and notably fast high-pitched *weerrrr weerrrr-whiiit* or *shree-shreee-shreee* trills that are shriller than Sick's Swift. [Vencejo Chico]

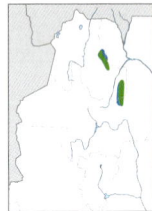

White-tipped Swift *Aeronautes montivagus* 12.5 cm

Resident in Yungas forest; seemingly lower in winter. Sometimes flocks with Rothschild's Swift. **ID** Adult: Blackish-brown with long, narrow wings and short, but appreciably forked tail when splayed, narrowly tipped white. White throat and upper breast. Blackish-brown below with white flank spots sometimes joining in a line across the belly. **Voice** Flight calls are very rapid trills terminating in short insect-like chatters. [Vencejo Montañés]

Andean Swift *Aeronautes andecolus* 14.5 cm

N and C Andes to N Patagonia, and C sierras. **ID** Adult: Brown with a broad white rump and usually a narrow white collar. Narrow white trailing edge to secondaries. White underparts contrast with brown underwing and blackish vent and undertail. **Voice** Flight calls involve a rapid series of very high-pitched, distinctly scratchy trills, sometimes descending somewhat in pitch. [Vencejo Blanco]

Sick's Swift

Chimney Swift

montivagus

Grey-rumped Swift

White-tipped Swift

andecolus

Andean Swift

Colibri Large, long-billed hummingbirds with glittering cheek flags. *Oreotrochilus* Sexually dimorphic hummingbirds with long, slightly drooped bills and green-throated males.

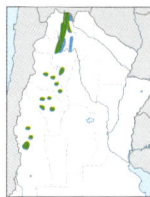

Sparkling Violetear *Colibri coruscans*   14 cm

Mainly in the NW Andes at 1500–3500 m. **ID Adult**: Glittering violet-blue throat and cheek flag. Blue belly patch, and undertail-coverts edged white. **Adult ♀** (not illustrated): Slightly smaller and duller. **Voice** Song is a monotonous series of metallic *teep* notes. Spectacular display flight with complex gurgling song, intermixed with electronic notes. [Colibrí Grande]

White-vented Violetear *Colibri serrirostris*   12.5 cm

Scarce in clearings and edge of dry chaco–Yungas forest intergrade; sparsely in the NE. **ID Adult**: Glittering magenta cheek flag, and glittering green throat. Bluish breast and white belly and vent. **Adult ♀** (not illustrated): Slightly smaller and duller. **Voice** Song is a tireless *tst-TSIT-tsu.... tst-TSIT-tsu...*, with variation; strongly recalls a warbling finch song. [Colibrí Mediano]

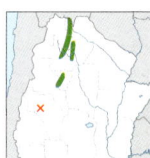

Blue-capped Puffleg *Eriocnemis glaucopoides*   10.8 cm

Yungas forest above 1000 m. **ID** Straight-billed hummingbird with prominent white leg puffs and forked blue-black tail. **Adult ♂**: Glittering blue forecrown. Glittering green throat and breast, and blue vent. **Adult ♀**: Cinnamon throat and breast. Turquoise vent. [Picaflor Frente Azul]

Andean Hillstar *Oreotrochilus estella*   13 cm

NW Andes above 2000 m. **ID Adult ♂**: Chestnut-brown stripe on abdomen (can look black depending on the light). **Adult ♀**: Throat spotting is thicker than White-sided Hillstar and tends to form lines. Outer tail feather is comparatively broader than White-sided. Local overlap with White-sided in the NW. [Picaflor Puneño]

White-sided Hillstar *Oreotrochilus leucopleurus*   12.5 cm

Locally throughout the Andes. **ID Adult ♂**: Blue-black stripe on abdomen. **Adult ♀**: Throat speckling finer than in Andean Hillstar. Narrow outer tail feather. Local overlap with Andean Hillstar in the NW. [Picaflor Andino]

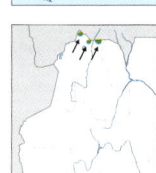

Wedge-tailed Hillstar *Oreotrochilus adela*   ♂ 13 cm; ♀ 12.5 cm

Very local in n. Jujuy. **ID Adult ♂**: Chestnut breast and belly a broad black neck tie. **Adult ♀**: Throat speckled green giving a streaked effect. Cinnamon underparts. [Picaflor Colorado]

Sparkling Violetear

coruscans

ad

Blue-capped Puffleg

ad ♀

ad ♂

White-vented Violetear

ad

ad ♂

ad ♀

Andean Hillstar

boliviana

ad ♂

ad ♀

White-sided Hillstar

ad ♂

ad ♀

Wedge-tailed Hillstar

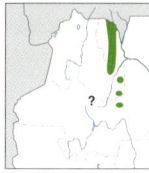

Yungas Speckled Hummingbird *Adelomyia [melanogenys] inornata*   8.6 cm

N Yungas forest. **ID** Small, straight-billed, brownish-green hummingbird. Sexes alike. Adult: Blackish mask and white post-ocular streak. Blue-speckled throat and sides of breast. Cinnamon tail tips. **Voice** Very fast monotonous series of shrill *tsip* notes inside forest. Tax note 25. [Picaflor Yungueño]

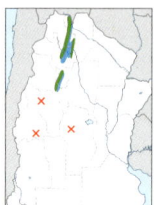

Slender-tailed Woodstar *Microstilbon burmeisteri*   ♂ 7.8 cm; ♀ 6.3 cm

Yungas forest. **ID** Tiny, straight-billed, forest hummingbird with a white flank spot. Insect-like flight. Adult ♂: Elongated outer tail feathers with a bulbous tip. Glittering magenta chin and pointed moustachial spikes. Adult ♀: Long buff post-ocular streak. Cinnamon underparts. Square-ended chestnut tail with green central tail feathers and a black subterminal band. **Voice** Nasal, mechanical *peenk* sound produced by tail in display flight. [Picaflor Enano]

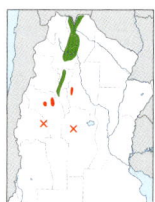

Yungas White-bellied Hummingbird *Elliotomyia [chionogaster] hypoleuca*   10.8 cm

Yungas forest to 1900 m. **ID** Rather nondescript forest hummingbird with a long, almost straight, bill. Sexes alike. Adult: Red lower mandible with a dusky tip. Bronze-brown crown contrasts with greener back. White below with green sides of the breast. White inner webs and tips of outer tail. **Voice** Endless series of monotonous notes from an exposed perch that accelerate and decelerate towards the end, *tsi tsi tsi tsi... tsitsitsitsi tsi... tsi tsi tsi.*. Tax note 26. [Picaflor Vientre Blanco]

Red-tailed Comet *Sappho sparganura*   ♂ 17.4 cm; ♀ 12.2 cm

N Andes and C sierras. **ID** A spectacular hummingbird with a long graduated, forked tail and short straightish bill. Adult ♂: Rosy-purple central mantle and rump. Uppertail glittering fiery-orange with black feather tips. Adult ♀: Green mantle and rosy-purple rump. Tail shorter than ♂ without black tips. Throat and breast spotted green. **Voice** Song is a short metallic, spluttered 'typewriter' chatter. Frequent call during chases is a series of high-pitched notes followed by a descending trill, *tee-tee-trreew...* [Picaflor Cometa]

Green-backed Firecrown *Sephanoides sephaniodes*   11 cm

Patagonian forest. **ID** Small hummingbird with a short, straight bill, feathered at the base. Adult ♂: Green with a speckled throat and breast. Glittering orange-red crown. Adult ♀: Brownish-olive crown. **Voice** Extremely high-pitched trill given in display. [Picaflor Rubí]

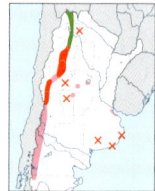

Giant Hummingbird *Patagona gigas*   20.5 cm

N and C Andes above 2500 m. **ID** Outstandingly large, slow-flying hummingbird with a long straight bill, feathered at the base. Adult ♂ *gigas*: Mainly brown, washed green above with contrasting white rump and vent. Adult ♀ (not illustrated): Lacks the green wash above, and can show a speckled throat. Adult *peruviana* (NW Andes; not illustrated) is slightly larger and rufous below. **Voice** Common call is a piercing *SIP* whistle given in flight or while perched. [Picaflor Gigante]

Yungas Speckled Hummingbird

ad

Slender-tailed Woodstar

ad ♂

ad ♀

ad

Yungas White-bellied Hummingbird

ad ♂

ad ♀

ad ♂

ad ♀

sappho

Green-backed Firecrown

Red-tailed Comet

ad ♂

gigas

Giant Hummingbird

Phaethornis Forest hummingbirds with long, slightly decurved bills, striped heads and long graduated tails. *Heliomaster* Large hummingbirds with straightish bills up to twice the length of the head.

Scale-throated Hermit *Phaethornis eurynome* 13.4 cm

Paraná forest. Overlaps with Planalto Hermit. **ID** Adult: Bill straighter than Planalto. Blackish centre of throat and drab grey underparts with a cinnamon belly and vent. Coppery rump. ♀ has somewhat shorter bill. **Voice** Song is a repeated, very high-pitched *swit cher-Wii*, sometimes transcribed as *keep... your-seat*, with variations. [Ermitaño Escamado]

Planalto Hermit *Phaethornis pretrei* 15.3 cm

Foothill Yungas forest and Misiones. **ID** Adult: Creamy throat and entirely rich cinnamon underparts. Contrasting rufous rump. ♀ has somewhat shorter bill. **Voice** Song is a repetitive, very high-pitched *Swit chu-wer*, with variations. [Ermitaño Canela]

Black-throated Mango *Anthracothorax nigricollis* 12 cm

Scarce in the NE. **ID** Chunky hummingbird with slightly decurved bill. Adult ♂: Black underparts bordered turquoise on sides of throat. Purple tail. Adult ♀: White underparts with black central stripe. Tail narrowly tipped white. [Picaflor Vientre Negro]

Long-billed Starthroat *Heliomaster longirostris* 11.6 cm

Unconfirmed in N lowlands; beware of moulting male Blue-tufted Starthroat. **ID** Adult ♂: Pale blue cap. Glittering violet throat highlighted by white malar. Adult ♀: Prominent white malar. Dusky speckled throat. White stripe on flanks and white tail corners. [Picaflor Picudo]

Stripe-breasted Starthroat *Heliomaster squamosus* 11.5 cm

Rare vagrant in the NE. **ID** Adult ♂: Glittering violet gorget and scarf. Sides of breast and flanks vivid green with contrasting narrow white median line reaching the vent. Long, deeply forked, dark green tail. Adult ♀: From ♀ Long-billed (no known overlap) by less prominent white malar streak, narrower white centre to the abdomen and pale green flanks. Outer tail feathers tipped white. [Picaflor Garganta Escamada]

Blue-tufted Starthroat *Heliomaster furcifer* 12.7 cm

N and C lowlands and foothills. **ID** Adult ♂: Iridescent turquoise-green crown. Glittering magenta throat extending into pointed blue cheek tufts. Deep blue underparts. Deeply forked, pointed tail. ♂♂ moulting into adult/breeding plumage may resemble Long-billed. Adult ♀: White post-ocular spot unlike ♀ woodnymphs (Plate 100). Forked green tail with black subterminal band and white tips. White below, broadly spotted green on sides of breast and flanks. [Picaflor de Barbijo]

Scale-throated Hermit

ad

Planalto Hermit

ad

araguayensis

ad ♂

ad ♀

nigricollis

Black-throated Mango

ad ♀

Long-billed Starthroat

longirostris

ad ♂

Blue-tufted Starthroat

ad ♀

ad ♀

ad ♂

ad ♂

ad ♀

ad ♂

Stripe-breasted Starthroat

Florisuga Unmistakable sturdy hummingbirds with straightish bills. *Thalurania* Forest hummingbirds with straightish bills and forked tails.

White-necked Jacobin *Florisuga mellivora* 10 cm

Rare vagrant known from ne. Buenos Aires. **ID Adult** ♂: Blue hood with a broad white nuchal band. Green flanks and upperparts. White outer tail feathers tipped black. **Adult** ♀: Drab green above often with the suggestion of a dusky facial mask. Throat, breast and vent greenish scaled white; contrasting unmarked white belly. Outer tail feathers tipped white. [Picaflor Nuca Blanca]

Black Jacobin *Florisuga fusca* 12 cm

Misiones (especially in winter) and sparse winter records in the NE. **ID Adult**: Mainly black with white flank patch and tail. **Juvenile/immature**: Browner, with broad rufous malar and less white in tail. [Picaflor Negro]

Swallow-tailed Hummingbird *Eupetomena macroura* ♂ 17.5 cm; ♀ 15.3 cm

Resident at Iguazú, n. Misiones; occurring sparsely southwards. Sexes similar. **ID** Outstandingly large slow-flying hummingbird with a very deeply forked tail. **Adult** ♀: Bluish-violet head and breast. Otherwise green with blue tail. **Adult** ♂: As ♀ but much longer tail. [Picaflor Tijereta]

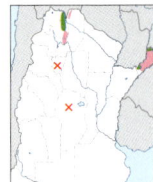

Fork-tailed Woodnymph *Thalurania furcata* 10.7 cm

Local in NW Andean foothills and Misiones. **ID Adult** ♂: Glittering green throat and violet-blue breast and belly. **Adult** ♀: Straightish bill. Brownish crown contrasts with green upperparts. Lacks green on the sides of the breast. Overlaps with Violet-capped in Misiones. [Picaflor Zafiro]

Violet-capped Woodnymph *Thalurania glaucopis* 10.9 cm

Paraná forest. **ID Adult** ♂: Glittering green throat and breast. Violet crown. **Adult** ♀: Straight bill. Green crown concolorous with back. Usually shows green on sides of breast. [Picaflor Corona Violácea]

Purple-crowned Plovercrest *Stephanoxis loddigesii* 9.5 cm

Paraná forest. **ID** Straight-billed hummingbird with white tail corners. **Adult** ♂: Spectacular, glittering violet-blue spike-like crest and patch on abdomen. **Adult** ♀: Green above and greyish-white below. Notable white tail corners and post-ocular spot which is often triangular. Compare with ♀ Glittering-bellied Emerald (Plate 101) and female woodnymphs. **Tax note 27**. [Picaflor Copetón]

White-necked Jacobin

ad ♀

ad ♂

Black Jacobin

ad

juv/imm

Swallow-tailed Hummingbird

ad ♀

macroura

Fork-tailed Woodnymph

ad ♂

ad ♀

Violet-capped Woodnymph

ad ♀

ad ♂

Purple-crowned Plovercrest

ad ♂

ad ♀

Glittering-bellied Emerald *Chlorostilbon lucidus* 9 cm

Widespread in the N and C lowlands and Yungas forest. **ID** Small hummingbird with straight bill; red with a black tip in ♂; more variable in ♀ but always with some pink at the base. Adult ♂ *lucidus* (illustrated; E of Paraná River and southwards): Entirely green with a forked blue tail. Adult ♂ *igneus* (W of Paraná River; not illustrated) differs by golden discs on the belly. Adult ♀: Bill with broader dusky tip than ♂ and often a dark upper mandible. Dusky mask and white post-ocular spot or streak. White tail tips. Greyish underparts. **Voice** A series of dry monotone, cricket-like trills; also softer brief chatters. [Picaflor Verde]

Gilded Sapphire *Hylocharis chrysura* 9.9 cm

N and C lowlands. Sexes alike. **ID** Small hummingbird with straight bill; red with a black tip in ♂; more variable in ♀ but always with some pink at the base. Adult: Orange-red bill with a black tip. Bronze-green above and brownish-olive below becoming buff on the belly. Rufous chin suggests ♀ Rufous-throated but tail is golden-bronze. **Voice** Very high-pitched rapid series of insect-like trills *tri-tri-tri-tri-tri…* [Picaflor Bronceado]

Rufous-throated Sapphire *Hylocharis sapphirina* 8.4 cm

Sparse records in NE forest. **ID** Small hummingbird with straight bill; red with a black tip in ♂; more variable in ♀ but always with some pink at the base. Adult ♂: Orange chin. Glittering violet-blue throat, green below. White leg puffs. Tail lacks white tips. Adult ♀: Pink lower mandible tipped dusky. Orange chin contrasts with mainly white underparts. Metallic chestnut tail with white tips on outer tail feathers. [Picaflor Cola Castaña]

White-chinned Sapphire *Chlorestes cyana* 8.2 cm

Vagrant to the NW (*conversa*; not illustrated). Hypothetical in the NE (*griseiventris*). **ID** Small hummingbird with straight bill; red with a black tip in ♂; more variable in ♀ but always with some pink at the base. Adult ♂: Glittering violet-blue head, throat and breast. Reddish uppertail-coverts. Chin speckled white. Adult ♀: Pink lower mandible. Greyish tail tips. Best distinguished from ♀ Glittering-bellied by reddish uppertail-coverts. [Picaflor Lazulita]

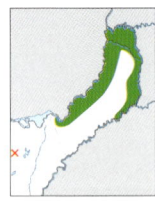

Versicoloured Emerald *Chrysuronia versicolor* 8.8 cm

Paraná forest. **ID** Small hummingbird with an almost straight bill, as long as the head. Sexes are very similar. Adult ♂: Reddish lower mandible tipped black. Glittering green throat, often mixed blue at the sides. White centre of abdomen (rarely up to chin) and olive sides of breast and flanks. Indistinct dusky subterminal tail band. Adult ♀ (not illustrated): Duller throat and indistinct greyish-brown tail corners. [Picaflor Esmeralda]

Sapphire-spangled Emerald *Chionomesa lactea* 9.5 cm

Rare in n. Misiones. **ID** Adult ♂: Pink lower mandible. Violet-blue throat and breast. Green flanks divided by white line. Adult ♀ (not illustrated): Breast mixed with grey, and greyish tail tips. [Picaflor Pecho Azul]

ad ♂

ad ♀

lucidus

Glittering-bellied Emerald

ad

Gilded Sapphire

ad ♂

ad ♀

Rufous-throated Sapphire

ad ♀

ad ♂

griseiventris

White-chinned Sapphire

ad ♂

Versicoloured Emerald

ad ♂

lactea

Sapphire-spangled Emerald

Festive Coquette *Lophornis chalybeus*
♂ 8.1 cm; ♀ 7.8 cm

Unconfirmed; sparse records in the NE. **ID** Tiny straight-billed hummingbird with a white or buffy rump band and insect-like flight. **Adult** ♂: Glittering green forecrown and large green scarf, spotted white. **Adult** ♀: Buffy-white malar streak. Underparts mixed with black. [Coqueta Verde]

Frilled Coquette *Lophornis magnifica*
*c.*6.9 cm

Unidentified ♀ coquettes recorded in Misiones could refer to this species. **ID** Tiny straight-billed hummingbird with a white or buffy rump band and insect-like flight. **Adult** ♀: Cinnamon forehead, lores and border to olive-speckled throat. Buff tail tips. **Adult** ♂ (not illustrated): Red bill tipped black, long erectile orange crest, scaled scarf with broad white feathers tipped black, and black-green throat and forehead. [Coqueta Naranja]

Titan Sphinx *Aellopos titan*
6.5cm

A widespread hawkmoth that looks and flies remarkably like coquettes. Hovers at flowers and makes fast darting flights.

Amethyst Woodstar *Calliphlox amethystina*
♂ 8.8 cm; ♀ 7.4 cm

Very rare (possible breeder) in n. Misiones. **ID** Small straight-billed hummingbird with a white flank spot. **Adult** ♂: Long blackish-purple tail with pointed tips. Amethyst gorget, bordered by a white pectoral band. **Adult** ♀: Square black tail with cinnamon tips. Lower throat speckled green. White post-ocular streak and pectoral band. Rich cinnamon flanks. [Picaflor Amatista]

White-throated Hummingbird *Leucochloris albicollis*
11.4 cm

NE lowlands; expanding to N Patagonia. **ID** Straight-billed lowland hummingbird. Often perches very high. Sexes alike. **Adult**: White throat and belly and broad green pectoral band. **Voice** Song is a rapid, springy and nasal series *zrrrriii zri-zri-zri-zri-zri* or *cheee che-che-che-che-che-che....* [Picaflor Garganta Blanca]

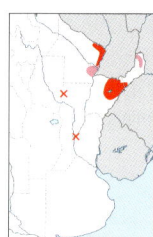

White-tailed Goldenthroat *Polytmus guainumbi*
10.7 cm

Scarce in the NE. **ID** Marsh hummingbird with slightly decurved bill and extensive white tail flashes. **Adult** ♂: Bronze-green above with a whitish post-ocular streak. Glittering green underparts. **Adult** ♀: Bronze-green above with a whitish post-ocular streak and malar. Buff below, spotted green on throat and breast. [Picaflor de Antifaz]

Ruby-topaz Hummingbird *Chrysolampis mosquitus*
9.2 cm

Vagrant to nw. Misiones. **ID** Small hummingbird with straightish bill, prominently feathered at the base. Fans the tail. **Adult** ♂: Brown with a chestnut tail. In favourable light shows a glittering ruby crown and orange throat and breast. **Adult** ♀: Brownish crown, olive mantle and off-white below. Tail has green central rectrices and chestnut outer tail feathers with a violet subterminal band and white tail corners. [Picaflor Topacio]

ad ♀

Festive Coquette

ad ♂

ad ♀

Frilled Coquette

chalybeus

Titan Sphinx

ad ♂

ad ♀

Amethyst Woodstar

ad

White-throated
Hummingbird

ad ♂

ad ♀

ad ♀

thaumantias

ad ♂

White-tailed Goldenthroat

Ruby-topaz
Hummingbird

Trogon Large robust, square-tailed sub-canopy birds with metallic plumage in males. Voices comprise loud resonant notes delivered in series.

Blue-crowned Trogon *Trogon curucui* 27.5 cm

N Yungas forest and locally in ne. Formosa where it overlaps with Surucua Trogon. **ID** Adult ♂: From ♂ Surucua by barred outer tail feathers and olive upper surface of tail, tipped black, and by whitish pectoral band. **Adult** ♀: From ♀ Surucua Trogon by finely barred outer tail feathers, black upper surface of central rectrices, and whitish pectoral band. **Voice** Very much like Surucua Trogon. [Surucuá Aurora]

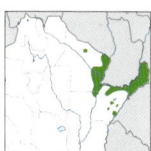

Surucua Trogon *Trogon surrucura* 27.5 cm

NE lowlands. Overlaps with Black-throated Trogon in Misiones and with Blue-crowned in ne. Formosa. **ID** Adult ♂: Turquoise central tail feathers and pure white outer tail feathers. **Adult** ♀: Dark brown central rectrices, tipped black, outer tail feathers edged and tipped white. **Voice** ♂ song is a long and fairly rapid series (4 notes per sec) of usually 15–20 tooting notes that starts softly and increases in volume. ♀ gives a softer and slower monotone series. [Surucuá Cola Blanca]

Black-throated Trogon *Trogon [rufus] chrysochloros* 28 cm

Paraná forest. Overlaps with Surucua Trogon. **ID** The only yellow-bellied trogon in Argentina. **Adult** ♂: Olive-bronze central tail feathers, tipped black. **Adult** ♀: Chestnut central tail feathers, narrowly tipped black. **Voice** ♂ song is a slow, mellow series (2 notes per sec) of 4–8 drawn-out whistles at similar volume; ♀ song is slightly higher-pitched. **Tax note 28**. [Surucuá Amarillo]

Notharcus and *Nonnula* Chunky, large-headed forest birds with stout bills, the former mainly in the canopy and the latter in midstorey. *Nystalus* Prefers open country and has a pale iris and narrow tail. Sexes alike.

Buff-bellied Puffbird *Notharcus swainsoni* 25 cm

Paraná forest. **ID** Large forest puffbird, mainly found in canopy. **Adult**: White throat, lores, ear-coverts and nuchal collar. Black pectoral band and buff belly. **Voice** Three common calls which can be strung together: 1) a long whistled series of upward-inflected *dWii* notes gradually becoming faster and higher-pitched; 2) a raptor-like *WEEU* whistle, often followed by 3) a manic and fast *wiPIRI-wiPIRI-wiPIRI*. [Chacurú Grande]

White-eared Puffbird *Nystalus chacuru* 21 cm

Clearings and savanna in Misiones and ne. Corrientes. **ID** Open-country puffbird with pale iris and narrow tail. **Adult**: Robust orange bill. White nuchal collar and cheek spot, surrounded by black wedge. Underparts often stained by red soil. **Voice** Song is a far-carrying series of up to 5 mellow, whistled *We Fó-LOw* phrases, slower and more quavering at the end. [Chacurú Cara Negra]

Spot-backed Puffbird *Nystalus maculatus* 19.5 cm

Mainly dry and sierran chaco. **ID** Open-country puffbird with pale iris and narrow tail. **Adult**: Tawny nuchal collar and rusty orange pectoral band. Upperparts spotted buff. White underparts, streaked black. **Voice** Song, often given in duet or sometimes trios, is a sudden rich, whistled *Cho-WEE chee CHwOOk*, while slowly wagging the tail. [Durmilí]

Rusty-breasted Nunlet *Nonnula rubecula* 15.5 cm

Paraná forest. **ID** Small inconspicuous midstorey forest puffbird. **Adult**: White lores and eye-ring. Breast varies from tawny-olive to brown with whitish belly and vent. **Voice** Song at dawn (or rarely during the day) is a steady series of 11–19 medium-pitched, whistled *weep* notes (*c.* 3 per sec), gradually rising and then falling in pitch. Recalls song of Squirrel Cuckoo, but faster-paced and higher-pitched. [Chacurú Chico]

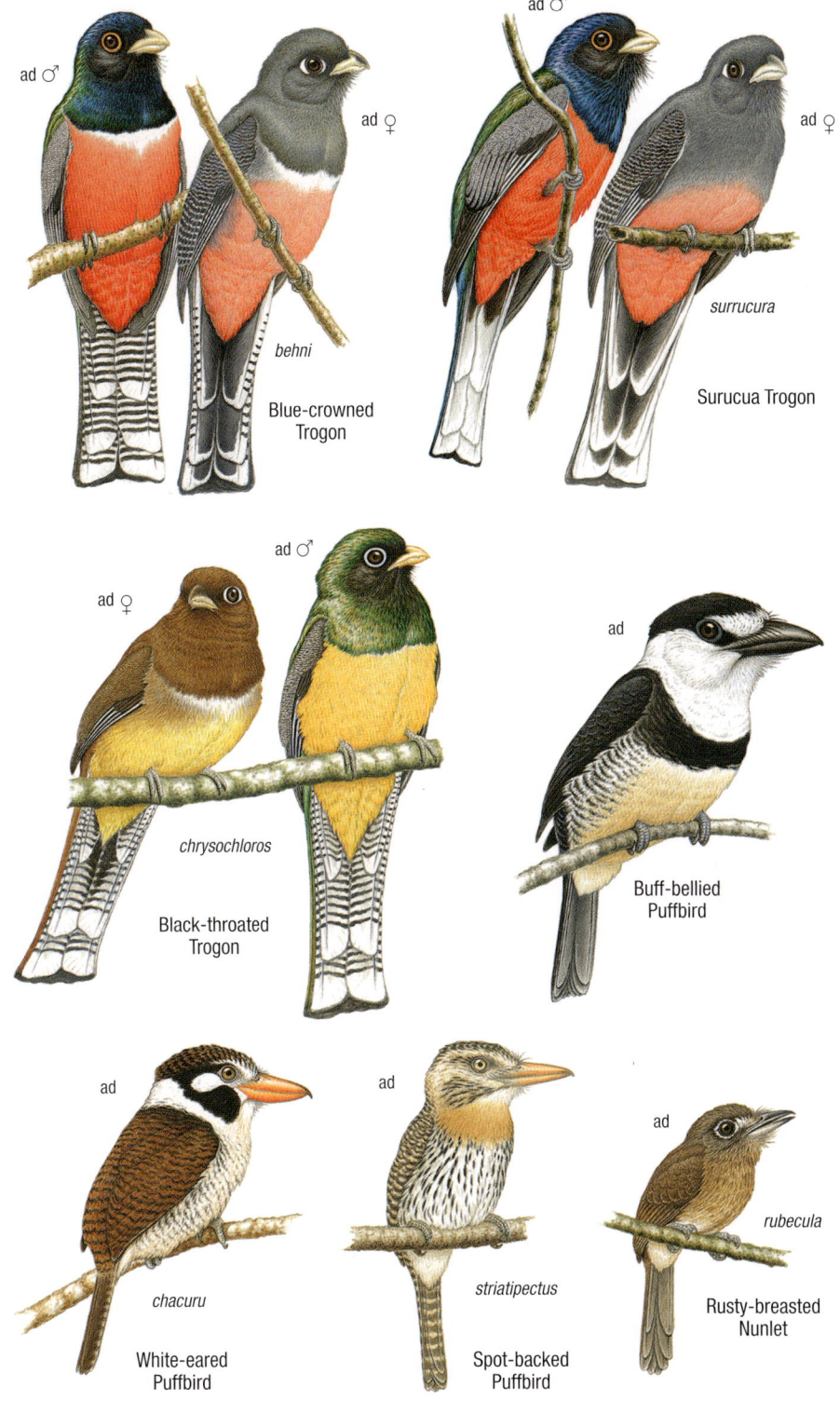

ad ♂

ad ♀

behni

Blue-crowned
Trogon

ad ♂

ad ♀

surrucura

Surucua Trogon

ad ♀

ad ♂

chrysochloros

Black-throated
Trogon

ad

Buff-bellied
Puffbird

ad

chacuru

White-eared
Puffbird

ad

striatipectus

Spot-backed
Puffbird

ad

rubecula

Rusty-breasted
Nunlet

Baryphthengus and *Momotus* Large, colourful (but mainly green) sub-canopy birds with long bill and long tail. Often inconspicuous in forest and best detected by voice. Sexes alike.

Rufous-capped Motmot *Baryphthengus ruficapillus* 40 cm

Paraná forest. **ID** Large motmot without tail rackets. **Adult**: Unmistakable. Often more rusty below due to red soil staining. **Voice** Crepuscular hollow rolling trills. [Yeruvá]

Amazonian Motmot *Momotus momota* 45 cm

Foothills in N Yungas forest. **ID** Large motmot with prominent tail rackets. **Adult**: Unmistakable. No overlap with Rufous-capped Motmot. **Voice** Short, often disyllabic, tremulous calls given at dawn and dusk. [Burgo]

Chloroceryle Green-backed kingfishers of variable size; two with flattened crests. *Megaceryle* Unmistakable, large crested kingfisher. Voices are rattles where pitch and speed match the size of the species; higher and faster the smaller the species.

Amazon Kingfisher *Chloroceryle amazona* 30 cm

N lowlands. **ID** Large crested kingfisher, heavier-billed than Green Kingfisher and without white spotting on the wings. **Adult** ♂: Broad chestnut pectoral band. **Adult** ♀: White underparts with a broken green pectoral band. [Martín Pescador Mediano]

Green Kingfisher *Chloroceryle americana* 21 cm

N lowlands and foothills. **ID** Small crested kingfisher with white spot-barring on the flight feathers. **Adult** ♂: White throat and belly, with contrasting chestnut breast band. **Adult** ♀: Pale buff throat and breast, grading to white on belly. Olive pectoral band. [Martín Pescador Chico]

American Pygmy Kingfisher *Chloroceryle aenea* 13 cm

Very rare in n. Misiones and cn. Corrientes. **ID** Tiny forest kingfisher. Lacks a crest and wing-spotting. **Adult** ♂: Bright chestnut underparts with a white centre to the belly. **Adult** ♀: Green pectoral band. [Martín Pescador Enano]

Ringed Kingfisher *Megaceryle torquata* 38–43 cm

Widespread. **ID** A large, notably crested, blue-backed kingfisher. **Adult** ♂ *stellata* (38 cm; Patagonian forest rivers and lakes): Plain chestnut breast and belly indicative of ♂. Dark bill. Upperparts spangled with white. **Adult** ♀ *torquata* (43 cm; N lowlands to 1500 m): Larger. Blue pectoral band indicative of ♀. Yellowish bill base, and unmarked back and wings. [Martín Pescador Grande]

Rufous-capped Motmot

ad

ad

Amazonian Motmot

nattereri

ad ♂

ad ♀

amazona

Amazon Kingfisher

ad ♀

ad ♂

mathewsii

Green Kingfisher

torquata

ad ♀

ad ♀

ad ♂

American
Pygmy Kingfisher

ad ♂

stellata

Ringed Kingfisher

Ramphastos Large, heavy billed toucans with a black back and contrasting throat, vent and rump. Slow croaking voices.

Toco Toucan *Ramphastos toco* 59 cm

N lowlands and foothills. **ID Adult**: Larger than Green-billed Toucan, with a black-tipped orange-yellow bill and white throat. **Voice** Slow croaking call is a loud snoring *WRRONG* or *WRRAH*, deeper than Green-billed Toucan. [Tucán Grande]

Green-billed Toucan *Ramphastos dicolorus* 49 cm

Paraná forest. **ID Adult**: Pea-green bill with a red cutting edge. Yellow throat and breast, becoming red on the abdomen. **Voice** Croaking call is a nasal frog-like *wreeh* or *reeah*, higher-pitched than Toco Toucan. [Tucán Pico Verde]

Selenidera Small sexually dimorphic toucan with a short graduated tail. Found alone or in pairs.

Spot-billed Toucanet *Selenidera maculirostris* 31 cm

N. Misiones. **ID** Short white bill with black tooth marks. **Adult** ♂: Black head and breast with golden head stripe. **Adult** ♀: Chestnut head and breast. **Voice** Remarkably deep, foghorn-like hoots *brrohw.. brrohw… brrohw…* [Arasarí Chico]

Pteroglossus Gregarious olive-backed toucans with a long bill (longer in ♂) and a long, narrow graduated tail.

Saffron Toucanet *Pteroglossus bailloni* 38 cm

Mainly n. Misiones. **ID Adult**: Crimson facial skin and base of yellow bill. Dull golden head and underparts. **Voice** Short, passerine-like *clí, quip* or *kek* notes, sometimes given in series. [Arasarí Banana]

Chestnut-eared Araçari *Pteroglossus castanotis* 46 cm

Paraná forest. **ID Adult**: Dark bill with an orange upper-mandible stripe. Red band across yellow belly. **Voice** Large variety of vibrant high-pitched yelps *cleeo, trlee, threeo, wheel…* [Arasarí Fajado]

ad

toco

Toco Toucan

ad

Green-billed Toucan

ad ♀

ad ♂

Spot-billed Toucanet

ad ♀

Saffron Toucanet

ad ♂

australis

Chestnut-eared Araçari

Picumnus Miniature, robust, and very agile woodpeckers which manoeuvre without using the tail for support. All deliver high-pitched trills and some drum rather loudly. ♂♂ have red foreheads; spotted white in ♀♀.

Mottled Piculet *Picumnus nebulosus*   10.5 cm

Local in gallery forest and *Guadua* bamboo on se. Misiones–ne. Corrientes border. Meets range of Ochre-collared and White-barred (ssp. *pilcomayensis*) Piculets. **ID** Bronzy-brown above, tawny below, thickly streaked dusky, becoming creamy on belly. Adult ♂: Red forehead. Adult ♀: Forehead and crown black finely spotted white. **Voice** Calls include 1–4 high-pitched notes, including trills. Drumming comprises 2–4 short hollow bursts in quick succession without syncopated pattern. [Carpinterito Ocráceo]

Ochre-collared Piculet *Picumnus temminckii*   10.5 cm

Paraná forest; often in bamboo. **ID** Warm brown mantle and tawny nuchal collar. Adult ♂: Red forehead. Adult ♀: Forehead and crown black finely spotted white. **Voice** Song is like White-barred Piculet but usually more monotonous and evenly-pitched. Drumming similar to White-barred but generally with longer bursts. [Carpinterito Cuello Canela]

White-wedged Piculet *Picumnus albosquamatus*   10.5 cm

Historical specimen from Yungas foothill forest of n. Salta. **ID** White underparts, scalloped with black fringing. Lacks barring on the flanks, but intermediates with ssp. *thamnophiloides* of White-barred occur along foothills. Adult ♂: Red forehead. Adult ♀: Forehead and crown black finely spotted white. [Carpinterito Escamado]

White-barred Piculet *Picumnus cirratus*   10.5 cm

Widespread and highly variable; overlaps with or meets range of all other piculets. **ID** Adult ♂ *tucumanus* (S and C Yungas): Underparts evenly barred black and white. Adult ♂ *pilcomayensis* (widespread in the N lowlands, except Misiones): Variable pattern of narrow black ventral barring, often wavy or broken. Adult ♀ *thamnophiloides* (C and S Yungas): White breast with black feather centres, and barred flanks. All three ssp. overlap with each other depending upon location, and variation is poorly understood. **Voice** Song is a high-pitched, fast and long trill, generally descending markedly in pitch. Compare with Ochre-collared Piculet. Drumming begins with a long (often faster) burst and is followed by 2–5 shorter (often slower) bursts creating a syncopated pattern. [Carpinterito Barrado]

Dryobates Small pied or brownish-green woodpeckers, with varied ventral markings and a strong or vestigial moustachial streak. ♂♂ show red on the crown or hindcrown. Forage at all heights. See also Plate 107.

Checkered Woodpecker *Dryobates mixtus*   16.5 cm

N and C lowlands. Mantle densely spotted white or yellowish buff. **ID** Adult ♂ *berlepschi* (C Sierras and Monte desert): Bill longer than other ssp. Complete or broken red band across hindcrown (orange-red patches restricted to sides of hindcrown in *mixtus*). Concolorous black crown and mantle. Fine black streaks on underparts. Adult ♀ *mixtus* (Mesopotamia and ne. Buenos Aires): Lacks streaking and red on crown. From Striped Woodpecker by finer streaking on underparts and shorter bill. Juvenile ♂: Crown finely streaked red. **Voice** Like Striped Woodpecker, and possibly not separable in the field. Compare with White-spotted Woodpecker. [Carpintero Bataraz]

Striped Woodpecker *Dryobates lignarius*   19 cm

Patagonian forest. **ID** Black mantle, distinctly barred white. Coarse streaking on underparts. Adult ♂: Black crown, finely streaked white. Red band across hindcrown. Adult ♀: Black crown and nape. **Voice** Song begins with an introductory *quip* followed by a sudden fast metallic rattle, very similar to Checkered, and possibly not safely distinguished in the field. [Carpintero Bataraz Patagónico]

Bolivian Woodpecker *Dryobates* [*lignarius*] *puncticeps*   19 cm

Not illustrated. Rare above Santa Victoria, n. Salta (2800–3000 m). **ID** Rather intermediate between Checkered and Striped. Browner-backed than both, while dorsal barring closer to Striped. Tail with sparse fine pale bars. Underparts washed buff; brown ladder streaks broadest on sides of breast. **Voice** Song is very similar to White-spotted Woodpecker and markedly different from Checkered and Striped Woodpeckers (no overlap with either). Tax note 29. [Carpintero Bataraz Boliviano]

Mottled Piculet

ad ♀

ad ♂

Ochre-collared Piculet

ad ♂

ad ♀

ad ♀

ad ♂

albosquamatus

White-wedged Piculet

ad ♀

ad ♂

tucumanus

ad ♂

pilcomayensis

thamnophiloides

White-barred Piculet

juv ♂

ad ♂

berlepschi

ad ♀

ad ♂

ad ♀

Checkered Woodpecker

mixtus

Striped Woodpecker

Dryobates Small pied or brownish-green woodpeckers, with varied ventral markings and a strong or vestigial moustachial streak. ♂♂ show red on the crown or hindcrown. Forage at all heights. See also Plate 106.

White-spotted Woodpecker *Dryobates spilogaster*   18 cm

Paraná and gallery forest in Mesopotamia. **ID** Narrow white supercilium and moustachial. Mantle, wing-coverts and tail barred yellowish. Underparts densely spotted white or yellow. **Adult** ♂: Fine red shaft streaks on blackish crown. **Adult** ♀: Fine white shaft streaks on crown. Overlaps with Little Woodpecker in n. Corrientes. **Voice** Song begins with an introductory *quip* note followed by a sudden slow rattle of slurred and churring notes. [Carpintero Oliva Manchado]

Little Woodpecker *Dryobates passerinus*   18 cm

Gallery forest, chaco and Espinal woodlands. **ID** Whitish moustachial. Sparse whitish spots on wing-coverts. Indistinctly barred below. **Adult** ♂: Red skullcap. **Adult** ♀: Dingy brown crown. Overlaps with White-spotted Woodpecker in n. Corrientes. **Voice** Song is a rapid, almost rattled *kwi-kwi-kwi-kwi-kwi-kwi...* with notes too fast to count. [Carpintero Oliva Chico]

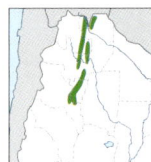

Dot-fronted Woodpecker *Dryobates frontalis*   17 cm

Yungas forest. **ID** Face grizzled whitish. Small whitish triangular flecks on wing-coverts. Underparts finely barred white. **Adult** ♂: Brown forehead, scarlet crown and nape. **Adult** ♀: Brown crown, finely spotted white. **Voice** A slow and emphatic *week week weeek weeek week week...* usually with 6–10 notes. [Carpintero Oliva Yungueño]

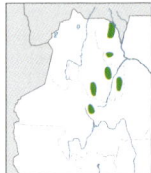

Smoky-brown Woodpecker *Dryobates fumigatus*   18 cm

Upper Yungas forest, generally in alder, in Jujuy and Salta. **ID** A small uniform woodpecker lacking ventral or dorsal markings. Contrasting pale face. **Adult** ♂: Crown speckled with scarlet. **Adult** ♀: Brown crown. **Voice** Song is a slow, low-pitched, dry grating rattle. Calls include *quick* and *tip* notes. [Carpintero Alisero]

Melanerpes Small to medium-sized, black-backed woodpeckers with a white stripe down the back and rump. White-fronted and White are highly gregarious. See also Plate 108.

Yellow-fronted Woodpecker *Melanerpes flavifrons*   19 cm

Misiones. **ID** Glossy blue-black upperparts. Bright yellow forehead and throat. Prominent yellow eye-ring. **Adult** ♂: Scarlet crown and large scarlet breast patch. Barred flanks. **Adult** ♀: Black crown and reduced scarlet breast patch. **Voice** Loud, rapid couplets or triplets *quip-quip-quip*, with short pauses between bouts. [Carpintero Arcoiris]

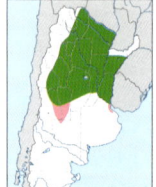

White-fronted Woodpecker *Melanerpes cactorum*   19 cm

Widespread in the north to 2800 m. **ID** Black with white forehead and wing-coverts. Yellow throat when breeding, otherwise white. Brownish-grey below. **Adult** ♂: Semi-concealed scarlet streak on centre of rear crown. **Adult** ♀: Lacks crown streak. **Voice** Soft *Week* or *weep-weep*. [Carpintero del Cardón]

ad ♀

ad ♂

White-spotted Woodpecker

ad ♀

olivinus

ad ♂

Little Woodpecker

ad ♂

ad ♀

Dot-fronted Woodpecker

fumigatus

ad ♂

ad ♀

Smoky-brown Woodpecker

ad ♂

ad ♀

Yellow-fronted Woodpecker

ad ♀

ad ♂

White-fronted Woodpecker

Piculus Medium-sized, bronze-olive woodpeckers with barred underparts and slightly peaked rear crowns. *Colaptes* Medium-sized midstorey woodpecker with barred underparts and a pale face.

Golden-green Woodpecker *Piculus chrysochloros*　　22.5 cm

Chaco woodlands and Yungas foothill forest. Midstorey to canopy. **ID** White or sky-blue iris. Yellow throat and long cheek-stripe continuing onto sides of neck. **Adult** ♂: Red crown and malar. **Adult** ♀: Olive crown and malar. **Voice**. Song consists of 4–5 extremely harsh screaming notes, *weeeah-weeeah-weeeah-weeeah*. [Carpintero Ojo Blanco]

Yellow-browed Woodpecker *Piculus aurulentus*　　22 cm

Paraná forest. Midstorey to canopy. **ID** Long yellow (or yellow and white) rear supercilium and moustachial. Yellow throat. **Adult** ♂: Red crown, nape and prominent malar. **Adult** ♀: Olive crown, red nape and dull red malar. **Voice** Song is a soft, higher-pitched introductory note followed by a strident series of 2–4 very loud and monotone whistles, *flee…TEW-TEW-TEW-TEW*. [Carpintero Dorado Verdoso]

Golden-olive Woodpecker *Colaptes rubiginosus*　　23 cm

Yungas forest. **ID** Grey forecrown and striking white lores and ear-coverts. **Adult** ♂: Red supercilium, hindcrown and malar. **Adult** ♀: Red restricted to nape. **Voice** An explosive and very loud *KÉU!* Also, an extended fast dry rattle. **Tax note 30**. [Carpintero Dorado Gris]

White Woodpecker *Melanerpes candidus*　　28.5 cm

Widespread in the N lowlands. **ID** Large black-and-white woodpecker with striking yellow orbital ring. Highly gregarious. Mainly white with blackish eye-stripe, mantle, wings and tail. Variable yellowish belly patch. **Adult** ♂: Variable yellow patch on nape; **Adult** ♀ (not illustrated): Lacks yellow nape. **Voice** Loud grating, tern-like *Creeeer*, in flight or when perched, audible from great distance. [Carpintero Blanco]

Celeus Large woodpeckers with yellow or cream heads and notable erectile crest, barred upperparts and a creamy or yellow rump and underwing-coverts.

Pale-crested Woodpecker *Celeus lugubris*　　27 cm

Palm savanna, chiefly in the humid chaco. **ID** Cream head and rump, brown body and rusty barring on remiges. **Adult** ♂: Broad red malar. **Adult** ♀: Broad dusky malar. **Voice** Explosive whistle *kuí kiú kiú* with 2–4 notes, each note slightly lower-pitched than the previous one and with a raspy quality. [Carpintero Copete Pajizo]

Blond-crested Woodpecker *Celeus flavescens*　　28.5 cm

Paraná forest. **ID** Yellow head and rump, blackish body and yellow barring on remiges. **Adult** ♂: Red malar. **Adult** ♀: Indistinct malar flecked with black. **Voice** Explosive series of evenly-pitched whistled notes *kuí kuí kuí kú kú*, sometimes with lower-pitched final notes; not as raspy as Pale-crested Woodpecker. [Carpintero Copete Amarillo]

Golden-green
Woodpecker

ad ♂

ad ♀

Yellow-browed
Woodpecker

ad ♂

ad ♀

ad ♀

tucumanus

ad ♂

Golden-olive
Woodpecker

ad ♂

White
Woodpecker

ad ♂

kerri

ad ♀

Pale-crested
Woodpecker

ad ♀

flavescens

ad ♂

Blond-crested
Woodpecker

Colaptes Large woodpeckers with barred upperparts and yellow primary shafts. Terrestrial and arboreal; often gregarious.

Campo Flicker *Colaptes campestris* 32 cm

Widespread. **ID** Black crown and nape, white face and contrasting orange-yellow sides of neck and breast. White rump. **Adult** ♂ *campestroides*: White throat. Dull red malar indicative of male. **Adult** ♀ *campestris* x *campestroides* (n. Misiones): Stippled or mostly black throat. Black malar indicative of female. [Carpintero Campestre]

Green-barred Woodpecker *Colaptes melanochloros* 29 cm

Paraná forest. Arboreal. **ID** Smaller and more yellow than Golden-breasted. Black crown and scarlet nape. Whitish face and spotted underparts. Mantle vivid green, barred brown. **Adult** ♂: Dark red malar. **Adult** ♀ (not illustrated): Black malar. [Carpintero Real Verde]

Golden-breasted Woodpecker *Colaptes [melanochloros] melanolaimus* 31 cm

Widespread in N lowlands to N Patagonia (two similar ssp: *nigroviridis* and *leucofrenatus*). Puna to 3400 m (*melanolaimus*). **ID** Pale dorsal barring and golden-orange breast. Whitish belly. **Adult** ♂ (not illustrated): Dark red malar. **Adult** ♀ *leucofrenatus* (Corrientes to Río Negro): Black malar. **Tax note 31.** [Carpintero Real]

Chilean Flicker *Colaptes pitius* 33 cm

Patagonian forest. **ID** Grey crown and nape; underparts buffy with heavy barring. **Adult** ♂: Narrow malar, finely streaked black. **Adult** ♀: Lacks a malar streak. [Carpintero Pitío]

Andean Flicker *Colaptes rupicola* 35 cm

Puna and Andean slopes. **ID** Long bill. Grey crown and nape; upperparts strongly barred. Underparts buffy, spotted on breast. **Adult** ♂: Grey and scarlet malar. Rarely shows red on lower nape. **Adult** ♀: Grey malar. [Carpintero Andino]

Campephilus Magellanic Woodpecker is the only large red-headed woodpecker in its range. See also Plate 110.

Magellanic Woodpecker *Campephilus magellanicus* 41 cm

Patagonian forest. **ID** Very large and mainly black. White tertials and tips to some flight feathers form a white stripe on the wing. No overlap with congeners. **Adult** ♂: Scarlet head with small, forward-pointing crest. **Adult** ♀: Black head with long recurved crest. Red lores, chin and tufts on forehead. [Carpintero Gigante]

campestris × campestroides

ad ♀

Campo Flicker

campestroides

ad ♂

Green-barred
Woodpecker

ad ♂

Golden-breasted
Woodpecker

ad ♀

leucofrenatus

ad ♂

ad ♀

cachinnans

Chilean Flicker

ad ♀

ad ♂

Andean Flicker

ad ♀

ad ♂

Magellanic
Woodpecker
(not to scale)

Celeus Helmeted Woodpecker is a medium-sized, red-crested woodpecker which overlaps with other red-headed species. *Dryocopus* and *Campephilus* Medium-sized to large, sexually dimorphic, red-crested woodpeckers. Territory defined by long drum rolls (*Dryocopus*) or drum-taps (*Campephilus*); single or double drum-taps (Robust, Cream-backed and Magellanic) or short sequence of deliberate loud taps (Crimson-crested). Foraging/contact calls in *Campephilus* are rich, nasal, metallic single or disyllabic notes. See also Magellanic Woodpecker on Plate 109.

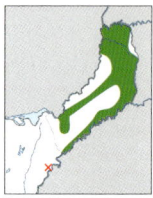

Helmeted Woodpecker *Celeus galeatus* 30 cm

Rare in Paraná forest. Forages on thick trunks and slender branches. **ID** Scarlet crown and upstanding shaggy crest, rarely folded. Cinnamon face and throat. Short stout bill. Whitish neck stripes and blackish upper breast. In flight, cinnamon wing-linings and wing-stripe. Note creamy rump. Adult ♂: Red malar. Adult ♀: Lacks a malar. Overlaps with Lineated and Robust Woodpeckers. **Voice** A far-carrying series of 4–10 slowly delivered (easy to count) *kweek* notes; always much shorter and more emphatic than Lineated Woodpecker. [Carpintero Cara Canela]

Black-bodied Woodpecker *Dryocopus schulzi* 31 cm

Local in the C sierras and chaco woodlands. **ID** White bill and dark iris. Pointed red crest. Ear-coverts light grey or whitish. Thick white neck stripes and black back. White throat and black underparts. In flight, underwing mainly white with a black carpal patch. Adult ♂ white-braced form: White braces. Red malar indicative of male. Typical adult ♀: Lacks white braces. Narrow black forehead. Black malar indicative of female. Overlaps with Lineated (hybridising in the humid chaco), Crimson-crested and Cream-backed Woodpeckers. **Voice** Very similar to Lineated Woodpecker. [Carpintero Negro]

Lineated Woodpecker *Dryocopus lineatus* 36 cm

N lowlands. **ID** Grey bill and white or yellow iris. Pointed red crest. Ear-coverts dark grey or black. Narrow white neck stripes and black back. White throat, streaked black. Upper breast black, barred below. In flight, white underwing-coverts. Typical adult ♂: Lacks white braces. Red malar indicative of male. Adult ♀ white-braced form: Black malar and forehead indicative of female. Overlaps with Helmeted, Black-bodied (hybridising), Crimson-crested, Robust and Cream-backed Woodpeckers. **Voice** A fast, far-carrying series of 8–20 *kweek* notes; always much longer and less emphatic than Helmeted Woodpecker. [Carpintero Garganta Estriada]

Crimson-crested Woodpecker *Campephilus melanoleucos* 34 cm

Local in gallery forest in the humid chaco and n. Corrientes. **ID** Pointed red crest. White neck stripe and converging braces. Black throat and upper breast. In flight, white underwing-coverts. Adult ♂: White lores, black-and-white cheek spot. Adult ♀: Black forehead to tip of crest. Broad white cheek stripe joining the neck stripe. Overlaps with Lineated, Cream-backed and possibly Black-bodied Woodpeckers. [Carpintero Garganta Negra]

Robust Woodpecker *Campephilus robustus* 38 cm

Paraná forest. **ID** Bushy red crest. Throat and neck scarlet. Creamy stripe down mantle and rump. In flight, creamy-buff underwing-coverts; flight feathers barred. Adult ♂: Black-and-white cheek spot. Adult ♀: Broad white cheek stripe, outlined in black. Overlaps with Helmeted and Lineated Woodpeckers. [Carpintero Grande]

Cream-backed Woodpecker *Campephilus leucopogon* 33 cm

N lowlands and Yungas forest to 1600 m. **ID** Creamy mantle. In flight, large cinnamon patch in base of primaries and a smaller patch on the carpal. Adult ♂: Scarlet head and pointed crest. Black-and-white oval cheek spot. Adult ♀: Thick white cheek stripe, bordered black. Black forehead extending to tip of crest. Overlaps with Black-bodied, Lineated and Crimson-crested Woodpeckers. [Carpintero Lomo Blanco]

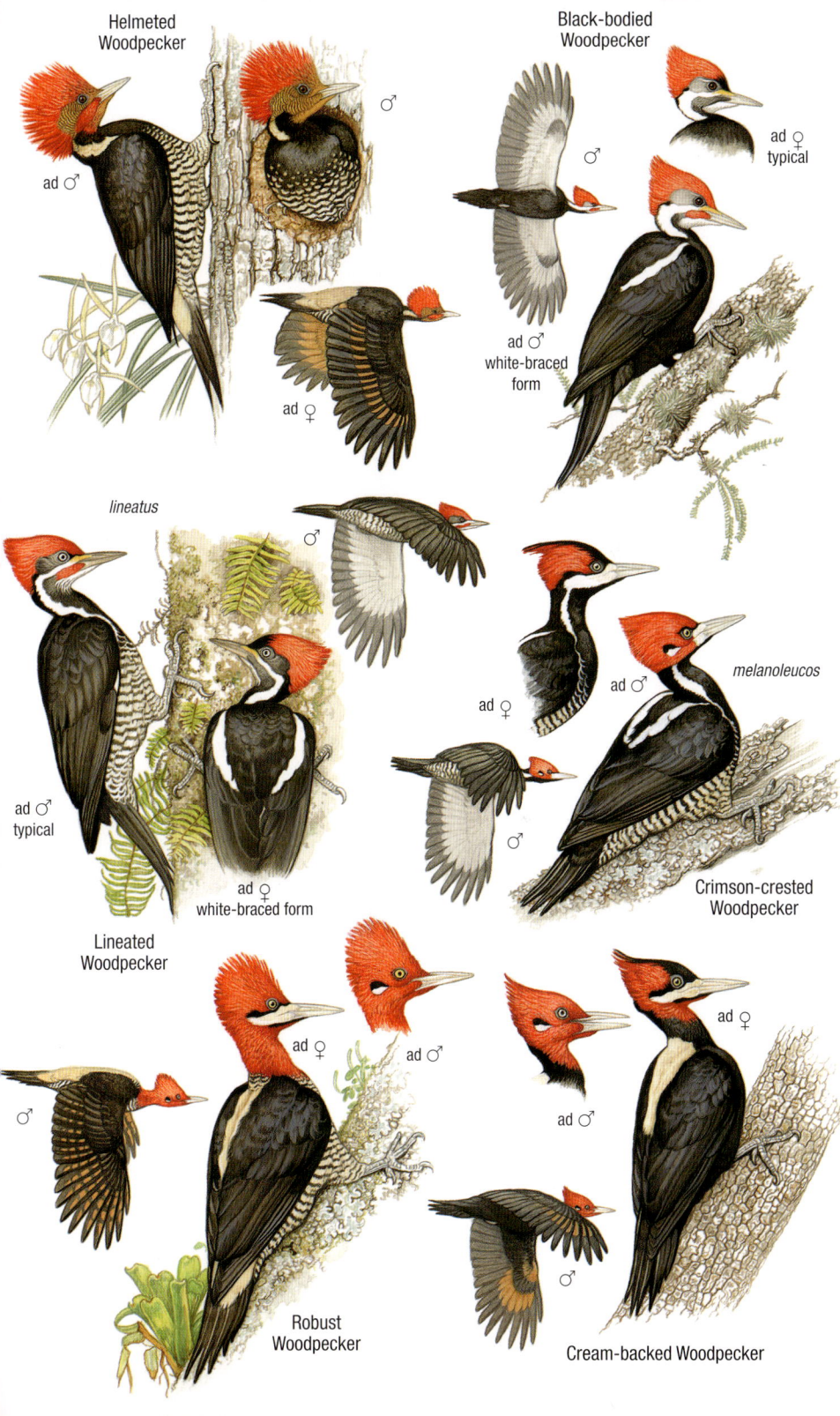

Helmeted Woodpecker
ad ♂
♂
ad ♀

Black-bodied Woodpecker
ad ♀ typical
♂
ad ♂ white-braced form

lineatus
♂
ad ♂ typical
ad ♀ white-braced form
Lineated Woodpecker

ad ♀
ad ♂
♂
melanoleucos
Crimson-crested Woodpecker

♂
ad ♀
ad ♂
Robust Woodpecker

ad ♀
ad ♂
♂
Cream-backed Woodpecker

White-bearded Antshrike *Biatas nigropectus* 18 cm

Rare in n. Misiones. **ID** Short-crested reclusive antshrike of tall dense *Guadua trinii* or *Merostachys* spp bamboo. **Adult** ♂: Black crown and breast. White cheeks, nuchal collar and supercilium of variable length. Striking concealed black and white patch on bend of wing. **Adult** ♀: Rufous crown, wings and tail. Narrow buff supercilium and tawny nuchal collar. **Voice** A mellow series of usually 8–10 whistles *keu-keu-keu-keu-keu...* with somewhat shorter final notes. May recall White-shouldered Fire-eye or Black-throated Trogon. [Batará Pecho Negro]

Great Antshrike *Taraba major* 21.5 cm

Widespread in N lowlands and foothills. **ID** Bicoloured antshrike with heavy bill and red iris. **Adult** ♂: Black and white plumage. **Adult** ♀: Rufous and white plumage. **Voice** A very long accelerating series of cackling notes, finishing with a drawn-out snarl, typically lasting 7–8 sec, *chew chew kew... ke-ke-ke-ke kekeke nyaaah.* Calls include long rattles and short churrs. [Chororó]

Tufted Antshrike *Mackenziaena severa* 24.5 cm

Paraná forest. **ID** Long-tailed, reclusive understorey antshrike; frequently in *Chusquea* bamboo. **Adult** ♂: Ashy-grey with prominent crest. **Adult** ♀: Upstanding rufous crest and barred plumage. **Voice** A slow rising series of usually 6–8 pure slurred whistles with progressively shorter notes, *weeeeo weee wee wee we...* Call is a dry piercing slightly descending *wrreeo.* [Batará Copetón]

Large-tailed Antshrike *Mackenziaena leachii* 28 cm

Paraná forest. **ID** Long-tailed, reclusive understorey antshrike. **Adult** ♂: Entirely black with white spotting on upperparts. **Adult** ♀: Brown with rusty crown and buff spotting throughout. **Voice** A fast, rising and falling series of 15–18 pure short penetrating whistles in 3–4 secs. Calls include a drawn-out wheeze. [Batará Pintado]

Spot-backed Antshrike *Hypoedaleus guttatus* 21.5 cm

Paraná forest. **ID** Heavy-billed, subcanopy antshrike, often encountered in vine tangles. **Adult** ♂: Spotted white above. Ochre restricted to lower belly and vent. **Adult** ♀: Black upperparts with buff teardrop-shaped spots. Ochraceous belly, flanks and vent. **Voice** A shrill, stuttered rising and falling trill with notes too fast to count, lasting 3–4 secs. Calls include a descending churr recalling trogon calls and a 'falling bomb' whistle, just like Sharpbill but shorter. [Batará Goteado]

Giant Antshrike *Batara cinerea* 36 cm

Paraná and Yungas forest. **ID** A highly reclusive, jay-sized, long-tailed antshrike with a flat bushy crest. **Adult** ♂ *cinerea* (Paraná forest): Black cap, grey face and nape. Black mantle, wings and tail, finely barred white. Whitish throat. **Adult** ♀ *cinerea*: Rufous forehead and black crest. Cinnamon barring on mantle and wings; rufous barring on tail. **Adult** ♀ *argentina* (Yungas forest): Entire crown rufous, tipped black. Both sexes are darker than *cinerea*, lack white on throat and show black markings on the bill. **Voice** An explosive, extremely loud, decelerating song that begins with a fast trill grading into longer notes, ending abruptly, *trrrrrcetetetetete-tew-tew-tew-tew-tew-teew-teew-teew-chuk.* Calls include a descending churr and a loud, slow rattle. [Batará Gigante]

White-bearded Antshrike

ad ♀

ad ♂

Great Antshrike

major

ad ♂

ad ♀

ad ♀

Tufted Antshrike

ad ♂

Large-tailed Antshrike

ad ♀

Spot-backed Antshrike

ad ♂

guttatus

ad ♀

ad ♂

ad ♀

ad ♀

cinerea

ad ♂

cinerea

Giant Antshrike

argentina

Thamnophilus Robust, compact antshrikes with stout hooked bills. ♀♀ lack tail spots. Both sexes sing similar songs in duet. Barred and Variable may exhibit clinal variation.

Barred Antshrike *Thamnophilus doliatus* 14 cm

Mainly humid chaco, very local in ne. Santa Fe and w. Entre Ríos (*radiatus*); local in dry chaco (*cadwaladeri*), Yungas foothills and n. Misiones (ssp.?). **ID** Adult ♂ *radiatus*: Yellow or white iris. Erectile black crest. Head finely streaked black and white. Mantle, wings and breast barred black and white. Only weakly barred below in the dry chaco, thickly barred in n. Misiones. Adult ♀ *radiatus*: Bright chestnut crown, mantle, wings and tail. Contrasting white face, finely streaked black. Buffy underparts. **Voice** An accelerating series becoming more nasal towards the end and ending in an emphatic note, *kah-kah-ka-ka-ka-kakakaka-wáh*. [Choca Listada]

Rufous-capped Antshrike *Thamnophilus ruficapillus* 16.5 cm

Yungas and Paraná forest and marshes in the NE. **ID** Adult ♂ *cochabambae* (NW): Reddish iris. Chestnut crown, buffy-white supercilium, olive-brown mantle and wings. Narrow black barring on breast. White tips visible on undertail. Adult ♂ *ruficapillus* (NE): Darker and less olive above than ♂ *cochabambae* with duller supercilium but more extensive breast barring. Adult ♀: Ssp. alike. Lacks breast barring of ♂ and shows duller crown, but brighter chestnut tail with no white tips. Compare with thornbirds (Plate 127). **Voice** A slow accelerating nasal series of monotone notes with an emphatic final note, *weer weer weer weer ke ke kekekekeke wáh*. Calls include monotone and inflected whistles and a soft harsh note. [Choca Corona Rojiza]

Variable Antshrike *Thamnophilus caerulescens* 16 cm

Widespread in N woodlands and forest. **ID** Adult ♂ *caerulescens* (n. Misiones): Black cap. Grey mantle, throat and breast, becoming white below. Blackish tail with white spots on underside. Adult ♂ *dinelli* (Yungas forest and chaco): Olive-grey mantle. Ochre wash over breast, belly and flanks; greyer breast in east. Adult ♂ *unnamed ssp.* (Sierra de Tartagal, n. Salta): Orange-buff flanks and white belly. Adult ♀: Ssp. similar. Warm brown cap and grey face. One or two narrow white wing bars and narrow greyish-white tips visible on underside of brown tail. Ochraceous belly and vent; extending onto the breast in the west. **Voice** Evenly-paced monotone series of usually 6–10 medium-pitched notes, *kaw-kaw-kaw-kaw-kaw-kaw*. Calls include nasal *kaaw* or *kow*, and short raspy notes. [Choca Corona Negruzca]

Dysithamnus Plump, short tailed antvireo of mid-strata. Notably large-headed appearance. Stout bill has a small hooked tip.

Plain Antvireo *Dysithamnus mentalis* 12 cm

Paraná forest. **ID** Adult ♂: Grey crown, blackish mask and olive back. White throat and pale yellow underparts. Adult ♀: Rufous-brown crown, dusky grey mask and narrow white eye-ring. Overlaps with Rufous-crowned Greenlet. **Voice** A slow rising and falling series, ending in a bouncing chatter, *pu-pu-pu-pu-pu-pu-chererere*. Calls include a rising series of 6–9 soft bouncing whistles, triple whistles and harsh calls. [Choca Amarilla]

Hylophilus Rufous-crowned Greenlet belongs in the Vireonidae family, but is most likely to be confused with the unrelated Plain Antvireo. See other vireos on Plate 163.

Rufous-crowned Greenlet *Hylophilus poecilotis* 12 cm

Paraná forest. **ID** Slender, long-tailed forest passerine of midstorey. Sexes alike. Adult: From unrelated adult ♀ Plain Antvireo by slender pointed bill, brighter rufous cap, brighter green upperparts and long narrow tail. **Voice** A series of evenly-paced, upward-inflected notes *fweet-fweet-fweet-fweet*. [Chiví Coronado]

Barred Antshrike

ad ♂

radiatus

ad ♀

Rufous-caped Antshrike

ad ♂

cochabambae

ad ♂

ruficapillus

ad ♀

ad ♀

Variable Antshrike

caerulescens

ad ♂

ad ♂

ad ♂

dinelli

unnamed subspecies

ad ♂

mentalis

ad ♀

Plain Antvireo

ad

Rufous-crowned Greenlet

Drymophila Small, slender, sexually dimorphic, understorey antbirds with long narrow graduated tails and semi-concealed black-and-white interscapular patches. *Myrmorchilus* Retiring understorey antbird (often terrestrial) with long slender bill. *Herpsilochmus* Capped antwrens with prominent white supercilia, white wing bars and slender graduated tails with white outer tail feathers.

Dusky-tailed Antbird *Drymophila malura* 15.5 cm

Paraná forest. **ID** Adult ♂: Head, throat and breast finely streaked black and greyish-white. Two narrow white wing bars. Adult ♀: Buff head, finely streaked black on crown and sides of neck. Buff wing bars. **Voice** Song is a sharp spluttered series of accelerating notes ending in a rattle, *CHew CHew CHew CH CH ch-ch-ch-chchchrrrrrrrrrrr*. Calls include an insect-like rasp with an emphatic metallic final note. [Tiluchi Estriado]

Bertoni's Antbird *Drymophila rubricollis* 15 cm

Bamboo in Paraná forest. **ID** Adult ♂: Black crown and eye-stripe. White supercilium and ear-coverts. Burnt orange below. Adult ♀: Pale brown crown and more washed-out underparts. **Voice** Song is a fast descending, and decelerating, stutter, *jee-ji-ji-ji-ji chew chew*, sometimes very raspy throughout. Calls include slow harsh notes. [Tiluchi Colorado]

Stripe-backed Antbird *Myrmorchilus strigilatus* 16.5 cm

Yungas foothills and dry chaco; local in humid chaco. **ID** Adult ♂: Rufous mantle, streaked black. Rufous tail, bordered black and white. Black throat and breast, bordered by broad white moustachial. Adult ♀: Lacks throat patch, instead finely streaked black on sides of breast. **Voice** Best detected by loud, high-pitched duet; ♂ gives a shrill *sheer-sheeRREO* and ♀ descending raspy notes *sheer-cheer-cheer-cheer-cherr-cherr-wurr*. Call is a piercing inflected whistle *feeuEEO*. [Batará Estriado]

Black-capped Antwren *Herpsilochmus atricapillus* 13 cm

N Yungas foothills. **ID** Midstorey to canopy, often in vine tangles. Adult ♂: Black crown and eye-stripe. Mainly white below, washed grey on breast. Adult ♀: Buff forehead and lores; remainder of crown black, streaked white; pale buff below. **Voice** Song is a slowing burry trill with 1–2 indistinct introductory notes, descending somewhat and duetted. Also delivers a nasal chatter, rising slightly towards the end; often in duet. [Tiluchi Plomizo]

Rufous-winged Antwren *Herpsilochmus rufimarginatus* 12 cm

Paraná forest. **ID** Midstorey to canopy, often in vine tangles. Adult ♂: Black cap and eye-stripe, striking rufous fringes to flight feathers, and yellow wash over belly. Adult ♀: Differs from ♂ by chestnut cap. **Voice** Slowing burry trilled song recalls Black-capped Antwren but with 2–3 long introductory notes. Call is a rich *who-who* or *who-who-who* and harsh chatters. [Tiluchi Ala Roja]

Streak-capped Antwren *Terenura maculata* 11 cm

n. Misiones. **ID** Tiny antwren mainly found in subcanopy vine tangles, often with Rufous-winged Antwren. Adult ♂: Black crown, streaked white. Bright rufous back. White throat and breast, finely streaked black; yellow wash over belly and vent. Adult ♀: From ♂ by buff crown streaking, duller back, and only vague breast streaking. **Voice** Song is a dry, fast, monotone piercing rattle. Frequent call is a syncopated series of spirited *tchew dee-dee*. [Tiluchi Enano]

Dusky-tailed Antbird

ad ♂

ad ♀

Bertoni's Antbird

ad ♂

ad ♀

ad ♀

ad ♂

suspicax

Stripe-backed Antbird

ad ♂

ad ♀

Black-capped Antwren

ad ♂

rufimarginatus

ad ♂

ad ♂

ad ♀

ad ♀

Streak-capped Antwren

Rufous-winged Antwren

White-shouldered Fire-eye *Pyriglena leucoptera* 18 cm

Paraná forest. **ID** A slender understorey antbird with a fine bill and red iris. **Adult** ♂: Black with white wing bars and interscapular patch. **Adult** ♀: Brown above with a contrasting black tail. **Voice** Song is a melancholic series of 5–8 evenly-spaced whistles, somewhat descending towards the end, *pew-pew-pew-pew-pew-pew-pew*; may recall White-bearded Antshrike from a distance. Call is a dry *pluck*. [Batará Negro]

Spotted Bamboowren *Psilorhamphus guttatus* 14 cm

Local in Paraná forest. **ID** Slender, long-tailed and long-billed tapaculo with a white iris. Mainly found in bamboo stands. Restless and often secretive. Flesh-coloured bill and tarsus. **Adult** ♂: Grey crown with fine ocellated spots at rear continuing over back and wing-coverts. White throat and breast, finely dotted black. **Adult** ♀: Grey restricted to forecrown, but spotted throughout upperparts. Underparts entirely buff. **Voice** Best detected by long series of ventriloquial, low-pitched vibrating notes *rü-drü-drü-drü-drü-...* [Gallito Overo]

Ochre-flanked Tapaculo *Eugralla paradoxa* 14.5 cm

Local in Andes of N Patagonia. Overlaps with Magellanic Tapaculo. **ID** Chunky understorey tapaculo with unique conical, broad-based bill. **Adult**: Mainly plumbeous-grey with blackish crown and white lores; whitish centre of belly and pale rufous flanks and vent. Tarsus chrome-yellow. **Juvenile**: Brown upperparts, finely vermiculated with blackish bars. Throat and breast greyish-buff, indistinctly scalloped with dark chevrons. Rusty flanks and vent. **Voice** Loud monotonous song is a rapid series *TE-TEA-TEW-TEW-TÉW-TÉW*, becoming more emphatic towards the end. Calls include a series of evenly-spaced *KIÚ* notes and single or doubled *KEE* or *KEE-KEE*. [Churrín Grande]

Scytalopus Small, mainly terrestrial, understorey tapaculos which forage restlessly and frequently cock tail. Best detected by loud strident songs. Sexes alike.

Planalto Tapaculo *Scytalopus pachecoi* 12 cm

Paraná forest. **ID Adult**: Mainly ash-grey with rusty rump, flanks and vent, barred blackish. Orange tarsus. **Juvenile**: Brown, finely vermiculated blackish throughout and with rusty rump; flanks and vent more coarsely barred. **Voice** Song is an endless series of harsh *chek* notes. Calls include sharp metallic calls and buzzes. [Churrín Plomizo]

White-browed Tapaculo *Scytalopus superciliaris* 11 cm

Endemic. Sub-Andes from Jujuy to La Rioja. **ID** Adult *superciliaris* (Salta, Tucumán and n. Catamarca): Extensive white throat and supercilium, grey underparts. Ochre tarsus. Adult *santabarbarae* (Sierra de Santa Bárbara, Jujuy and Cresta del Gallo, Salta): Considerably darker than *superciliaris*. Birds from s. Catamarca and La Rioja (undescribed ssp) are paler than *superciliaris* with a white centre to the breast. **Juvenile** (not illustrated): Scaled throughout but also shows a white throat. **Voice** Disyllabic song, *ti-CHRRR*, with last note descending and churring, very different from Zimmer's Tapaculo. Sonogram on p. 446. [Churrín Ceja Blanca]

Zimmer's Tapaculo *Scytalopus zimmeri* 11 cm

Not illustrated. Andes of Salta and Jujuy. **ID** Adult Like pale ssp. of White-browed Tapaculo and only safely separated by voice. **Voice** Syncopated 3–8 note song, compared to 2 notes in White-browed. Sonogram on p. 446. [Churrín de Zimmer]

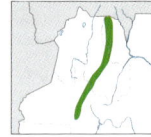

Magellanic Tapaculo *Scytalopus magellanicus* 10 cm

C and S Andes. Overlaps with Ochre-flanked Tapaculo. **ID** White-crowned adult: Dark grey with browner flanks, vent and rump, barred blackish and with striking silvery white forehead, variable in extent. **Dark-crowned adult**: As white-crowned adult but lacks white on crown. Birds from high Andes of Mendoza (undescribed ssp) are darker, lack barring and rarely show white on the crown. **Juvenile**: Essentially brown, barred throughout with broader barring on rump, flanks and tail. **Voice** Call is a syncopated *cha-chook* or *tow-chee*, sometimes repeated for minutes on end. Main call is a short, fast, dry descending trill. [Churrín Andino]

White-shouldered Fire-eye

ad ♀

ad ♂

ad ♂

ad ♀

Spotted Bamboowren

juv

Planalto Tapaculo

juv

ad

ad

Ochre-flanked Tapaculo

ad
white-crowned
form

ad
dark-crowned
form

ntabarbarae

ad

juv

superciliaris

ad

Magellanic Tapaculo

White-browed Tapaculo

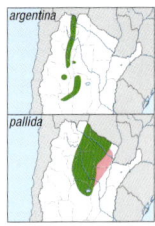

Olive-crowned Crescentchest *Melanopareia maximiliani* · 15.5–17 cm

Widespread. **ID** Slender, long-tailed skulking passerine of grass and scrub. Monotonous voice reveals presence. Adult *argentina* (15.5 cm; NW Andes and C Sierras): Olive upperparts with black mask and pectoral band. Ochraceous supercilium and throat. Rich chestnut below. Adult *pallida* (17 cm; edge of dry chaco and local in the humid chaco, especially in grassy clearings): Paler and more washed-out than smaller *argentina* (no overlap) with a narrower pectoral band and white supercilium. **Voice** *argentina*: Rapid trill, descending slightly at end; *pallida*: lower-pitched song than *argentina*; a much slower monotone series of 2 notes per sec. **Tax note 32**. [Gallito de Collar]

Crested Gallito *Rhinocrypta lanceolata* · 21.5 cm

Dry chaco, monte desert and S espinal. **ID** Large crested terrestrial and arboreal tapaculo. Cocks tail when on the ground. Perches high to sing. Adult: Chestnut-brown head and erect spiky crest, streaked white. Extensive rufous flanks. **Voice**. Gives single loud resounding *t-chok* note and grating series of calls. [Gallito Copetón]

Sandy Gallito *Teledromas fuscus* · 17.5 cm

Endemic. Monte desert; especially scrubby sand dunes. **ID** Cryptic desert tapaculo. Uses bushes as song posts. Cocks tail while running at surprisingly high speed. Adult: Sandy-brown above with short whitish supercilium and blackish tail. Creamy below with buff flanks. **Voice** Series of 7–10 resonant *chuck* notes, often accompanied by wing-rotating display. Also, long series of flicker-like whistles. [Gallito Arena]

Chucao Tapaculo *Scelorchilus rubecula* · 18 cm

Patagonian forest, often in bamboo. **ID** Small chunky understorey tapaculo with fairly long narrow cocked tail. Adult: Greyish crown, ear-coverts and flanks. Orange throat, breast and supercilium. White crescentic markings on abdomen. **Voice** Explosive loud ringing voice *popocatepetl*, sometimes preceded or followed by a low-pitched grunt. Calls include a querulous *kuoo* and a wooden rattle *beereo*. [Chucao]

Black-throated Huet-huet *Pteroptochos tarnii* · 24 cm

Humid Patagonian forest. **ID** Large and plump, but small-headed, understorey tapaculo with fairly long broad tail, usually held cocked; prominent white eye-ring, and very large claws used for scraping. Adult: Slaty-black upperparts, throat and breast. Rufous-chestnut cap, rump and belly. **Voice** Long series of hollow resonating notes which can be slow, descending and fading *kwók* notes or fast, ascending *whoop* notes, increasing in volume. Low-pitched *hwed* calls reveal presence. No overlap with Chestnut-throated Huet-huet. [Huet-huet]

Chestnut-throated Huet-huet *Pteroptochos castaneus* · 24 cm

Very local in dry forest of nw. Neuquén. **ID** Structure and behaviour much like Black-throated Huet-huet. Adult: Broader eye-ring than Black-throated. Rufous forehead, supercilium and entire underparts. Scalloping on underparts mixed with white. Slaty-brown upperparts with narrow buff wing bars. **Voice** Long series of hollow resonating notes and *hwed* calls are faster or higher-pitched than Black-throated Huet-huet. Contact call is a nasal *wehk*, recalling a squeaky toy. [Huet-huet Castaño]

Moustached Turca *Pteroptochos megapodius* · 23 cm

Arid shrubby slopes at c.2500 m, sw. Mendoza-Chilean border. **ID** Greyish-brown above with a rusty rump. Narrow white postocular streak and broad moustachial. Warm brown lower throat and breast; chestnut belly and vent, irregularly barred white. **Voice** Song begins with resonant *quip* series, like dripping liquid, well-spaced and descending into whooping cries. [Turca]

Olive-crowned Crescentchest

pallida

ad

ad

argentina

Crested Gallito

ad

ad

Sandy Gallito

ad

rubecula

Chucao Tapaculo

ad

Black-throated Huet-huet

megapodius

ad

ad

Moustached Turca

Chestnut-throated Huet-huet

Conopophaga Small chunky, short-tailed, reclusive, understorey passerine.

Rufous Gnateater *Conopophaga lineata*
13 cm

Bamboo in Paraná forest. **ID** Adult: ♂: Brown above and rufous below with whitish centre of belly and spot in centre of breast. Grey supercilium terminates in a concealed slivery-white streak which can be flared out. Adult ♀: Very similar to ♂ but supercilium shorter and greyer. **Voice** Fast ringing series of rising notes that ends abruptly, lasting 4–6 sec. Raspy sneeze-like contact call. Audible wing whir. [Chupadientes]

Grallaria Reclusive, long-legged, understorey birds with upright postures. Relatively large bills. *Hylopezus* Smaller and more slender-billed than *Grallaria* species.

Variegated Antpitta *Grallaria varia*
23 cm

Paraná forest. **ID** Adult: Intricately patterned but whitish lores and broad moustachial stand out in forest interior. Grey crown and notable buff spotting on wing-coverts. **Voice** Crepuscular song from high perch is a low-pitched hollow resonant series of 8–12 haunting notes *wu wu wu Wu WU WU WU Wu wu wu...*, rising and falling in pitch. [Chululú Pintado]

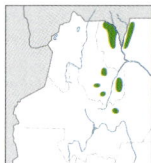

White-throated Antpitta *Grallaria albigula*
22 cm

N Yungas forest. **ID** Adult: Bright chestnut crown. White throat contrasts with grey breast. **Voice** Song from low perch is a rich two-note whistle with higher-pitched second note *WU WOO* with variations, heard throughout the day. Calls include a soft quavering *clu-clu-clu-clu-clu* with a toad-like quality, and an inquisitive *WEEOH*. [Chululú Cabeza Rojiza]

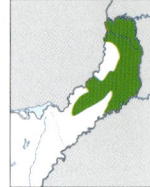

Speckle-breasted Antpitta *Hylopezus nattereri*
14 cm

Bamboo in Paraná forest. **ID** Adult: Buffy lores and eye-ring. Ochraceous below with strongly spotted breast and flanks. **Voice** Song is a pleasant rising series of 8–11 whistles in 2 secs, with a skipping cadence, *de-dee-dee-see-see-DEE-DEE-di.* Call is a rich short and monotone liquid chatter. [Chululú Chico]

Chamaeza Solitary understorey passerines which spend much of the time walking on the forest floor. Best detected by long whistled series of notes.

Short-tailed Antthrush *Chamaeza campanisona*
22.5 cm

Paraná forest. **ID** Adult: Rufescent crown contrasts with dull olive mantle. Blackish subterminal tail band tipped white. Broad chevron markings below. **Voice** Song is a medium-pitched, pulsing series of rising notes terminating in 4–8 dog-like descending yaps lasting 10–20 secs, *cu-cu-cu-cu-cu..... ke ke ke ke...* [Tovaca Parda]

Rufous-tailed Antthrush *Chamaeza ruficauda*
19 cm

Local in steep terrain in ne. Misiones. **ID** Adult: From Short-tailed Antthrush by smaller blackish bill, rufous rump and unbanded tail; denser ventral markings and distinctly barred undertail-coverts. **Voice** Song is a fast rising, resonant, tremulous trill of 4–5 secs, rather different and higher-pitched than Short-tailed. Call is a short rising or flat slow trill *brrrep.* [Tovaca Colorada]

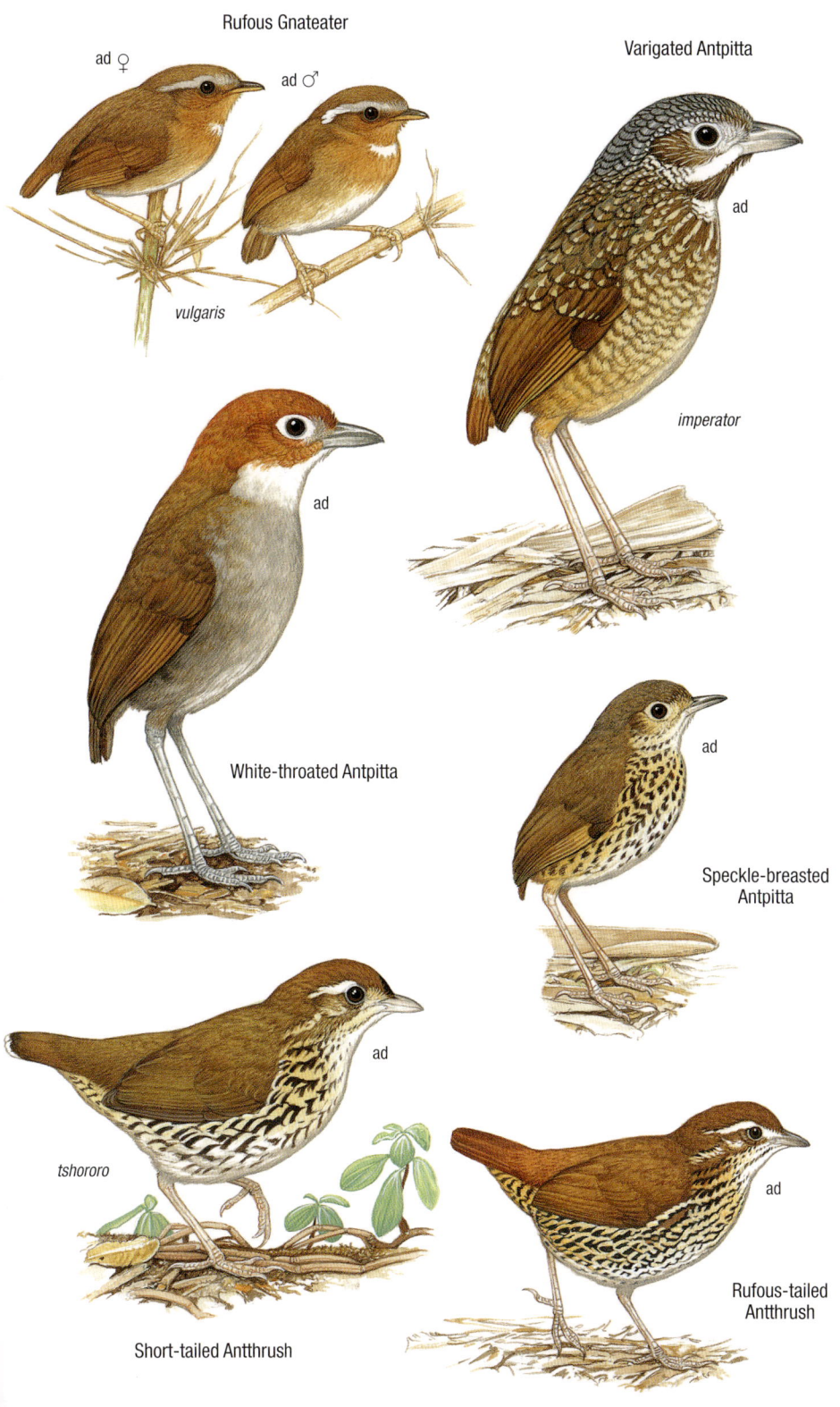

Rufous Gnateater

ad ♀

ad ♂

vulgaris

Varigated Antpitta

ad

imperator

ad

White-throated Antpitta

ad

Speckle-breasted Antpitta

ad

tshororo

Short-tailed Antthrush

ad

Rufous-tailed Antthrush

Dendrocolaptes Medium-sized with straight or slightly decurved bills with a hooked tip. *Xiphocolaptes* Large, stout bills with strongly curved culmen. *Campylorhamphus* Large with extremely long, decurved bills.

Black-banded Woodcreeper *Dendrocolaptes picumnus*   28 cm

Yungas forest and gallery forest in e. Formosa. **ID Adult**: Bill whitish or horn. Rufescent-brown above, brightest on wings and tail. Indistinct buff supercilium. Shaft streaks restricted to upper mantle unlike Planalto with which it overlaps in e. Formosa. **Voice** Song is an explosive, descending, liquid whinny that slows towards the end, lasting 2 secs. Long song is a slower monotone series, sometimes endless. [Trepador Colorado]

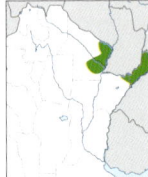

Planalto Woodcreeper *Dendrocolaptes platyrostris*   27.5 cm

NE forests. **ID Adult**: Black bill straighter and narrower than White-throated Woodcreeper. Whitish supercilium and streaking on ear-coverts; usually shows a narrow black moustachial. Shaft streaks cover entire mantle. Fine white streaks on breast, outlined in black. Overlaps with White-throated in Misiones and Corrientes, and with Black-banded in e. Formosa. **Voice** Song is a slow, monotone, liquid series of notes that slows towards the end, usually lasting 3–5 secs. Long song is a slower monotone series, sometimes endless. [Trepador Oscuro]

White-throated Woodcreeper *Xiphocolaptes albicollis*   31 cm

Paraná forest. **ID Adult**: Whitish supercilium and stripe across lower ear-coverts, bordered by thick black moustachial. Shaft streaks restricted to upper back and most of breast. White or creamy throat. Overlaps with Planalto Woodcreeper. **Voice** Syncopated crepuscular song is a slow, descending series of usually 6 disyllabic notes *k-TEW k-TEW k-TEW...*, sometimes preceded by a soft cat-like meow with a sneeze. [Trepador Garganta Blanca]

Great Rufous Woodcreeper *Xiphocolaptes major*   34 cm

Chaco woodlands and Yungas foothill forest. **ID Adult**: Stout pale bill. Rich rufous, often with a pale throat. **Voice** Syncopated crepuscular song is a slow, descending series, of usually 6 disyllabic notes *KI-ku KI-ku KI-ku...*, sometimes preceded by a soft cat-like meow with a sneeze. [Trepador Gigante]

Red-billed Scythebill *Campylorhamphus trochilirostris*   27 cm

Chaco woodlands and n. Corrientes. **ID Adult**: Red or orange-red bill. Rufous back, wings and tail. **Voice** Long song is a fast series of 7–10 rich, melancholic up-slurred whistles, with the final note clearly inflected downwards, lasting 2–3 secs. Short song is a rich whistled phrase that rises in pitch and ends in an inquisitive stutter, *flweee fwee wee wi ki-ki-k.* [Picapalo Colorado]

Black-billed Scythebill *Campylorhamphus falcularius*   28 cm

Paraná forest; mostly in bamboo. **ID Adult**: Black bill. Crown and face black, streaked creamy. Contrasting chestnut wings and tail. **Voice** A series of usually 5–8 well-spaced similar raspy notes that slowly increase in volume and length, *shre shre shrEE SHREE SHREEE shreee.* Also, an insect-like, dry, metallic, piercing phrase of 5–6 notes, *Klipzikzikzikzik!* [Picapalo Oscuro]

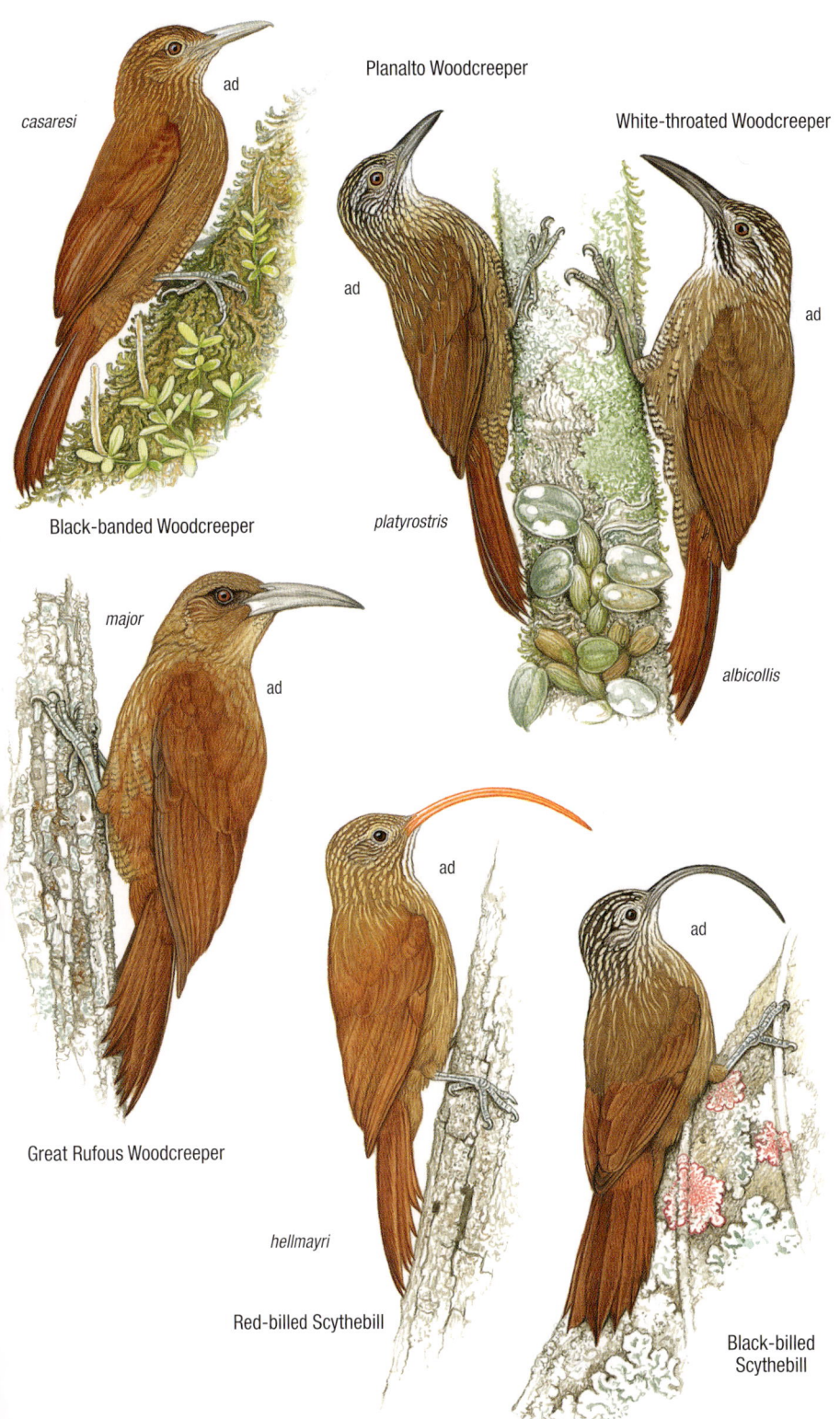

casaresi

ad

Planalto Woodcreeper

White-throated Woodcreeper

ad

ad

Black-banded Woodcreeper

platyrostris

albicollis

major

ad

Great Rufous Woodcreeper

ad

ad

hellmayri

Red-billed Scythebill

Black-billed Scythebill

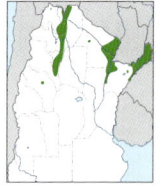

Olivaceous Woodcreeper *Sittasomus griseicapillus* 15.5 cm

Widespread. **ID** Tiny and agile with a short straight bill. **Adult** *griseicapillus* (Yungas forest and chaco woodlands): Greyish-olive body with contrasting rufous wings, rump and tail. **Adult** *sylviellus* (Paraná forest): Bronze-olive upperparts and rich olive-buff underparts. **Voice** *griseicapillus*: song is a long, slow, series of rising and falling whistled notes; *sylviellus*: song is a slow series of high-pitched, sharp, rising notes progressively lower-pitched and fading, typically only first 3 notes given. Both ssp. also differ in their long, sometimes endless, trills. **Tax note 33.** [Tarefero]

Plain-winged Woodcreeper *Dendrocincla turdina* 20 cm

Paraná forest. **ID** Medium-sized with uniform plumage and short, straight bill. **Adult:** Uniform olive-brown above, more tawny below with a paler olive-buff throat. Often shows an indistinct dusky malar streak. Contrasting dark rufous-chestnut tail. **Voice** A very long, monotone, series of dry insect-like chips (4–6 per sec) that indiscriminately speed up or slow down, sometimes endlessly. [Arapasú]

Lesser Woodcreeper *Xiphorhynchus fuscus* 16.5 cm

Paraná forest. **ID** Smaller than Narrow-billed and Scaled Woodcreepers and with a straighter bill. **Adult:** Bill almost straight. Buff or white eye-ring. Throat and underparts buffy, with diffuse streaking edged pale olive-brown. Overlaps with Scaled Woodcreeper. **Voice** Song is a falling and rising series of accelerating/decelerating chip notes; may recall Southern Bristle Tyrant, but less metallic and lower-pitched. [Chinchero Enano]

Scaled Woodcreeper *Lepidocolaptes squamatus* 20.5 cm

Paraná forest. **ID** Small, with slender, notably decurved bill and scaled underparts. **Adult:** Pure white throat. Breast white becoming buff on belly, boldly scaled blackish throughout. Overlaps with smaller Lesser Woodcreeper. **Voice** A sudden, tremulous note given alone or in loose series, very similar to one call of Black-capped Foliage-gleaner but lower-pitched and less shrill. Less often, a series of 2–4 melancholic rising and falling whistles. **Tax note 34.** [Chinchero Escamado]

Narrow-billed Woodcreeper *Lepidocolaptes angustirostris* 21.5 cm

Widespread. **ID** Small, with slender decurved bill and streaked underparts. Blackish crown with broad white supercilium and blackish eye-stripe. Rufous tail. **Adult** *praedatus* (most widespread ssp. throughout most of N, also C and E lowlands): Pale, slightly decurved bill. Underparts diffusely streaked. **Adult** *angustirostris* (humid chaco, n. Corrientes and N sectors of dry chaco): Straight bill; much shorter than *praedatus*. Back and wings distinctly rufous; more distinct ventral streaking. **Voice** Series of clean whistles or tremulous notes that speed up and descend towards the end, lasting 2–3 secs. Common call is a loud, explosive, tremulous whistle. [Chinchero Chico]

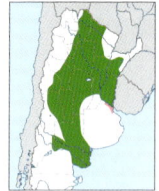

Scimitar-billed Woodcreeper *Drymornis bridgesii* 31 cm

N and C lowlands. **ID** Terrestrial and arboreal with very long decurved bill. **Adult:** Shaggy crown; bold white supercilium and moustachial. Underparts densely streaked white. Rufous tail. **Voice** Song is a long, powerful, springy, metallic laughter that accelerates towards the end and finishes with a fast chatter; often abbreviated to the first notes. Call is a dry, disyllabic *CHITIK*. [Chinchero Grande]

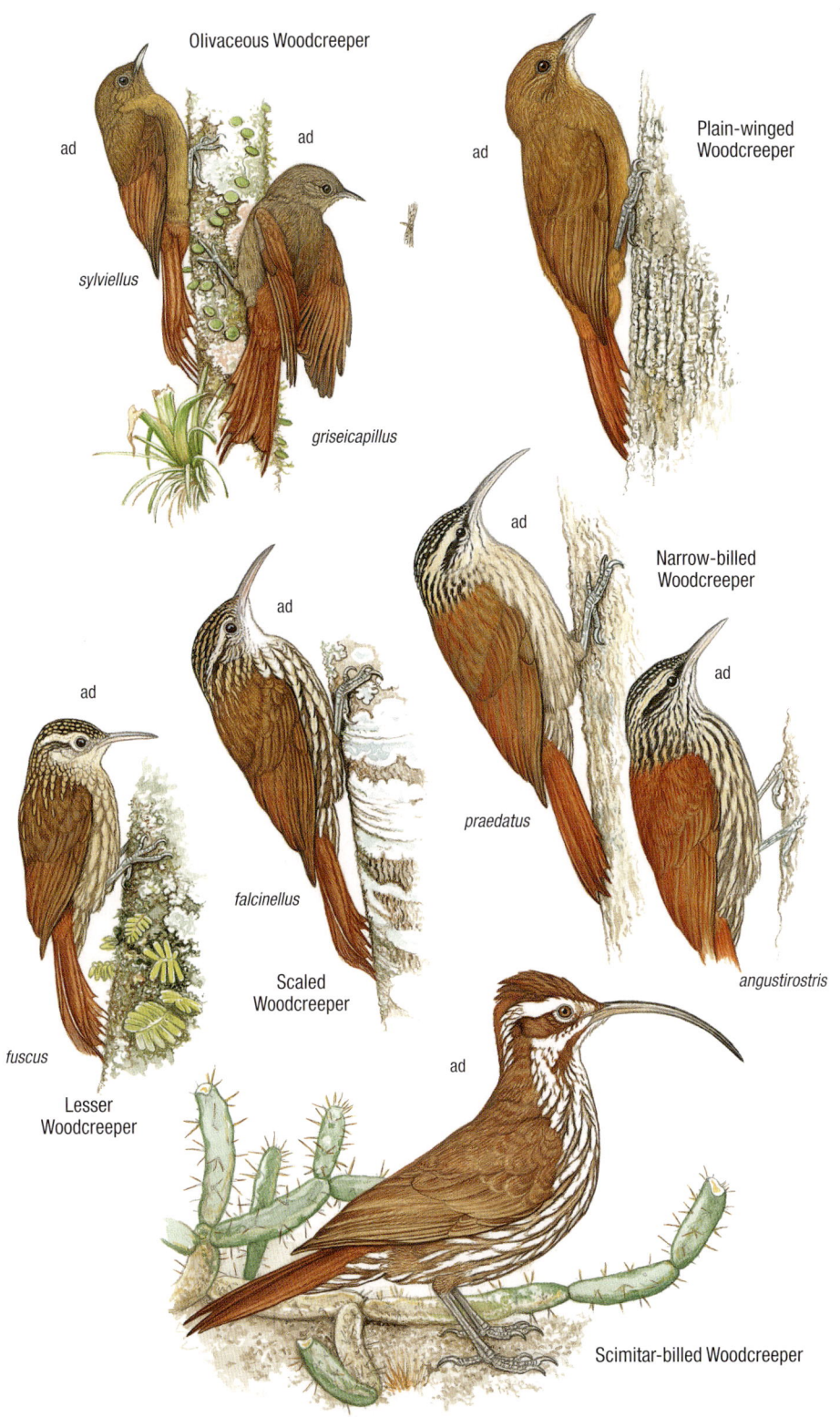

Olivaceous Woodcreeper

ad

ad

sylviellus

griseicapillus

Plain-winged Woodcreeper

ad

Narrow-billed Woodcreeper

ad

ad

ad

praedatus

falcinellus

angustirostris

ad

fuscus

Lesser Woodcreeper

Scaled Woodcreeper

ad

Scimitar-billed Woodcreeper

Geositta Short-tailed, terrestrial furnariids which wag the hindbody while foraging. They differ by wing pattern, rump colour, bill shape and length, and presence of streaking or mottling on the breast.

Short-billed Miner *Geositta antarctica*   15 cm

S Patagonia to C Andes in winter. **ID Adult:** Bill short, almost straight. Breast mottled brown outlining whitish spots, which can suggest streaking. Overlaps with nominate ssp. of Common Miner. In flight, wings mostly indistinct pale cinnamon. **Voice** A spluttered series of short rachet-like notes, lasting 1–2 secs. Compare with nominate ssp. of Common Miner. [Caminera Patagónica]

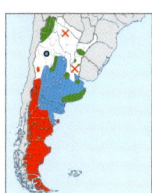

Common Miner *Geositta cunicularia*   15 cm

Widespread. **ID Adult** *cunicularia*: The only miner with prominent breast streaking. Overlaps with Short-billed and Trilling Miners. **Adult** *titicacae* (N Andes): Paler above than *cunicularia* with indistinct breast streaks. Overlaps with Puna, Buzzing and Slender-billed Miners. In flight, wings mostly rufous with a dusky subterminal bar. Contrasting creamy rump and white edges to tail. **Voice** *cunicularia*: a steady series of complaining couplets *we-TUK we-TUK we-TUK...* lasting 2–6 secs, while perched or in display flight. Compare with Short-billed Miner. *titicacae*: vocally little known; flight call is a spluttered trill followed by a squeaky stutter *chrrrrr kwe-kwe-kwe-kwe-kwe-kwe...* **Tax note 35.** [Caminera Estriada]

Puna Miner *Geositta punensis*   15.5 cm

Puna. **ID Adult:** Short black bill, upright stance and warm sandy upperparts. Clean white below. In flight, wings mostly rufous with dusky subterminal bar. Cinnamon rump and tail edges. Overlaps with ssp. *titicacae* of Common Miner and Buzzing Miner. **Voice** Song is a long, slow series of variable chipping notes. Call is a sharp, piercing *PEEW*. [Caminera Puneña]

Buzzing Miner *Geositta [rufipennis] rufipennis*   17.5 cm

C (below 2400 m) and N Andes, C sierras, mesetas in N Patagonia. **ID Adult:** Warm brown upperparts and cleaner underparts than Trilling Miner. Overlaps with ssp. *titicacae* of Common Miner and Puna Miner. In flight, wings mostly rufous with prominent blackish subterminal bar. Rufous rump, blackish subterminal tail band. **Voice** Very distinctive, long, slow, gravelly buzzing series of notes *GRRI-GRRI-GRRI-GRRI...*, given while perched. Call is a descending, vibrant *prrew.* **Tax note 36.** [Caminera Zumbadora]

Trilling Miner *Geositta [rufipennis] fasciata*   17.5 cm

C (above 2400 m) and S Andes. **ID Adult** *giaii* (S Andes): Short straight bill. Grey-brown above, plain unmarked breast, and rufous on flanks. Overlaps with Short-billed, Common (nominate), Buzzing and Creamy-rumped Miners. Plumage pattern identical to Buzzing in flight. **Voice** Unlike Buzzing, fast high-pitched trill, punctuated by rapid piping notes *chi-ch-ch-ch-ch-ch-PPPPPPprrrr, chi-ch-ch-ch-ch-ch-PPPPPP prrrr, chi-ch-ch-ch-ch-ch-PPPPPPprrrr,* in flight display. Calls include short *chuip* and *choc* notes. **Tax note 36.** [Caminera Trinadora]

Slender-billed Miner *Geositta tenuirostris*   18.5 cm

N and C Andes. **ID Adult:** Long, slender, decurved bill suggesting an earthcreeper. Short rufous tail with a dusky T-mark. Brown scalloping on breast, sometimes indistinct. In flight, rufous wings lack the dusky subterminal bar of other miners in range. Brown rump and rufous sides to tail. Wings proportionately broader and tail proportionately shorter than Buzzing Miner. Overlaps with ssp. *titicacae* of Common Miner, as well as Buzzing, Trilling and Creamy-rumped Miners. **Voice** A slow, unevenly patterned, series of dry, resonant, piping notes *khe khe khe khe khe...,* given in flight display. Call is a complaining *WEEO* or *WEEK.* [Caminera Picuda]

Creamy-rumped Miner *Geositta isabellina*   17.5 cm

Mainly C Andes (2600–5000 m); w. La Rioja, San Juan and Mendoza. **ID Adult:** Fairly long bill, often with a yellow base. Head and sometimes throat washed rusty. Unmarked breast. In flight, dull rufous wings and contrasting creamy white rump. Overlaps with ssp. *titicacae* of Common Miner, as well as Trilling and Slender-billed Miners. **Voice** Song is a long, somewhat descending, series of fast, squeaky ringing notes. [Caminera Grande]

Short-billed Miner

ad

ad

ad

ad

Common Miner

ad

cunicularia

titicacae

Puna Miner

ad

ad

ad

Buzzing Miner

ad

ad

Trilling Miner

giaii

ad

Creamy-rumped Miner

ad

tenuirostris

ad

Slender-billed Miner

Ochetorhynchus: Terrestrial tail-cocking earthcreepers of the Andes and Patagonia with a straightish bill. *Tarphonomus*: Terrestrial and arboreal earthcreepers of dry scrub. *Lochmias* Small dark furnariid of forest rivers and streams.

Rock Earthcreeper *Ochetorhynchus andaecola* 19 cm

NW Andes (above 2400 m). **ID Adult:** Bill slightly decurved. Tail rufous-chestnut. Broad whitish dagger-like streaks on sides of breast and flanks, highlighted with brown. Overlaps with Bolivian Earthcreeper; closely approaches Straight-billed but keeps to shrubby valleys and gullies, generally at lower altitude. **Voice** Song is a series of squeaky, shrill, introductory notes dropping in pitch to a final chatter *weez, weez, weez, weet, weet, weet WI-WITITITtttt-TEEEE*. Common call is a drawn-out, strident, pure whistle *SWEEET*. Also gives long trills and shrill calls. [Bandurrita Cola Castaña]

Straight-billed Earthcreeper *Ochetorhynchus ruficaudus* 18 cm

Andes. **ID Adult:** Bill almost straight. Tail rufous with blackish inner webs on all but central rectrices. Underparts flammulated with white. Usually on flatter terrain above Rock Earthcreeper in the NW. **Voice** Song is very similar to Rock Earthcreeper, but with faster introductory notes and lower-pitched, squeakier, shorter notes throughout. Common call is a shorter, flatter, whistled *QUEEP*, but very similar to Rock Earthcreeper. Also gives long trills and shrill calls. [Bandurrita Pico Recto]

Band-tailed Earthcreeper *Ochetorhynchus phoenicurus* 18.5 cm

S Andean foothills and Patagonia. **ID** Arboreal and terrestrial furnariid with a fairly long straight bill; recalls a large canastero. Cocks tail showing diagnostic pattern. **Adult:** Chestnut tail with black T-shaped mark. Narrow whitish supercilium; flammulated white below. Nest on Plate 201. **Voice** Harsh, fast, low-pitched rattled trill followed by up to 8 well-spaced shrill whistles *brrrrrrrrlip CLIP… CLIP… CLIP… CLIP…*. Common call is a very fast, high-pitched, monotone, metallic trill *ZRRRRIII* ending abruptly. [Bandurrita Patagónica]

Bolivian Earthcreeper *Tarphonomus harterti* 16 cm

Rare in Andes of extreme n. Salta (2500–2650 m). **ID Adult:** Yellowish base to almost straight black bill. Blackish lores and eye-stripe. Rufous supraloral and narrow cinnamon supercilium. Short, squarish, bright rufous tail. Creamy throat; drab brownish-buff below. Compare with Rock Earthcreeper and Streak-fronted Thornbird (Plate 127). **Voice** Perhaps all vocalisations are indistinguishable from Chaco Earthcreeper in the field, although long song may be shriller. [Bandurrita Boliviana]

Chaco Earthcreeper *Tarphonomus certhioides* 18 cm

N and C lowlands and foothills. **ID Adult:** Dull rufous forehead and narrow supercilium. Tail dull rufous-brown with brown central rectrices. White throat contrasts with grey-brown underparts. No overlap with Bolivian Earthcreeper; compare with Rufous Hornero (Plate 128). **Voice** Short song is a very loud, steady series of 6–10 monotone *chit* notes, sometimes fading off and ending in more spaced, lower-pitched *CHEP* notes. Long song is a somewhat descending series of flat or inflected short, shrill whistles *ZWIT ZWIT ZWIT zwit zwit…* Call is a nasal, hoarse *TZÜ*. [Bandurrita Chaqueña]

Sharp-tailed Streamcreeper *Lochmias nematura* 14.5–15.5 cm

Paraná and Yungas forest. **ID** Small, short-tailed, dark brown furnariid, found exclusively along forest rivers and streams. **Adult *nematura*** (14.5 cm; mainly Paraná forest): Narrow whitish supercilium; underparts heavily spotted white. **Adult *obscuratus*** (15.5 cm; local in Yungas forest): Larger and longer-billed than *nematura*, but lacks a supercilium and shows only sparse spotting on throat and breast. **Voice** Song is a fast, metallic rattle that increases in volume, usually lasting 5 secs. Common call is a succession of scolding, metallic couplets or triplets *ti-rip* or *zi-ri-rip* (rarely with more notes). Tax note 37. [Macuquito]

Rock Earthcreeper

montanus
ad

Straight-billed Earthcreeper

ad

Band-tailed Earthcreeper

certhioides
ad

ad

Bolivian Earthcreeper

Chaco Earthcreeper

ad

ad

obscuratus

nematura

Sharp-tailed Streamcreeper

Cinclodes Terrestrial, semi-aquatic furnariids. Best identified by wing and tail pattern. See also Plate 122.

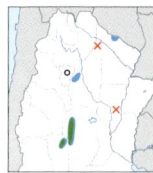

Córdoba Cinclodes *Cinclodes comechingonus* 17.5 cm

Endemic. C sierras; some winter in N lowlands. **ID** Adult: Yellow bill base. Rufous wing panel, primary-coverts and tail corners. In flight, striking rufous wing-stripe. Overlaps with Grey-flanked (ssp. *olrogi*) and White-winged (ssp. *schocolatinus*) Cinclodes, and with Buff-winged in winter. **Voice** Song is composed of alternating series of twittering and rapid, scratchy trills that fluctuate upscale and downscale. [Remolinera Serrana]

Cream-winged Cinclodes *Cinclodes albiventris* 16.5 cm

NW Andes; lower in winter. **ID** Adult: Overlaps with and resembles a miniature White-winged Cinclodes but buff carpal patch and tail corners, and less rufescent above. In flight, wing-stripe buff, becoming white distally. **Voice** Song is composed of alternating series of very high-pitched ticking notes and very rapid, dry trills that fluctuate upscale and downscale. Tax note 38. [Remolinera Acanelada]

Buff-winged Cinclodes *Cinclodes fuscus* 18 cm

Patagonia to the N lowlands in winter. **ID** Adult: Grey-brown or brown above, and greyish-buff below. In flight, cinnamon-buff wing-stripe; tail corners buffy to buffy-white. Overlaps with Grey-flanked (ssp. *oustaleti*) and Dark-bellied Cinclodes, and with Córdoba, Cream-winged, Grey-flanked (ssp. *olrogi*) and White-winged (ssp. *schocolatinus*) in winter. **Voice** Song is composed of alternating series of very high-pitched ticking notes and rapid, flat, dry trills, sometimes including distinctive buzzy notes. Short song is a springy stutter *ch ch ch ch chdritit-chdritit-chdritit* repeated at short intervals. [Remolinera Parda]

Grey-flanked Cinclodes *Cinclodes oustaleti* 16.5–18 cm

Widespread. **ID** From congeners by small white centre to abdomen. Adult *oustaleti* (17.5 cm; C Andes): Brown above with a greyer crown. Buff tail corners. Narrow white supercilium. Light brown breast, speckled whitish. Overlaps with Buff-winged. In flight, wing-stripe entirely buff. Adult *hornensis* (18 cm; S Andes): Crown and especially back greyer than *oustaleti* and *olrogi*. Largest ssp. In flight, wing-stripe buff at the base and white distally. Overlaps with Dark-bellied Cinclodes. Adult *olrogi* (16.5 cm; C sierras): Similar to *oustaleti* but smaller with warmer brown back and flanks. In flight, bicoloured wing-stripe as *hornensis*. Overlaps with Córdoba and White-winged Cinclodes, and with Buff-winged in winter. **Voice** Identical to Cream-winged Cinclodes. Tax note 38. [Remolinera Chica]

Dark-bellied Cinclodes *Cinclodes patagonicus* 20 cm

S Andes and Tierra del Fuego; often by forest rivers, lakes and seashore. **ID** Adult: Overlaps with Grey-flanked Cinclodes, but bill longer, supercilium broader, throat and sides of neck pure white, and more heavily streaked below. Often appears crested. In flight, buff wing-stripe and tail corners. **Voice** Song is a rising and falling trill with two introductory notes, *wi-chi-chitrrrrrrrrrrrrrrrrrrr*, overall slower and lower-pitched than congeners. [Remolinera Araucana]

White-winged Cinclodes *Cinclodes atacamensis* 21 cm

N Andes and C sierras. **ID** Adult *atacamensis* (N Andes): Chocolate-brown above, greyish breast, and warm brown flanks and vent. White wing-stripe, carpal patch and tail corners. Overlaps with Cream-winged Cinclodes. Adult *schocolatinus* (C sierras): Less rufescent above and greyer on the breast than *atacamensis*. Overlaps with Córdoba and Grey-flanked Cinclodes (ssp. *olrogi*), and with Buff-winged in winter. **Voice** Common call is a mellow whistle followed by a slow rattle *WEE TR-R-RR*, recalling the song of Cliff Flycatcher but less burry. [Remolinera Castaña]

Córdoba Cinclodes

ad

Cream-winged
Cinclodes

ad

ad

Buff-winged
Cinclodes

oustaleti

ad

ad

oustaleti

Grey-flanked
Cinclodes

ad

olrogi

hornensis

ad

hornensis

ad

schocolatinus

patagonicus

ad

ad

ad

Dark-bellied
Cinclodes

atacamensis

White-winged
Cinclodes

Cinclodes Terrestrial, semi-aquatic furnariids. Best identified by wing and tail pattern. See also Plate 121.

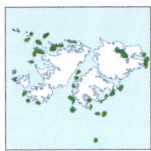

Tussacbird *Cinclodes* [*antarcticus*] *antarcticus* 19.5 cm

Endemic to coast and steppe on offshore islands of the Falklands. **ID** Adult: Short dark bill. Brown with a pale buff throat and indistinct supercilium. In flight, ill-defined warm brown wing-stripe. Contrasting pale central tail feathers. **Voice** Song is a fast monotone ringing trill while standing or in flight. Calls are mostly *chip* notes. Vocally similar to Buff-winged Cinclodes. **Tax note 39**. [Alt. Blackish Cinclodes] [Remolinera Malvinera]

Fuegian Cinclodes *Cinclodes* [*antarcticus*] *maculirostris* 21 cm

Scarce at pinniped and shag colonies of s. Fuegian, Hornean and Isla de los Estados islets. **ID** Adult: Yellow bill with a black upper mandible. Sooty-brown, mottled on the throat. In flight, uniformly dark with no wing-stripe. **Voice** Vocally unknown. **Tax note 39**. [Alt. Blackish Cinclodes] [Remolinera Fueguina]

Upucerthia Mainly terrestrial furnariids with a proportionately long decurved bill and long tail which is often cocked. Perform wing-raising and wing-rotating displays and often sing in flight.

Patagonian Forest Earthcreeper *Upucerthia saturatior* 20 cm

Patagonian forest edge and clearings. **ID** Adult: Bill shorter and blacker than Scale-throated Earthcreeper; also upperparts much darker brown and breast scalloping poorly defined. **Voice** Song is a series of stuttered ticking couplets or triplets *p-p-tirik-tirik... p-p-tirik-tirik-tirik*, often from a treetop. Call is a dry chipping *pep*. **Tax note 40**. [Bandurrita de Bosque]

Scale-throated Earthcreeper *Upucerthia dumetaria* 21.5 cm

Widespread. **ID** Adult *dumetaria* (Patagonian steppe to N lowlands in winter): Long decurved bill. Breast prominently scalloped black. Cinnamon tail corners. Adult *hypoleuca* (Resident in NW; not illustrated): Warmer above. **Juvenile**: More extensive ventral scaling. Pale streaks on nape and upper mantle. **Voice** Song is a continuous series of similar ringing notes that indiscriminately rise and fall and change speed *pli-pli-pli-pli-pli-pli...* Call is a sharp *keep*. [Bandurrita Esteparia]

Buff-breasted Earthcreeper *Upucerthia validirostris* 18–21 cm

NW Andes and C sierras. **ID** Adult *validirostris* (21 cm): Heavily decurved bill. Warm brown above and rich buff below. Conspicuous rufous patch in flight feathers. Adult *pallida* (18 cm; Bolivian border to s. Jujuy and n. Salta): Smaller than *validirostris*. Pinkish-buff or sandy underparts. Bill tends to be smaller than *validirostris*. **Voice** Song begins with a stutter that accelerates into a strident rising and falling rattled trill, sometimes longer and indiscriminately oscillating in frequency, given from a rock or boulder top. Also gives a brief staccato *tk-t* call note. **Tax note 41**. [Bandurrita Andina]

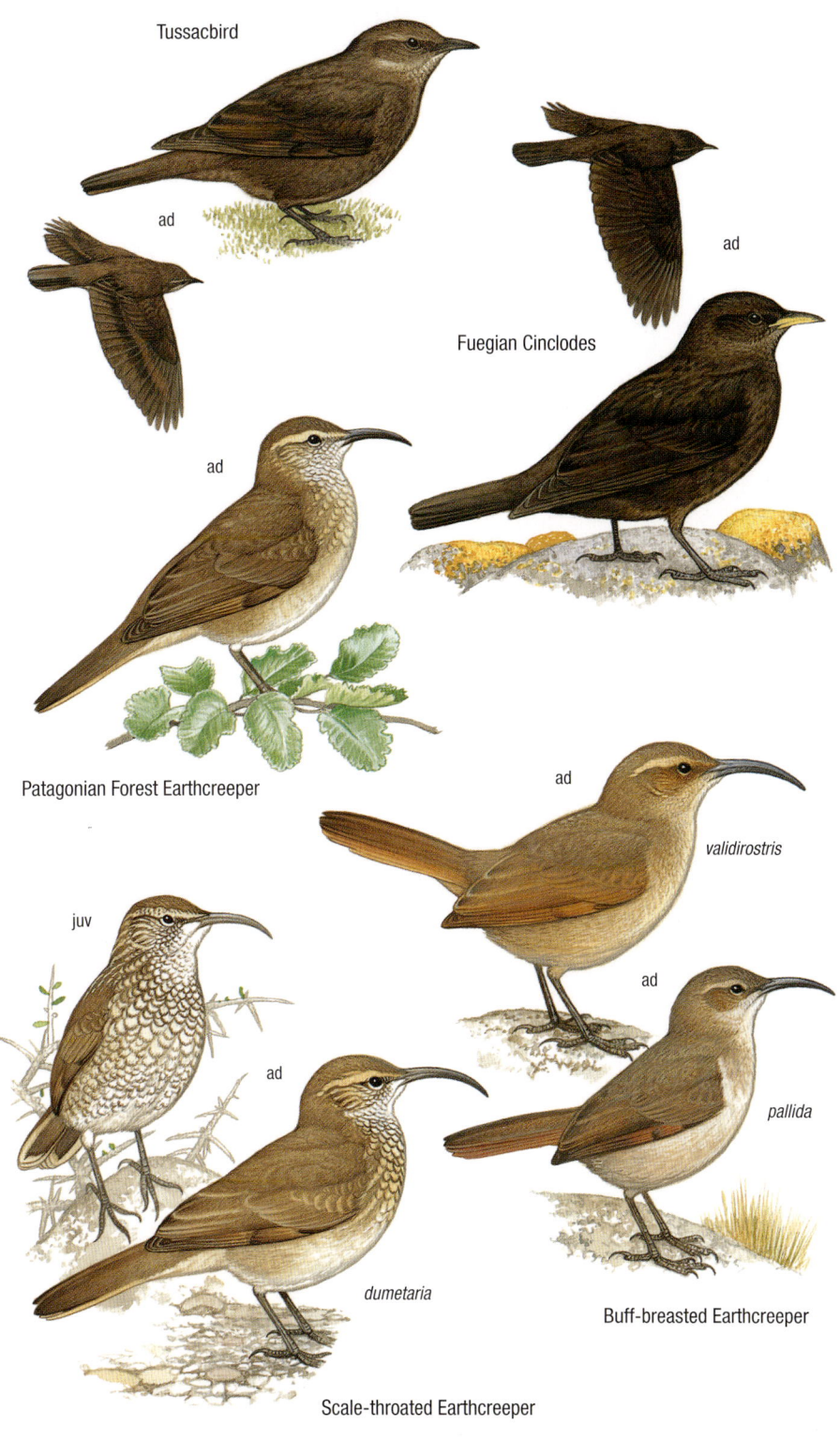

Tussacbird

ad

ad

Fuegian Cinclodes

ad

Patagonian Forest Earthcreeper

ad

validirostris

juv

ad

ad

pallida

dumetaria

Buff-breasted Earthcreeper

Scale-throated Earthcreeper

Buff-browed Foliage-gleaner *Syndactyla rufosuperciliata*   17.5 cm

Widespread in NE and Yungas forest. **ID** Understorey foliage-gleaner with slightly upturned lower mandible. **Adult** *acrita* (NE, illustrated): Pink bill base. Narrow buff supercilium. White throat, yellowish spots on breast becoming flammulated below. Overlaps with Sharp-billed Treehunter and White-browed Foliage-gleaner. **Adult** *oleaginea* (Yungas forest, not illustrated): Very similar but with a slightly less olivaceous back. **Voice** A variety of song-types; most commonly a squeaky, rising and falling rattle that accelerates and becomes harsher. Also, a monotone slow rattle with well-spaced introductory notes, *KEW…KEW… KEW-KEW-KEW-KEW…* and a harsh monotone stutter. Calls include a variety of sharp, metallic notes and screeches. **Tax note 42**. [Ticotico Estriado]

White-browed Foliage-gleaner *Anabacerthia amaurotis*   15.5 cm

Rare in ne. Misiones understorey. **ID** Small foliage-gleaner with a relatively short, stout, straight bill. **Adult**: Prominent buffy white supercilium from eye, bordered by blackish eye-stripe. Plain back. Blurry spots on breast. Compare with Buff-browed Foliage-gleaner and Sharp-billed Treehunter. **Voice** A high-pitched, insect-like, sharp series of *spit* notes; may speed into a fast chatter. [Ticotico Ceja Blanca]

Sharp-billed Treehunter *Heliobletus contaminatus*   13 cm

Paraná forest. **ID** Small midstorey furnariid with slender bill. **Adult**: Ochraceous supercilium from eye. Yellowish-buff throat, sides of neck and streaks on upper mantle. Overlaps with Buff-browed and White-browed Foliage-gleaners. **Voice** Song is a metallic, monotonous, medium-pitched, spluttered trill, sometimes with a higher-pitched introduction. Call is a sharp *pit* note, usually repeated. [Picolezna Estriado]

Pearled Treerunner *Margarornis squamiger*   15 cm

Very rare in Yungas forest. **ID** Small arboreal furnariid with tail spines. Favours mossy limbs in cloud forest. **Adult**: Mostly rich rufous above. Narrow white supercilium. Yellowish throat and tear-drop-shaped spots on underparts. Compare with Streaked Xenops. **Voice** A very high-pitched, shrill twittering. [Ticotico Goteado]

Streaked Xenops *Xenops rutilans*   12.5 cm

Paraná forest and Yungas foothills (ssp. similar). **ID** Small arboreal furnariid with upturned lower mandible and white malar streak. **Adult** *rutilans* (Paraná Forest, illustrated): Rufous mantle and unmarked rufous tail. Streaked underparts. **Voice** Song is a fast series of 5–10 dry, lisping notes, *shi-shi-shi-shi-shi-shi*, sometimes fading away and dropping in pitch. Calls comprise similar but isolated notes. [Picolezna Rojizo]

Plain Xenops *Xenops minutus*   11.5 cm

Paraná forest; mainly in riparian vines. **ID** Small arboreal furnariid with upturned lower mandible and white malar streak. **Adult**: Brown mantle and cinnamon tail with a symmetrical black pattern. Unstreaked underparts. **Voice** Song is a steady series of 5–10 upward-inflected pure notes, *swip swip swip swip swip*, sometimes followed by a fast chatter. Calls comprise similar but isolated notes. [Picolezna Chico]

acrita
ad

Buff-browed Foliage-gleaner

ad

White-browed Foliage-gleaner

contaminatus

ad

Sharp-billed Treehunter

ad

squamiger

Pearled Treerunner

ad

rutilans

Streaked Xenops

ad

minutus

Plain Xenops

Ochre-breasted Foliage-gleaner *Anabacerthia lichtensteini* 17.5 cm

Paraná forest. Midstorey to canopy. **ID** Small foliage-gleaner with a relatively short, stout, straight bill. Adult: Olive-grey lores, crown and nape. Dull rufous wings and slightly notched tail. Compare with Buff-fronted Foliage-gleaner. **Voice** Song is a steady series of 5–12 dry, insect-like, metallic notes *clip clip clip clip clip*, higher-pitched and slower than Buff-fronted Foliage-gleaner. Call is a brief shrill insect-like trill *ziririp!* [Ticotico Ocráceo]

Buff-fronted Foliage-gleaner *Philydor rufum* 19.5 cm

Paraná forest. Midstorey to canopy. **ID** Unstreaked foliage-gleaner with stout bill and notched tail. Usually in mixed flocks. Adult: Ochraceous lores and forehead; rest of crown, nape and eye-stripe grey. Fairly bright rufous wings. Tail notably forked. Compare with Ochre-breasted Foliage-gleaner. **Voice** Song is a stuttered sharp metallic chipper *kekekekekecliclicliclicliclicliclicli*, lower-pitched and notably faster than Ochre-breasted Foliage-gleaner. Calls include a brief rasp and isolated metallic notes. [Ticotico Grande]

Black-capped Foliage-gleaner *Philydor atricapillus* 17 cm

N Paraná forest. Understorey to midstorey. **ID** Unstreaked foliage-gleaner with stout bill and notched tail. Usually in mixed flocks, often with Red-crowned Ant Tanager. Adult: Striking black crown, moustachial and eye-stripe. Bright rufous sides of neck. **Voice** Song is a harsh, 5 sec-long, fast churring trill that descends gradually and ends abruptly. Common call is a loud rich upward-inflected series of 2–4 *chWEE* notes. Also, a tremulous note given alone or in a loose series, *trreew*, very similar to Scaled Woodcreeper but higher-pitched and shriller. [Ticotico Cabeza Negra]

White-eyed Foliage-gleaner *Automolus leucophthalmus* 20 cm

Paraná forest. **ID** Noisy understorey foliage-gleaner with broad rounded tail. Often found in bamboo. Adult: White iris. Contrasting puffed-out white throat. Broad rufous tail. **Voice** A fast, querulous *clee-cly-cly-cly-cly-cly-cly* (song?). Alternates with a noisy series of almost woodpecker-like, excitable *KIK-werk* phrases. [Ticotico Ojo Blanco]

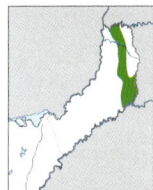

Canebrake Groundcreeper *Clibanornis dendrocolaptoides* 22.5 cm

Local in n. Misiones. **ID** Chunky, raucous, reclusive forest furnariid of bamboo (especially *Merostachys*) and vine tangles. Adult: Robust bill. Rufous-chestnut crown and broad tail stand out in dark forest interior. Long whitish or grey supercilium. White throat, speckled dusky at sides. **Voice** Song is a two-part scolding strident series, emphatic at first and followed by a higher-pitched chatter *chok chok tchuk chok ke-ke-ke-ke-ke-ke*, occasionally duetted; sings from a perch in shady understorey. Calls include a loud, sneezing *SCREW* and a disyllabic or trisyllabic *TA-tk*. [Tacuarero]

Rufous-breasted Leaftosser *Sclerurus scansor* 18.5 cm

Paraná forest. **ID** Mainly terrestrial furnariid with long slender bill and short tail. Nests in banks. Adult: Dark brown with rufous-chestnut rump and black tail. Mature adult shows rufous wash on breast. **Voice** Song is a cascading, metallic chipper trill lasting 3 secs. Call is a metallic, piercing *TIK*, sometimes doubled or in decelerating series. [Raspahojas]

Ochre-breasted
Foliage-gleaner

ad

ad

Buff-fronted
Foliage-gleaner

rufum

Black-capped
Foliage-gleaner

ad

ad

White-eyed
Foliage-gleaner

sulphurescens

Rufous-breasted
Leaflosser

ad

ad

scansor

Canebrake Groundcreeper

Leptasthenura Slender, arboreal furnariids with long graduated tails, with spiked central rectrices. Open woodlands and arid scrub (except Araucaria Tit-Spinetail). *Sylviorthorhynchus* Furtive, slender-tailed furnariids with a fine bill and rufous forehead.

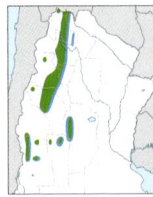

Brown-capped Tit-Spinetail *Leptasthenura fuliginiceps* 15 cm

N Andes and C sierras. **ID** Adult: Rusty crown with bushy crest; rufous wings and tail contrast with pale brown mantle and greyish-buff underparts. Overlaps locally with Tawny Tit-Spinetail. **Voice** Song is a high-pitched, weak, monotone chipper (or stuttered) trill with audibly separate notes; sometimes long, complex and decelerating. Contact call is a high-pitched *swip* note. [Coludito Canela]

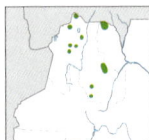

Tawny Tit-Spinetail *Sylviorthorhynchus yanacensis* 16 cm

Local in *Polylepis* groves in Jujuy and Salta. **ID** Adult: From Brown-capped Tit-Spinetail by lack of crest; small rufous spot on forehead, long fine bill; longer more spiny tail and ochraceous underparts. **Voice** Song is a series of fairly loud, sharp *chip* notes accelerating into variable metallic trills. Also, weak twittering trills which may rise and fall in pitch; may recall Maquis Canastero (Plate 130), but higher-pitched. [Coludito de Queñoal]

Tufted Tit-Spinetail *Leptasthenura platensis* 16.5 cm

Widespread in lowlands and foothills. **ID** Adult: Prominent crest, tawny-buff (not white) edges to tail. Notable streaking on throat and upper breast. Some overlap with Plain-mantled, and possibly with Puna Tit-Spinetail. **Voice** Song is a high-pitched, short series ending in a short, scratchy trill *tsee-tee-tee-te-t-tttttt*, lasting only 2–3 secs. May recall song of Straneck's Tyrannulet (Plate 135), which overlaps [Coludito Copetón]

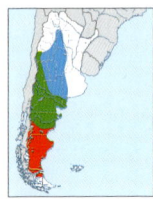

Plain-mantled Tit-Spinetail *Leptasthenura aegithaloides* 16 cm

Patagonia to C lowlands in winter. **ID** Adult *pallida*: Similar to Puna Tit-Spinetail but crown streaked buff; colder, greyer mantle and paler underparts. Differs from Tufted Tit-Spinetail by white edges to tail and reduced crest. Adult *aegithaloides* (NW Patagonian forest; not illustrated): Darker overall. **Voice** Song is a fluctuating medium-pitched, long, twittering chatter, punctuated by squeaky notes *chititititititi-pwe-chititi-pwe-chititititi-pwe...*, lasting 4 secs or more. Commonest call is a very fast, short series of buzzy, metallic, high-pitched notes *zrri-zri-zri-zri-zri-zrree-zrree* in less than 1 sec. Also, a variety of soft trills. [Coludito Cola Negra]

Puna Tit-Spinetail *Leptasthenura [aegithaloides] berlepschi* 15 cm

NW Andes. **ID** Adult: Slightly crested, blackish crown, streaked rufous; whitish streaks on nape and sides of neck; olive-brown mantle; tail edged white. Fine mottling on throat and breast but clean white in centre; buff below. **Voice** A flat, dry, fast ticking of evenly spaced notes lasting 3–4 secs. Also, a variety of soft trills. Tax note 43. [Coludito del Altiplano]

Araucaria Tit-Spinetail *Leptasthenura setaria* 18 cm

Restricted to Paraná Pine (*Araucaria angustifolia*) forest and plantations in Misiones. **ID** Adult: Blackish crest, finely streaked white; russet mantle and rump; rufous tail with very long spines; throat white, finely speckled brown; light buff below. **Voice** Song is a fast descending metallic chatter ending in a faster trill. May resemble song of *Cranioleuca* spinetails but faster and higher-pitched. Long song comprises many chipper phrases ending with the trill. Contact call is a *Tic* note. [Coludito de los Pinos]

Des Murs's Wiretail *Sylviorthorhynchus desmursii* 23 cm

Patagonian forest, often in bamboo understorey. **ID** Elongated central tail feathers; only six filamentous rectrices. Adult: Long fine bill, rufous forehead, buff below. **Voice** Song is a very rapid sequence of 5–7 disyllabic springy notes, *wa-kili wa-kili wa-kili...* Also delivers a harsh, rapid monotone trill, and a nasal *kük*, often in a well-spaced series. [Colilarga].

Brown-capped Tit-Spinetail

paranensis

ad

Tawny Tit-Spinetail

ad

Plain-mantled
Tit-Spinetail

ad

ad

pallida

ad

Tufted Tit-Spinetail

ad

Puna Tit-Spinetail

Araucaria Tit-Spinetail

ad

Des Murs's Wiretail

Straight-billed Reedhaunter *Limnoctites rectirostris* 16.5 cm

Local in Paraná Delta. **ID** Restricted to marshes with serrucheta (*Eryngium* spp). Very secretive. **Adult:** Bill long, slender and straight with pinkish-grey lower mandible, tipped dark; crown and nape tinged grey; long narrow whitish supercilium. Juvenile: Shorter bill; warm buff underparts and supercilium; lacks grey on head. Compare with Sulphur-bearded Reedhaunter with which it is nearly always found. **Voice** Song is a series of sharp notes accelerating into a rattled trill *chit chit chit chit cht cht-trt trt-trt-trrrrrrrrrrrrrrr*. Call is a sharp *stip* note. [Pajonalera Pico Recto]

Sulphur-bearded Reedhaunter *Limnoctites sulphuriferus* 15.5 cm

Local in Mesopotamia and Pampas. **ID** Adult: Whitish supercilium; tawny panel in primaries; lemon-yellow throat stripe; underparts white, finely streaked grey. Compare with Yellow-chinned Spinetail and vocally very similar Straight-billed Reedhaunter. **Voice** Song has an initial rising rattle and a chortled series of churring monotone couplets that speeds into a descending rattle *trrrrrep chr chr chr chr chr chr chr chr ch-t-t-trrrrew*; lower-pitched than Straight-billed Reedhaunter. Call is a brief *chú*, higher-pitched than Curve-billed Reedhaunter. [Curutié Ocráceo]

Curve-billed Reedhaunter *Limnornis curvirostris* 17 cm

NE marshes; mainly e. Buenos Aires. **ID** Chunky, marsh-dwelling furnariid with broad, rounded rufous tail. Often bold and inquisitive. **Adult:** Heavy, slightly decurved bill with whitish lower mandible. Broad white supercilium. Nest on Plate 201. **Voice** Song has an initial rising rattle ascending into a stuttering series of chipping notes that speeds into a descending rattle, *trrrrrep te te TE TE te te te-t-t-trrrrew*; similar to Sulphur-bearded Reedhaunter but more obviously rising and falling, and without couplets. Sometimes followed by a long machine-gun rattle punctuated with squeaky notes. Call is a brief metallic *Tk*. [Pajonalera Pico Curvo]

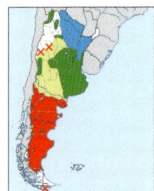

Wren-like Rushbird *Phleocryptes melanops* 14.5–17 cm

Widespread. **ID** Chunky, short-tailed marsh furnariid, mainly restricted to rushbeds. Adult *melanops* (14.5 cm; widespread): Bold supercilium; tawny nuchal collar; dark brown mantle, streaked whitish; extensive rufous in wings. Adult *schoenobaenus* (17cm; Puna marshes, not illustrated): Much larger, with a rufescent rump and warmer buff breast. Nest on Plate 201. **Voice** Monotonous ticking voice and a sporadic, 'cloth-ripping', harsh *zeeep*. [Junquero]

Yellow-chinned Spinetail *Certhiaxis cinnamomeus* 16.5 cm

Widespread in north. **ID** Marsh-dwelling spinetail with bicoloured plumage. Adult: Rufous crown, wings and tail. Underparts mainly white; yellow chin spot. Nest on Plate 201. **Voice** A variable dry metallic rattle, remarkably like that of Rufous-sided Crake (Plate 54), but never as long. Also, a monotone accelerating rattle and a brief nasal *chrree*. [Curutié Colorado]

Bay-capped Wren-Spinetail *Spartonoica maluroides* 14.5 cm

Local in Pampas and Mesopotamia; some moving north in winter. **ID** Saltmarsh-dwelling furnariid with streaked upperparts. **Adult:** Pale iris. Orange-rufous crown. Heavily streaked upperparts. Juvenile: Buffy below with white throat; lacks rufous crown and pale iris. Compare with Chotoy Spinetail (Plate 131). **Voice** Simple song is a very rapid, dry, shrill, grasshopper-like trill lasting 1.5 secs. Complex song is a fast series of alternating higher- and lower-pitched trills and raspy sounds. Call is a high-pitched *tsi* or *tip*. [Espartillero Enano]

ad

Straight-billed Reedhaunter

juv

ad

Sulphur-bearded
Reedhaunter

ad

Curve-billed
Reedhaunter

melanops

ad

Wren-like Rushbird

ad

juv

Bay-capped
Wren-Spinetail

russeola

ad

Yellow-chinned Spinetail

Phacellodomus Mainly arboreal furnariids with conspicuous stick nests. Rather similar loud descending songs that speed up, often duetted. Pitch and speed match the size of species. Notes are very nasal in Rufous-fronted, whistling in Little and resounding in Greater. Also, shorter rising-falling songs. Ticking calls are very similar.

Rufous-fronted Thornbird *Phacellodomus rufifrons*   18 cm

Dry chaco and dry rain shadow Yungas foothills. **ID Adult:** Rufous forecrown; grey supercilium broadens behind eye. Outer tail feathers tinged rufous. Overlaps with Little, Greater and Spot-breasted Thornbirds. Nest on Plate 200. [Espinero Frente Rojiza]

Streak-fronted Thornbird *Phacellodomus striaticeps*   18 cm

Rocky or shrubby gullies in high Andean steppe and pre-Puna (usually above 2000 m). **ID Adult:** Rufous forehead; greyish streaks on central crown. Indistinct whitish supercilium. Rufous shoulder. Blackish tail with rufous outer tail feathers, tipped dusky. [Espinero Andino]

Little Thornbird *Phacellodomus sibilatrix*   15 cm

Widespread in N thorn woodland. **ID Adult:** Plain grey face with capped effect. Tiny amount of rufous on forehead; less than others of genus. Rufous shoulder spot and outer tail feathers. Overlaps with Rufous-fronted, Freckle-breasted and Greater Thornbirds. Compare with Short-billed Canastero (Plate 129). Nest on Plate 200. [Espinero Chico]

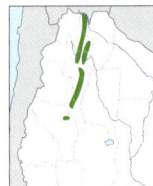

Spot-breasted Thornbird *Phacellodomus maculipectus*   17.5 cm

Clearings and shrubby slopes in Yungas forest (400–2900 m), south to La Rioja. **ID Adult:** Dark iris. Rufous crown with whitish shaft streaks. Short, broad whitish rear supercilium. Rufous outer tail feathers. White throat contrasting with fairly bright rufous wash over malar region and breast, mottled with whitish spots. Possible overlap with Streak-fronted Thornbird at treeline, and with Rufous-fronted Thornbird in dry Yungas ecotone. Nest on Plate 200. [Espinero Serrano]

Freckle-breasted Thornbird *Phacellodomus striaticollis*   17.5 cm

Shrubs in Pampas marshes. **ID Adult:** Amber or dull yellow iris. Crown rufous with small area of brown on forehead. Dull whitish supercilium. Dull rufous outer tail feathers. Breast washed rufous, with tiny white shaft streaks. Overlaps with Rufous-fronted and Greater Thornbirds. Nest on Plate 200. [Espinero Pecho Manchado]

Greater Thornbird *Phacellodomus ruber*   21 cm

Marshes and riversides in chaco and Mesopotamia. **ID Adult:** Bright yellow iris. Whitish lores but no supercilium. From others of genus by rufous crown, wing-coverts, remiges and tail. Overlaps with Rufous-fronted, Little and Freckle-breasted Thornbirds. [Espinero Grande]

Orange-breasted Thornbird *Phacellodomus ferrugineigula*   18 cm

Marshy clearings in or near Paraná forest in ce. Misiones. **ID** No other thornbird in range. **Adult:** Dark iris. Brown above with contrasting rufous crown and outer tail feathers. Orange-rufous throat and breast. **Voice** Song is a loud monotone, almost flicker-like *clu KWEEP WEEP WEEP WEEP WEEP*, or more disyllabic *clu KILIP KILIP KILIP KILIP*, sometimes in duet. [Espinero Pecho Naranja]

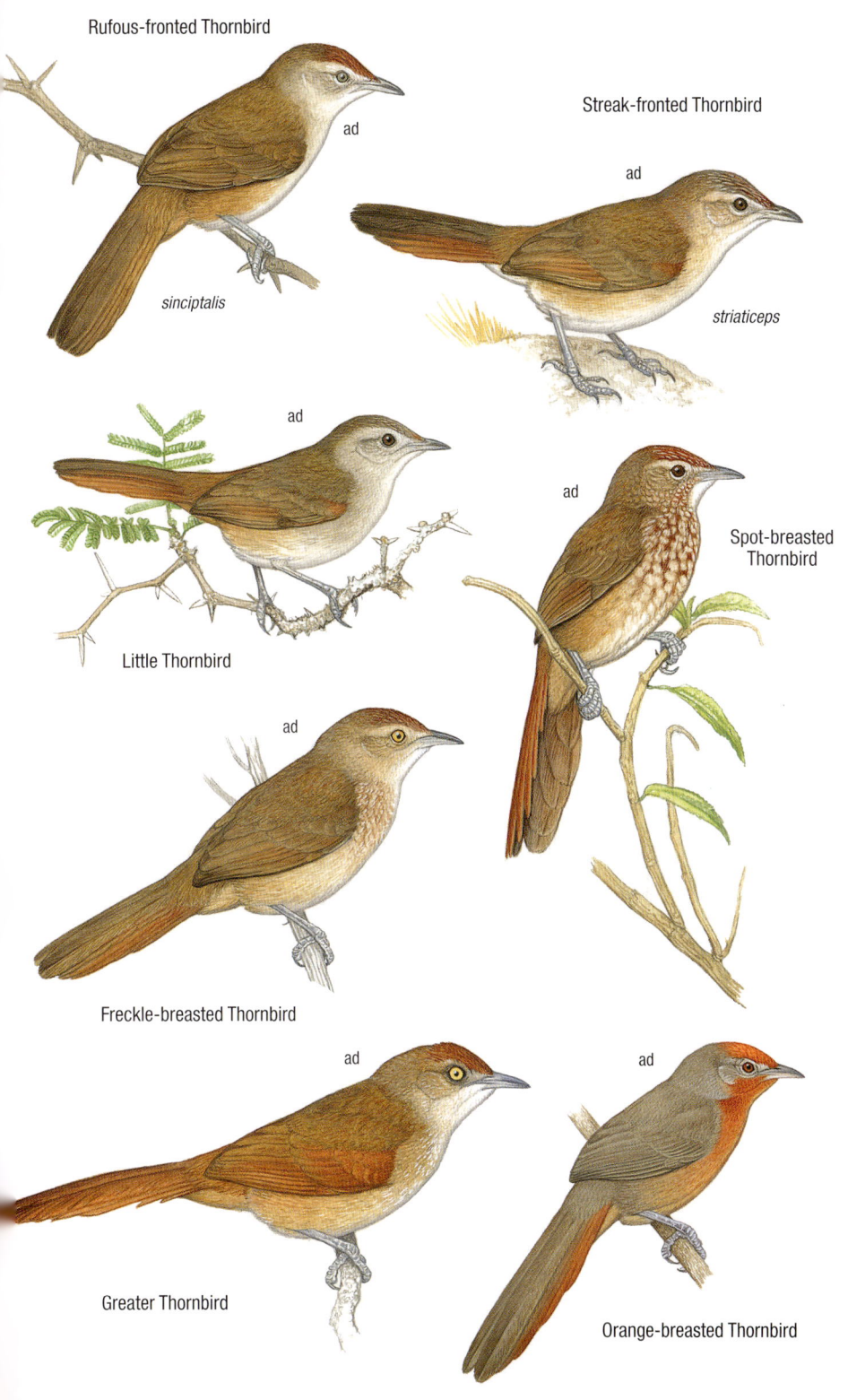

Rufous-fronted Thornbird

ad

sinciptalis

Streak-fronted Thornbird

ad

striaticeps

ad

Little Thornbird

ad

Spot-breasted
Thornbird

ad

Freckle-breasted Thornbird

ad

Greater Thornbird

ad

Orange-breasted Thornbird

Rufous Hornero *Furnarius rufus*   18–22 cm

Widespread. **ID** Terrestrial and arboreal furnariid with contrasting rufous tail and characteristic strutting gait. Conspicuous oven-shaped mud nest (Plate 200). **Adult** *rufus* (22 cm; Mesopotamia and C lowlands south): Rufous wash on nape and forecrown. Buffy lores. White throat, centre of belly and vent. **Adult** *paraguayae* (18 cm; NC lowlands): Smaller than *rufus* with more contrasting rufous nape. Compare with Chaco Earthcreeper (Plate 120). **Voice** Loud duetting of accelerating and decelerating notes, sometimes in short renderings by solitary individuals. Calls include monotone rattles that increase in volume, and *KEE* notes, isolated, or in a long series. [Hornero]

Crested Hornero *Furnarius cristatus*   16 cm

Dry chaco and C Sierras. **ID** Plumage and behaviour as Rufous Hornero. Smaller mud nest. **Adult:** Pointed crest. White lores, throat and centre of abdomen and vent. Overlaps with ssp. *paraguayae* of Rufous Hornero. **Voice** Song and calls are similar to Rufous Hornero but higher-pitched and faster. Distinctive call is an upswept note followed by a rattle of semi-musical metallic notes, *KWEE dedededede* or *KWEE bereberebereberebere...* [Hornerito Copetón]

Lark-like Brushrunner *Coryphistera alaudina*   16.5 cm

N and C lowlands. **ID** Gregarious, terrestrial and arboreal furnariid with upstanding blackish crest. Nest on Plate 201. **Adult:** White subocular crescent and chestnut ear-coverts. Rufous tail bases. **Voice** Common voice is a weak, muffled yet metallic *zhheerrrrr*, sometimes followed by a fast, tinny trilll *trrleeeeeeee*. Excited groups can deliver a mixture of rasps, chatters and soft whistles [Crestudo]

Firewood-gatherer *Anumbius annumbi*   19.5 cm

Widespread in the north, south to N Patagonia. **ID** Slender, mainly arboreal furnariid which builds enormous stick nest (Plate 200). **Adult:** Indistinct rufous forecrown. White throat outlined with blackish flecks. Graduated, pointed tail with white tips; best seen in flight. **Voice** Song is an accelerating series of 2–5 chipping introductory notes ending in a dry rattle, *che che che-che-che chrrrrrrree*. Also, a long accelerating and decelerating rattle and ticking calls. [Leñatero]

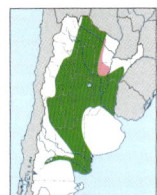

Brown Cacholote *Pseudoseisura lophotes*   27 cm

Widespread in the N and C. **ID** The largest, most raucous furnariid. Terrestrial and arboreal. Huge stick nests (Plate 200). **Adult:** Yellow iris. Prominent crest. Contrasting rufous-chestnut throat, nape and tail. **Voice** Song is an unmistakable cacophony of very loud raucous grating notes that become faster, shorter and lower-pitched towards the end; often a second bird duets with piping or short raspy notes, lasting 12–16 secs. Call is a loud cackling *cluck*, sometimes in series. [Cacholote Castaño]

White-throated Cacholote *Pseudoseisura gutturalis*   25–26 cm

Endemic. Monte desert and Patagonia. **ID** Large raucous furnariid. Terrestrial and arboreal. Huge stick nests. **Adult** *gutturalis* (25 cm; S Monte desert and Patagonia, illustrated): Greyish-brown with slight bushy crest. White throat with black smudge at base. **Adult** *ochroleuca* (26 cm; N Monte desert): differs by sandy-brown upperparts and sandy-buff underparts. **Voice** Song is a far-carrying duet starting with 4–6 soft raspy notes that switch into short whistled notes, grading into disyllabic cackles while speeding up and descending in pitch for 12–16 secs. Sings much faster in the north. [Cacholote Pardo]

Rufous Hornero

ad

paraguayae

ad

rufus

Crested Hornero

ad

ad

alaudina

Lark-like Brushrunner

ad

Firewood-gatherer

ad

argentina

Brown Cacholote

ad

gutturalis

White-throated Cacholote

Asthenes and *Pseudasthenes* Small brown terrestrial and arboreal furnariids of open-country and light woodland, with long tails, rufous wing-stripes and coloured gular patches. Best distinguished by combination of bill shape, colour of gular patch, shape and amount of rufous in tail, behaviour, especially tail-cocking or pumping, and habitat. All *Asthenes* (except Short-billed and Rusty-vented) give similar double trills. See also Plate 130.

Cordilleran Canastero *Asthenes modesta*   17 cm

Rocky areas in the Andes, Sierras and Patagonia. **ID** Adult *modesta* (NW Andes): Long, slender, pointed bill. Shows more rufous in the tail than any other canastero, with blackish-brown restricted to inner webs. Adult *australis* (Patagonia): Darker grey-brown above than *modesta* with prominent streaking on sides of throat and breast. **Voice** Song is a short, fast, rising liquid trill *triririririririRRI*, lasting 1.5–2 secs. Recalls Puna Canastero (Plate 130) but lacks the descending rattle. Call is a sharp *TÉU* or *PIP*. [Canastero Pálido]

Patagonian Canastero *Pseudasthenes patagonica*   15 cm

Endemic. Brush-steppe in Patagonia and Andean foothills north to s. San Juan. **ID** Adult: Short bill. Blackish tail with rufous on outer web of outermost rectrix; rarely visible. Blackish gular patch, spotted white. Greyish breast and pale cinnamon flanks, belly and vent. Nest on Plate 201. **Voice** Song is a long, dry, rattled trill that gradually descends, *TWRRRRRrrrrreeeeeeeeeew*, lasting 3–5 secs. Also, a series of chipping notes that slows towards the end, *tiptiptiptiptiptip-tip-tip tip tip tip*, recalling Chaco Earthcreeper (Plate 120). Calls include slow grating rasps and short churrs, *prr*, often in series. [Canastero Patagónico]

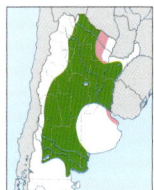

Short-billed Canastero *Asthenes baeri*   15 cm

Light woodlands in N lowlands, Andean foothills and C sierras to N Patagonia. **ID** Short stubby bill. Grey supercilium, broadening behind eye. Relatively short tail. Blackish central tail feathers and rufous outer tail feathers. Orange gular patch. Overlaps with Sharp-billed (especially in winter), Steinbach's and Patagonian Canasteros. Compare also with Little Thornbird (Plate 127). Nest on Plate 201. **Voice** Short song during the day is a very fast reeling trill of sharp notes, speeding in the middle and fading towards the end, lasting 2 secs. Long song at dawn is a long string of higher-pitched fast trills and lower-pitched slower trills, *trrrrreeeeRRRRRrreeeRRRRrreeeRRRRrrreeeeRRRR…* lasting 4 secs. Long descending metallic trills beginning with spaced notes (may recall a *Cranioleuca* spinetail). Also, soft raspy notes. [Canastero Chaqueño]

Steinbach's Canastero *Pseudasthenes steinbachi*   17.5 cm

Endemic. Monte desert and badland canyons from Salta to n. Chubut, and Sierra de las Quijadas, n. San Luis. **ID** Adult: Relatively long attenuated tail. Brown crown and mantle with grey wash on crown. Rufous rump and entirely rufous outer pair of tail feathers; rest of tail feathers blackish, fringed rufous on both webs. Usually lacks a gular patch. Overlaps with Short-billed Canastero in the south, and with Rusty-vented Canastero in the north. **Voice** Song is a long, slow, monotone, rich rippling chatter *webereberebereberebere*; lasting 4–6 secs. Call is a dry nasal buzz, recalling Southern Martin (Plate 160). [Canastero Castaño]

Rusty-vented Canastero *Asthenes dorbignyi*   15.5 cm

Woodland and shrubland in NW Andes. **ID** Adult: Brown crown (usually tinged rufous) and mantle, becoming rufous on rump. Relatively short, square-ended, blackish tail which shows little rufous in outer tail feathers. Dark red gular patch; may appear blackish. Nest on Plate 201. **Voice** Much like Short-billed Canastero, but slightly slower and lower-pitched. [Canastero Rojizo]

Dusky-tailed Canastero *Pseudasthenes humicola*   17 cm

Historical specimen record from Andes of Mendoza. **ID** Adult: Brown above, slightly warmer on crown and rump. Prominent creamy supercilium. Blackish tail with rufous on outer web of outermost rectrix. Rufous shoulder. White throat, stippled black; breast flammulated greyish. May overlap with Patagonian Canastero. **Voice** Song is a squeaky rising or rising and falling series of syncopated disyllabic notes that speed up into a final trill, *we ka we ka we-ka we-ka weka-wekawetidldldee* and variations, lasting up to 3 secs. Also, a long descending metallic trill. Calls include ticking *STIT* and *STUT* notes often given in series. [Canastero Estriado]

modesta

ad

Cordilleran Canastero

australis

ad

ad

Patagonian Canastero

ad

baeri

Short-billed Canastero

ad

Steinbach's Canastero

ad

Rusty-vented Canastero

ad

Dusky-tailed Canastero

Asthenes See notes on Plate 129. Maquis and Sharp-billed have distinctive long tails that are pumped while foraging.

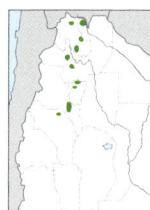

Maquis Canastero *Asthenes heterura*   18 cm

Local in NW Andean brush-steppe. **ID** Pumps long tail. **Adult:** Slender, pointed bill. From Sharp-billed Canastero by more pointed, brown central tail feathers, more rufous wing-coverts and more extensive rufous wing-stripe. **Voice** Song is a steady, squeaky, rising or rising-falling *chu chui chui CHUI CHUI-chi-chir-chu-chu*. Also, an excited dry descending or oscillating trill of variable length and intensity *rrreeeee*. [Canastero Quebradeño]

Sharp-billed Canastero *Asthenes pyrrholeuca*   16 cm

Breeds in Patagonia; winters in marshy habitats in N lowlands; local breeding populations in C and NW. **ID** Pumps long tail. **Adult:** Slender straight bill with chisel-shaped tip. Tail has blackish central rectrices. **Voice** Song is a rapid, squeaky rising or rising and falling *du-dui-dui-duiduiDUIDUIDUIDUIdid idididdidiu*. Call is a piercing, metallic upswept *SUIT*, often given on the wintering grounds. [Canastero Coludo]

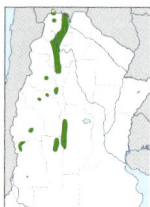

Puna Canastero *Asthenes sclateri*   17.5–18.5 cm

Pre-Puna, Puna and Sierran rocky grasslands. **ID** Adult *lilloi* (17.5 cm; Tucumán to La Rioja): Brown upperparts, coarsely streaked black. Tail edged and tipped cinnamon. Cinnamon-buff below; orange gular patch. Overlaps with Scribble-tailed Canastero. **Adult** *sclateri* (18.5 cm; Sierras of Córdoba–San Luis): Greyer above than *lilloi* with denser streaking; greyish-brown below. **Voice** Song is a short, fast, rising dry trill that ends in a faster, slightly descending rattle. Recalls Cordilleran Canastero Plate 129) but not so liquid and with a different ending. Calls include ticking notes, often in series. **Tax note 44.** [Espartillero Serrano]

Scribble-tailed Canastero *Asthenes maculicauda*   15 cm

Local in steep patches of tall grasslands in NW Andes. **ID** Adult: Short bill. Striking rufous forecrown. Mantle streaked black and white. Relatively short, graduated pointed tail feathers with diagonal brown barring and buff tips to penultimate pair. Lacks gular patch. Overlaps with Puna Canastero (ssp. *lilloi*). Compare with Grass Wren (Plate 162). **Voice** Song is a raspy, fairly high-pitched, slightly rising *zree zree zree sri-sri-sri-sri titrrrrrrrr*, with a rapid final descending trill, from the top of a *Festuca* grass clump. Call is a high-pitched whistled *swee-IP* or *whoo-IT*, inflected upwards on last syllable; ♀ responds with higher-pitched whistle. [Espartillero Estriado]

Austral Canastero *Asthenes anthoides*   16 cm

W and S Patagonia. **ID** Adult: Grey-brown upperparts, heavily streaked blackish. White fringes to tertials. Tail tipped and edged cinnamon-buff; conspicuous in flight. Pale orange gular patch. Compare with Grass Wren (Plate 162). **Voice** Song is a short, fast, dry metallic monotone or very slightly rising trill of 1 sec. Recalls Cordilleran Canastero, but faster and higher-pitched. Call is a high-pitched metallic *stik* note, often in couplets. [Espartillero Austral]

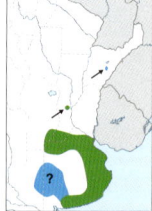

Hudson's Canastero *Asthenes hudsoni*   17.5 cm

Pampas marshes and Paraná Delta. **ID** Adult: Whitish fringes on streaked mantle and tertials. Large dull orange gular patch; dark streaks on flanks. Compare juvenile Bay-capped Wren-Spinetail (Plate 126) and Long-tailed Reed Finch (Plate 183). **Voice** Song begins with 1–4 pure whistles that ascend in pitch, running into a trill that descends at the end, *see swee SWEE SWEE zizizizirrrrrrrrr*. Call is a dry muted *ch-t* or *t-t*. [Espartillero Pampeano]

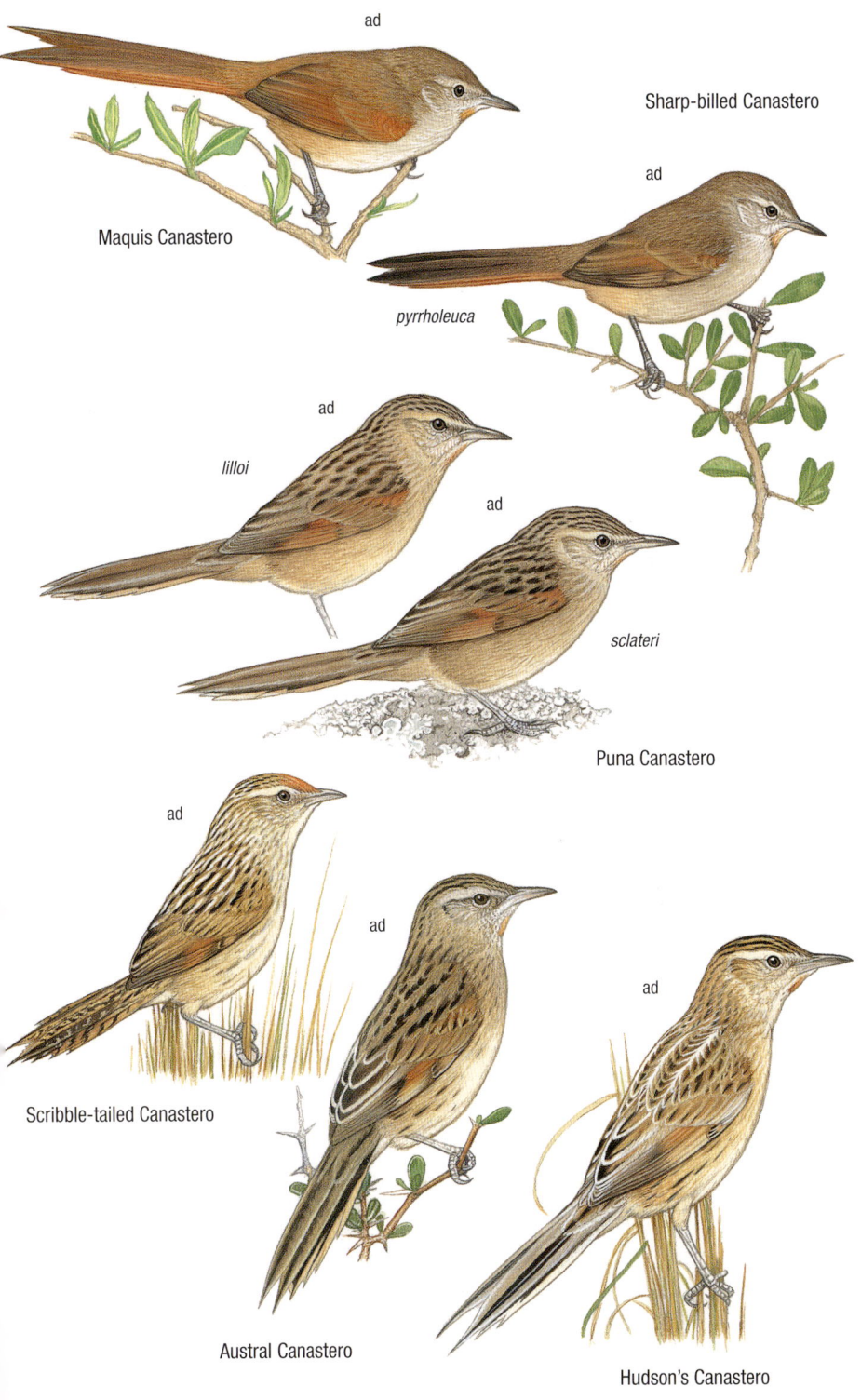

ad

Sharp-billed Canastero

Maquis Canastero

ad

pyrrholeuca

lilloi

ad

ad

sclateri

Puna Canastero

ad

ad

ad

Scribble-tailed Canastero

Austral Canastero

Hudson's Canastero

Chotoy Spinetail *Schoeniophylax phryganophilus*   20 cm

Widespread in N lowlands. **ID** Stocky, open-country spinetail with long graduated tail and streaked upperparts. **Adult:** Rufous cap and shoulders; yellow and black throat; orangey breast. **Juvenile:** Lacks rufous cap and throat markings. Compare with juvenile Bay-capped Wren-spinetail (Plate 126). Nest on Plate 200. **Voice** Unmistakable loud, throaty chortling song ending in a churring rattle lasting up to 5 secs, often in duet. Also a cackling series ending in a rattle, recalling Dark-billed Cuckoo. [Chotoy]

Olive Spinetail *Cranioleuca obsoleta*   14.5 cm

Paraná and gallery forest in NE. Possible overlap with Stripe-crowned Spinetail. **ID Adult:** Brownish-olive above; narrow whitish or buff supercilium; dull brownish-olive below, whiter on throat and browner on flanks. Note rufous shoulder and tail. **Voice** A fairly fast, cascading song of short, tinny, lisping notes with variation. Call is a short, trilled *chiririp* or *trreep*. [Curutié Oliváceo]

Stripe-crowned Spinetail *Cranioleuca pyrrhophia*   15.5 cm

Widespread in N and C (except Misiones). **ID Adult:** Crown streaked whitish; restricted or lacking in some birds. Upperparts olive-brown; broad white supercilium; white or whitish below becoming dull buff on flanks and vent. Note rufous shoulder and outer tail-feathers. Nest on Plate 201. **Voice** Song is highly variable, but generally begins with spaced-out stuttered metallic lisping notes that speed into a descending trill. Can recall the descending song of Short-billed Canastero. Call is a short, trilled *chiririp* or *trreep*. [Curutié Blanco]

White-throated Treerunner *Pygarrhichas albogularis*   16 cm

Patagonian forest. **ID** Chunky arboreal furnariid with chisel-shaped bill. **Adult:** Brown upperparts becoming rufous on wings, rump and spiny tail. Blackish eye-stripe. White throat, black below spotted white. **Juvenile:** Buff streaks on mantle and spots on crown. Blackish fringes on throat. **Voice** Delivers loud ticking calls, often in couplets. [Picolezna Patagónico]

Thorn-tailed Rayadito *Aphrastura spinicauda*   14 cm

Patagonian forest; in tussock grass in the extreme S. **ID** Chiefly arboreal, inquisitive and generally abundant furnariid with notable tail spines. **Adult:** Striking combination of black, rufous, buff and white. **Voice** Complex reeling, twittering and trilling, often in a long frantic series. [Rayadito]

Synallaxis Understorey spinetails. Knowledge of range and voice greatly aids ID. Extent of gular patch highly variable. Note combination of rufous on crown, wings and tail; presence of brown forehead; and coloration of underparts. Typically short disyllabic songs and also complex multi-note songs in most species. Six more species on Plate 132.

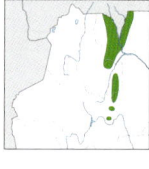

Ochre-cheeked Spinetail *Synallaxis scutata*   14.5 cm

Yungas foothills. **ID Adult:** White supercilium turning buff at rear; dark eye-stripe. Chestnut wings and tail. Large black gular patch. Face, sides of neck, breast and flanks ochraceous. **Voice** Song is a pure upswept sweet *su-íp* or *sweet sueet*; very like allopatric Grey-bellied Spinetail (Plate 132). Call is a single piercing note. **Tax note 45.** [Pijuí Canela]

juv

Chotoy Spinetail

phryganophilus

ad

ad

Olive Spinetail

ad

pyrrhophia

Stripe-crowned Spinetail

juv

ad

ad

spinicauda

Thorn-tailed Rayadito

whitii

ad

Ochre-cheeked
Spinetail

White-throated
Treerunner

Synallaxis Understorey spinetails. Knowledge of range and voice greatly aids ID. Extent of gular patch highly variable. Note combination of rufous on crown, wings and tail; presence of brown forehead; and coloration of underparts. Typically short disyllabic songs and also complex multi-note songs in most species. One further species on Plate 131.

Azara's Spinetail *Synallaxis azarae* 17 cm

Yungas forest. **ID** Overlaps with Sooty-fronted Spinetail. Adult *superciliosa*: Olive-brown forehead, bright rufous cap, buff supercilium, rufous wings, chestnut tail and white central underparts. **n. Salta form**: White supercilium. **Voice** A repeated *pi-hwee*, slightly slower and lower-pitched than Sooty-fronted. Call is a disyllabic *k-wee*. [Pijuí Ceja Canela]

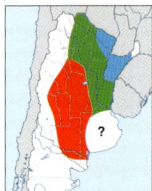

Austral Spinetail *Synallaxis [albescens] australis* 15 cm

Widespread in N and C; southern birds move north in winter. **ID** Overlaps with Azara's, Spix's, Rufous-capped and Grey-bellied Spinetails. **Adult**: Olive-brown forehead; cap and wing-coverts rufous but tail brown and proportionately short for genus. Indistinct whitish supercilium; underparts greyish-white (whiter than other *Synallaxis*). **Voice** A rapid *peer-shick*; like Sooty-fronted Spinetail but with rasping, not whistled, first note. Common call is a pure, sharp, whistled *tiú*; often during winter. **Tax note 46**. [Pijuí Cola Parda]

Spix's Spinetail *Synallaxis spixi* 16.5 cm

NE scrub and light woodlands. **ID** Overlaps with Sooty-fronted, Austral, Rufous-capped and Grey-bellied Spinetails. **Adult**: Entire crown and wing-coverts rufous; tail brown; face and underparts darker grey than Sooty-fronted and Austral. **Voice** A chattering *whit tututut*. [Pijuí Plomizo]

Sooty-fronted Spinetail *Synallaxis frontalis* 15.5 cm

Widespread in north. **ID** Overlaps with Azara's, Austral and Spix's Spinetails. **Adult**: Brown forehead; cap, wings and tail rufous but browner on central rectrices. Underparts darker than Azara's and Austral, but paler than Spix's. Nest on Plate 201. **Voice** A repeated *chi-clee*, more liquid and strident than Azara's Spinetail and without the raspy first note of Austral Spinetail. Call is a piercing *twíp*. [Pijuí Frente Gris]

Grey-bellied Spinetail *Synallaxis cinerascens* 15 cm

Paraná forest. **ID** Overlaps with Austral, Spix's and Rufous-capped Spinetails, all of which have rufous crowns. **Adult**: Olive-brown upperparts; wings rufous; tail chestnut-brown; face and underparts smoky grey. **Voice** Song is a pure, upswept sweet *pu-ít*; very like allopatric Ochre-cheeked Spinetail (Plate 131). Call is a single piercing note. [Pijuí Negruzco]

Rufous-capped Spinetail *Synallaxis ruficapilla* 15.5 cm

Paraná forest. **ID** Overlaps with Austral, Spix's and Grey-bellied Spinetails. **Adult**: Orange-rufous crown, short orange supercilium, dusky mask and white throat; rufous wings and tail. **Voice** A rapid squeaky chatter ending abruptly in an emphatic nasal note *jäjäjäjäjäjäjä kwéd*. Call is a scolding rattle *CHrrrrrrrr*. [Pijuí Corona Rojiza]

N Salta form

ad

Azara's Spinetail

ad

superciliosa

ad

Austral Spinetail

ad

Spix's Spinetail

ad

frontalis

Sooty-fronted Spinetail

ad

Grey-bellied Spinetail

ad

Rufous-capped Spinetail

Elaenia Mainly frugivorous flycatchers with stubby bill and pink or pale base. Best recognised by voice, knowledge of range and seasonal status, presence and extent of a coronal patch or semi-erectile crest, number, colour and width of wing bars, and general plumage tones especially throat and breast colour. Songs are mainly heard at dawn and dusk (except Slaty), but calls greatly aid identification. See also Plate 134.

Yellow-bellied Elaenia *Elaenia flavogaster*   16 cm

Resident in the NE lowlands except subtropical forest; local in the NW. **ID** Adult: Semi-erectile crest usually obvious. White coronal patch often visible. Wing bars as Large Elaenia. Greyish-white throat with an olive wash across breast; yellow below. Proportionately short-tailed and short-billed compared to Large Elaenia. **Voice** All calls are drawn-out and wheezy. Dawn song is a rapid *zri-weee-zreeu-weee*. Common call is a long hoarse *zweeehrr*, sometimes in excited duets including higher- and lower-pitched versions. [Fiofío Copetón]

Large Elaenia *Elaenia spectabilis*   18 cm

N lowlands and Yungas forest to 1500 m (Oct–Apr). **ID** Adult: Recalls a *Myiarchus* flycatcher. White coronal patch and small erectile crest rarely visible. From all congeners by grey throat and breast. 2–3 wing bars with a broad median-covert bar. Proportionately long-tailed. **Voice** Dawn song is a very fast, gravelly *bridbree-doBRÍ*, lasting 0.5 sec. Recalls Small-billed but accentuated in final syllable. Calls include a loud *WEEO* and an explosive *p-CHÚ*. [Fiofío Grande]

White-crested Elaenia *Elaenia albiceps*   15 cm

Patagonian forest (Oct–Apr) and Yungas forest (Aug–May); local breeder in the C Sierras?; passage migrant throughout the N lowlands. **ID** Breeding adult: Distinctly crested with a large white coronal patch. Long primary projection. Non-breeding adult looks very like Small-billed and Olivaceous Elaenias, although slightly duller above than both, with slightly broader wing bars (always two), and prominent eye-ring, but caution is strongly advised from Apr to Oct. Juvenile: Lacks a coronal patch. Browner above than adult, and wing bars distinctly tinged buff. **Voice** Dawn song is a very short disyllabic note repeated rapidly, *chbrew* or *tebree*. Calls include a drawn-out *weeeo* (higher-pitched than Large Elaenia), a burry *beerruh* and a pure *wee-uu*. Tax note 47. [Fiofío Silbón]

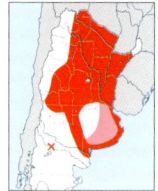

Small-billed Elaenia *Elaenia parvirostris*   14.5 cm

Mainly Oct–May throughout the N lowlands and C sierras. **ID** Adult: Concealed white coronal patch. Short primary projection. Olive above with relatively narrow wing bars. A third wing bar is often visible, and distinguishes from White-crested. **Voice** Dawn song is a fairly high-pitched *chidi-WÓRR-free*, lasting 0.5 sec; sometimes with high-pitched *swip* notes between songs. Recalls Large Elaenia but accentuated in the middle syllable. Calls include a dry *pk* or *pik* and a descending *péeuuu*. [Fiofío Pico Corto]

Highland Elaenia *Elaenia obscura*   18.5 cm

Resident in Yungas forest (800–1700 m; descending somewhat in winter). **ID** Adult: Slightly larger, duller and more brownish-olive than Small-headed Elaenia (no overlap) with dull yellowish-white wing bars. Pink bill, tipped black. Overlaps with White-crested, Large and Small-billed Elaenias. **Voice** Dawn song is a relatively slow and complex *ee-ooh WIORR-rohe*, lasting 1.5 sec. Calls include harsh burry *RREEE* or *REEARR*, and hoarse whistles. [Fiofío Oscuro]

Small-headed Elaenia *Elaenia sordida*   18 cm

Resident in Misiones and ne. Corrientes. **ID** Adult: Tiny bill with pink base. Lacks a coronal patch. Appears spectacled and round-headed, not crested like Yellow-bellied and Large Elaenias. Bright olive primary fringes and relatively long tail. Two broad white wing bars. Whitish throat. Greyish-olive breast. **Voice** Dawn song is a very fast buzzy nasal *wizur-drä*, repeated rapidly. Calls include a toad-like, low-pitched *brrrp*. [Fiofío Paranaense]

Yellow-bellied Elaenia

ad

flavogaster

Large Elaenia

ad

ad

juv

ad br

Small-billed Elaenia

chilensis

White-crested Elaenia

Highland Elaenia

ad

ad

Small-headed Elaenia

Elaenia Mainly frugivorous flycatchers with stubby bill and pink or pale base. See generic text on Plate 133.

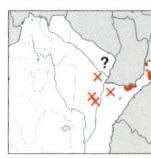

Olivaceous Elaenia *Elaenia mesoleuca*   14.5 cm

Sep–Jan in Misiones. Very local in n. Corrientes, e. Chaco and e. Formosa. **ID Adult:** Lacks a coronal patch. Brighter olive above than sympatric Small-billed Elaenia and a more obvious olive wash over the breast. Overlaps briefly with migrating White-crested Elaenia. **Voice** Dawn song is a very rapid *wit-zreebrroh*, with first note sharp and inflected and the remainder burry and buzzy. Calls include *burrr* or *brr* (higher-pitched and shorter than Small-headed Elaenia) and an ascending, whistled *whip*. [Fiofío Oliváceo]

Lesser Elaenia *Elaenia chiriquensis*   14 cm

Oct–Mar in n. Misiones (no recent records). **ID Adult:** Smallest member of the genus. Concealed white coronal patch and short semi-erectile crest. Distinctly browner above than White-crested, Small-billed and Olivaceous Elaenias with 1–2 narrow white wing bars; prominent on median wing-coverts. **Voice** Dawn song is a deliberate *chil-JAHR*. Calls include variable harsh buzzy notes. [Fiofío Belicoso]

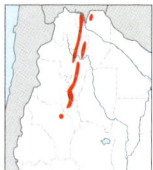

Slaty Elaenia *Elaenia strepera*   16 cm

Oct–Mar in Yungas forest (900–1800 m). **ID Adult** ♂: Orange lower mandible. Slate-grey with a white coronal patch. Contrasting white centre of abdomen and vent. Two whitish wing bars. **Adult** ♀: Differs from adult ♂ by an olive wash on the back, rump and breast, and yellow wash on lower underparts. Contrasting grey crown. Narrow ochraceous wing bars. **Voice** A mechanical sounding, ratchet-like series *te te tek t-t-t-TTTRRR-re rek*, sometimes followed by a rapid ticking *ki-ki-ki-ki-ki-ki...* Call is a very fast cicada-like *churr zr-ZRRRRRR*. [Fiofío Plomizo]

Myiopagis Small midstorey and subcanopy flycatchers with a concealed coronal stripe. Recall large tyrannulets but stockier.

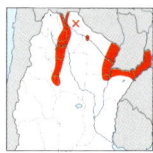

Greenish Elaenia *Myiopagis viridicata*   13.5 cm

Yungas foothills, humid chaco and Paraná forest (Sep–Apr). **Adult:** Pinkish lower mandible, tipped black. Dull olive above with a broad golden coronal patch; rarely visible. Short whitish supercilium and eye-ring, broken in front of eye. No wing bars, but yellow-fringed secondaries and tertials. Greyish throat, olive breast and sulphur-yellow abdomen. **Voice** Dawn song is a repetitive series of two well-spaced segments, *beer-wee-o... chiw-o*. During the day, a buzzy, upward-inflected *beer-wit*, confusingly similar to Euler's Flycatcher (Plate 141), but faster and higher-pitched. Call is a piercing, shrill *tsee*. [Fiofío Corona Dorada]

Grey Elaenia *Myiopagis caniceps*   12.5 cm

Resident in N Yungas foothills and Paraná forest. **ID Adult** ♂ *cinerea* (Andean foothills of Salta and Jujuy to 1400 m): Ash-grey above with a narrow white coronal stripe. Two white wing bars and fringes to remiges. Pale grey below, becoming whiter on the abdomen. **Adult** ♂ *caniceps* (Paraná forest; not illustrated): Can resemble adult ♀ or is otherwise duller grey above with a slight olive dorsal wash. **Adult** ♀: Ssp. alike. Grey crown with a pale yellow coronal stripe. Vivid olive back and rump. Two yellow wing bars and fringes to remiges. Pale grey throat and breast, usually becoming pale yellow on the abdomen. **Juvenile:** As adult ♀ but with reddish-brown wing bars. Mixed or blotched reddish over the nape and upper back. **Voice** Often in overlapping duet, with one sex giving a slow rattle punctuated with semi-musical notes, *tetetetete-TEW-tetete-TEW-tetete-TEW-tete-TEW-te-TEW* and variations, and the other sex giving a rising and falling (or falling) trill, *twe-twe-twe-teeerrrrrrrrrr*. Calls include a high-pitched *SWEE*, similar to that of Rough-legged Tyrannulet (Plate 137) but shorter and more piercing. [Fiofío Ceniciento]

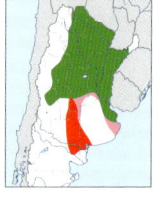

Suiriri Flycatcher *Suiriri suiriri*   15 cm

N and C lowlands. **ID** Widespread, medium-sized flycatcher of midstorey with a burry voice. **Adult:** Stout black bill. Grey above with blackish lores. Two whitish wing bars with prominent median-covert bar. Black tail, edged white. White below with a grey wash over the breast. Compare with White-crested Tyrannulet (Plate 135). **Adult NE form** (e. Formosa): From normal adult by a contrasting olive back and rump, and yellow wash over the belly and vent. Compare with Southern Scrub Flycatcher (Plate 141). **Juvenile:** Brown upperparts, flecked white. Olive-fringed secondaries. Pale yellow belly. **Voice** Song is a distinctive fast nasal chatter that rises and falls without a definitive pattern. Call is an explosive, burry and buzzy *PEEWR*. [Suirirí Gris]

Olivaceous Elaenia
ad

Lesser Elaenia
albivertex
ad

ad ♀

ad ♂

Slaty Elaenia

viridicata
ad

Greenish Elaenia

juv

cinerea
ad ♂

ad
NE form

ad ♀

juv

Grey Elaenia

ad

suiriri

Suiriri Flycatcher

Southern Beardless Tyrannulet *Camptostoma obsoletum* 12 cm

N lowlands and foothills. **ID** A robust, midstorey to subcanopy tyrannulet with a distinctly peaked crown. **Adult:** Stubby bill with pinkish lower mandible. Grey crown and white supraloral. Dull olive above with two cinnamon wing bars. Pale greyish throat, washed olive across the breast, and pale yellow below. **Voice** Dawn song is a short, relatively flat whinny followed by a short trill, *we-te-te-te-drriu*. Day song is a faster, longer, often descending whinny. Commonest call is a single plaintive whistle *feeeeit*, with variations; may resemble calls of Purple-throated Euphonia and Eastern Wood Pewee. [Piojito Silbón]

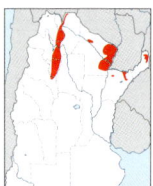

Mouse-coloured Tyrannulet *Phaeomyias murina* 12 cm

Local in the N lowlands and foothills (Sep–Apr). **ID** A brown-backed tyrannulet of midstorey and subcanopy. **Adult:** Pink base to lower mandible. Broad whitish supercilium. Dark brown wings with two broad whitish wing bars and tertial fringes. White throat, washed grey on the sides of the breast, and drab yellowish below. **Voice** Dawn song is a very fast *Elaenia*-like, burry *two-beer-or-bleed*. Day song is a rapid rising series of buzzy notes, often with some squeaky notes at the end, *brbrbrbrbrBRR-BRR TEW-TU-TU*. Calls include 2–4 short staccato *TEW* notes, sometimes alone. [Piojito Pardo]

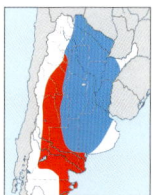

Straneck's Tyrannulet *Serpophaga griseicapilla* 10 cm

Breeds in Monte desert and N Patagonia (Oct–Mar), winters in N and E lowlands. **ID** Small tyrannulet with slender bill and a semi-erectile crest. **Adult:** Best told from White-crested Tyrannulet (ssp. *subcristata*) and Plain Inezia by voice. Also note grey crown with a few blackish streaks and concealed white feather bases (no definitive coronal stripe), shorter tail, and usually has a paler yellow belly. **Juvenile** (not illustrated): Browner above with rusty wing bars. **Voice** Song is a soft high-pitched fast trill, generally preceded by 1–4 flat, high-pitched introductory notes *TWIE-TWI-tirrrrrrrrrrrrrrrr*. Call is *teee-trrrrrrr* with a clear but short break between introduction and fairly descending fast trill (mostly in winter, when overlaps with Plain Inezia). Sonogram on p. 447. **Tax note 48**. [Piojito Trinador]

White-crested Tyrannulet *Serpophaga subcristata* 11 cm

Widespread. **ID** Small tyrannulet with slender bill and a semi-erectile crest. **Adult** *subcristata* (NE lowlands, wintering mostly in the humid chaco. Mostly overlaps with Straneck's in Apr–Oct): Differs by its white coronal stripe, complete white supercilium above the eye, richer yellow belly and voice. **Adult** *munda* (Andean foothills south to Mendoza across the Monte desert to coastal Rio Negro, moving north in winter): Resembles *subcristata* but back and rump uniform grey and lacks yellow belly (intermediates occur). **Juvenile** (not illustrated): Browner above with rusty wing bars. **Voice** Song is a loud succession of 5 or 6 (sometimes more) whistled notes, *tsil-tsil-tsil-tsil-tsil*, followed by 7–18 (sometimes many more) notes *cliclicl-cl-cl-cliclicli...* Calls include a rapid metallic rattle *TIRIRIP*, and a high-pitched, very commonly heard, syncopated phrase *chip che-rip chep*. Sonogram on p. 447. **Tax note 49**. [Piojito Tiquitiqui]

Plain Inezia *Inezia inornata* 10 cm

Dry and humid chaco south to n. Cordoba (Sep–Mar); local in n. Corrientes and s. Misiones. **ID** Resembles *Serpophaga* tyrannulets but more thickset and larger-headed with a stouter bill. **Adult:** From Straneck's and White-crested Tyrannulet (ssp. *subcristata*) by white supercilium ending at eye in a broad, but broken, eye-ring; black loral line extends behind eye; yellow on sides of breast and flanks contrasts with paler yellowish-white belly. Lacks the white coronal streak of White-crested Tyrannulet. Some show diagnostic grey breast flammulations. **Voice** Song is a rising and falling trill, often followed by an emphatic whistle, *trrrrrrEEEEu-tee*. Call is *teeeewrrrrrrrrrrr...*, without a break between introduction and evenly-pitched fast trill (mostly in winter, when overlaps with Straneck's Tyrannulet). Sonogram on p. 447. [Piojito Picudo]

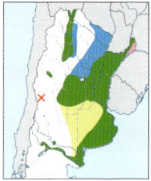

Sooty Tyrannulet *Ridgwayornis nigricans* 12 cm

Mostly resident in N lowlands and Andean foothills. **ID** Chunky, water-loving tyrannulet; frequently fans tail. **Adult:** Grey above with a slight olive tinge on the back. Concealed white coronal stripe. Two indistinct pale brown wing bars. Contrasting blackish tail. Pale grey below with a whitish lower belly. **Voice** Song is a dry, thin and very fast monotone trill, slowing towards the end, often in duet. Calls include a gravelly repeated *ch-vlip* or *dj-wit*, and a more scolding *stip* note. [Piojito Gris]

Southern Beardless Tyrannulet

ad

ad

ignobilis

Mouse-coloured Tyrannulet

ad

Straneck's Tyrannulet

ad

subcristata

Plain Inezia

ad

ad

munda

White-crested Tyrannulet

ad

Sooty Tyrannulet

Yellow-billed Tit-Tyrant *Anairetes flavirostris*   10.5 cm

N Andes to 3400 m, C sierras, arid W lowlands and N Patagonia, wintering north into the Pampas. **ID** Small, restless tyrant with a filamentous crest, concealed white coronal patch and tiny bill. **Adult:** Pink or orange base to lower mandible. Dull iris. Straightish crest and streaky face, grizzled with white. Two bold white wing bars and tertial fringes. Thick black breast streaks. **Juvenile:** Lacks the crest and has buffy wing bars. **Voice** Song has a short sharp introduction and a clearly inflected, reeling trill *T-DU-reeeeee-reeeo*. Also a rapid series of short purring trills, *pwrr-pwrrr-pwrrr…* and a rising and falling rapid dry trill. [Cachudito Pico Amarillo]

Tufted Tit-Tyrant *Anairetes parulus*   11 cm

Widespread in Andes; 3 ssp. **ID** Small, restless tyrant with a filamentous crest, concealed white coronal patch and tiny bill. **Adult** *parulus* (S Andes to 1100 m, from Neuquén to Tierra del Fuego): Black bill. Yellow or white iris. Fine recurved crest. Short white supraloral. Usually lacks wing bars, otherwise very faint. Very fine breast streaking. **Adult** *patagonicus* (Andean slopes to 2630 m, from Catamarca to ne. Neuquen, C sierras and N Patagonia): Coarser breast streaking than *parulus*, narrower wing bars than Yellow-billed Tit-Tyrant. **Adult** *aequatorialis* (above 2300 m in the Andes of Jujuy and Salta; not illustrated): Resembles *patagonicus*, but is darker above with little white crest streaking. **Juvenile:** Lacks the crest and has buffy wing bars. **Voice** Song is a short, upswept ratchet-like *wi-drrik*. Trills are slower and flatter than in Yellow-billed Tit-Tyrant. Complex vocal repertoire not well understood. [Cachudito Pico Negro]

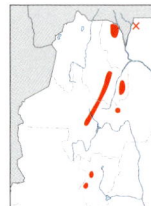

Cinnamon Flycatcher *Pyrrhomyias cinnamomeus*   13 cm

Yungas forest to 1700 m (Nov–Mar). **ID** An upright-perching flycatcher which sallies from midstorey or subcanopy snags. **Adult:** Brown crown with concealed golden-yellow coronal patch. Brownish-olive back with contrasting ochre rump band. Two rufous wing bars and broad wing stripe. Rufous below, becoming ochre on belly. **Voice** A high-pitched, dry, spluttered trill *tzeerrrrrt*. [Birro Chico]

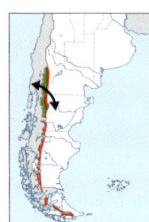

Patagonian Tyrant *Colorhamphus parvirostris*   13 cm

Patagonian forest to 1800 m; southern birds move north in winter, while northern birds descend lower. **ID** Inconspicuous midstorey flycatcher with a short slender bill. Perches upright. **Adult:** Grey face with a dusky smudge behind eye. Brown above with 1–2 orange or cinnamon wing bars. Grey throat and breast and creamy-buff below. **Voice** Song is a thin, shrill, pure, short double whistle *peeeeee ti-siú*, sometimes without the second part. Call is a melancholic descending *pseeeeuu*. [Peutrén]

White-browed Chat-Tyrant *Ochthoeca leucophrys*   15 cm

NW Andes (1700–4000 m). **ID** Chunky, unobtrusive flycatcher with fairly long slender bill and a bold white supercilium. Generally in wetter places than D´Orbigny's Chat-Tyrant, including upper Yungas forest. **Adult:** Two rufous wing bars and grey underparts. **Voice** Common call is a strident whistled *SPEE* or *PSIU*. [Pitajo Gris]

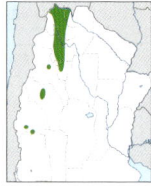

D'Orbigny's Chat-Tyrant *Ochthoeca oenanthoides*   15.5 cm

NW Andes (1700–4000 m). **ID** Structure and behaviour as White-browed Chat-Tyrant. Found alone or in pairs in high Andean or Puna shrub and cactus zones, and *Polylepis* woodland. Generally in drier areas than White-browed Chat-Tyrant. **Adult:** 1–2 cinnamon wing bars. Rich cinnamon-orange below with a greyish throat. **Voice** Song is a series of squeaky *Pi-chu* couplets, often in syncopated duets, with male and female giving slightly different versions. [Pitajo Canela]

flavirostris

ad

parulus

Tufted Tit-Tyrant

ad

patagonicus

ad

Yellow-billed Tit-Tyrant

ad

cinnamomeus

Cinnamon Flycatcher

ad

Patagonian Tyrant

ad

tucumana

White-browed Chat-Tyrant

ad

oenanthoides

D'Orbigny's Chat-Tyrant

Phyllomyias Chunky tyrannulets of midstorey and subcanopy. Shorter and broader-tailed than *Phylloscartes*. More sluggish than *Mecocerculus* and *Phylloscartes*. **Phylloscartes** Small, sleek, restless tyrannulets with a long, narrow tail, pinched at the base. **Mecocerculus** Two midstorey and subcanopy tyrannulets with slender bills which are otherwise very different and seemingly unrelated.

Rough-legged Tyrannulet *Phyllomyias burmeisteri* 12.5 cm

Yungas forest to 1700 m (locally higher) and Paraná forest. **ID** Chunky tyrannulet of midstorey and subcanopy. **Adult:** Stouter bill than Sclater's with an entirely pink/orange lower mandible. Vivid olive above with a slightly darker crown. Whitish supercilium and eye-ring. Dull wing bars and creamy-yellow tertial fringes; broader and whiter on the innermost. Greyish-white chin, washed olive across the breast, and pale yellow below. Often perches upright. Compare with Greenish and Planalto Tyrannulets (Plate 138) in Misiones. **Voice** Delivers astonishingly loud pure whistles; isolated or in a series. May resemble calls of Grey Elaenia, but longer and lower-pitched. Sonogram on p. 446. [Mosqueta Pico Curvo]

Sclater's Tyrannulet *Phyllomyias sclateri* 13 cm

Yungas foothill forest to 1500 m. **ID** Chunky tyrannulet of midstorey and subcanopy. **Adult:** Slender black bill with a pinkish base. Olive above with a dull grey wash on the crown. Short white supercilium. Two yellow or yellowish-white wing bars and fringes to remiges. Whitish throat. Pale grey breast, washed olive at the sides. Pale yellow below, especially flanks. **Voice** Distinctive series of well-spaced, emphatic raspy, churring notes. Sonogram on p. 446. [Mosqueta Corona Gris]

Mottle-cheeked Tyrannulet *Phylloscartes ventralis* 12 cm

Yungas forest to 1700 m, Paraná forest south to ne. Buenos Aires. **ID** Small, sleek, restless tyrannulet with a long narrow tail, pinched at the base. **Adult** *tucumanus* (Yungas forest; illustrated): Long, slender blackish bill with a pink base. Short whitish supercilium. Dusky lores and grizzled face. Bright olive above with two pale yellow or whitish wing bars. Tertials tipped with white or yellowish-white spots. Greyish-white throat and yellow below. **Adult** *ventralis* (Paraná forest; not illustrated): Shorter tail. **Voice** Call is a soft *chek*, sometimes running into a brief, liquid accelerating–decelerating trill. Sonogram on p. 446. [Mosqueta Carasucia]

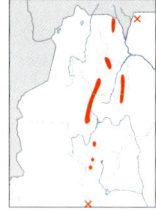

Buff-banded Tyrannulet *Mecocerculus hellmayri* 10.5 cm

Yungas forest, especially alder and *Podocarpus* (1200–2200 m); mainly Sep–May. **ID** Midstorey and subcanopy tyrannulet with a slender bill. **Adult:** Small delicate black bill with a fleshy lower mandible. Slate-grey crown and long white supercilium broadening behind eye. Vivid olive back and dull buff rump. Two cinnamon-buff wing bars. Short brown tail. Greyish-white throat, pale yellow belly and flanks. **Voice** Common call is a series of up to 5 high-pitched, shrill whistles *tsee tsee tsee*, with slight pitch variation. May recall Purple-throated Euphonia, but higher-pitched and more piercing. Also delivers metallic chatters. [Piojito de los Pinos]

White-throated Tyrannulet *Mecocerculus leucophrys* 14 cm

Yungas forest to 2400 m, south to ne. La Rioja. **ID** Midstorey and subcanopy tyrannulet with a slender bill. **Adult:** Large and conspicuous. Olive-brown above with two broad buffy wing bars. Striking white throat (often puffed out) contrasts with grey-brown breast and pale yellow belly. **Voice** Dawn song is a rapidly repeated short phrase *chi-d-dt*. Calls include a fast bubbly chatter, a short slow series of variable squeaky notes, *chup chup cheet cheet cheet…* Also loud wing rattles. [Piojito Gargantilla]

Tawny-rumped Tyrannulet *Phyllomyias uropygialis* 11 cm

Unconfirmed in Yungas forest of n. Salta. **ID** Chunky tyrannulet of midstorey and subcanopy. **Adult:** Dusky cap and lores, and short, narrow greyish supercilium. Olive-brown back and tawny rump. Black wings with two broad cinnamon wing bars and fringes to remiges. Greyish throat and breast, washed yellow below. **Voice** Common call is a fast series of high-pitched, shrill whistles *tsee tsee-tsiu*, with a similar quality to Buff-banded Tyrannulet. [Mosqueta Rabadilla Canela]

Rough-legged Tyrannulet

ad

Sclater's Tyrannulet

ad

sclateri

tucumanus

ad

Mottle-cheeked Tyrannulet

ad

Buff-banded Tyrannulet

ad

leucophrys

White-throated Tyrannulet

ad

Tawny-rumped Tyrannulet

Greenish Tyrannulet *Phyllomyias virescens*   13 cm

Misiones and ne. Corrientes. Midstorey to subcanopy, often in riparian forests. **ID** Chunky with a fairly broad tail. More sluggish than *Phylloscartes* tyrannulets. Horizontal stance with slightly cocked tail and drooped wings is typical. **Adult:** Stubby bill with a mostly pink lower mandible. Fairly bright olive above. Bold white or yellowish supercilium to just behind eye. Broken white eye-ring. Two broad, yellow or whitish wing bars. Flight feathers and tertials fringed yellowish. Greyish-white throat, washed olive across the breast, and pale yellow below. Compare with Planalto Tyrannulet, and Rough-legged and Mottle-cheeked Tyrannulets (Plate 137). **Voice** Song is a loud, fast spluttered trill of raspy notes. Sonogram on p. 446. [Mosqueta Corona Oliva]

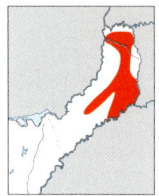

Planalto Tyrannulet *Phyllomyias fasciatus*   12.5 cm

n. Misiones. **ID** Chunky with a fairly broad tail. More sluggish than *Phylloscartes* tyrannulets. Perches fairly upright in all strata. **Adult:** Fairly short black bill with a small pink base rarely visible. Drab brownish-olive above; usually duller and browner on the crown. Indistinct whitish supraloral. Two indistinct buffy wing bars, and fringes to secondaries and tertials. Proportionately short brown tail. Greyish throat with a dull olive wash across the breast, and pale yellow below. **Voice** Song consists of 3–4 slow, drawn-out, mellow, rising whistles *weee-weee-weee-ít.* Calls include a long series of notes and a raspy chatter, often in duet. Sonogram on p. 446. [Mosqueta Olivácea]

São Paulo Tyrannulet *Phylloscartes paulista*   10 cm

Rare in n., c. and sw. Misiones. Found in all strata. **ID** Small, sleek, agile tyrannulet with a small, slender bill and narrow tail, pinched at the base. **Adult:** Mostly pink lower mandible and pinkish tarsus. Olive above with a yellow face and narrow white to yellowish supercilium, curving behind the large black ear-crescent. Wing bars lacking or very narrow yellowish-white. Yellow below, sometimes washed olive across the breast. Compare with Southern Bristle Tyrant and Eared Pygmy Tyrant (Plate 139). See also widespread and common Mottle-cheeked Tyrannulet (Plate 137) and Sepia-capped Flycatcher. **Voice** Song is a fast, metallic *swi-di-di-di,* sometimes in an even faster rollicking series *swidibre-swidbre-swidbre...* [Mosqueta Oreja Negra]

Sepia-capped Flycatcher *Leptopogon amaurocephalus*   14 cm

Yungas, humid chaco and Paraná forest. **ID** Perches upright in midstorey. Performs single wing-lifts. **Adult:** Brown crown and dusky ear-patch highlighted by buffy lores. Dark olive nape and back. Cinnamon wing bars. Compare with São Paulo Tyrannulet. **Voice** Drawn-out scolding chatter, often with emphatic final note, *wed-de-de-de-de-de-de-dijk,* with much variation. [Mosqueta Corona Parda]

Southern Bristle Tyrant *Pogonotriccus eximius*   11.5 cm

n. c. and sw. Misiones. **ID** Recalls a *Phylloscartes* tyrannulet, although it has a shorter, stouter bill and shorter, squarer tail, and usually perches upright in midstorey to subcanopy, performing stylised wing-lifts. **Adult:** Slate-grey crown. Intricate facial pattern with long broken white supercilium, black ear-crescent, loral spot and subocular spot. Fairly bright olive back. No wing bars. White throat and bright golden-yellow below, often with a pale olive wash over the upper breast. Compare with São Paulo Tyrannulet and Eared Pygmy Tyrant (Plate 139). **Voice** Song is a fast, steady chatter that rises slightly and falls markedly towards the end while slowing. [Mosqueta Media Luna]

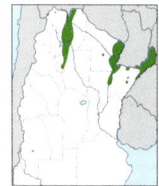

Yellow-olive Flycatcher *Tolmomyias sulphurescens*   14.5 cm

Northern forests. **ID** Resident in midstorey and subcanopy. Robust with a flattened bill. **Adult** *sulphurescens* (Paraná forest; illustrated): Grey crown (rarely with olivaceous markings) and olive back. White supraloral. Two yellow wing bars. Whitish chin. Yellow below; breast clouded olive. **Adult** *pallescens* (Yungas foothill; not illustrated): similar to *sulphurescens* but paler. Ssp. *grisescens* of humid chaco appears identical to *pallescens*. **Voice** *sulphurescens*: 3–4 upward-inflected rasps, *schweet schweet schweet. pallescens*: 3–4 upward-inflected whistles, *sweet sweet sweet.* **Tax note 50.** [Alt: Yellow-olive Flatbill] [Picochato Grande]

Planalto Tyrannulet

ad

Greenish Tyrannulet

ad

brevirostris

São Paulo Tyrannulet

ad

ad

amaurocephalus

Sepia-capped Flycatcher

ad

Southern Bristle Tyrant

ad

sulphurescens

Yellow-olive Flycatcher

Drab-breasted Bamboo Tyrant *Hemitriccus diops* 11.5 cm

Mainly n. and c. Misiones, nearly always in bamboo. **ID** Small, plump flycatcher of bamboo understorey. Slender bill and large iris. Best detected by ticking calls. **Adult:** Pinkish lower mandible. Fairly bright olive above with a prominent yellowish or white supraloral and eye-ring. No wing bars, but notable yellow carpal. Ashy-grey below with a white spot on the lower throat, white centre to the belly and yellowish wash on the flanks. **Voice** Vocally diverse. Common calls include often doubled, shrill fast ringing trills *krrreee krrri*, sometimes in longer renderings, isolated or rapid triplets of *KIK* notes, and upswept *wip* whistles. [Mosqueta de Anteojos]

Brown-breasted Bamboo Tyrant *Hemitriccus obsoletus* 11 cm

Local in n. Misiones; mainly in *Chusquea tenella* bamboo. **ID** Structure and behaviour much as Drab-breasted Bamboo Tyrant. **Adult:** Pink lower mandible. Dull brownish-olive above. Buff loral spot. Unmarked olive wings. Creamy-brown throat and breast; latter sometimes flammulated grey. Whitish centre to the belly. Yellowish wash on flanks and vent. **Voice** A stuttering series of usually 6–8 sharp notes *ki-ki-ki-ki-ki-kik*. Calls include single, doubled or tripled *KIK* notes. All voices are slower, lower-pitched and less shrill than Drab-breasted Bamboo Tyrant. [Mosqueta Pecho Pardo]

Eared Pygmy Tyrant *Myiornis auricularis* 8.5 cm

Misiones and ne. Corrientes. **ID** Tiny, upright-perching flycatcher of midstorey and subcanopy. Fairly long slender bill and short narrow tail. **Adult:** Olive above with a brownish crown. Grey sides of the neck. Cinnamon around eye. Broad black facial crescent. One or two narrow yellow wing bars. White throat and upper breast with blurry blackish streaks. Fairly bright yellow below with blurry olive streaks on the breast. **Voice** Song is a fast, soft ringing ascending trill *brrreee*. Calls include unpatterned series of ticking *pic* or *puik* notes, sometimes preceding the song. [Mosqueta Enana]

Bay-ringed Tyrannulet *Phylloscartes sylviolus* 11 cm

n. and c. Misiones. Mainly in subcanopy. **ID** Small, sleek, agile tyrannulet with a small, slender bill and narrow tail, pinched at the base. **Adult:** White iris and rusty spectacles. Fairly bright olive above without wing bars. Clean white below, washed yellow on the throat and undertail-coverts. See ♀ Chestnut-vented Conebill (Plate 173). See also widespread and common Mottle-cheeked Tyrannulet (Plate 137). **Voice** Call is a springy, metallic *swi´d´d´du* with a variable number of notes. The infrequently heard song is a series of shrill notes that speeds into a slightly descending chatter, *see swee swee see see see see si-si-si-sisisisisi*. [Mosqueta Cara Canela]

White-throated Spadebill *Platyrinchus mystaceus* 10 cm

Understorey in Misiones and ne. Corrientes. **ID** Relatively large-headed compact flycatcher with an extremely broad-based flattened bill, long rictal bristles, distinctive facial pattern and large concealed coronal patch (usually lacking in ♀♀). **Adult:** Creamy lower mandible. Brownish-olive above with a golden coronal patch. Buffy lores and eye-ring. White cheek spot. Flight feathers can be fringed rufous, like Russet-winged Spadebill. Stubby tail unlike Russet-winged. White or creamy throat. Ochraceous-buff below with a paler centre to the belly. **Voice** Song is a rich, burry, liquid, cascading trill *wee-urrrrrrrrrrr*, lasting 2 secs. Call is a usually doubled *wee-kip* or *kwi-kwi*, like a squeaky toy. [Picochato Enano]

Russet-winged Spadebill *Platyrinchus leucoryphus* 13 cm

Unconfirmed in Misiones. Midstorey of shady forest. **ID** General structure like White-throated Spadebill, but notably longer tail. **Adult:** Pink lower mandible. Olive above with browner crown and white coronal patch. White or buff supraloral, eye-ring, post-ocular streak and cheek spot. Wings broadly-fringed rufous. Fairly long brown tail. Off-white or yellowish below. Buff or cinnamon breast band; olive at the sides. **Voice** Song is an unmistakable explosive fast trill that rises in pitch and volume, ending in an emphatic whistled note *weeeeeeeiiiiiiirrrrrrrrr-CLEU*, lasting 3 secs. Call is a loud disyllabic whistle *Pluh*. [Picochato Chico]

Drab-breasted Bamboo Tyrant

ad

Brown-breasted Bamboo Tyrant

ad

zimmeri

ad

Eared Pygmy Tyrant

ad

Bay-ringed Tyrannulet

ad

mystaceus

White-throated Spadebill

ad

Russet-winged Spadebill

Tawny-crowned Pygmy Tyrant *Euscarthmus meloryphus* 11 cm

N lowlands and foothills; mainly Oct–Mar, but winters in the humid chaco. **ID** Small understorey flycatcher of thorny woodland. **Adult:** Olive-brown above. Buffy lores and around eye. Concealed cinnamon-orange coronal patch. Some show cinnamon wing bars. Greyish-white throat and breast; washed pale yellow below. **Voice** Song comprises an introductory trill, a squeaky rapid phrase and a short final trill, *teeerrrrrrrrr ick-trri-kitik trrr*, repeated over and over. Calls include rattled trills and a dry rapid *pídidrr*. [Barullero]

Ochre-faced Tody-Flycatcher *Poecilotriccus plumbeiceps* 10 cm

Andean foothills to 1600 m in Jujuy and Salta, Paraná forest and locally in the chaco. **ID** Resident understorey flycatcher with a slender flattened bill. **Adult** *viridiceps* (NW; illustrated): Slate-grey crown. Cinnamon throat and face with a dusky ear-spot. **Adult** *plumbeiceps* (NE; not illustrated): More prominent dusky ear-spot. **Voice** A dry, mechanical, hollow trill *brraaap* or *prrreep*; like someone breaking wind. Pitch varies and sometimes given in fast series. [Mosqueta Cabeza Canela]

Pearly-vented Tody-Tyrant *Hemitriccus margaritaceiventer* 11 cm

N lowlands and Andean foothills. **ID** Small resident, short-tailed, understorey flycatcher with a very slender bill. **Adult:** Pale yellow iris with a reddish eye-ring. Grey crown; olive back and wings. Yellow bend of wing, but indistinct wing bars. White below, flammulated grey on the breast. **Juvenile** (not illustrated): As adult, but drabber and with dark eye. **Voice** Song is a soft ticking series that rises, followed by a soft ringing trill *tuc tic-tic-tic-tic-treew*; sometimes lacks the trill. Diverse array of calls includes a slow descending scolding series, a long flat accelerating trill, inflected chuckling notes and a very nasal, harsh rasp. [Mosqueta Ojo Dorado]

Grey-hooded Flycatcher *Mionectes rufiventris* 13.5 cm

n. and c. Misiones. **ID** Chunky with a fairly long black bill. Perches upright at all levels of shady forest. **Adult:** Slate-grey hood, fairly bright olive back and rich cinnamon-orange below. **Voice** Throaty antshrike-like song at leks is an accelerating series of nasal notes ending abruptly in a final isolated note, *daao daao daao-dao-daodadadada dao* lasting 4 secs. Call is an unevenly patterned series of raspy emphatic *reee* or *reeah* notes. [Ladrillito]

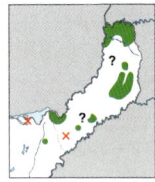

Common Tody-Flycatcher *Todirostrum cinereum* 10.5 cm

Misiones, expanding south. **ID** Slender-bodied flycatcher of modified habitats. Flattened bill with a ridged culmen and fairly long graduated tail. **Adult:** Unmistakable with black foreface, yellow iris, bright olive back and bright yellow below. Variable yellow lores (sometimes extensive). Black tail with white vane to outermost feather and white tips. **Voice** A variety of short, high-pitched ticking trills and steady series of ticking notes. [Mosqueta Pico Pala]

Southern Antpipit *Corythopis delalandi* 14 cm

Paraná forest and local in N Yungas foothills. **ID** A terrestrial and arboreal forest understorey flycatcher. **Adult:** Dull olive-brown above. Whitish lores and eye-ring. White below with a broad black pectoral band and streaked flanks. Local form in nc. Salta (undescribed ssp?) has a narrow pectoral band, less extensive flank streaking and a slightly more olive back. **Voice** A pleasant, rich series of alternating whistles and rippling notes, *peee-wree pe-wreee-blibli*, lasting 2 secs. An atypical variant is a whistled *qwee qwi-qwi-qwi*, with second part lower-pitched. Often detected by bill snapping. [Mosquitero]

Greater Wagtail-Tyrant *Stigmatura budytoides* 15 cm

Monte desert to 2500 m, C sierras, chaco and N Patagonia. **ID** Bold, noisy member of mixed-species flocks in thorn woodland. Long, graduated tail. **Adult:** Dull olive above with yellowish supercilium, whitish median wing-covert bar, fringes to greater coverts and tertials. White tips to all but central tail feathers. Pale yellow below. **Voice** Dawn song is a short burry *djrjrjrr-brrt*. Diurnal song is a long series of rollicking frantic notes, *peru-witch-you* or *pir-pin-to*, often in duet, lasting up to 7 secs. Call is a burry *berr*, often in series. [Calandrita]

Tawny-crowned Pygmy Tyrant

ad

meloryphus

Ochre-faced Tody-Flycatcher

ad

viridiceps

Pearly-vented Tody-Tyrant

ad

margaritaceiventer

ad

Grey-hooded Flycatcher

ad

coloreum

Common Tody-Flycatcher

delalandi

ad

Southern Antpipit

ad

flavocinerea

Greater Wagtail-Tyrant

Yellow Tyrannulet *Capsiempis flaveola* 11.5 cm

NE forests, mainly in Misiones. **ID** A slender tyrannulet with a particularly narrow tail, recalling *Phylloscartes* tyrannulets. Mainly in *Chusquea* and *Merostachys* bamboo understorey to midstorey. **Adult:** Fairly bright olive above. Bright yellow supercilium and underparts. Indistinct olive wing bars. **Voice** Song in overlapping duets is a fast series of rollicking notes that become louder and faster, lasting 4–7 secs. Call is a purring trill, generally ascending slighty in pitch *brrrrrreeeee*, repeated at short intervals. [Mosqueta Ceja Amarilla]

Large-headed Flatbill *Ramphotrigon megacephalum* 14 cm

Misiones. **ID** Chunky midstorey flycatcher with a broad flattened bill. Sits upright in *Merostachys* and *Guadua* bamboo. **Adult:** Pale base to lower mandible. Fairly bright olive above. Short yellowish supercilium. Broad black lateral crown stripes. Two ochre wing bars. Pale olive below with a pale yellow central abdomen. **Voice** Song is a melancholic whistle with two well-spaced notes, the second one shorter and lower-pitched, *whee...cuh*; often gives only the first note. Calls include drawn-out whistles *weeuu* or *wuuuu*, recalling *Myiarchus* flycatcher whistles. [Picochato Cabezón]

Southern Scrub Flycatcher *Sublegatus modestus* 14 cm

Sep–Apr in the N lowlands and foothills. **ID** Upright-perching, peaked crown when singing. **Adult:** Pink base to lower mandible. Narrow white supraloral and broken eye-ring. Dull brownish-olive above. 2–3 whitish wing bars; broadest on median-coverts. Whitish wing panel. Greyish face, throat and upper breast; pale yellow below. **Juvenile** (not illustrated): Distinctive white crown spots. **Voice** Dawn song is a series of alternating inflected pure whistles *wi-chu didi-cleeu*. Diurnal song is an ascending series of stuttered notes ending in a drawn-out whistle, *ti-di-di-d-d-dddd SFEEW*. [Suirirí Pico Corto]

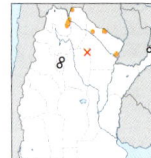

Alder Flycatcher *Empidonax alnorum* 13.5 cm

Along rivers in foothill Yungas, chaco and Paraná forest (Nov–Mar). **ID** Upright-perching, boreal migrant. Broad bill and rather short tail. **Adult:** Pink or orange lower mandible. Dull olive above with peaked, slightly darker crown. Contrasting whitish lores. Two broad, whitish wing bars. Dull olive breast and flanks; pale yellow below. **Voice** Song is a fast gravelly buzz *free-bree-o*. Common call is a sharp *pit* that may resemble voices of Small-billed Elaenia and Pearly-vented Tody-Tyrant. Rarely delivers inflected whistles. [Mosqueta Boreal]

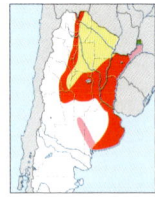

Bran-coloured Flycatcher *Myiophobus fasciatus* 12.5 cm

Sep–Apr in the N lowlands and Yungas foothills. **ID** Inhabits understorey. Slender, broad-based bill and concealed golden or rufous coronal patch. **Adult** *flammiceps* (widespread; illustrated): Rusty-brown above with dark wing-coverts. Two cinnamon, buffy or white wing bars. Whitish below with fine brown breast streaking, washed yellow on belly. **Adult** *auriceps* (Yungas foothills; not illustrated): Indistinct breast streaking and lacks the yellow belly. **Voice** Dawn song is a vireo-like series of well-spaced warbles *che-wep... chu-do-it...ch-wip*. Common call is a rippling chatter *weeberererererererere*. [Mosqueta Estriada]

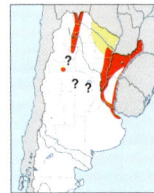

Euler's Flycatcher *Lathrotriccus euleri* 13 cm

N woods and forests to 1600 m. **ID** Understorey and midstorey. **Adult** *argentinus* (NW): Pinkish or white lower mandible. Brown to warm brown above with darker wings. Two fairly broad buff wing bars and wing panel. Whitish throat, olive-brown wash on breast. Greyish-white below. **Adult** *euleri* (NE Argentina; not illustrated): Pale yellow below. **Voice** Dawn song is a repetitive series of two well-spaced segments, *beer-widu... pewurr*. Diurnal song is a buzzy, upward-inflected *beer-whit*, confusingly similar to Greenish Elaenia, but slower and lower-pitched. Calls include 1–3 burry notes, *speer-speer-speer...* or a fast burry chatter *speer-chchchchch*. [Mosqueta Parda]

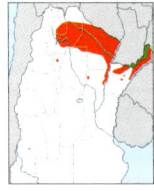

Whistling Fuscous Flycatcher *Cnemotriccus* [*fuscatus*] '*bimaculatus*' 15.5 cm

Mainly Sep–Apr in the chaco, n. Corrientes and Paraná forest. Midstorey. **ID** Fairly long, stout black bill. **Adult:** Brown above with a warmer rump. Whitish or pale buff supercilium and dusky lores. Two cinnamon wing bars and fringed remiges. Whitish below, washed brown on the breast; often pale yellow on the belly. **Voice** Dawn song is a series of two well-spaced burry notes and a whistle ending in a chuckle, *beerro... peeaRR... peeeoCHK*. Calls include a pure plaintive rising whistle *pweeee* and a fast raspy series, *dje-dje-dje-dje-dje-dje-dje-dje* that may recall Euler's Flycatcher. **Tax note 51.** [Mosqueta Ceja Blanca]

ad

flaveola

Yellow Tyrannulet

ad

megacephalum

Large-headed Flatbill

Southern
Scrub Flycatcher

ad

brevirostris

ad

ad

flammiceps

Bran-coloured Flycatcher

ad

Alder Flycatcher

ad

ad

argentinus

Euler's Flycatcher

Whistling Fuscous Flycatcher

Contopus Upright-perching flycatchers, often on a dead snag, with notable crests, broad-based bills and pale lores. *Tyrannus* An overshooting boreal migrant of open and edge habitats. *Sirystes* Chunky, bold, noisy subcanopy resident with a stout bill and bushy crest.

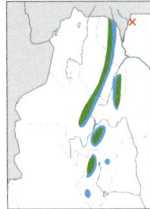

Smoke-coloured Pewee *Contopus fumigatus* 17.5 cm

Humid Yungas forest to 1800 m. **ID** Tall bushy crest. **Adult:** Mostly dusky grey, washed olive dorsally. Pale grey wing bars. Indistinct pale lores. Whitish throat and centre of belly. **Voice** Dawn song is a series of inquisitive, plaintive liquid whistles *pee-wili pe-weeeear*, sometimes followed by a long series of rapid ethereal metallic notes. Call is a whistled *pwip*, often in series. [Burlisto Copetón]

Olive-sided Flycatcher *Contopus cooperi* 19 cm

Known from one vagrant record. **ID Adult:** Greyish above with contrasting white patch on sides of rump. White throat; white line down central breast and belly contrasts with greyish breast-sides, flammulated dusky. **Voice** Song is a rapid whistled *quick THREE-BEERS*. Call is very similar to Smoke-coloured Pewee. [Burlisto Oliváceo]

Eastern Wood Pewee *Contopus virens* 12.5 cm

Rare in NW (Sep–Apr). **ID Adult:** Little or no pale colour on lores. Very narrow eye-ring, usually restricted to rear of eye. Crown darker than back. Primaries reach one third to half of tail length. Green tinge on breast band. Yellow on belly stronger than Tropical Pewee. Worn and moulting birds may look very drab, and lack the distinctive long primary projection. **Voice** Drawn-out, rich whistled *P-weeeee*. May resemble calls of Southern Beardless Tyrannulet. [Burlisto Boreal]

Tropical Pewee *Contopus cinereus* 14.5 cm

Widespread. **ID Adult** *pallescens* (Drier rain-shadow Yungas forest to 2000 m): Pale olive-grey with striking pale lores. Indistinct wing bars. Throat and belly paler. **Adult** *cinereus* (Paraná forest): Ashy-grey, tinged yellow on belly. Wing bars are highly variable, yet generally indistinct or lacking. **Juvenile** (not illustrated): Rusty wing bars and scaling across crown, back and rump. **Voice** Song has a raspy note and a descending whistle *tri-youuu*. Call is like Smoke-coloured Pewee but higher-pitched. [Burlisto Chico]

Eastern Kingbird *Tyrannus tyrannus* 20 cm

Overshooting migrant (Oct–Mar) with widespread records. **ID** Open and edge habitats. Slender bill and peaked crown. **Adult:** Blackish head with concealed orange coronal patch. Slate-grey back. Narrow whitish wing bars. Relatively short, glossy-black, square-ended tail, edged white with a white terminal band. White below, washed grey on sides of breast and flanks. Compare with Sibilant Sirystes and Fork-tailed Flycatcher (Plate 146). **Voice** Usually silent. Call is a high-pitched *dzee*. [Suirirí Boreal]

Sibilant Sirystes *Sirystes sibilator* 19.5 cm

Paraná forest. **ID** Chunky, bold, noisy subcanopy resident. Often in mixed flocks. Stout bill and bushy crest. **Adult:** Blackish head contrasts with pale greyish-olive back, vaguely streaked brown. Whitish wing-fringes. Pale grey below, some washed yellow on the belly. Compare with Eastern Kingbird. **Voice** An excited series of fast, medium-pitched whistles *we-PI-PI-PI-PI-PEW*, sometimes in long series. [Suirirí Silbón]

ad

brachyrynchus

Smoke-coloured Pewee

ad

Olive-sided Flycatcher

Tropical Pewee

ad

pallescens

ad

Eastern Wood Pewee

ad

cinereus

ad

ad

tyrannus

sibilator

Eastern Kingbird

Sibilant Sirystes

Pseudocolopteryx All species occur in marshes or shrubbery near water. Up to four species may be found at a single locality. Males have black bills and gapes. Females have orange gapes, some orange on the lower mandible, and pale supercilia. Juveniles are highly variable and difficult to identify. *Polystictus* and *Culicivorus* Small tyrants of open country.

Crested Doradito *Pseudocolopteryx sclateri*　　　　　　　11.5 cm

Resident and nomadic partial austral migrant in the NE. **ID** Adult ♂: Striking vertical blackish crest with golden and rusty streaks. Bright yellow underparts. Adult ♀: Black crown, streaked yellow, and white supercilium. Two yellow, buff or white wing bars. Juvenile (not illustrated): Resembles ♀ but with broader buff or whitish wing bars and pale yellow underparts mixed with white. **Voice** Song, given while perched or in flight, is a series of metallic bill snaps ending in a syncopated mixture of high-pitched vocalisations and bill snaps, *t t t t t tík-t-tík-sí*. [Doradito Copetón]

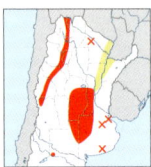

Subtropical Doradito *Pseudocolopteryx acutipennis*　　　　　　12 cm

NW Andes, local in C sierras and N and C lowlands (Nov–Mar). **ID** Adult ♂: Upperparts are brighter and more vivid green than congeners, and crown is noticeably peaked. Underparts bright lemon-yellow. Adult ♀ (not illustrated): Less peaked crown; shows a short yellowish supercilium and pale pinkish lower mandible. **Voice** Song is a series of introductory buzzy notes, followed by a short circular display with mechanical wing-whirring, *zek zek zek whi-crrru*. [Doradito Oliváceo]

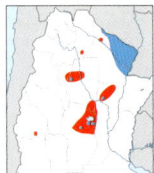

Dinelli's Doradito *Pseudocolopteryx dinelliana*　　　　　　　12 cm

Endemic breeder. Sep–Mar in NC lowlands; some winter in the humid chaco. **ID** Adult ♂: Indistinct rusty forecrown. Diffuse greyish supercilium. Drab brownish-olive above with two whitish or buff wing bars. Yellow to golden-yellow below. Adult ♀: Lacks rusty forecrown and shows buff supercilium which can join across forehead. **Voice** Song is a very complex succession of segments of fast, accelerating, short buzzy notes endings with distinctive raspy, nasal or whistled notes, *cht-cht-chtcht WHZ… cht-cht-cht-chtcht WEE… cht-cht-cht-chtcht TSINK*. Final whistles can be given in display flight. [Doradito Pardo]

Warbling Doradito *Pseudocolopteryx flaviventris*　　　　　　13 cm

Breeds in Pampas marshes; winters to the north. **ID** Adult ♂: Chestnut-brown crown contrasts with olive-brown back, and yellow underparts. Extensive black lores suggest a mask. Adult ♀ (not illustrated): Lacks the mask and shows a greyish supercilium. Juvenile: Presence and colour of wing bars variable, crown and mantle usually warm brown, and underparts mixed yellow and white. **Voice** Song is a series of soft introductory nasal notes, followed by a slow syncopated musical phrase, *ke ke ke ke KE-WI KE KE-LIP*, with variations. Sonogram on p. 446. [Doradito Pampeano]

Ticking Doradito *Pseudocolopteryx citreola*　　　　　　　13 cm

CW and N Patagonia; passes through the range of Warbling Doradito on migration. **ID** Like Warbling Doradito and only safely separated by voice. **Voice** Does not tick. Song is a gravelly series of insect-like introductory notes that then rise in pitch, speed and volume in a final phrase, *jep jep jep jep jep JEP TRIP TRRiu*. Sonogram on p. 446. Tax note 52. [Doradito Limón]

Bearded Tachuri *Polystictus pectoralis*　　　　　　　　11 cm

N and C lowlands (Oct–Apr); winters in the humid chaco. **ID** Small insectivorous tyrant of open-country, especially numerous along overgrown verges. Adult ♂: Crown, with erectile crest, face and throat finely streaked black and white. Rest of plumage as ♀. Adult ♀: Nondescript brown flycatcher with a short buff supercilium, tawny rump, two buff wing bars and an ochraceous wash on the flanks. Sometimes found with Sharp-tailed Grass Tyrant. **Voice** Perched song comprises alternating series of pure whistles followed by a mechanical wing-whirr, *peeee wididiPRRRT… wididididiPRRRT…* In display flight, a longer rising series of introductory notes. [Tachurí Canela]

Sharp-tailed Grass Tyrant *Culicivorus caudacuta*　　　　　　11 cm

Local in NE savanna. **ID** Small grass tyrant with a narrow tail and fairly stout bill. Adult: Striking broad white supercilium. Streaked crown and back with a tawny rump. Ochraceous wash on ear-coverts, sides of breast and flanks. Sometimes found with Bearded Tachuri. **Voice** Male song is a monotonous steady series of rising nasal buzzy notes, *wreeh*; female intermixes short chatters, *pirirrip*. [Tachurí Coludo]

ad ♂

Crested Doradito

Subtropical Doradito

ad ♂

ad ♀

Dinelli's Doradito

ad ♂

ad ♀

ad ♂

ad ♀

pectoralis

Bearded Tachuri

ad ♂

juv

Sharp-tailed Grass Tyrant

ad

Warbling/Ticking Doradito

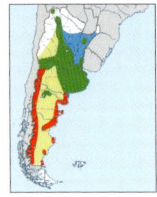

Many-coloured Rush Tyrant *Tachuris rubrigastra*   11.5 cm

Widespread. **ID** Tiny, restless, rush-dwelling tyrant with delicate bill. **Adult** *rubrigastra* (widespread, illustrated): Striking olive and yellow flycatcher with yellow supercilium, blue ear-coverts, red coronal tuft, incomplete black pectoral bar, white wing bar and panel, and white outer tail feathers. **Adult** *alticola* (NW Andes; not illustrated): Differs from *rubrigastra* by dark iris. **Juvenile:** Like washed-out version of adult, but with white supercilium and lacking pectoral bar. Wings and tail as adult. **Voice** Song is a rich resonant series ending in a rattle, *K'IUP PEERH K'IUP K'IUP-CHRddd*. [Tachurí Sietecolores]

White-headed Marsh Tyrant *Arundinicola leucocephala*   14.5 cm

Local in the NE. **ID** Short-tailed marsh tyrant, always encountered close to water's surface. Pale lower mandible. **Adult ♂:** Black with contrasting white head and throat. **Adult ♀:** Greyish hindcrown and back. Brown wings and blackish tail. White foreface and underparts. **Voice** Rarely heard song is a short ticking followed by a high-pitched metallic whistle, *tikik TSI-IK*. [Lavandera]

Black-backed Water Tyrant *Fluvicola albiventer*   14.5 cm

Erratic austral migrant in N marshes. **ID** Pied marsh or river tyrant. Sexes alike. Often perches in midstorey. **Adult:** Black with white face and underparts and two narrow white wing bars. Narrow white rump band. **Voice** Song is an explosive raspy note *SPÉwr*. Call is a liquid *pid*. [Viudita Lomo Negro]

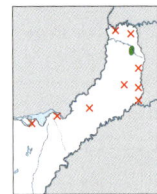

Masked Water Tyrant *Fluvicola nengeta*   15 cm

Rare in forest edge of rivulets and ponds in Misiones (expanding southwards). **ID** Pied marsh or river tyrant. Sexes alike. **Adult:** White with narrow black eye-stripe. Greyish back, unmarked black wings and tail. Broad white rump and tail tips. **Voice** Song is a gurgling chatter with grating notes. Call is a medium-pitched *PIK*, with much variation. [Viudita Enmascarada]

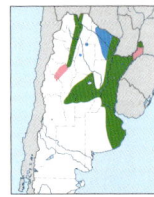

Yellow-browed Tyrant *Satrapa icterophrys*   16 cm

N lowlands and Andean slopes to 2700 m (mainly Oct–May). **Adult:** Vivid olive upperparts with a striking lemon-yellow supercilium and underparts. Extensive white wing-fringing and at least one wing bar. **Voice** Dawn song is a fast series of rising liquid notes, *wudubibiribirí*. Call is a weak, short, rising whistle *fuít*. [Suirirí Amarillo]

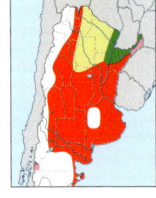

Vermilion Flycatcher *Pyrocephalus rubinus*   13.5 cm

Resident in the N lowlands; austral migrant south to N Patagonia. **ID** Small dimorphic flycatcher of open woodlands. **Adult ♂:** Vermillion crown and underparts. **Adult ♀:** Brown above with white wing-fringing but lacks wing bars. White below with extensive blurry brown streaking, yellow to yellow-orange undertail coverts. Immature ♂♂ have pink undertail-coverts. **Voice** In undulating display flight, a very fast rising, accelerating stutter ending in a metallic trill, *du-du-dededididddd-rrink*. Call is a weak, short, rising whistle *fee*. [Churrinche]

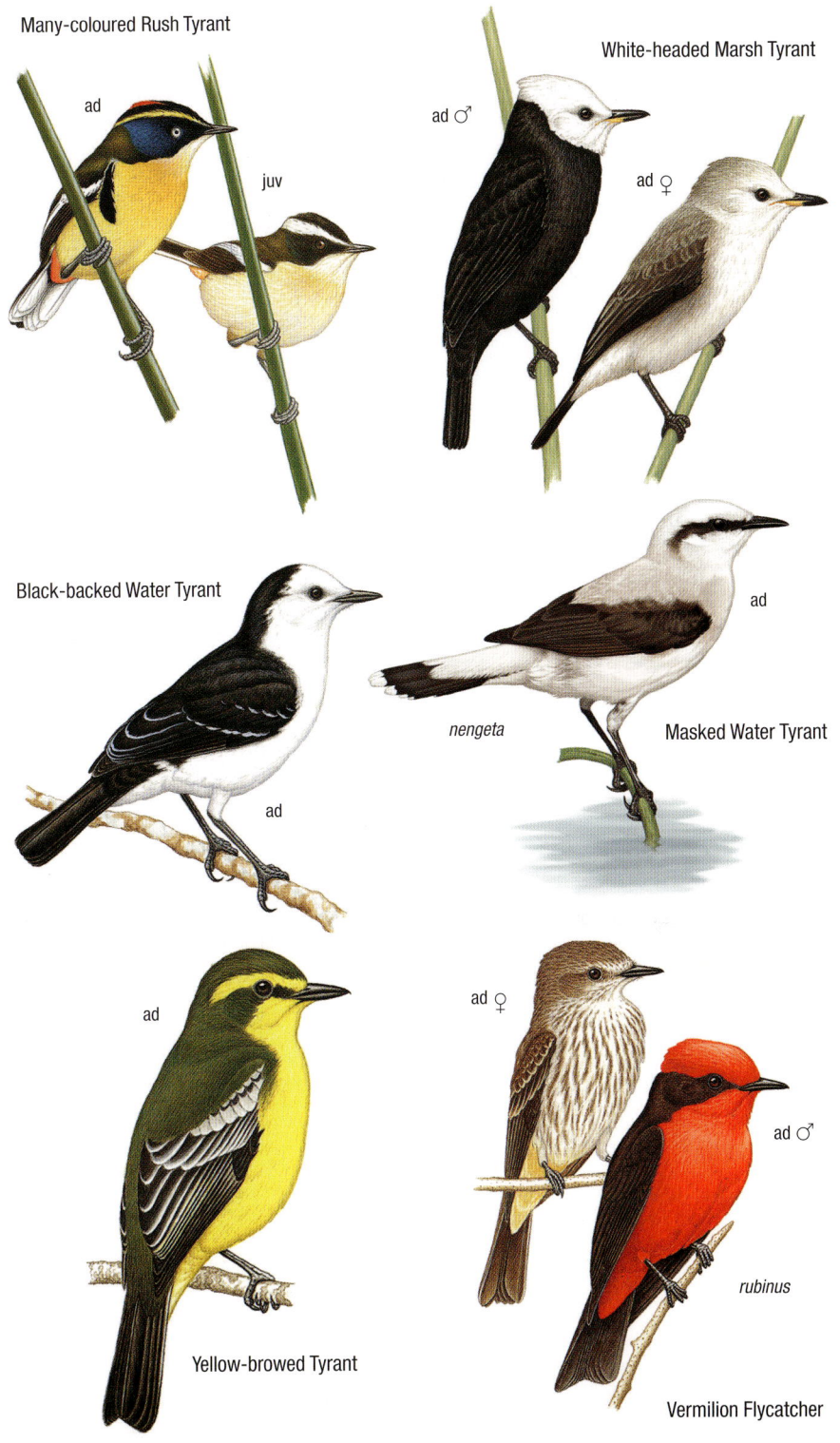

Many-coloured Rush Tyrant

ad

juv

White-headed Marsh Tyrant

ad ♂

ad ♀

Black-backed Water Tyrant

ad

ad

nengeta

Masked Water Tyrant

Yellow-browed Tyrant

ad

ad ♀

ad ♂

rubinus

Vermilion Flycatcher

Knipolegus Sexually dimorphic, often inconspicuous, tyrants of scrub and forest. ♂♂ perform cartwheel display flights with mechanical clicks at apex. Most are partial or austral migrants. *Hymenops* Sexually dimorphic marsh tyrant with an eye wattle. ♂♂ perform high cartwheel displays.

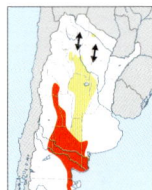

Hudson's Black Tyrant *Knipolegus hudsoni*   17 cm

Endemic breeder. Local in woodlands of N Patagonia (mainly Oct–Jan); passage migrant in the dry chaco (Oct). **ID Adult** ♂: Makes a short display jump while switching perches (unlike White-winged). Black or dull chestnut iris. Proportionately larger-headed than White-winged and bill deeper. White on lower flanks sometimes obvious, but not safe to distinguish from White-winged. **Adult** ♀: From adult ♀ White-winged by dense blotchy breast streaking, creamy buff belly, contrasting cinnamon undertail-coverts, and less rufous in rump and tail. Never shows grey on the crown, unlike some White-winged. **Voice** In display flight, series of hollow low-pitched notes followed by a quick succession of higher-pitched vocal and mechanical notes. [Viudita Chica]

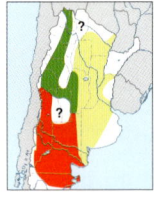

White-winged Black Tyrant *Knipolegus aterrimus*   17.5 cm

Widespread in the west, south to N Patagonia. **ID Adult** ♂: From Hudson's Black Tyrant by high vertical display flight always landing on the same perch. Iris tends to be redder than in Hudson's, head smaller and bill finer. In display flight, broad white wing-stripe. White on lower flanks reduced and difficult to see; not useful to distinguish from Hudson's. **Adult** ♀: Highly variable. Greyish-olive to dark brown above with contrasting rufous rump and tail base. White, buff or cinnamon wing bars. Cinnamon below, often whiter on throat or belly; some with indistinct breast streaking. **Voice** In display flight, short high-pitched trill followed by two mechanical clicks. [Viudita Trinadora]

Blue-billed Black Tyrant *Knipolegus cyanirostris*   17 cm

NE, chiefly gallery forest in Mesopotamia. **ID Adult** ♂: Bright red iris and blue bill. Glossy black with a white wing-stripe visible in flight. **Adult** ♀: Dull red iris. Rusty crown and nape. Two broad cinnamon wing bars. Rufous rump. Broad blackish breast streaking. Overlaps with Cinereous Tyrant in the humid chaco in winter. **Voice** Display flight is little known. Call is a sharp *tsip*. [Viudita Pico Celeste]

Cinereous Tyrant *Knipolegus striaticeps*   14 cm

Chaco woodlands. **ID Adult** ♂: Orange-red iris. Entirely grey with a blackish mask. **Adult** ♀: Orange-red iris. Rusty crown, nape and rump. Two pale wing bars. Blurry breast flammulations. **Voice** In high vertical display flight, a rising liquid series of soft notes followed by sharp vocal notes and a mechanical *clock*. [Viudita Chaqueña]

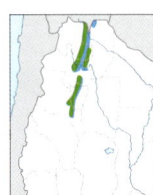

Plumbeous Tyrant *Knipolegus cabanisi*   16.5 cm

Yungas forest. **ID Adult** ♂: Entirely plumbeous-grey with an orange-red iris. **Adult** ♀: Dull olive or olive-brown above with two buff or cinnamon wing bars. Rufous rump and tail base. Dull greyish-olive below with whitish throat and belly and buff undertail-coverts. **Voice** In vertical display flight, a fast dry wing-whirr followed by a mechanical smack. Tax note 53. [Viudita Plomiza]

Spectacled Tyrant *Hymenops perspicillatus*   16.5 cm

Widespread. **ID Adult** ♂: Black with yellow bill, iris and eye wattle. White primaries visible on closed wing. In flight, striking white distal primaries. **Adult** ♀: Iris and smaller wattle as adult ♂. Pale bill. Dark brown crown and broad pale supercilium. Cinnamon wing bars and chestnut primaries, visible on closed wing. White or buffy below, streaked brown on breast. In flight, wings mostly bright chestnut. **Juvenile** (not illustrated): Much like ♀, but with dark eyes and usually lacks the bare wattle. **Voice** In high cartwheel display flight, a low mechanical humming interrupted by a sharp metallic click at the apex. [Pico de Plata]

Hudson's Black Tyrant

ad ♂

ad ♀

White-winged Black Tyrant

ad ♂

♂ display flight

ad ♀

aterrimus

ad ♀

ad ♂

ad ♀

ad ♂

Cinereous Tyrant

Blue-billed Black Tyrant

ad ♀

ad ♂

ad ♀

ad ♀

ad ♂

perspicillatus

Plumbeous Tyrant

Spectacled Tyrant

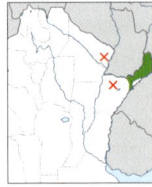

Long-tailed Tyrant *Colonia colonus* 25 cm

Paraná forest. **ID** Canopy tyrant with narrow elongated central rectrices. Perches vertically on exposed twigs. **Adult**: Black with white forehead, and grey to white cap. Square white rump patch. **Juvenile**: Somewhat browner than adult. White develops gradually on forehead. No streamers. **Voice** Dawn song is a series of two well-spaced, rich whistled phrases, *we vi-ew... wee-wrrrr*. Call is a pure rising and falling, pewee-like whistle *weeou*. [Yetapá Negro]

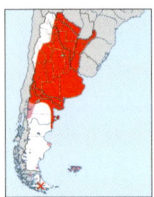

Fork-tailed Flycatcher *Tyrannus savana* 34 cm

Widespread austral migrant. **ID** Large open-country tyrant with extremely long forked tail and semi-concealed bright yellow coronal patch. **Adult**: Glossy black hood, grey mantle and long black tail. Pure white underparts. In flight, tail often spread open. **Juvenile**: Brown crown and ear-coverts. Cinnamon fringing on wing-coverts and brown tail; considerably shorter and less graduated than adult. Compare juvenile with Eastern Kingbird (Plate 142). **Voice** Song is a variable twittering trill that speeds up into a short cricket-like rattle and drops in pitch. Also a gravelly series of monotone burry notes. [Tijereta]

Shear-tailed Grey Tyrant *Muscipipra vetula* 22.5 cm

Clearings in Paraná forest; mainly in winter. **ID** Local, slender tyrant of forest edge with deeply notched tail. Undulating flight. **Adult**: Soft blue-grey plumage with dusky ear-coverts, dark brown wings and black tail. In flight, tail spread open forming distinctive fork; only occasionally spread while perched. **Juvenile** (not illustrated): Similar but with white fringes on wing-coverts. **Voice** Common call is a disyllabic, hollow and wooden *pu-PAK*, sometimes in longer series. [Viudita Coluda]

Streamer-tailed Tyrant *Gubernetes yetapa* 42 cm

Boggy grasslands in s. Misiones and ne. Corrientes. **ID** Spectacular marsh tyrant with extremely long and narrow forked tail. **Adult** ♂: Mainly grey with white supercilium and throat, deep reddish-brown ear-coverts extending into pectoral band. Tail averages shorter in ♀. In flight, striking cinnamon patch in flight feathers. White underwing-coverts. **Juvenile**: Buff crown and nape, finely speckled brown. Chestnut-brown ear-coverts. Buffy-white throat. Comparatively short brown notched tail, edged whitish and with blackish terminal band. Cinnamon patch in primaries as adult. **Voice** Duet is a series of burry warbled notes with simultaneous whistles and wing-snaps during bowing/wing-raising display, creating a bizarre ensemble. Call is a far carrying, burry *WEERRR*. [Yetapá Grande]

Strange-tailed Tyrant *Alectrurus risora* ♂ 31 cm; ♀ 20 cm

Local in humid grasslands of NE, mainly Corrientes. **ID** Sexually dimorphic tyrant restricted to natural grasslands. ♂ has unique pair of tail flags with vertical vanes, important in tail jerking displays. **Adult** ♂ **breeding**: Black head and upperparts. Bare scarlet throat skin, bordered by broad black pectoral band. White lower underparts and scapulars. Long attenuated black tail flags. In flight, tail flags held straight out and used as rudders. White scapulars and grey rump. **Adult** ♂ **non-breeding**: Feathered white throat. Black areas of upperparts and pectoral band fringed with brown. Tail flags usually worn. **Adult** ♀ **non-breeding**: Brown upperparts with contrasting white face and underparts. Complete buffy pectoral band and buff wash on flanks. Elongated shafts of outer tail feathers with slender brown rackets at tip. Breeding ♀ has darker pectoral band. **Voice** High-pitched weak whistles during display flight, with audible wing sounds. Call is a liquid *pwit*. [Yetapá de Collar]

Cock-tailed Tyrant *Alectrurus tricolor* ♂ 14.5 cm; ♀ 12.5 cm

Extinct in s. Misiones savanna? **ID** Sexually dimorphic tyrant restricted to natural grasslands. ♂ has unique modified tail, important in display. **Adult** ♂ **breeding**: Black crown, nape, back, patches on sides of breast and wedge-shaped tail flags. White face, underparts and scapulars. **Adult** ♀: Brown crown and mantle, becoming rusty on rump. Square blackish tail. Contrasting white face and underparts with brown spurs on sides of breast and buff wash on flanks. **Voice** Generally silent. [Yetapá Chico]

Long-tailed Tyrant

ad

juv

ad

ad

juv

savana

Fork-tailed Flycatcher

colonus

ad ♂

ad

ad

Shear-tailed Grey Tyrant

juv

Streamer-tailed Tyrant

♂ non-br

♂ br

♀ non-br

Cock-tailed
Tyrant

ad ♀

♂ br

♂ br

Strange-tailed Tyrant

Pyrope, Nengetus, Neoxolmis, Xolmis and *Heteroxolmis* Conspicuous and approachable arboreal and terrestrial tyrants, usually found in pairs. Sexes alike, although somewhat brighter in the ♂♂ of Salinas and Rusty-backed Monjitas. Austral migration is pronounced in Black-crowned and Rusty-backed Monjitas. Whistled songs given mostly at dawn. See also Plate 148.

Fire-eyed Diucon *Pyrope pyrope* — 21.5 cm

Patagonian forest. **ID** Perches upright, mostly in subcanopy or canopy, but forages at all levels. **Adult:** Sooty-grey with contrasting white throat and black wings. Scarlet iris is striking at close range. **Voice** Song starts with soft introductory notes, followed by a loud pure whistle *whip… whip… wu-EEO*. In high, slow, fluttering display flight, a medium-pitched, bold *TJU* repeated every 1.5 secs. Calls include a variety of *pic* notes. [Diucón]

Grey Monjita *Nengetus cinereus* — 23 cm

Partial austral migrant in NE grasslands. **ID Adult:** Grey with contrasting white supraloral streak and throat, red iris and black moustachial. In flight, inner primaries entirely white, forming striking white panel and contrasting with black distal primaries. Broad whitish or pale brown terminal band on tail. **Voice** Song is a slurred, melancholic whistle *feee-fio-weeeh*. Calls include high-pitched *SWEE* and *SWI-LI*. [Monjita Gris]

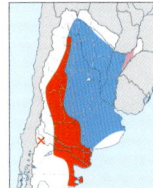

Black-crowned Monjita *Neoxolmis coronatus* — 21.5 cm

Endemic breeder in Monte desert; winters in N lowlands. **ID Adult:** White forehead and diadem surrounding black skullcap. Contrasting black post-ocular streak. Broad greyish bars across wing-coverts. In flight, blackish wings with broad white wing-stripe and narrow trailing edge. **Voice** Song is a far-carrying, rich, mellow falling and rising disyllabic whistle *chuwo-weeee*. Call is a short, explosive *PWÉ*. [Monjita Coronada]

White-rumped Monjita *Xolmis velatus* — 20 cm

Sporadic or invading in N lowlands, especially in the NE. **ID Adult:** Pale grey crown and nape. White forehead and indistinct supercilium. Greyish-brown back concealing white panel in secondaries. Black wings and tail tip. Compare with Black-and-white Monjita. In flight, contrasting broad white rump and tail base. Thick white stripe across base of flight feathers. **Voice** Song is a burry, descending *DJERR*. Calls include a variety of short rich whistles. [Monjita Rabadilla Blanca]

White Monjita *Xolmis irupero* — 18.5 cm

Widespread. **ID Adult:** Entirely white with black primaries and tail tip. In flight, short slightly notched tail; wings mostly white, tipped black. Compare with Black-and-white Monjita in Mesopotamia. **Voice** Song comprises two well-spaced mellow, upward-inflected whistles, one higher and one lower-pitched, *weeee… woeee*. Calls include short single whistles. [Monjita Blanca]

Black-and-white Monjita *Heteroxolmis dominicanus* — 22 cm

Local in NE grasslands, principally Mesopotamia. **ID** Slender-bodied, long-tailed marsh tyrant with unusual dichromatic wing pattern. Sexually dimorphic, perhaps nomadic and occasionally gregarious. **Adult ♂:** White body with contrasting black wing-coverts, flight feathers and tail. **Adult ♀:** Crown, nape and mantle pale brown, highlighting white face and scapulars. In flight, both sexes show black wings, tipped white. **Voice** Song is a nasal, squeaky series of upward-inflected whistles with intermixed short notes, *week co-co queeek*, with much variation. Calls include pure upward-inflected whistles. [Monjita Dominica]

Fire-eyed Diucon

pyrope

ad

Grey Monjita

ad

Black-crowned Monjita

ad

White-rumped Monjita

ad

White Monjita

irupero

ad

Black-and-white Monjita

ad ♂

ad ♀

Neoxolmis **(continued)** See generic notes on Plate 147.

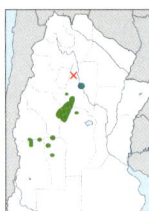

Salinas Monjita *Neoxolmis salinarum*

18 cm

Endemic to saltpans of C Argentina. **ID** Adult ♂: Smaller than Rusty-backed Monjita (no definite overlap), differing by supercilium continuing across hindcrown, and in some individuals a second whitish nuchal collar; also by striking white upperwing-coverts and rump, and sparsely streaked breast. Adult ♀: Lacks rusty cap of ♂. Note white wing-coverts and rump in flight. **Voice** Much like Rusty-backed Monjita. [Monjita Salinera]

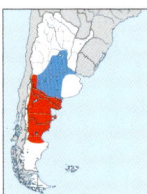

Rusty-backed Monjita *Neoxolmis rubetra*

21 cm

Endemic breeder in N Patagonian shrub-steppe, winters to N lowlands. **ID** Adult ♂: Rusty crown and warm brown back and black wings. Contrasting white supercilium and underparts, finely streaked black across breast. Adult ♀: Lacks rusty cap. Blackish primaries with cinnamon flashes and narrow white trailing edge in flight. Possible overlap with Salinas Monjita in winter. **Voice** Rarely heard song is a disyllabic, low nasal note followed by a rising low whistle *AANH-WOOO*. Calls include soft short whistles and *chek* notes. [Monjita Castaña]

Chocolate-vented Tyrant *Neoxolmis rufiventris*

25 cm

Patagonian grass-steppe to N lowlands in winter. **ID** A robust, mainly terrestrial, open-country tyrant; gregarious in winter. Adult: Grey head, back and breast with contrasting rusty belly and vent, whitish patch on wing-coverts and dusky mask. In flight, contrasting rufous bases to flight feathers and broad white trailing edge. Juvenile: Indistinct whitish supercilium, mottled upperparts and breast, and cinnamon belly. **Voice** Song is a two-note, falling and rising, pure whistle *PEEEW-WEEE*. Calls include a rapid, ringing *pi-di-dit*. [Monjita Chocolate]

Lessonia Two similar small, terrestrial, open-country flycatchers. Often segregated by sex outside the breeding season when gregarious.

Andean Negrito *Lessonia oreas*

13.5 cm

Puna marshlands. **ID** Adult ♂: Black with contrasting cinnamon saddle. In flight, contrasting white inner webs of primaries. Adult ♀: Similar to ♂ but black plumage replaced by dark brown. **Voice** Display at dawn and dusk consists of short high-pitched notes while standing, followed by ticking notes in flight and a wing-whirr, with a long final high-pitched note *tsi...tsi...tsi...tsi... tic tic tic tic trrrrrrr-psie*. Calls include high-pitched short notes. [Sobrepuesto Andino]

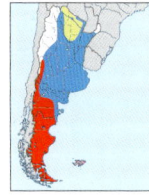

Austral Negrito *Lessonia rufa*

12.5 cm

Breeds in Patagonia, winters throughout N lowlands. **ID** Adult ♂: Entirely black with contrasting chestnut saddle. In flight, black flight feathers. Adult ♀: Biscuit-brown crown and upperparts; somewhat warmer on mantle. White throat contrasts with pale brown underparts. **Voice** Display flight has not been described. Calls include high-pitched short notes. [Sobrepuesto Austral]

Cnemarchus A high-Andean tyrant, recalling a shrike-tyrant but slimmer-bodied with a very slender bill.

Rufous-webbed Bush Tyrant *Cnemarchus rufipennis*

22 cm

Local in NW Andes above 2600 m, often in *Polylepis* woodlands. **ID** Perches upright and frequently hunts by hovering. Adult: Ashy-grey with a contrasting creamy-white belly and vent. Cinnamon undertail with a dusky tip. In flight, underwing mostly cinnamon with paler coverts and dusky trailing edge. Tail cinnamon with a dusky T-shape. **Voice** Common call is a shrill, quavering descending whistle *sreeu*, with much variation. [Birro Gris]

ad ♀

ad ♂

Salinas Monjita

Rusty-backed Monjita

ad ♀

ad ♂

Chocolate-vented Tyrant

ad

ad

juv

Andean Negrito

ad ♀

ad ♂

ad

ad

ad ♀

ad ♂

bolivianus

Austral Negrito

Rufous-webbed Bush Tyrant

Muscisaxicola Terrestrial tyrants of the Andes and Patagonia, mostly distinguished by head pattern. Sexes alike. Juveniles lack or have subdued crown patches. Most species occur together, and some form mixed flocks. See also Plate 150.

Spot-billed Ground Tyrant *Muscisaxicola maculirostris* 15 cm

Partial austral migrant throughout much of Andes. **ID** Adult: Smallest ground tyrant with pale yellowish base to short lower mandible visible at close range. Two narrow rusty wing bars when fresh. Grey wash on breast, sometimes flammulated. **Voice** Song in display flight is a series of ticking notes, ending in a whistle during a static hover with open wings, *pic...pic pic pic-pic-pididiSWEEO*; alternatively, only delivers whistles in quick succession. [Dormilona Chica]

Dark-faced Ground Tyrant *Muscisaxicola maclovianus* 16.5–18.5 cm

Widespread. **ID** Adult *mentalis* (16.5 cm; Breeds in S Andes and Tierra del Fuego, winters in C sierras and N lowlands): Blackish lores and forehead form distinct dark foreface but beware juvenile Cinnamon-bellied (Plate 150). Adult *maclovianus* (18.5 cm; Endemic to the Falklands): From *mentalis* by warm brown crown, whiter underparts, shorter bill and larger size. [Dormilona Cara Negra]

Cinereous Ground Tyrant *Muscisaxicola cinereus* 17 cm

NW and C Andes. **ID** Adult: Uniform brownish-grey upperparts but no skullcap or rufous tinge on crown and whitish supercilium indistinct. Ssp. essentially alike. Compare Spot-billed, White-browed, Puna and Rufous-naped Ground Tyrants. [Dormilona Cenicienta]

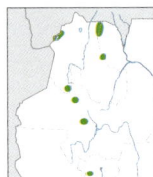

Puna Ground Tyrant *Muscisaxicola juninensis* 17 cm

NW Andes. **ID** Adult: Very similar to White-browed Ground Tyrant (possible overlap in winter) but skullcap less distinct, upperparts somewhat greyer, and slightly more white in outer tail. Compare also with Cinereous and Rufous-naped Ground Tyrants. [Dormilona Puneña]

White-browed Ground Tyrant *Muscisaxicola albilora* 18 cm

Austral migrant in the S Andes, recorded on passage in the C sierras and NW Andes. **ID** Adult: Grey-brown above with rufescent skullcap. Primaries extend halfway along tail. [Dormilona Ceja Blanca]

Ochre-naped Ground Tyrant *Muscisaxicola flavinucha* 18.5 cm

Austral migrant on Andean slopes and Tierra del Fuego. **ID** Adult: Large, long-winged ground tyrant with prominent white lores extending into supercilium, and yellowish-buff skullcap. [Dormilona Fraile]

Rufous-naped Ground Tyrant *Muscisaxicola rufivertex* 17 cm

N Andes and C sierras. **ID** Adult *achalensis* (C Sierras; illustrated): Ash-grey upperparts, paler and greyer than congeners, and with contrasting black wings and tail. Rufous skullcap. White underparts. Adult *pallidiceps* (Andes; not illustrated): Brownish-black wings and tail, and less clean underparts, often with a greyish wash. [Dormilona Gris]

Spot-billed Ground Tyrant

maculirostris

ad

Dark-faced Ground Tyrant

ad

maclovianus

ad

mentalis

argentina

Cinereous Ground Tyrant

Puna Ground Tyrant

ad

ad

Ochre-naped Ground Tyrant

ad

flavinucha

ad

White-browed Ground Tyrant

achalensis

Rufous-naped Ground Tyrant

Muscisaxicola **(continued)** Terrestrial tyrants of Andes and Patagonia, mostly distinguished by head pattern. Sexes alike. Most species occur together, and some form mixed flocks. See also Plate 149.

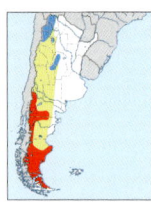

Cinnamon-bellied Ground Tyrant *Muscisaxicola capistratus*   17.5 cm

Breeds in Patagonia; winters in C Sierras and N Andes. **ID Adult:** Black forecrown, chestnut skullcap and cinnamon flanks and vent. **Juvenile:** Dusky crown, rarely with chestnut at rear. Buff fringes to wing-coverts and tail edged cinnamon. Vent washed cinnamon. Compare with Dark-faced Ground Tyrant (ssp. *mentalis*) (Plate 149). [Dormilona Canela]

Black-fronted Ground Tyrant *Muscisaxicola frontalis*   18.5 cm

Mainly high Andes of Mendoza and San Juan; local in Jujuy and Salta. **ID Adult:** Long, slightly decurved black bill. White supraloral and contrasting black forehead, terminating in a point. [Dormilona Frente Negra]

Streak-throated Bush Tyrant *Myiotheretes striaticollis*   22 cm

NW Andes. **ID** Large, stout-billed Andean flycatcher of forest edge, cliffs and tree-line. **Adult:** Brown upperparts and dull cinnamon underparts with contrasting white throat streaked dusky. Undertail rufous with dusky terminal band. In flight, wings mostly rufous. Tail rufous with black T-shape visible from above. **Voice** Common call is a far-carrying, yet relatively soft, low-pitched mellow whistle *pweeee.* Less often, a higher-pitched disyllabic *su-vit* or *swee-vit.* [Birro Grande]

Cliff Flycatcher *Hirundinea ferruginea*   18.5 cm

N Andes, C sierras; rare in Misiones and n. Corrientes. **ID** Compact flycatcher with peaked crown and notably aerial habits. **Adult** *pallidior* (N Andes and C Sierras, illustrated): Brown above with rufous fringing on wing-coverts. Bright rufous panel in closed primaries. Rufous rump and tail, tipped brown. Rufescent underparts. In flight, extensive rufous in wings and tail. **Adult** *bellicosa* (NE, not illustrated): Slightly darker than *pallidior.* **Voice** Common call is a mellow whistle followed by a slow burr, *WEE-RRRRR*, recalling the song of White-winged Cinclodes but less rattled. **Tax note 54.** [Birro Colorado]

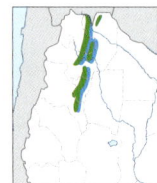

Black Phoebe *Sayornis nigricans*   18.5 cm

NW Andean rivers. **ID** Fairly long-tailed, rock-dwelling flycatcher of montane rivers. **Adult ♂:** Black with two white wing bars and tertial fringes, and whitish centre of belly. **Adult ♀** (not illustrated): Somewhat paler. **Voice** Dawn song is composed of two alternating shrill notes, *tshreee pdi-swee.* Call is a sharp, metallic *pit.* [Viudita de Río]

Cinnamon-bellied Ground Tyrant

juv

ad

Black-fronted Ground Tyrant

ad

ad

pallidus

ad

Cliff Flycatcher

Streak-throated
Bush Tyrant

ad ♂

latirostris

Black Phoebe

Agriornis Large brown flycatchers with streaked throats and hooked bills, most notable in largest species. Inhabit open terrain in the high Andes, Patagonia and C sierras. Best identified by tail pattern and bill shape and colour.

Lesser Shrike-Tyrant *Agriornis murinus*   18.5 cm

Endemic breeder in monte desert and N Patagonia; winters in the N lowlands. **ID** Adult: Small pink bill with a black culmen and tip. Buffy wing bars; sometimes white. Brown tail, edged buffy-white. Streaked throat, diffuse when worn. Brownish breast (whitish when worn) with warm buffy flanks. Compare with Grey-bellied Shrike-Tyrant. **Voice** Perched calls include pure or shrill whistles *chewit chí-weeet*. In high display flight, stalls and free-falls tail first, in silence or giving a short sharp descending whistle every 2–3 secs. [Gaucho Chico]

Black-billed Shrike-Tyrant *Agriornis montanus*   23–26 cm

The most widespread shrike-tyrant. **ID** Distinguished from congeners by its straight, slender black bill with little or no hook. Tail mostly white with brown central tail feathers, varying slightly between ssp. Adult *leucurus* (26 cm; C and S Andes, and Patagonia): Grey-brown above, fairly dark brown breast, washed cinnamon on the belly, and whitish undertail-coverts. Overlaps with Lesser, Great, and nominate Grey-bellied Shrike-Tyrants. Adult *montanus* (23 cm; NW Andes): Smaller and paler than *leucurus* with no cinnamon on belly and indistinct throat streaking. Overlaps with Grey-bellied (ssp. *andecola*) and White-tailed Shrike-Tyrants. Juvenile (not illustrated): Yellow base to lower mandible and dark iris. **Voice** Song is a loud, rich, pure rising and falling whistle *WEEÉ-YOU*, given mostly at dawn. Silent, high, undulating display flight with static pauses on outstretched wings. [Gaucho Serrano]

Great Shrike-Tyrant *Agriornis lividus*   30 cm

Patagonian shrub-steppe and forest edge along the S Andes. **ID** Adult: Much larger than other shrike-tyrants. Huge hooked bill. Strong cinnamon wash on belly and vent. Brown tail, edged white with a whitish tip when fresh. Overlaps with Black-billed (ssp. *leucurus*) and Grey-bellied (ssp. *micropterus*) Shrike-Tyrants. **Voice** A quiet, somewhat metallic *wit* or *pit* note given sporadically. [Gaucho Pardo]

Grey-bellied Shrike-Tyrant *Agriornis micropterus*   24–25.5 cm

Widespread. **ID** Adult *micropterus* (25.5 cm; breeds in C Andean foothills, to 1300 m, and much of Patagonia, except Tierra del Fuego; winters in the N lowlands): Heavy bill with a pink base; variable hooked tip. Crown darker brown than back. Tail fringed white. Overlaps with Lesser, Great and Black-billed (ssp. *leucurus*) Shrike-Tyrants. Adult *andecola* (24 cm; local resident in Puna): Smaller, paler and more buffy-brown than *micropterus* with a less capped appearance; tends to show weaker throat streaks. Overlaps with Black-billed (ssp. *montanus*) and White-tailed Shrike-Tyrants. **Voice** A quiet, dry *kip* note given sporadically. [Gaucho Grande]

White-tailed Shrike-Tyrant *Agriornis albicauda*   25.5 cm

Rare above 3400 m. in Jujuy, Salta and Tucumán. **ID** Adult: Always outnumbered by very similar Black-billed (ssp. *montanus*) which tends to be slimmer. Tail patterns are very similar and throat streaking is variable. Best distinguished by very broad pink-based bill, usually with an obvious hook. From similar Grey-bellied (ssp. *andecola*) by tail pattern. **Voice** A quite dry *chit* note given sporadically. [Gaucho Andino]

ad

Lesser Shrike-Tyrant

ad

leucurus

ad

montanus

Black-billed Shrike-Tyrant

ad

fortis

Great Shrike-Tyrant

ad

ad

albicauda

ad

andecola

micropterus

White-tailed Shrike-Tyrant

Grey-bellied Shrike-Tyrant

Myiarchus Upright-perching flycatchers with a bushy crown, grey throat and breast, and yellow belly, inhabiting light woodland or forest edge in midstorey or subcanopy. Beware of confusion with Large Elaenia (Plate 133) which has a similar ventral pattern but a stubby bill. Voices greatly aid identification, although repertoires are extensive and only the commonest calls are described here. Juveniles have rusty wing bars and fringes to the flight feathers and tail, although the tertials are fringed white in Swainson's and Short-crested Flycatchers, and the crown is black in Dusky-capped.

Swainson's Flycatcher *Myiarchus swainsoni* 19–20 cm

Widespread. **ID** Two very different ssp, although intermediates occur in Mesopotamia. Adult *swainsoni* (19 cm; Paraná forest, Sep–Apr): Less crested than congeners. Stout, dark brown bill with a small pink base. Dark brownish-olive head and upperparts. 1–2 indistinct greyish-brown wing bars. Dull whitish fringes to secondaries and tertials. Adult *ferocior* (20 cm; throughout the N and C lowlands, Andean foothills to 1050 m and N Patagonia, Sep–Mar): Paler brown bill than *swainsoni* with an extensive pink base. Much paler above than *swainsoni* with obvious broad whitish wing bars, and prominent secondary and tertial fringing. **Voice** Dawn song is a mournful whistle, followed by a short rollicking phrase, *waaah widi-clew*. Call is a soft drawn-out *meeoow*. [Burlisto Pico Canela]

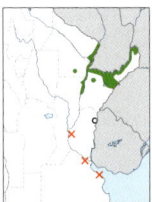

Short-crested Flycatcher *Myiarchus ferox* 18.5 cm

Resident in Misiones and ne. Corrientes. Sparse Sep–Mar records from the humid chaco, Entre Ríos and ne. Buenos Aires. **ID** Adult: Slender black bill, shorter than congeners (some show pale bill base, mostly in winter?). Slight erectile dusky crest. Fresh birds have narrow pale brown wing bars and warm brown-fringed secondaries and tertials; worn birds lack these markings. Tail proportionately shorter than congeners. **Voice** Commonest call is a low-pitched, rippling, whistled trill *whirrr* or *beeurrr*, with much variation. [Burlisto Pico Negro]

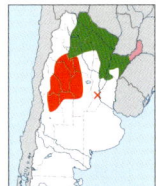

Brown-crested Flycatcher *Myiarchus tyrannulus* 20 cm

Resident in the N lowlands and foothills to 1200 m, south to e. Mendoza, n. San Luis, w. Córdoba and n. Santa Fe; southern birds withdraw north in Apr. **ID** Adult: Stout black or brown bill, usually with a small pink base. Distinctly crested with a warm brown crown. Two whitish wing bars and rufous-fringed primaries, striking in flight. Secondaries and tertials fringed white. Fairly long brown tail with rufous inner webs; visible when folded. Beware of juveniles of Swainson's and Short-crested Flycatchers. **Voice** Dawn song is a sharp note followed by a rolling trill *KWIP-piruborr*. Commonest call is a thrush-like *PWIK* or *WHIP*, sometimes also integrated into dawn songs. [Burlisto Cola Castaña]

Dusky-capped Flycatcher *Myiarchus tuberculifer* 18.5 cm

Yungas forest interior and edge to 2000 m (Sep–Apr). **ID** Adult: Stout black bill. Readily identified by the notable bushy black crown. Only overlaps with Brown-crested Flycatcher, and conceivably Swainson's (ssp. *ferocior*). **Voice** Dawn song is a melancholic series of whistles, with short interspersed notes, *wee-wiu wit weeo*. Meowing call is very similar to Swainson's. [Burlisto Corona Negra]

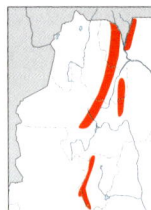

Rufous Casiornis *Casiornis rufus* 18 cm

N lowlands and foothills. **ID** Solitary, rufous, *Myiarchus*-like flycatcher of midstorey. Round-headed although somewhat crested. Adult ♂: Stout black bill with pink base. Chestnut upperparts, brightest on crown, wing-coverts, rump and tail. Buffy lores, ear-coverts and throat, becoming ochraceous on breast and creamy-yellow on abdomen. Adult ♀: Somewhat duller. **Voice** Song is an unpredictable series of 1–3 progressively lower-pitched, shrill, descending whistles *kleeu-kleeu-kleeu*. Call is a piercing flat shrill *KLEE*. [Burlisto Castaño]

Rufous-tailed Attila *Attila phoenicurus* 19.5 cm

Rare in Paraná forest. **ID** Chunky flycatcher of mid- and upper strata at forest edge and interior. Adult: Long black bill, sometimes with a pink base. Bright rufous mantle with paler rump and bright rufous-chestnut tail. Contrasting ashy-grey hood, suggestion of a rufous pectoral band and ochraceous yellow belly and vent. **Voice** Song is a very loud rising series of pure whistles ending in a lower-pitched note *FWEE-FEE-FIFU*, recalling a hawk-eagle. Call is a disyllabic, lower-pitched, drawn-out melancholic whistle *peee-uu*. [Burlisto Cabeza Gris]

Swainson's Flycatcher

swainsoni

ad

ad

ferocior

ad

australis

Short-crested Flycatcher

tyrannulus

ad

Brown-crested Flycatcher

ad

atriceps

Dusky-capped Flycatcher

ad ♂

Rufous Casiornis

ad

Rufous-tailed Attila

Three-striped Flycatcher *Conopias trivirgatus*   16 cm

Resident in Paraná forest edge. **ID** A small subcanopy flycatcher with a fairly slender black bill. **Adult:** Brown crown, narrow blackish mask and long white supercilium. Olive back and unmarked brown wings. Yellow underparts (including the throat); usually washed olive on the sides of the breast. **Voice** A frenetic series of alternating buzzes and fast chatters, *pewrr pddd pewrr pewrr pddd pewrr*, with much variation. Alternatively delivers a series of buzzes. [Benteveo Chico]

Social Flycatcher *Myiozetetes similis*   17.5 cm

Resident in Paraná forest. **ID** Small, robust, stub-billed flycatcher with a concealed golden or orange coronal patch. Perches low or in midstorey of edge habitats and secondary growth. **Adult:** Short stout black bill. Dusky face and broad white supercilum. Olive back and brown wings with indistinct rusty fringing and usually a single buffy wing bar. Bright yellow below with a contrasting white throat. **Voice** Dawn song is a whistle followed by a rapid phrase *fee fi-tri-dri-d*. Call is a repeated whistle *SHRILL*, or a pure flatter *SWlh*. [Benteveo Mediano]

Lesser Kiskadee *Philohydor lictor*   17 cm

Rare vagrant. **ID** Recalls *Pitangus* but with a relatively long, narrow, slender bill. **Adult:** Considerably smaller than Great Kiskadee with a slender body, different bill shape and tends to show more rusty fringes in wings and tail. Usually found near water. **Voice** Common call is a short rising rasp *KREEE*. [Benteveo Pico Fino]

Great Kiskadee *Pitangus sulphuratus*   24 cm

Andean foothills and C Sierras to 1600 m, N lowlands south to N Patagonia, and similar ssp. in the high NW Andes to 2950 m and Paraná forest. **ID** Conspicuous, large flycatcher of open country and forest edge. Stout bill and concealed golden coronal patch. **Adult** *argentinus* (widespread): Black head with a white diadem and throat. Otherwise brown above and yellow below. Wings and tail as Boat-billed Flycatcher. Fans open crest when displaying. **Voice** Highly vocal. Local name *bicho-feo* and *kiskadee* are onomatopoeias for the diurnal song, often given in duet. Diverse repertoire includes a loud *WAAAAH*, querulous notes and rattles. [Benteveo]

Boat-billed Flycatcher *Megarynchus pitangua*   23 cm

Humid chaco, n. Corrientes and Paraná forest; local in the dry chaco. **ID** Kiskadee-like flycatcher of forest edge with a shovel-shaped bill. Concealed yellow or chestnut coronal patch. **Adult:** From Great Kiskadee mainly by bill shape, but crown tends to be browner. Fresh wings and tail fringed rusty. **Voice** Dawn song is a very fast, harsh rolling *clear-ree-rip*. Diurnal song is a long series of harsh whistles and trills, often in duet. Commonest call is a complaining, nasal, grating phrase *queerrearrear*, sometimes with a distinctive ending. [Pitanguá]

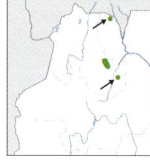

Golden-crowned Flycatcher *Myiodynastes chrysocephalus*   21 cm

Local along N Yungas rivers. **ID** An upright-perching flycatcher with a broad, stout bill and concealed coronal patch. **Adult:** Blackish lores, brown crown and ear-coverts with an indistinct whitish supercilium. Olive back and brown wings and tail, both with rufous fringing. White throat with broad blackish malar streak. Yellow below with thick olive flammulations over the breast and flanks. **Voice** Common call is an emphatic disyllabic nasal *chu-WI*, often in couplets. [Benteveo de Barbijo]

Three-striped Flycatcher

ad

trivirgatus

Social Flycatcher

ad

pallidiventris

Lesser Kiskadee

ad

fanning
crest during
display

ad

argentinus

Great Kiskadee

ad

pitangua

Boat-billed Flycatcher

chrysocephalus

ad

Golden-crowned Flycatcher

Empidonomus, *Myiodynastes* and *Legatus* Upright-perching, migratory tyrants, mostly with striped heads, concealed yellow coronal stripes and streaked breasts. *Legatus* nests inside hanging icterid nests. Most tend to perch high except Crowned Slaty Flycatcher.

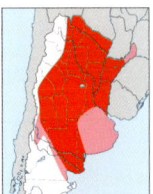

Crowned Slaty Flycatcher *Empidonomus aurantioatrocristatus* 18.5 cm

Sep–Mar in the arid N lowlands. **ID** Adult: Black crown and indistinct eye-stripe. Concealed yellow coronal patch. Grey face and underparts; brownish wings and tail. **Juvenile** (not illustrated): Very similar to juvenile Variegated Flycatcher but with yellow wash over belly and vent, and blacker crown. **Voice** Dawn song is a high-pitched short series, that starts with buzzy notes and ends in a slightly rising sequence of flat metallic whistles, *zrr zeet zi-zi-zi-zi-zi*, with variation. Calls include springy, non-musical notes *zrrp*. [Tuquito Gris]

Variegated Flycatcher *Empidonomus varius* 18.5 cm

Sep–Apr in N forests and Andes to 1300 m. **ID** Adult: Stout black bill. Somewhat mottled back and blurry steaks on the underparts. Compare with Solitary and Piratic Flycatchers. **Juvenile**: Plain brown crown and back. Wing-coverts fringed white and tail fringed dull rufous. Greyish below. **Voice** Dawn song is a high-pitched, accelerating series of shrill raspy metallic notes ending in a dry trill, *zreeu zreeu zree-zree-zre-zrrrrr*. Calls include shrill rasps. [Tuquito Rayado]

Solitary Flycatcher *Myiodynastes [maculatus] solitarius* 22 cm

Sep–Apr in the N lowlands and Andes to 1350 m. **ID** Adult: Differs from Variegated Flycatcher by larger bill with a pink base, heavily streaked on the back, finely streaked throat and more prominent breast streaking. **Voice** Dawn song is a loud squeaky whistled phrase *WEE-CHWI-CLU* or *we-believe-you* or *we-squeeze-you*. Call is a loud, dry hiccuping *WEK* or *WHEP*. **Tax note 55**. [Benteveo Rayado]

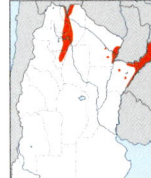

Piratic Flycatcher *Legatus leucophaius* 16 cm

Sep–Mar in N forests and Andes to 1500 m. **ID** Adult: Bill much shorter than Variegated Flycatcher, breast streaking less distinct, belly more obviously pale yellow, lacks wing bars and rufous in the tail. **Juvenile**: Wing-coverts and tail fringed rusty. Compare with Variegated Flycatcher. **Voice** Song is a loud drawn-out, rising and falling whistle followed by a fast series of 3–5 short whistles, *see-WEEEe…piwiwiwi*. Also, a slower, richer *whi-whi-whi-whi-whi*, sometimes in a long series. [Tuquito Chico]

Tyrannus and *Machetornis* Two yellow-bellied flycatchers of open areas; Tropical Kingbird is highly aerial while Cattle Tyrant is terrestrial and mostly perches up high to sing and preen.

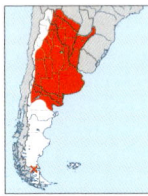

Tropical Kingbird *Tyrannus melancholicus* 22 cm

N lowlands and Andean foothills to 1700 m, south to N Patagonia (Sep–Apr). **ID** Adult: Grey head with dusky lores and dark grey ear-coverts suggesting a mask. Olive back and brown wings fringed white to buff. Grey throat, olive breast and sulphur-yellow belly. **Voice** Dawn song is a series of short shrill segments, each progressively higher-pitched and rising notably in the last phrase, *ziridi-ziridi-ziridiii*. Call is a fast rising and falling metallic trill, *swiririri*, in less than 0.5 sec, resembling the song of Plain Inezia but faster and shorter. [Suirirí Real]

Cattle Tyrant *Machetornis rixosa* 19.5 cm

Widespread in the N lowlands and Andes to 2000 m, south to N Patagonia. **ID** Arboreal and mainly terrestrial tyrant with a fairly long slender bill, long tarsus and concealed red coronal patch. Sometimes takes insects from the backs of domestic or wild mammals. **Adult**: Dull reddish iris. Soft grey head with a dusky eye-stripe. Pale olive-brown above, fringed pale buff. White throat, greyish breast and yellow below. **Voice** Song is a thin, tinny, rapid tinkling *tsi tsi-wid-si*, with much variation. [Picabuey]

ad

ad

varius

Variegated Flycatcher

juv

juv

ad

aurantioatrocristatus

Crowned Slaty Flycatcher

ad

Solitary Flycatcher

leucophaius

Piratic Flycatcher

ad

Cattle Tyrant

ad

melancholicus

rixosa

Tropical Kingbird

Oxyruncus Plump, subcanopy passerine of uncertain affinities with a stout, conical bill and short squarish tail.

Sharpbill *Oxyruncus cristatus*   18 cm

Rare (resident?) in N Paraná forest, mostly in places with Palmito Palm (*Euterpe*) and emergent Palo Rosa (*Aspidosperma*). **ID** Adult: Greyish bill sharply pointed. Face and throat grizzled with fine blackish barring. Vivid olive above and pale yellow below spotted brown on the breast and flanks. Orange iris and concealed coronal patch only visible at close range. **Voice** Song is a shrill descending 'bomb-drop' whistle lasting 4–5 secs. Call is a shorter, lower-pitched version, recalling Spot-backed Antshrike. [Picoagudo]

Phytotoma Sexually dimorphic cotingas of scrub, light woodlands and forest edge. Orange iris. Stout bill with a serrated cutting edge and bushy crown. Both species have distinctive 'mechanical' reeling voice.

White-tipped Plantcutter *Phytotoma rutila*   18.5 cm

Widespread in the lowlands and Andes (to 3500 m) south to n. Chubut. **ID** Adult ♂: Rather bright rufous forecrown and underparts. Grey above with a broad white median-covert bar and tail tips. Adult ♀: Greyish-brown above, densely streaked dark brown. Unmarked greyish rump. Narrow white wing bars. Tail as ♂. Buffy below, streaked dark brown. **Juvenile** ♂: Dark or orange iris. Brown above, streaked dusky. Wing and tail markings as adult. Dull cinnamon below with blurry streaking on the sides of the breast. **Voice** Song is a bizarre, reeling, drawn-out sound like a creaking door with 2–6 phrases, *WRREEEH wreeeh wreeeh...* Calls include a short dry reeling note *reek* and longer rasps. [Cortarramas]

Rufous-tailed Plantcutter *Phytotoma rara*   18.5 cm

Scrub and Patagonian forest edge from nw. Mendoza to sw. Santa Cruz (Sep–Apr). Overlaps with White-tipped Plantcutter in Mendoza. **ID** Adult ♂: Chestnut crown, blackish ear-coverts, short white moustachial and broken supercilium. Streaked back and large white bar on median-coverts. Blackish tail with rufous base. Rich rufous below. Adult ♀: Resembles ♀ White-tipped but tends to be more buffy overall with finer ventral streaking, and with rufous tail pattern of adult ♂. **Voice** Song is a stuttering gravelly series of 6–14 short, raspy notes followed by a gargled reeling churr, *djé-djé-djé-djé-djé-djé-GRRREEH*. Call is a short dry raspy note *jek* and longer rasps. [Rara]

Swallow-tailed Cotinga *Phibalura flavirostris*   21.5 cm

Erratic, rare winter visitor to Paraná forest; especially the NE sierras. **ID** A sexually dimorphic subcanopy cotinga with a deeply forked tail and stubby pink bill. Adult ♂: Glossy blue-black mask. Concealed orange-chestnut coronal patch. Yellow-olive above with blue-black barring. Blue-black wings and tail. Bright yellow throat. White breast, barred with black crescents. Yellow below with small black chevrons on the flanks. Adult ♀: From ♂ by greyish face and crown, smaller coronal patch, olive wings and tail, and much paler underparts. **Voice** Generally silent. [Tesorito]

Red-ruffed Fruitcrow *Pyroderus scutatus*   43 cm

Paraná forest. **ID** Very large forest cotinga of midstorey and canopy with an anvil-shaped head. Adult: Glossy black with a fiery crimson throat and upper breast. Variable chestnut spots on the abdomen. **Voice** Song is an extremely low-pitched, deep, curassow-like booming series of drawn-out notes, like blowing over a huge bottle. [Yacutoro]

Bare-throated Bellbird *Procnias nudicollis*   27 cm

Rare winter visitor to Paraná forest. **ID** Sexually dimorphic, frugivorous canopy cotinga with a very broad bill. Adult ♂: Plumage entirely white. Bare greenish facial skin and gular sac. Only turns white after 4 years, with various intermediate plumages. **3rd-year immature** ♂: White back, mottled brown. Brown tail and wings with olive-fringed secondaries. White below with variable indistinct olive streaks. Adult ♀: Blackish head and throat with sparse whitish streaks. Olive-green above, and blurry pale yellow with olive streaks below. **Voice** One of the loudest birds in the world. Song from canopy is a series of metallic echoing clangs like hammering an anvil, followed by an extremely powerful explosive note *glin glin glin glin glin glin... KWONG*. [Pájaro Campana]

Sharpbill

ad

White-tipped Plantcutter

rutila

ad ♂

juv ♂

ad ♀

ad ♂

ad ♀

Rufous-tailed Plantcutter

ad ♂

ad ♀

Swallow-tailed
Cotinga

Red-ruffed
Fruitcrow

ad

scutatus

ad ♂

imm ♂
3rd-yr

ad ♀

Bare-throated Bellbird

Chiroxiphia, *Manacus* and *Pipra* Males perfom spectacular lekking displays; females are drab and inconspicuous.

Yungas Manakin *Chiroxiphia boliviana*   13 cm

N Yungas forest block. **ID** Chunky, sexually dimorphic, subcanopy blue and black manakin. **Adult** ♂: Black with sky-blue saddle and red skullcap. In display ♂ splays open an extraordinary red fan-like crest. **Adult** ♀: Olive with brownish wings and tail. Dull tarsus. **Voice** Very loud whistles include a disyllabic *CHIWEEO*, and up to 3 medium-pitched resounding *PÍU* notes, sometimes in fast series. An upward-inflected whistle *wu-eee* may recall Rufous-capped Antshrike. At height of lek display, a nasal, drawn-out whirring ending in mechanical clicks, *nyaaah… nyaaah… cli-cli-tic-tic.* [Bailarín Yungueño]

Swallow-tailed Manakin *Chiroxiphia caudata*   15.5 cm

Paraná forest. **ID** Chunky, sexually dimorphic, subcanopy blue and black manakin. **Adult** ♂: Cobalt-blue with scarlet crown, and black head and wings. **Adult** ♀: Fairly vivid green with pink bill and tarsus. **Juvenile** ♂: as ♀, but crown scarlet. **Voice** Very loud whistles include a disyllabic *CHI-RRU*, sometimes shortened to the first part, and medium-pitched resounding *WHIO* or *WHIP* notes. At height of lek display, a nasal, drawn-out whirring ending in mechanical clicks, *wawawawa… wawawawawa… cli-cli.* [Bailarín Azul]

White-bearded Manakin *Manacus manacus*   11 cm

N Paraná forest. **ID** Small, sexually dimorphic, understorey manakin with a dark iris. **Adult** ♂: Black and grey with a striking white throat and nuchal collar. **Adult** ♀: Rich olive upperparts; duller below with pale belly and vent. Dark iris and orange tarsus. **Voice** Highly varied repertoire. In display arena, loud, often doubled, sharp wing-snaps followed by a dry mechanical rattle. Isolated wing-snaps sometimes intermixed with a purring *chrrr* and a short shrill whistle. [Bailarín Blanco]

Band-tailed Manakin *Pipra fasciicauda*   11.5 cm

Paraná forest. **ID** Small, short-tailed, sexually dimorphic, understorey and midstorey manakin with a pale iris. **Adult** ♂: Striking crimson, yellow and black plumage. White band across base of tail. **Adult** ♀: Dull olive upperparts. Carpal, belly and vent yellowish. Note whitish iris and pinkish-brown tarsus. **Voice** Song is a strange-sounding, weak, nasal, downslurred *yeeeuur.* Call is an ascending pure whistle *fweee-e.* [Bailarín Naranja]

Piprites Chunky, compact midstorey and subcanopy forest passerines with loud, far-carrying vocalisations.

Wing-barred Piprites *Piprites chloris*   14 cm

Paraná forest. **ID** Adult: Olive above with a yellow eye-ring, greyish ear-coverts, two yellow or white wing-bars and tertial spots. Lores often cinnamon. Yellowish or olive throat and breast, becoming paler yellow below. **Voice** Song is a memorable, rhythmical, medium-pitched series of wooden whistles, sounding like Morse Code, *wep wep wep-wep pededep wép wép*, with some variation. [Bailarín Verde]

Black-capped Piprites *Piprites pileata*   12.5 cm

Breeding population at one remote black laurel (*Ocotea*) forest in ce. Misiones. **ID** Adult ♂: Stout yellow bill. Black crown and dark chestnut upperparts; ochraceous throat and upper breast; yellow below. **Adult** ♀: From adult ♂ by olive mantle and wing-coverts. **Voice** Song is a series of rich complaining whistles ending in syncopated couplets, *PIU WEWEWEWEWE-KIÚ wer-KIÚ wer-KIÚ wer-kiú.* Common call is a loud *PIU* or *PEW* in sparse series. [Bailarín Castaño]

ad ♂

ad ♀

Swallow-tailed Manakin

juv ♂

Yungas Manakin

♂ display

ad ♂

ad ♂

ad ♀

ad ♂

ad ♀

gutturosus

scarlatina

Band-tailed Manakin

White-bearded Manakin

ad ♀

ad

ad ♂

ad ♀

Wing-barred Piprites

Black-capped Piprites

Pachyramphus Three sexually dimorphic and one monomorphic species of midstorey and subcanopy in forests and woodlands. Often in mixed-species flocks.

Green-backed Becard *Pachyramphus viridis* 15 cm

Widespread in N lowlands and foothills. **ID** Adult ♂: Glossy black cap contrasts with white lores. Vivid green back, wings and tail. Yellow pectoral band. Adult ♀: Differs from ♂ by olive cap and chestnut shoulders. **Voice** Song is a fast, rising and decelerating series of whistles that increase in volume *w´w´w´weweweweWEWEWE-WI-WI-WII*, lasting 2 secs. Call is a repeated loud rich whistle *chu-WEE*. [Anambé Verdoso]

Chestnut-crowned Becard *Pachyramphus castaneus* 15.5 cm

Paraná forest. **ID** Sexes alike. Adult: Sky-blue bill with dark culmen. Chestnut upperparts. Contrasting grey mask joining across nape. Ochraceous underparts with suggestion of darker pectoral band. Compare with Rufous Casiornis and Rufous-tailed Attila (Plate 152) and ♀ Crested Becard; also Ochre-breasted and Buff-fronted Foliage-gleaners (Plate 124). **Voice** Common calls include a series of 3–5 flat, thin, progressively lower-pitched, plaintive whistles *SEEU-SEEU-SEEU-SEEU…* and a fast high-pitched ringing trill that rises at the end *trreeeerrrrrrwerrr*. [Anambé Castaño]

White-winged Becard *Pachyramphus polychopterus* 16 cm

Widespread in N lowlands and foothills. **ID** Adult ♂: Sky-blue bill. Mainly sooty-black with glossy blue-black crown, white wing-fringing, broad median-covert bar and tail tips. Adult ♀: Olive with contrasting chestnut fringing on wings, median-covert bar and narrow tips to tail. Pale yellow below with a greyish-olive breast; some brighter than others. **Voice** Song is a series of short descending cackling whistles that either accelerates throughout and fades-out, or accelerates and decelerates, *CLU CLUCLUCLU-CLU-CLU*. [Anambé Negro]

Crested Becard *Pachyramphus validus* 17.5–19 cm

Widespread. **ID** Bushy crest in both sexes. Adult ♂ *audax* (19 cm; NW Andes and C Sierras): Heavy blue-grey bill with a hooked tip. Glossy black crown with bushy crest. Blackish upperparts and ashy-grey below. Adult ♀ *audax*: Ashy-grey crown. Chestnut upperparts and buff underparts, warmest on upper breast. Adult ♂ *validus* (17.5 cm; NE lowlands): From ♂ *audax* by greyish-buff underparts. Adult ♀ *validus* (not illustrated): Differs from ♀ *audax* by its smaller size. **Voice** Dawn song is a repeated, high-pitched, short disyllabic whistle *shi-svit*. Long diurnal voice is a complex rapid series including a nasal note, an ascending chatter and a drawn-out whistle *ank-drrweee-si-wueeee*, sometimes with long chatters and multiple whistles. Call is a high-pitched shrill *SHREEW*. [Anambé Grande]

White-naped Xenopsaris *Xenopsaris albinucha* 13 cm

Widespread but very local in N lowlands. **ID** Small, becard-like passerine with distinctive voice. Adult ♂: Glossy black cap. Grey or white nuchal collar. Grey upperparts with white-fringed wing-coverts. Underparts entirely white. Adult ♀: Glossy black cap with brown forehead. **Juvenile:** Rusty crown, barred black. Mantle and wing-coverts fringed cinnamon. **Voice** Extraordinarily high-pitched rising song is a sequence of thin inflected whistles with a terminal quavering stutter, *swip sweee-eee swi ititititi*. Also, a fast series of rising twittering notes may recall Tropical Kingbird. [Tijerilla]

Greenish Schiffornis *Schiffornis virescens* 16 cm

Interior of Paraná forest. **ID** Solitary, stocky midstorey and understorey passerine. Adult: Generally dull dark olive with warm brown wings and prominent yellowish eye-ring. **Voice** Loud drawn-out melodic whistled series of emphatic notes, *CHI-you… chu-WÍP chu-WÍP*, with much variation. [Flautín]

Green-backed Becard

ad ♀

viridis

ad ♂

ad

Chestnut-crowned
Becard

White-winged Becard

ad ♀

ad ♂

spixii

ad ♂

ad ♂

ad ♀

validus

audax

Crested Becard

audax

ad ♀

ad ♂

juv

albinucha

White-naped Xenopsaris

Greenish Schiffornis

ad

Tityra Large, upright perching, black-and-white, cotinga-like canopy passerines. Unusual toad-like vocal phrases.

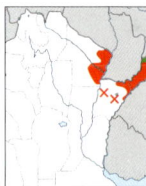

Black-crowned Tityra *Tityra inquisitor* 20 cm

NE forests. **ID** Lacks bare facial skin and bill hook. **Adult** ♂: Mantle, scapulars and tertials greyer than Black-tailed Tityra; lacks a black chin. **Adult** ♀: Distinctive black crown and cinnamon face. Back streaked with brown. Black wings and tail. **Voice** Song is a cacophonous series of harsh 'white-noise' notes with grating segments, *schrrraah schrraah schrraah...* or *zhrek zhrek zhrek...* Calls include harsh *SHHP* or *SHUP* notes. [Tueré Chico]

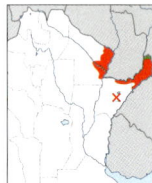

Black-tailed Tityra *Tityra cayana* 21.5 cm

NE forests. **ID** Reddish bare ocular skin and base of hooked bill. **Adult** ♂: Black hood, wings and tail; greyish-white mantle, scapulars and tertials; chin black; white below. **Adult** ♀: Crown and back streaked with brown. Underparts lightly streaked. Black wings and tail. **Voice** Song is a fast series of nasal toad-like croaks, *wrre-dek wree-dek...* Also delivers a series or isolated *wrre* notes. [Tueré Grande]

Masked Tityra *Tityra semifasciata* 21 cm

n. Misiones, expanding southwards. **ID** Reddish bare ocular skin and base of hooked bill. **Adult** ♂: Recalls Black-tailed Tityra but black restricted to foreface forming a mask. Tail tipped white. **Adult** ♀: Similar to Black-tailed Tityra but is grey above and lacks the brown dorsal streaking. Tail tipped pale grey or white. **Voice** Song and calls resemble Black-tailed Tityra, but much shorter and higher-pitched. [Tueré Enmascarado]

Cyanocorax Gregarious, chunky long-tailed birds of forest and light woodlands. Sexes alike. Far-carrying distinctive voices.

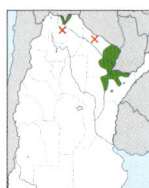

Purplish Jay *Cyanocorax cyanomelas* 36.5 cm

Yungas foothill forest, humid chaco and gallery forest from ne. Salta to n. Corrientes. **ID** Adult: Brownish-mauve with a black hood and breast, and narrow violet tail. Often together with Plush-crested Jay. **Voice** Simple vocal repertoire. Calls include rapidly repeated series of short parrot-like *KRE* notes, or longer, harsher, nasal crow-like *AARH* notes. [Urraca Morada]

Plush-crested Jay *Cyanocorax chrysops* 33.5 cm

Widespread in the north. **ID** Adult: Black head with bushy, flat-topped crest. Sky-blue eye-crescents and nape. Blue back and creamy or yellow belly. Tail, broadly tipped cream. **Voice** Complex vocal repertoire. Common calls include a high-pitched *BIP BIP*, a laser-gun *TIÚ TIÚ* and a nasal descending *NNYAAAH*; also screeches and rattles. Imitates other birds, most notably *Accipiter* hawks and *Micrastur* forest falcons. [Urraca Criolla]

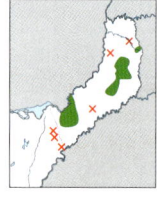

Azure Jay *Cyanocorax caeruleus* 39 cm

Rare in Paraná forest; mostly in *Araucaria* forest of ce. Misiones. **ID** Adult: Vivid azure-blue with a black hood, breast and small frontal crest. Relatively short, very broad tail. **Voice** Moderate vocal repertoire. More raucous than congeners. A repeated series of lapwing- or parrot-like *KREYRR*, sometimes without raspy quality. Short, disyllabic whistled *KA-KU*. Drawn-out descending *WEEEH*. [Urraca Azul]

Black-crowned Tityra

inquisitor

ad ♂

ad ♀

ad ♂

ad ♀

Black-tailed Tityra

braziliensis

ad ♂

ad ♀

Masked Tityra

Plush-crested Jay

ad

ad

chrysops

Purplish Jay

ad

ad

Azure Jay

Tachycineta Compact blue-backed swallows with forked tails and white rumps. Complex songs and calls include trills, twitterings, liquid trills, rasps and buzzes.

White-rumped Swallow *Tachycineta leucorrhoa* 14 cm

N lowlands to NW foothills (Sep–Apr; some overwinter in the NW). **ID** Adult: Glossy blue upperparts, tinged green; narrow white supraloral streak. Primaries extend beyond tail. Note deeply forked tail in flight. Proportionately larger white rump than Chilean Swallow. Juvenile (not illustrated): Non-glossy brown upperparts, reduced white supraloral streak. Some show a diffuse dark breast-band. [Golondrina Ceja Blanca]

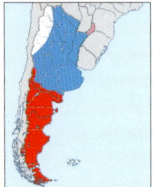

Chilean Swallow *Tachycineta leucopyga* 13.5 cm

Patagonia (Sep–Apr), migrating to the N lowlands in winter. **ID** Adult: Glossy royal-blue upperparts. Lacks supraloral of White-rumped Swallow. Primaries usually extend just beyond tail. Note slightly forked tail in flight. Proportionately smaller white rump than Chilean Swallow. Juvenile (not illustrated): Much like White-rumped but generally with very reduced or no white supraloral streak. [Golondrina Patagónica]

White-winged Swallow *Tachycineta albiventer* 14.5 cm

Local resident in Misiones and n. Corrientes. **ID** Adult: Glossy blue-green upperparts; conspicuous white area on wings, and white rump. In flight, striking white greater wing-coverts and secondaries. Note moderately forked tail. [Golondrina Ala Blanca]

Pygochelidon Bicoloured swallows with forked tails and dark rumps. Vocalisations include gravelly twitterings, electrical buzzes and short whistles.

Blue-and-white Swallow *Pygochelidon cyanoleuca* 13 cm

Widespread. **ID** Adult *cyanoleuca* (N lowlands): Glossy dark blue upperparts; white below with blue-black vent and undertail-coverts. Primaries extend just beyond forked tail. In flight, note deeply forked tail. Adult *patagonica* (Patagonia, Sep–Mar, and Andes, resident in NW): As *cyanoleuca* but only undertail-coverts blue-black. Note contrast between sooty underwing-coverts and underside of remiges compared to Andean Swallow. [Golondrina Barranquera]

Black-collared Swallow *Pygochelidon melanoleuca* 13 cm

Mainly n. Misiones on rapids where perhaps sporadic. **ID** Adult: Glossy blue-black upperparts and pectoral band contrast with white underparts. Tail streamers extend well beyond wingtips. In flight, note very deeply forked tail. [Golondrina de Collar]

Orochelidon Compact swallow with notched or squarish tail and dark rump.

Andean Swallow *Orochelidon andecola* 13.5 cm

Local resident in NW Andes, above the tree-line. **ID** Adult: Upperparts brown, tinged dull blue; primaries extend just beyond tail. Throat dirty brown; white below. In flight, note notched tail, brownish sides of breast and flanks, and uniform brown underwing. Compare with Blue-and-white Swallow. **Voice** Delivers gravelly liquid trills and a mellow *uip* or *bruip* call in flight. Generally more liquid and lower-pitched than Blue-and-white Swallow. [Golondrina Andina]

White-rumped Swallow

ad

ad

Chilean Swallow

cyanoleuca

ad

ad

White-winged Swallow

patagonica

Blue-and-white Swallow

ad

andecola

ad

Black-collared Swallow

Andean Swallow

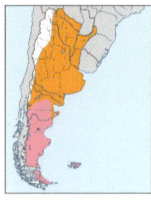

Sand Martin *Riparia riparia* 12.5 cm

Lowlands south to Patagonia (Aug–Mar). **ID** Non-breeding boreal migrant brown hirundine with notched tail. **Adult:** Uniform brown upperparts including rump. White below with brown pectoral band. In flight, proportionately shorter and broader wings and faster, more erratic flight than Brown-chested Martin. Wings always held horizontal. **Voice** Generally silent. Gives a short churr while perched or in flight. [Alt: Bank Swallow] [Golondrina Zapadora]

Progne Large, chunky, mostly sexually dimorphic hirundines with forked tails (except Brown-chested). Identification of some species can be challenging without good views. Note that hybrids between Grey-breasted and Southern Martin may occur where their ranges overlap. Vocally very similar (except Brown-chested). Song is a series of rich gurgles and musical trills. Calls include buzzy churrs and rasps.

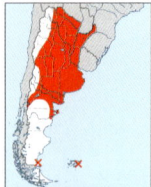

Brown-chested Martin *Progne tapera* 18 cm

Lowlands south to N Patagonia (late Aug–Apr). **ID Adult:** Sexes alike. Brown upperparts. White throat and sides of neck, sometimes forming nuchal collar. Brown pectoral band, and brown spots in centre of white abdomen. In flight, note shallow notched tail. Characteristic flight on downwardly-angled wings. **Voice** Song is composed of shorter and more glassy notes than congeners. Diagnostic call is a clearly cascading series of sharp liquid notes. [Golondrina Parda]

Grey-breasted Martin *Progne chalybea* 19 cm

N lowlands (late Aug–late Mar). **ID Adult** ♂: Glossy steel-blue upperparts. Greyish-brown throat becoming mottled on breast. In flight, note deeply forked tail. **Adult** ♀: Small streaks on flanks, lacking in ♂. [Golondrina Doméstica]

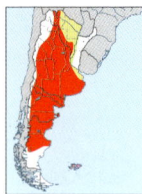

Southern Martin *Progne elegans* 18.5 cm

Mainly Patagonia, C sierras and NW (late Aug–early Apr). **ID Adult** ♂: Entirely glossy purple-blue. In flight, tail generally more forked than Grey-breasted and Purple Martins. **Adult** ♀: Glossy purple-blue upperparts. Underparts brown, mottled with pale brown crescentic markings; crescents most prominent on undertail-coverts. Deeply forked tail. [Golondrina Negra]

Purple Martin *Progne subis* 19.5 cm

Rare vagrant known from sparse records in the north. **Adult** ♂ (not illustrated): Not safely separable in the field from ♂ Southern Martin although, in the hand, the external webs of the rectrices are fine and narrow. **Adult** ♀: Greyish forehead and nape; rest of upperparts glossy purple. Pale brown throat and breast; mottled with white on breast. Fine brown shaft streaks over centre of white abdomen. In flight, greyish forehead and nuchal collar (blue in Grey-breasted Martin) extending to sides of neck. Brown underwing-coverts with white fringing on marginal wing-coverts (entirely blackish in Grey-breasted). [Golondrina Purpúrea]

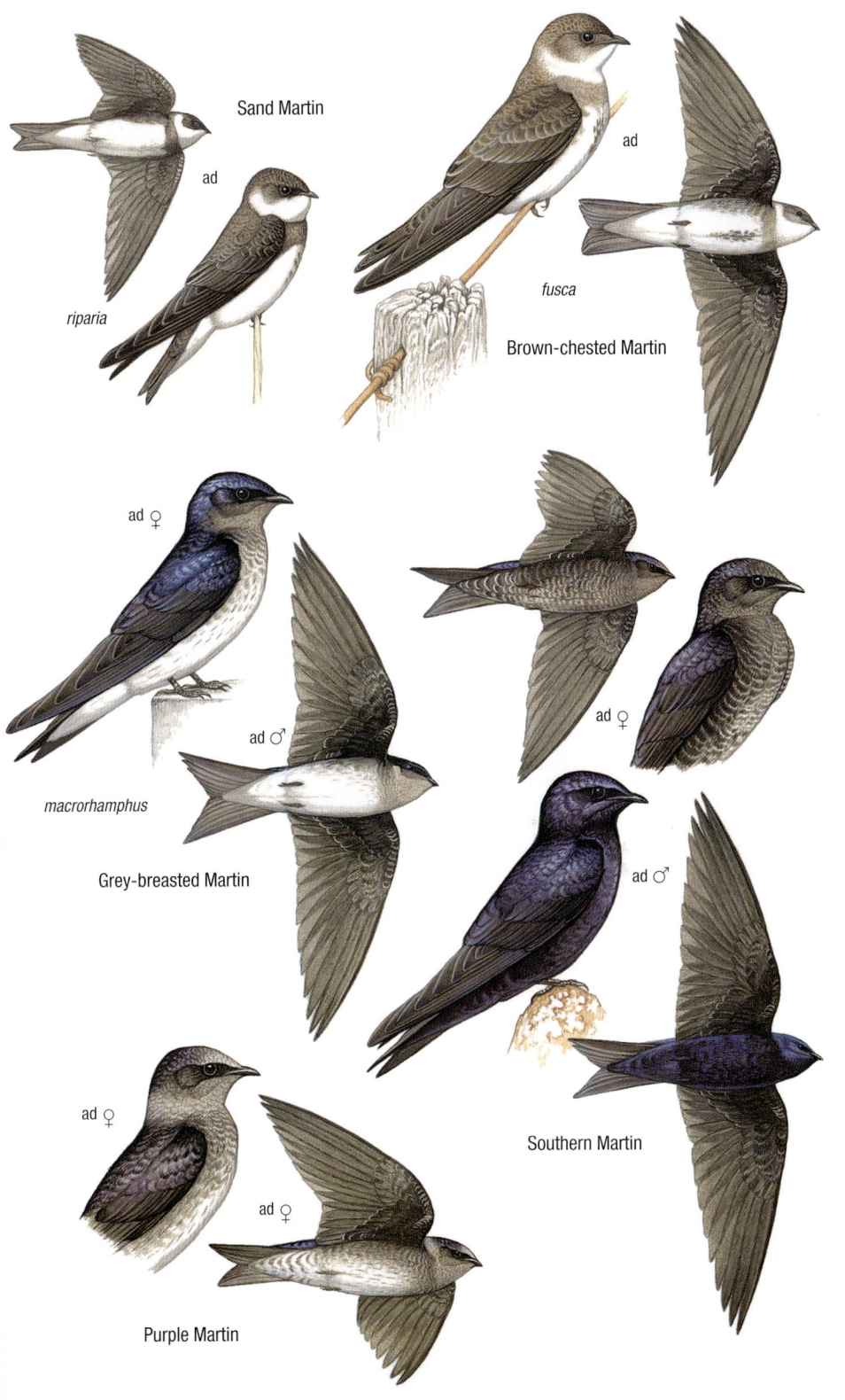

Sand Martin

ad

riparia

ad

fusca

Brown-chested Martin

ad ♀

ad ♀

ad ♂

macrorhamphus

Grey-breasted Martin

ad ♂

Southern Martin

ad ♀

ad ♀

Purple Martin

Tawny-headed Swallow *Alopochelidon fucata*   12 cm

N lowlands and foothills (late Aug–Apr). **ID** A small, compact, brown austral migrant swallow with a slightly notched tail. **Adult:** Brown crown and dusky lores; cinnamon supercilium, ear-coverts, narrow nuchal collar, throat and upper breast; white below. In flight, uniform upperparts; cinnamon nuchal collar, throat and upper breast usually evident. Squarish sillhouette. **Voice** Song comprises gravelly notes followed by a short rich warble, repeated over. [Golondrina Cabeza Rojiza]

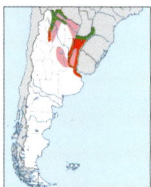

Southern Rough-winged Swallow *Stelgidopteryx ruficollis*   14 cm

Andean foothills and Mesopotamia (Sep–Mar); many are resident in extreme north. **ID** Large brown swallow with a notched tail. Modified barbules on outer web of first primary in ♂♂. **Adult:** Brown upperparts with paler brown rump and whitish fringing on tertials. Cinnamon-rufous throat, grey-brown wash over breast and flanks; yellowish or white below. In flight, obvious pale rump and indistinct pale nuchal collar. **Voice** Common call is a pleasant raspy *greet-reet*, sometimes varied to more musical upslurred couplets. [Golondrina Ribereña]

Barn Swallow *Hirundo rustica*   15–17 cm

Boreal migrant almost throughout (early Sep–early Apr), also widespread expanding breeder in C provinces. **ID** Narrow-winged swallow with distinctive long tail streamers in breeding plumage. **Adult breeding:** Glossy blue upperparts and extensions onto sides of breast. Chestnut forehead and throat; rest of underparts and underwing-coverts cinnamon. White subterminal tail spots; outermost rectrices extended as streamers. **Adult non-breeding:** Forehead and throat dull cinnamon. Upperparts glossed blue. Underparts cinnamon, mixed with white. Lacks tail streamers. **Juvenile:** Similar to non-breeding adult. Brown upperparts lack blue gloss; lower underparts clean white. **Voice** Song is a quick succession of harsh, squeaky, semi-musical notes and monotone ticking rattles. Common flight call is an upswept metallic *ch-veet*. [Golondrina Tijerita]

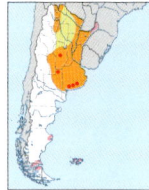

American Cliff Swallow *Petrochelidon pyrrhonota*   13.5 cm

Boreal migrant to the N lowlands (early Oct–early Apr), and breeding locally in the CE (perhaps becoming established?). **ID** Compact and broad-winged swallow with a square-ended tail. **Adult *pyrrhonota*** (boreal migrant and local breeder in CE provinces): White or pale cinnamon forehead and glossy dark blue cap and mantle with whitish fringes; chestnut ear-coverts and throat. **Adult *melanogaster*** (Boreal migrant mainly to NE lowlands): Differs from *pyrrhonota* by chestnut forehead. Note cinnamon rump and brownish nuchal collar in flight. **Voice** Diverse repertoire of flight calls includes a nasal descending *weeáh*, a churring *brrrr*, metallic upslurred notes and electrical twitterings. [Golondrina Rabadilla Canela]

Tawny-headed Swallow

Southern
Rough-winged Swallow

ruficollis

ad

ad

ad

melanogaster

ad br

erythrogaster

juv

ad

Barn Swallow

pyrrhonota

ad non-br

American Cliff Swallow

Troglodytes Small, compact, rather nondescript wrens with relatively short tails.

House Wren *Troglodytes aedon* 12 cm

Throughout. **ID** All ssp. show black barring on wings and tail. **Adult** *musculus* (Paraná forest): Brightest ssp. with a rufous wash over the breast. **Adult** *rex* (Widespread ssp. of NW Andes and CN lowlands): Rather plain brown with rufescent rump, flanks and undertail-coverts. **Voice** Song is a pleasant rapid string of warbles and trills ending abruptly. Common call is a nasal rasp. [Ratona]

Cobb's Wren *Troglodytes cobbi* 13.5 cm

Endemic to tussock grass on mostly off-islands of the Falklands where House Wren has only occurred as a vagrant. **ID** Adult: Differs by longer dark bill, paler brown upperparts and a proportionately shorter tail. **Voice** Song and call resemble those of House Wren but are somewhat lower-pitched and slower. [Ratona Malvinera]

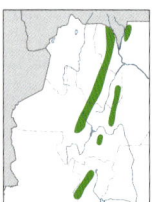

Mountain Wren *Troglodytes solstitialis* 10.5 cm

Yungas forest; at all heights. Frequently gregarious and in mixed-species flocks. **ID** Adult: Small brown forest wren with striking white rear supercilium. **Voice** Song is a rapid string of short, high-pitched, tinkling whistles and raspy descending churrs. Call is a repeated, piercing, high-pitched wheezy trill *zreeeu*, sometimes with a raspy quality. [Ratona Ceja Blanca]

Grass Wren *Cistothorus platensis* 12 cm

Widespread, from sea coasts to Puna grasslands. Several similar species may be involved. **ID** Inconspicuous (unless singing) in marsh grass, sedges and rushes. Holds tail cocked. **Adult:** Black back, streaked with white. Wings and tail barred black. Compare with Wren-like Rushbird and Bay-capped Wren-Spinetail (Plate 126). **Voice** Song is generally long and comprises clearly separated repetitive phrases (sometimes with short soft introductions), including churrs, rasps and whistles, *ch-ch-ch-ch krr-krr-krr-krr wee-wee-wee-wee…* Call is a scolding series of short harsh *jep* notes. **Tax note 56.** [Ratona Aperdizada]

Thrush-like Wren *Campylorhynchus turdinus* 20.5 cm

NE lowlands expanding southwards. **ID** Large wren of palms or palm forest. **Adult:** Fairly long, slightly decurved bill with pink lower mandible. Long creamy supercilium. Pale brown above and buff below. Long graduated tail with barred undertail-coverts. **Voice** Unmistakable explosive song, often in duet, is a very loud *CHIRU-CHOROP CHOP-CHOP… CHOP-CHOROP CHOP-CHOP…* Call is a short dry rasp *rrah*. [Ratona Grande]

Cinclus Plump, pot-bellied river bird with relatively large tarsus.

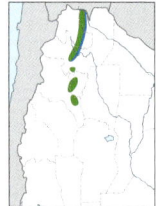

Rufous-throated Dipper *Cinclus schulzii* 15.5 cm

Yungas forest (especially alder) and Andean steppe rivers (600–3200 m). **ID** Adult: Plumbeous-grey with contrasting pale orange patch on throat and upper breast. Bobs and flicks wings downwards to expose large white wing flashes. **Voice** Rarely heard song is a springy warble of whistled and raspy notes. Common call, often given in flight, is a loud, dry nasal rasp *RREK* or *REG*. [Mirlo de Agua]

House Wren

musculus

ad

rex

ad

Cobb's Wren

ad

ad

Mountain Wren

auricularis

ad

platensis

Grass Wren

ad

unicolor

Thrush-like Wren

ad

bobbing

Rufous-throated Dipper

Cyclarhis is an upright-perching passerine of light woodland and forest with a proportionately very stout bill. *Vireo* Typical vireo of midstorey to canopy in light woodland and forest, with a fairly long slender bill. *Hylophilus* Rufous-crowned Greenlet is also in the Vireonidae family but is depicted on Plate 112.

Rufous-browed Peppershrike *Cyclarhis gujanensis* · 16.5 cm

N and C lowlands, Andes to 1650 m. **ID** Adult *viridis* (widespread, except extreme NW and extreme NE; illustrated): Orange iris. Rusty supercilium and grey face. Bright olive back, wings and tail. Buffy-white below with a broad yellow pectoral band. Compare with Green-backed Becard (Plate 157). **Voice** Varied repertoire. Song is a mellow quavering double warble (local onomatopoeic name is Juan Chiviro), given 4–5 times before switching to another rendering. Calls include a slow series of falling and rising whistles, each progressively lower-pitched, fading in volume (may recall Rufous-bellied Thrush), and short triple whistles, *fiu-fiu-fuiiiiii*. [Juan Chiviro]

Chivi Vireo *Vireo chivi* · 14 cm

N and C lowlands, Andes to 1800 m, Aug–Apr. **ID** Adult *chivi* (widespread; illustrated): Dark iris. Grey crown bordered black. Creamy supercilium and dusky eye-stripe. Otherwise olive above and white below with pale yellow undertail-coverts. Adult *diversus* (Paraná forest; not illustrated): Brighter with a yellow wash on the sides of the breast and flanks. Compare with ♀ Chestnut-vented Conebill (Plate 173). **Voice** Song is a slow series of well-spaced musical couplets, triplets or warbles *chi-vi… vireo… wiri-wiri… dididi… chi-vi….* Call is a raspy nasal scold *nyaaah*. Tax note 57. [Chiví-chiví]

Polioptila Slender, agile, warbler-like arboreal passerines with long narrow tail.

Creamy-bellied Gnatcatcher *Polioptila lactea* · 11 cm

Scarce in Paraná forest, also gallery forest in ne. Corrientes. **ID** Adult ♂: Black cap, rest of upperparts blue-grey. Plain white face and underparts, often tinged yellow on the belly. Adult ♀: As ♂ but without black cap. Some overlap with Masked Gnatcatcher in south of range. **Voice** Song is a fast, short series of high-pitched upward-inflected or flat whistles *seeseeseeseeseesee….* May recall renderings of song of Fawn-breasted Tanager, but always lower-pitched and generally slower. [Tacuarita Blanca]

Masked Gnatcatcher *Polioptila dumicola* · 12 cm

Widespread in N lowlands. **ID** Adult ♂: Black mask through eye extending across forehead, bordered below by a silvery moustachial streak. Blue-grey upperparts. Grey throat and breast becoming white below. Adult ♀: White lores and eye-ring with dark rear supercilium. Grey ear-coverts. Rest of plumage as ♂. **Voice** Vast repertoire includes many accurate imitations. Songs include pure rapid whistles, twitterings, rasps and ticking trills. Call is a nasal *nyaaah*, sometimes recalling House Wren and Chivi Vireo. [Tacuarita Azul]

Donacobius Large, slender marshbird with bright yellow inflatable pouches on sides of neck.

Black-capped Donacobius *Donacobius atricapilla* · 24 cm

NE Marshes. **ID** Adult: Yellow iris. Mostly black above and ochraceous below with extensive white fringing on long graduated tail. In flight, shows striking white stripe over base of primaries. **Juvenile** (not illustrated): Narrow ochre supercilium, turning white behind the eye. **Voice** Song, in duet, comprises a slow series of 4–8 loud, medium-pitched ascending whistles *WUUUEE-WUUUEE-WUUEE-WUUUEE…* each overlapping with a harsh cicada-like rasp *rehh-rehh-rehh-rehh*. Also gives a faster series of descending or ascending whistles and trills. Calls include sequences of soft, parrot-like drawn-out churrs and harsh wren-like rasps. [Angú]

viridis   ad

Chivi Vireo

ad

chivi

Rufous-browed Peppershrike

ad ♀

Creamy-bellied Gnatcatcher

ad ♂

ad ♀

ad ♂

dumicola

Masked Gnatcatcher

ad

atricapilla

Black-capped Donacobius

Turdus Large, mostly resident thrushes found in all habitats, with forest species generally more shy. Some species are strongly sexually dimorphic while in others sexes are alike. Bills may be duller and darker in winter. See also Plates 165 and 166.

Yellow-legged Thrush *Turdus flavipes* — 21 cm

Extremely rare in n. Misiones. **ID Adult** ♂: Bright yellow bill and legs. Narrow yellow eye-ring. Glossy black head, throat, breast, wings and tail contrasting with grey back, rump, belly and vent. Back sometimes scaly, chin often spotted white and centre of belly often whitish. **Adult** ♀: Buffy lores and fine buff flecking on the ear-coverts. Uniform brown above with a contrasting grey rump. Cinnamon underwing-coverts. Throat mottled buffy-white. Pale brown below with warmer flanks and a greyish centre to the abdomen. Compare with ♀ Eastern Slaty and Creamy-bellied Thrushes (Plate 165). **Voice** Song is a metallic mix of high-pitched, shrill, repeated notes, slow mellow warbles and other tinkling notes or chatters. Successive songs differ markedly. Call is a single or double metallic rasp. [Zorzal Azulado]

Eastern Slaty Thrush *Turdus subalaris* — 21 cm

Paraná forest; mainly in leafy canopy. **ID Adult** ♂: Orange-yellow bill. Narrow yellow eye-ring. Dull orange-yellow legs. Blackish hood and slate-grey back, rump and wing-coverts. Brown wings and tail. White throat, streaked black. Narrow white gorget, pale grey below with a whitish centre of belly. **Adult** ♀: Yellowish-brown bill. Yellow legs. Olive-brown above with warmer flight feathers. White throat, streaked brown at the sides. Soft grey-brown below with a whitish centre to the belly and white vent. Overlaps with Yellow-legged, Creamy-bellied, White-necked and Pale-breasted Thrushes. **Voice** Very distinctive song (usually Oct–Nov), is composed of a large variety of well-spaced short phrases of repeated thin metallic notes, *tink-tink-tink… see-see-see-see… ch-lit ch-lit ch-lit… deer-deer-deer… zeer-zeer…* Call is a high-pitched *seeee*. [Zorzal Campana]

Pale-breasted Thrush *Turdus leucomelas* — 23.5 cm

NE forests. **ID Adult**: Brown bill, rarely with yellow. Dull red iris. Grey-brown crown and nape; variable. Contrasting warm brown back, wings, rump and tail. White throat, streaked brown. Soft pale brown below with a whitish centre to the belly. In flight, rather bright chestnut-orange underwing-coverts. **Voice** Song is a mockingbird-like series that repeats notes several times before switching to another type. Short phrases may sometimes recall Creamy-bellied Thrush but are more continuous. Common call is a rapid series of very harsh notes, *SCHRR-SCHRR-SCHRR-SCHRR-SCHRR*, often with just 1–2 notes. [Zorzal Sabiá]

White-necked Thrush *Turdus albicollis* — 22 cm

Paraná forest and Yungas foothills. **ID Adult** *paraguayensis* (Paraná forest): Yellow bill with a black culmen. Narrow orange eye-ring. Pinkish-brown legs. Dark brown hood and tail; warmer back, wings and rump. Black throat, sometimes barely streaked white. Contrasting white gorget. Brownish-grey breast, whitish belly and vent. Contrasting orange flanks. In flight, orange underwing-coverts. **Adult** *contemptus* (Rare in Yungas foothill forest of Salta and Jujuy): Differs from *paraguayensis* by being greyer overall with brown on flanks and buffy underwing-coverts. **Voice** Song is a long slow series of drawn-out, medium-pitched, pure or slurred melancholic whistles with distinctive inflections, *chi-wo-oh, ch-wee, wee-berr, chi-wo-ueh….* Calls include a short *WÁH*. [Zorzal Collar Blanco]

Cocoa Thrush *Turdus fumigatus* — 25 cm

Invading in ne. Misiones; secondary growth and towns. **Adult**: Large dark bill. Can show grey around eye. Warm brown above and fuscous below; often paler on belly. Fine dusky throat streaking. Sometimes fairly uniform chocolate-brown throughout, including throat. From similar Pale-breasted Thrush by stouter, darker bill and darker, more uniform plumage. **Voice** Song principally includes a mellow repetitive phrase of 5–6 short flat whistles, each successively lower-pitched, followed by 1–2 higher-pitched whistles. Calls include a rapid whistled *tíu-tíu-tíu-tíu-tíu…* or harsh clipped notes *KIKIKIKIKI*. [Zorzal Chocolate]

Yellow-legged Thrush

ad ♂

flavipes

ad ♀

Eastern Slaty Thrush

ad ♂

ad ♀

ad

ad

leucomelas

Pale-breasted Thrush

paraguayensis

ad

fumigatus

ad

contemptus

ad

Cocoa Thrush

White-necked Thrush

Turdus Large, mostly resident thrushes found in all habitats, with forest species generally more shy. Some species are strongly sexually dimorphic while in others sexes are alike. Bills may be duller and darker in winter. See also Plates 164 and 166.

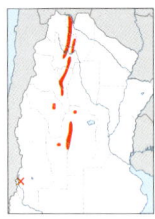

Andean Slaty Thrush *Turdus nigriceps*   21 cm

Yungas forest to 2300 m, C Sierras. **ID** Adult ♂: Bright yellow bill. Yellowish legs and narrow eye-ring. Sooty-black above with a black hood, becoming greyer on the rump. Black tail. White throat, thickly streaked black. Ash-grey below, becoming white on the centre of belly. Adult ♀: Brownish bill with dull yellow base. Dull yellow legs and indistinct eye-ring. Dark brown above; darkest on the crown and tail. Whitish throat, densely streaked dark brown. Brown below with a whitish centre to the belly. Overlaps with Creamy-bellied and White-necked Thrushes. **Voice** Song, throughout the day, is a loud metallic mix of strident whistles, high-pitched shrill notes and other tinkling or chatters. Call is a disyllabic metallic *CH-WINK, TZZD-WEE* or a simple high-pitched *TZEE*. [Zorzal Cabeza Negra]

Creamy-bellied Thrush *Turdus amaurochalinus*   23.5 cm

Complex migratory patterns, but many remain N lowlands, C sierras and Andean foothills to 1300 m. **ID** Adult: Yellow bill, often with a dusky tip. Brown above with a darker tail. Conspicuous blackish lores. White throat, thickly streaked dark brown, sometimes with a small white gorget in the centre of the lower throat. Off-white below with a pale brown wash over the breast. In flight, cinnamon underwing-coverts. Overlaps with most other *Turdus* species. **Voice** Song is a slow series of well-spaced, often slurred, rich whistles, warbles or churrs that are frequently doubled, imparting a distinctive temporal pattern with interruptions (unlike Rufous-bellied Thrush). Calls include a hollow *puk*, a liquid *uip* and a rapid series of raspy upswept whistles. [Zorzal Blanco]

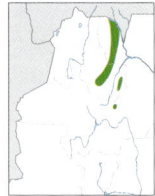

Glossy-black Thrush *Turdus serranus*   23.5 cm

Yungas forest of Salta and Jujuy at 1250–2500 m. **ID** Adult ♂: Bright orange bill and eye-ring; slightly paler legs. Entirely glossy black with slightly browner flight feathers. Adult ♀: Dull yellow or black bill. Yellowish legs and narrow eye-ring. Dark sooty-brown above; slightly warmer below and on the lores. Overlaps with Chiguanco Thrush; in general, sleeker and proportionately longer-tailed. **Voice** Song is composed of short phrases that begin with similar short whistles but which differ in their metallic, squeaky or trilled endings, *we-keep kwong… we-keep quick… we-keep chrrr….* Calls include a low-pitched, rapid liquid *kuikuikuikuikui* and a gurgling *GWRRE*. [Zorzal Negro]

Chiguanco Thrush *Turdus chiguanco*   27 cm

Andes to 3600 m, south to Neuquén, C sierras to 2200 m, sparsely in N Patagonia. **ID** Adult ♂: Stout orange bill, yellow eye-ring and pale orange legs. Entirely dark sooty brown; often paler below. Adult ♀ (not illustrated): Paler and browner throughout. Duller bill and eye-ring. **Voice** Song is a steady phrase of pure, musical, thin whistles with much variation in pitch and generally at least one downslurred or upslurred quavering or trilled note. The same phrase is repeated over and over. Calls include loud clipped whistles *ZIK* or *WEAK*, often in couplets or triplets, and low chuck sounds. **Tax note** 58. [Zorzal Chiguanco]

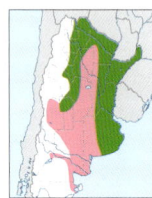

Rufous-bellied Thrush *Turdus rufiventris*   23.5 cm

Yungas forest to 1700 m and N lowlands south to N Patagonia. **ID** A large conspicuous resident thrush. Adult: Mainly olive-brown with a striking orange-rufous belly. In flight, orange-rufous underwing-coverts and axillaries. Juvenile: Recalls adult, but breast spotted brown and wing-coverts tipped orange. **Voice** Song is a melodic, leisurely series of mellow undulating whistles with a few slurred notes, creating a continuous temporal pattern (onomatopoeic Guaraní name is Corochiré). Calls include an upward-inflected, liquid, drawn-out *chuu-wip* (at dusk), a *whâ* and a rapid series of pure, upswept whistles (harsher in the NW). [Zorzal Colorado]

Andean Slaty Thrush

ad ♂

ad ♀

Creamy-bellied Thrush

ad

Glossy-black Thrush

serranus

ad ♂

ad ♀

Chiguanco Thrush

anthracinus

ad ♂

ad

rufiventris

ad

juv

Rufous-bellied Thrush

Turdus See Plates 164 and 165 for other species. *Catharus* Small thrushes with spotted breasts of shady understorey and midstorey. Two species are boreal migrants. *Hylocichla* Habits and plumage very like *Catharus*, although a little plumper and more pot-bellied.

Austral Thrush *Turdus falcklandii* 25.5 cm

s. Mendoza, Río Negro and s. Buenos Aires, south throughout Patagonia. **ID** The only thrush over most of its range. Adult ♂ *magellanicus*: Yellow legs, eye-ring and brighter bill. Blackish hood, contrasting with a warm brown back and grey rump. Thickly streaked throat, and mostly ochre-buff belly. In flight, note ochre-buff underwing-coverts. Adult ♀: Browner hood than ♂ and paler underparts. Adult *falcklandii* (Falkland Is; not illustrated): Similar to *magellanicus* but apparently larger. Juvenile: Brown above with buff shaft streaks. Indistinct throat streaking. Breast and belly ochraceous with large blackish-brown spots. **Voice** Song is a notably slow series of diverse, often metallic or wheezy, slurred whistles interspersed with odd churrs. Calls include short, harsh rasps, sometimes in a fast series, and a *WÉH* note. [Zorzal Patagónico]

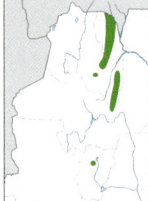

Sclater's Nightingale-Thrush *Catharus maculatus* 18.5 cm

Resident in Yungas forest to 1700 m in Salta and Jujuy. **ID** Adult: Bright red bill and eye-ring. Orange legs. Black hood and olive-grey upperparts. Mainly yellowish-buff below with a whitish central throat, and dusky breast spotting. **Voice** Song recalls a slowly swinging rusty gate and is made up of well-spaced couplets of flat, metallic, pure or raspy whistles, second note higher- or lower-pitched, and often alternating. Occasionally interrupted by raspy trills and bouts of bill snapping. **Tax note 59.** [Zorzalito Overo]

Veery *Catharus fuscescens* 17 cm

Rare vagrant in Misiones. **ID** Adult: Reddish-brown above with grey lores. Indistinct eye-ring. Greyish-white below. Narrow malar and small sparse brown spotting on upper breast. **Voice** Generally silent, but can give a descending buzz *ZEEWR* or *ZZEOU*. [Zorzalito Colorado]

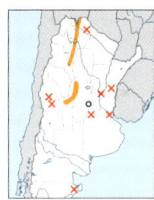

Swainson's Thrush *Catharus ustulatus* 17.5 cm

Yungas forest to 2500 m and C sierras (Oct–Mar); rare in the N lowlands. **ID** Adult: Olive-brown above. Creamy supraloral streak and eye-ring. Greyish-white throat, becoming whiter below with a buff wash over the upper breast. Narrow blackish malar streak and breast spotting. **Voice** Song is a series of strange, soft, flute-like notes that rise and fall in pitch creating an overall ascending melody. Calls include a resounding liquid *quip* or *puip*, and a long, nasal burry *kwi-burrrrrr*. [Zorzalito Boreal]

Wood Thrush *Hylocichla mustelina* 18 cm

Vagrant; once on the Falklands. **ID** Adult: Contrasting rufescent crown and nape. Rest of upperparts warm brown. Narrow white eye-ring, white-streaked face and blackish malar streak. White below with rounded black spots. **Voice** Calls include liquid *quip* notes and descending raspy chatters. [Zorzalito Rojizo]

Austral Thrush

ad ♂

magellanicus

ad ♀

juv

ad

Sclater's Nightingale-Thrush

ad

Veery

ad

swainsoni

Swainson's Thrush

ad

Wood Thrush

Mimus Thrush-like but with fairly long graduated tails. Worn birds can look subdued, lack some features and present uncertainty over their identity, although structure aids in identification. Large repertoire of complex songs consists of the rapid repetition of a single note-type before switching to another, including whistles, rasps, trill, churrs and imitation of birds and other sounds. Calls are harsh rasps of variable length.

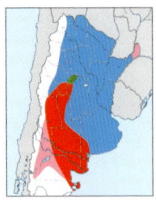

White-banded Mockingbird *Mimus triurus* 23 cm

Endemic breeder in Monte desert of N Patagonia (Oct–Feb), C sierras; winters through N lowlands and Andes to 2800 m. **ID Adult**: Greyish crown and back with a contrasting rufous-brown rump. Striking white panel on folded wing. Black tail with white outer tail feathers. In flight, black primaries and central tail feathers contrast with white secondaries, carpal and outer tail feathers. Can show a black trailing edge. Some overlap with Brown-backed Mockingbird in winter. [Calandria Real]

Brown-backed Mockingbird *Mimus dorsalis* 25.5 cm

Resident above 2400 m mainly in Salta and Jujuy. **ID Adult**: Chocolate-brown above, becoming rufous on the rump. White supercilium and thick blackish eye-stripe. Black wings with white primary-coverts and broad tertial fringes. Tail as White-banded Mockingbird. In flight, recalls White-banded Mockingbird but with less white in the wing; restricted to the primary bases, primary wing-coverts and tips to greater coverts. [Calandria Castaña]

Chalk-browed Mockingbird *Mimus saturninus* 24.5 cm

N lowlands and Andes to 2500 m, south to N Patagonia. **ID Adult**: Similar to Patagonian Mockingbird but vaguely streaked dark brown above; sometimes with a rufescent rump. Restricted flank streaks. In flight, long brown tail broadly tipped and narrowly fringed white. Overlaps with Patagonian Mockingbird in the NW and N Patagonia. [Calandria Grande]

Patagonian Mockingbird *Mimus patagonicus* 22.5 cm

Arid NW Andes to 3000 m, and throughout Patagonia except Tierra del Fuego. **ID Adult**: Ear-coverts sometimes tinged rufous. Sandy brown above; usually warmer on the rump. Note shorter, blacker tail than Chalk-browed Mockingbird. Can show vague flank streaks. In flight, relatively short black tail, tipped and fringed white. White trailing edge. Overlaps with Chalk-browed and Chilean Mockingbirds. [Calandria Mora]

Chilean Mockingbird *Mimus thenca* 27 cm

Sub-Andean Neuquén to nw. Chubut. **ID Adult**: Similar to Chalk-browed Mockingbird but longer-tailed and with less white in the tail. Note broad dusky malar streak, and long, crisp, dark flank streaks. [Tenca]

White-banded Mockingbird

ad

ad

Brown-backed Mockingbird

modulator

ad

Chalk-browed Mockingbird

ad

ad

Chilean Mockingbird

Patagonian Mockingbird

Anthus Best distinguished by song and the presence or lack of a malar streak, scapular braces, ventral coloration, and extent and shape of ventral markings. Worn and fresh plumages differ greatly. The length and curvature of the hindclaw (usually visible with a scope) greatly aids identification. Outer rectrices are white in lowland species, but help to distinguish Puna, Paramo, Hellmayr's and Correndera Pipits in the NW Andes. Here we emphasise diagnosis of display songs based on structural patterns, including spacing and quality of buzzes and other details, rather than describing individual songs. See also Plate 169. See Appendix 3 (p. 444) for identification keys and Sonograms (p. 448) to use with sound descriptions.

Short-billed Pipit *Anthus furcatus*　　　　15 cm

N and C lowlands. **ID** Short curved hindclaw. **Adult worn:** Fairly uniform brown above with little contrast. No scapular braces. Primary tips do not extend beyond the long tertials. Whitish below with little buff wash across the breast. Smaller, neater breast spotting. **Adult fresh:** Back thickly blotched blackish and fringed pale brown. White outer rectrices. Indistinct buffy eye-ring and supercilium. Open-faced with little or no streaking. Buffy throat with narrow blackish malar. Buff wash over the breast with dense brown spotting or inverted chevrons, well demarcated from the greyish-white belly. A few brown flank streaks. **Voice** The display flight includes a monotone buzz and 1–2 couplets or triplets of short notes, then a second buzz which ascends, followed by 1–2 couplets or triplets, and this pattern is repeated over. Sonogram on p. 448. [Cachirla Uña Corta]

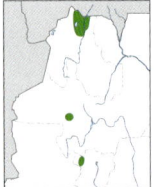

Puna Pipit *Anthus [furcatus] brevirostris*　　　　15 cm

Not illustrated. Replaces Short-billed Pipit in N Puna. **ID Adult:** Extremely similar to Short-billed Pipit; differs in fresh plumage by a more ochre wash over the breast, and some rusty dorsal fringing. **Voice** The display flight includes a long and descending churring buzz followed by a short monotonous series of similar notes or couplets. Sonogram on p. 448. **Tax note 60.** [Cachirla Puneña]

Yellowish Pipit *Anthus lutescens*　　　　13.5 cm

N lowlands and foothills; often on marshy ground. **ID** Long, arched hindclaw. **Adult worn:** Upperparts like worn Short-billed Pipit. White or greyish-white below with small fine breast streaks and unmarked flanks. **Adult fresh:** Dark brown above with pale fringing creating a mottled effect. No scapular braces. Mostly white outer rectrices. Indistinct malar, but short, thick, brown submoustachial streak. Washed sulphur-yellow below with fine black breast streaking (sometimes blurry) and indistinct flank streaks. **Voice** During display flight or while standing, gives a brief ticking introduction followed by an unmistakable, long, raspy, nasal descending note, *tic-tic dzeeeeeaahhhh*. Sonogram on p. 448. [Cachirla Chica]

Ochre-breasted Pipit *Anthus nattereri*　　　　14.5 cm

Resident in Corrientes savanna. **ID** Very long, slightly arched hindclaw. **Adult worn:** Brown above with blackish feather centres and whitish braces. Mainly white below, sometimes washed buff on breast and flanks, finely streaked black on breast and flanks. Shows a malar unlike Pampas and Hellmayr's Pipits. **Adult fresh:** Fairly bright chestnut above, thickly streaked blackish. Buffy-white braces. Strong black centres to the median and greater wing-coverts. Outer two rectrices buffy-white. Indistinct buff supercilium. Ear-coverts finely flecked black. Narrow blackish malar. Rich ochre-yellow over the breast and flanks; paler and whiter in the centre of the belly. Sparse black breast streaking and longer flank streaks. **Voice** The high display flight comprises a musical introduction of fast jumbled notes recalling *Spinus* siskins, followed by 3–6 very nasal flat buzzes, each successively lower in pitch. Sonogram on p. 448. [Cachirla Dorada]

Short-billed Pipit

ad worn

ad fresh

furcatus

ad fresh

ad worn

ad fresh

ad fresh

ad worn

Ochre-breasted Pipit

ad worn

ad fresh

ad worn

lutescens

Yellowish Pipit

Anthus Best distinguished by song and the presence or lack of a malar streak, scapular braces, ventral coloration, and extent and shape of ventral markings. Worn and fresh plumages differ greatly. The length and curvature of the hindclaw (usually visible with a scope) greatly aids identification. Outer rectrices are white in lowland species, but help to distinguish Puna, Paramo, Hellmayr's and Correndera Pipits in the NW Andes. Here we emphasise diagnosis of display songs based on structural patterns, including spacing and quality of buzzes and other details, rather than describing individual songs. See also Plate 168. See Appendix 3 (p. 444) for identification keys and Sonograms (p. 448) to use with sound descriptions.

Pampas Pipit *Anthus chacoensis*   13.5 cm

Endemic breeder. Mainly in crops in the Pampas and sierran foothills; reaches the humid chaco in winter. **ID** Short hindclaw. **Adult fresh:** Cold brown above, densely streaked blackish, but no scapular braces. Pure white outer rectrices. Very indistinct or no malar streak. Greyish-white below with a dull buff wash across the breast which is sparsely streaked blackish and with longer streaks on the flanks. **Voice** In extremely high display flight, the very distinctive ringing song lacks any buzz and comprises rapid repetitions of distinct note-types given in long series that tend to ascend in pitch. Call is a very high-pitched, short shrill note. Sonogram on p. 448. [Cachirla Trinadora]

Paramo Pipit *Anthus bogotensis*   16.5 cm

Scarce in the NW Andes above 2700 m; favours humid rocky grasslands. **ID** Long, slightly arched hindclaw. **Adult fresh:** Pale brown to buffy above, thickly streaked blackish. Creamy-white outer tail feathers. Indistinct buffy supercilium. Plain ear-coverts. No malar but short dusky submoustachial. Mostly buff below, usually whiter on the throat. Fine dark brown streaks over breast. Some show a few longer brown flank streaks. **Voice** In low display flight (usually just one song per flight), a short, high-pitched buzz is followed by thin leisurely musical notes recalling a *Leistes* meadowlark. Call is an ethereal, high-pitched cascading trill. Sonogram on p. 448. [Cachirla Andina]

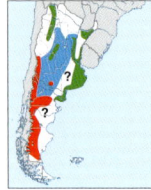

Hellmayr's Pipit *Anthus hellmayri*   15.5 cm

Widespread. Three similar ssp: *hellmayri* (NW Andes to 2000–4000 m), *dabbenei* (C and S Andes to 1500 m), *brasilianus* (Pampas and Mesopotamia; illustrated). **ID** Fairly long hindclaw. **Adult worn:** Cold grey-brown with blackish feather centres and indistinct whitish braces. Note short primary extension; two tips extend beyond tertials. Greyish-white below with a necklace of fine streaks, and fine flank streaks. No malar streak. **Adult fresh:** Pale to warm brown above, thickly streaked blackish-brown with buffy braces. Pure white outer rectrices and distal half of penultimate rectrices (buff in *hellmayri*). Whitish eye-ring and variable supercilium. Streaked ear-coverts. Vague malar streak; usually lacking. Buff below with a whitish throat. Necklace of fine blackish streaks, and broader flank streaks. **Voice** The display flight contains short, nasal buzzes separated by long series of musical notes (unlike Correndera Pipit). Short flight-song is a series of well-spaced, varied couplets. Sonogram on p. 448. [Cachirla Pálida]

Correndera Pipit *Anthus correndera*   15.5 cm

Widespread. Five fairly similar ssp. cover the region. **ID** Fairly long, arched hindclaw. **Adult worn:** Dull brown above, sometimes with indistinct pale scapular braces. Outer pair of rectrices mostly white; tinged buff in the altiplano (*calcaratus*). Lacks the buff wash over breast, and shows smaller ventral spots which can appear streaked. **Adult fresh:** Thickly streaked blackish above with chestnut or ochre fringes. Prominent whitish or creamy braces. Short white or buffy supercilium and eye-ring. Streaked ear-coverts. Narrow black malar and submoustachial. Whitish throat, washed buff on the breast and flanks, and greyish-white below. Thick black spots on breast, and more elongated oval flank spots. **Voice** The display flight includes long, grating buzzes separated by a few quick short notes (unlike Hellmayr's Pipit). Sonogram on p. 448. [Cachirla Goteada]

Pampas Pipit

ad fresh

ad fresh

shiptoni

Paramo Pipit

ad worn

ad fresh

ad worn

Hellmayr's Pipit

brasilianus

ad worn

ad fresh

ad fresh

ad worn

Correndera Pipit

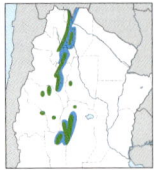

Brown-capped Whitestart *Myioborus brunniceps* 13 cm

NW Andes (to 2400 m) south to ne. San Juan; C Sierras (to 1900 m). **ID** Inquisitive and restless while fanning tail in midstorey to subcanopy. Often in mixed-species flocks. **Adult:** Slate-grey above with white spectacles, chestnut cap and green back. Shows white outer tail feathers. Bright yellow below. **Juvenile:** Drabber above than adult, lacking the cap and green back. Dusky throat and upper breast; paler yellow below. **Voice** Song is a fast, very high-pitched, shrill ascending trill lasting 3–4 secs. Call is an audible *tic*, similar to Tropical Parula but fuller and lower-pitched. [Arañero Corona Rojiza]

Southern Yellowthroat *Geothlypis [aequinoctialis] velata* 13.5 cm

Andes (to 2000 m), N and C lowlands (mainly Sep–Apr; winters in the humid chaco). **ID** A sexually dimorphic migratory breeding warbler of marshes, lush grasslands and edge of light woodlands. **Adult ♂:** Olive above and golden below. Grey crown. Drooping black mask. **Adult ♀:** Black upper mandible and fleshy lower mandible. Olive above and duller yellow below than ♂. Narrow yellow supraloral and eye-ring. Compare with doraditos, which lack pink legs (Plate 143). **Voice** Varied repertoire. Typical diurnal song is a long, spirited, pleasant warble with some metallic trills or whistles. Call is a sharp buzzy chatter, *TZEW ZED-ZED-ZED-ZED*, sometimes with just the initial note. **Tax note 61.** [Arañero Cara Negra]

Northern Waterthrush *Parkesia noveboracensis* 15 cm

Vagrant. **ID** Boreal migrant. Recalls a miniature thrush. Bobs rear body and tail. Mainly terrestrial in wooded marshes or edges of pools. **Adult:** Brown above. Buff or white supercilium. Black eye-stripe. White or yellowish below with linear streaking. **Voice** Call is a loud, somewhat metallic *SPIK*, repeated over. [Arañero de Agua]

Setophaga Tropical Parula is a resident midstorey and canopy warbler. Restless and often in mixed-species flocks. The other three species are arboreal non-breeding boreal migrants of light woodland or forest edge, at any height but mostly midstorey; all give a high-pitched sharp note, often repeated.

Tropical Parula *Setophaga pitiayumi* 11 cm

N Andes, C sierras and N lowlands, rarely south to N Patagonia. **ID Adult ♂:** Grey-blue above with a green mantle and two white wing bars. Yellow below with an orange throat and breast. **Adult ♀** (not illustrated): Lacks orange on the yellow breast. **Juvenile** (not illustrated): Lacks wing bars and is duller below. **Voice** Song is a very fast rising, high-pitched trill ending abruptly in a sharp note or a rapid succession of differently pitched warbles and trills, also ending abruptly. Call is a sharp *tic*, similar to Brown-capped Whitestart but thinner and higher-pitched. [Pitiayumí]

Yellow Warbler *Setophaga [petechia] aestiva* 12.5 cm

Vagrant. **ID Adult ♂:** Foreface and underparts bright yellow. Broad chestnut streaks on abdomen. **Adult ♀:** A few fine chestnut streaks on the breast and flanks. Wing-coverts and tertials fringed yellow unlike ♀ Southern Yellowthroat. **Tax note 62.** [Arañero Petequia]

Blackpoll Warbler *Setophaga striata* 13 cm

Vagrant. **ID** Note orangish legs in all plumages. **Adult ♂ breeding:** Black cap and contrasting white cheeks. Olive-grey above, streaked black. Two white wing bars. White below with black malar streak and spot-streaking on sides of breast and flanks. **Adult ♂ non-breeding:** Olive above, finely streaked black. Two white wing bars. Short yellowish supercilium. Dull yellow-olive below with white undertail-coverts. **Adult ♀ breeding** (not illustrated): Lacks the black cap and malar of breeding ♂. Crown like back and sparsely streaked below. [Arañero Estriado]

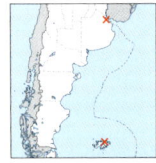

American Redstart *Setophaga ruticilla* 13 cm

Vagrant. **ID** Fans tail, pirouettes and droops wings. **Adult ♂:** Black upperparts and breast; white belly. Striking orange wing and tail flashes, and patch on sides of breast. **Adult ♀:** Grey head and olive-brown back. Darker wings and tail with yellow flashes. White below. **Immature:** Like ♀ but browner above. Yellow restricted to the tail. Yellow patch on sides of breast. [Arañero Yanqui]

Brown-capped Whitestart

ad

juv

Southern Yellowthroat

velata

ad ♂

ad ♀

Tropical Parula

pitiayumi

ad ♂

Northern Waterthrush

ad

Yellow Warbler

aestiva

ad ♀

ad ♂

American Redstart

ad ♂

imm

ad ♀

Blackpoll Warbler

♂ br

♂ non-br

Basileuterus and *Myiothlypis* Olive-backed forest warblers; mainly in understorey. All are resident. Sexes alike. Note head pattern and song.

Golden-crowned Warbler *Basileuterus culicivorus*   12.5 cm

Yungas forest to 1200 m (rarely to 1600 m); humid chaco and NE lowlands south to ne. Buenos Aires. Understorey and mid-strata. **ID** Adult: Thick brown lateral crown stripes and orange coronal patch. Whitish or creamy supercilium and grizzled area below eye. Overlaps with Pale-legged, Two-banded and Flavescent Warblers. **Voice** Song is a springy musical series of sweet warbles with characteristic cadence, rising and falling abruptly at the end, *su-su-su-swi-SWI-SÍ-SIÚ*. Calls include a short flat buzzy rasp, *chr* or *shreep*, sometimes in a steady series Tax note 63. [Arañero Coronado Chico]

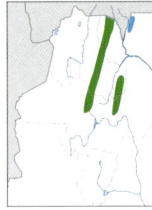

Pale-legged Warbler *Myiothlypis signata*   14 cm

Yungas forest understorey (900–1900 m) of Jujuy and Salta. **ID** Adult: Blackish lores and short indistinct lateral crown stripes, bordering olive crown. Broad bright yellow supercilium terminating just behind the eye. Overlaps with Two-banded Warbler to 1700 m and with Golden-crowned Warbler mostly below 1200 m. **Voice** Song is a fast, explosive steady series of liquid whistles, starting with a stutter and ending in a trill that rises and falls, e.g. 40 notes in 5 secs. Alternatively, a shorter rising and falling or falling series of descending whistles becoming more emphatic towards the end. Call is a sharp *stit* note, often doubled or repeated in a rapid series. [Arañero Ceja Amarilla]

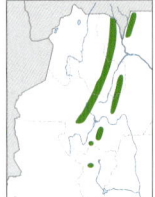

Two-banded Warbler *Myiothlypis bivittata*   15 cm

Yungas forest understorey (450–1400 m) of Jujuy and n. Salta. **ID** Adult: Blackish eye-stripe and thick lateral crown stripes, bordering orange or yellow coronal patch. Short yellow supercilium and notable crescent below eye. More olive on flanks than Pale-legged Warbler. Overlaps with Golden-crowned, and with Pale-legged at 900–1600 m. **Voice** Song is a fast strident jumbled duet, with one bird giving a syncopated series of differently pitched raspy couplets and the other a low-pitched rising and accelerating stutter of excited burry notes, *ti-ku ti-ku chrr-chrr kip-it kip-it … du-du-di-di-wi-wie-ZIRRIRRI*. Call is a resounding *whik* or *whit*, sometimes in a rapid series. [Arañero Coronado Grande]

Flavescent Warbler *Myiothlypis flaveola*   14 cm

Local in gallery forest in ne. Formosa. **ID** Adult: Pale bill. Short, broad, bright yellow supercilium and dusky lores. Brighter yellow below than congeners. Olive wash on the sides of the breast and flanks. Overlaps with Golden-crowned Warbler. **Voice** Song is a short rapid explosive series ending in 3 (rarely 4–5) distinctive notes, *susísusísusí-CHU-CHU-CHU*. Call is a sharp *tíc*. [Arañero Pico Pálido]

White-rimmed Warbler *Myiothlypis leucoblephara*   14.5 cm

Understorey of riparian woodland in humid chaco and NE lowlands south to ne. Buenos Aires. **ID** Adult: Grey crown, face and flanks. White supraloral patch, eye crescents and throat. Bright olive back, wings and tail. Yellow undertail-coverts. **Voice** Song is a very distinctive long, cascading series of piercing high-pitched pure whistles, e.g. 20 notes in 5 secs. Call is a high-pitched *psee*. [Arañero Silbón]

Riverbank Warbler *Myiothlypis rivularis*   14.5 cm

Pools, streams and rivers in Paraná forest. **ID** Adult: Slate-grey crown and nape. Blackish eye-stripe and lateral crown-stripe, divided by narrow buff supercilium. Rest of upperparts mainly olive. White throat and mainly buff below. **Voice** Song is a powerful, relatively low-pitched, rich, barely descending, series of whistles that begins with 2–3 *tew* notes and continues into an explosive trill, with more emphatic and slower final notes, e.g. 22 notes in 3 secs. Abbreviated songs may lack the trilled portion and sometimes ascend slightly in pitch. Call is a scolding *chit*, often doubled or tripled. [Arañero Ribereño]

azarae

Golden-crowned Warbler

ad

flavovirens

ad

Pale-legged Warbler

argentinae

ad

Two-banded Warbler

ad

Flavescent Warbler

leucoblephara

ad

White-rimmed Warbler

rivularis

ad

Riverbank Warbler

Plushcap *Catamblyrhynchus diadema* 13.5 cm

Very rare in mid-elevation Yungas forest (winter visitor?). **ID** Small understorey passerine with a stout arched bill and bristle-like feathers on the forecrown. Often in *Chusquea* bamboo. **Adult:** Contrasting yellow forehead, slaty upperparts and deep chestnut below. **Immature:** Yellowish supraloral, and dull olive-grey upperparts. Underparts paler and more washed-out than adult. **Voice** Song is a springy, fast, jumbled series of high-pitched squeaky notes (unlikely to be heard in Argentina). Call is a weak, thin, high-pitched *psee* or *tsit-tsit-tsit.* [Diadema]

Grey-bellied Flowerpiercer *Diglossa carbonaria* 13 cm

Hypothetical at 3450 m on Bolivian border in Jujuy. **ID** Hook-tipped upper mandible and upturned lower mandible for piercing flower bases. Generally in pairs in Andean shrubbery, sometimes with mixed flocks. **Adult:** Black upperparts, throat and breast. Contrasting blue-grey shoulders, rump and belly. Chestnut undertail-coverts. [Payador Vientre Gris]

Giant Conebill *Conirostrum binghami* 15.5 cm

Very rare in *Polylepis tomentella* woodland above 3000 m in Salta and Jujuy. **ID** Slender conical bill. Unobtrusive, usually in upper strata. Often in mixed flocks. **Adult:** Blue-grey above and chestnut below. Contrasting white forehead and cheeks. **Voice** Song (at dawn?) consists of the repetition of two fairly high-pitched, thin, syncopated phrases *chúit chew-it SWEEP… chúit chew-it SWEEP…,* with variation. When excited, gives a longer, jumbled high-pitched twittering with squeaky notes. [Saí Gigante]

Rusty Flowerpiercer *Diglossa sittoides* 11.5 cm

Andean meadows and Yungas forest edge at 1150–3400 m; mostly 1600–2600 m. **ID** Hook-tipped upper mandible and upturned lower mandible for piercing flower bases. Generally in pairs in Andean shrubbery, sometimes with mixed flocks. **Adult** ♂: Grey-blue above with a dusky mask. Buff below. **Adult** ♀: Olive-brown above with two narrow wing bars. Drab yellow below with dusky breast flammulations. **Voice** Song is a rapid series of high-pitched, metallic, ringing or buzzy notes often ending in a churr or short trill *zlin-zlin-zlin-zlin chrrrr*; with much variation, sometimes sputtered or with slower, more musical notes. Call is a very thin, high-pitched *chit chit.* [Payador Canela]

Rufous-bellied Mountain Tanager *Pseudosaltator rufiventris* 22.5 cm

Local in Andes of Jujuy and Salta (2450–3400 m). **ID** Unobtrusive, chunky mountain tanager, found very locally in the NW Andes. **Adult** ♂: Blue-grey upperparts, throat and breast. Long narrow white supercilium. Brick-red belly and vent. **Adult** ♀ (not illustrated): Greyer, less blue above, and paler below. Iris duller, not so strikingly red. **Voice** Rarely heard (dawn?) song is a short, rapid yodelling and a higher-pitched, strident whistle, *wu-luli-luli-TCHEEE.* Also gives a rapid high-pitched springy series, *chi wi-cher wi-cher wi-cher.* Most frequent calls are varied single, nasal, gruff notes *gwäo, khweo, kauh* or *cao,* recalling Mourning Sierra Finch. [Pepitero Colorado]

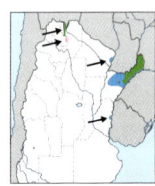

Black-goggled Tanager *Trichothraupis melanops* 17 cm

Paraná forest, south to ce. Entre Ríos in winter; local population in Yungas foothills. **ID** Slender-billed and slender-tailed tanager which forages at all levels. Striking white wing-stripe, only visible in flight. **Adult** ♂: Black lores and mask. Semi-concealed golden-yellow coronal patch. Dull brownish-olive above with blackish wings and tail. White underwing-coverts. Uniform buff below, with a paler belly. **Adult** ♀: Brownish-olive above. Dark brown wings and tail. Buff below. **Voice** Song is a thin, slow whistled series with two alternated phrases, *cha-wer SWII-SIA-wer CHurr, …cha-wer SWII-Cha-wa,* repeated over and over. Call is a dry, scolding *chhitt* note, often repeated. [Frutero Corona Amarilla]

Chlorospingus Conspicuous and often gregarious compact passerine. Mainly in midstorey where it is often in mixed flocks.

Common Chlorospingus *Chlorospingus flavopectus* 14 cm

Yungas forest. **ID Adult:** Olive above with a brownish hood and contrasting white postocular triangle. Yellow pectoral band and greyish-white belly. **Voice** Dawn song is an unmusical repeated phrase of chipping couplets, *chip chip-it chip-it chip-it…,* sometimes varied to a *chip chuet-chuet-chuet-chuet…* Large repertoire of calls includes long, shrill, twittering and high-pitched notes. [Alt: Common Bush Tanager] [Frutero Yungueño]

Plushcap

imm

ad

citrinifrons

Grey-bellied Flowerpiercer

ad

Giant Conebill

ad

ad ♀

Rufous-bellied
Mountain Tanager

ad ♂

sittoides

ad ♂

Rusty Flowerpecker

ad

ad ♂

ad ♀

argentinus

Common Chlorospingus

Black-goggled
Tanager

Thlypopsis Smaller tanagers found mostly in midstorey with white or yellow underwing-coverts. *Hemithraupis* Canopy tanager, often with mixed flocks; white underwing-coverts diagnostic if seen.

Chestnut-headed Tanager *Thlypopsis pyrrhocoma* 14.5 cm

Paraná forest, often in *Chusquea* bamboo. **ID Adult** ♂: Grey with a contrasting chestnut hood. Extensive black lores. **Adult** ♀: Dull rusty-brown crown. Often a diffuse yellowish supercilium. Olive back, wings and tail. Buff or whitish throat, becoming olive on breast and flanks. Creamy-buff belly and vent. **Voice** Song is a distinctive, high-pitched, shrill whistled series, *tsee-tsee-tsee swee-swee*, sometimes ending with 3 notes, which are always distinctly lower-pitched than the faster introduction. Calls include a ticking *tik* and a shrill *tzz*. [Pioró]

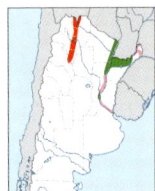

Orange-headed Tanager *Thlypopsis sordida* 14 cm

Yungas forest to 1500 m; humid chaco, n. Corrientes and s. Misiones. **ID Adult** *sordida*: Chestnut crown grading to yellow on face and throat. Olive back. Buff breast, becoming whiter on belly. **Adult** *chrysopis* (NE; not illustrated): More rusty crown and greyish back. **Juvenile**: Olive above. Yellow forehead, face, throat and breast, becoming buffy below. White centre to belly. **Voice** Song is a very high-pitched thin, slowly delivered *tsit... tsit... tsit-tsit-tsit-Síeu* with emphasis on the final note, sometimes with just one or no introductory notes. Calls include a thin *tseet*, *tsip*, and a harsh note. [Tangará Gris]

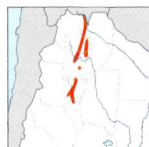

Rust-and-yellow Tanager *Thlypopsis ruficeps* 14 cm

Yungas forest 1700–2600 m (Oct–Mar). **ID Adult**: Bright chestnut crown and ear-coverts. Vivid olive back and wings. Golden-yellow below, sometimes washed olive on sides of breast. **Juvenile**: Dull olive above. Buff forehead and face. Pale yellow below with an olive wash on the sides of the breast and flanks. **Voice** Song is a shrill, high-pitched, springy *sit-sa-Si, sit-sa-Si, sit-sa-Si* series, or a faster spluttered series. Call is a short, high-pitched rasp *ship*. [Tangará Alisero]

Guira Tanager *Hemithraupis guira* 13.5 cm

Yungas foothill forest to 1250 m and Paraná forest. **ID Adult** ♂ *fosteri* (Paraná forest): Black face encircled by yellow supercilium. Orange breast. **Adult** ♂ *boliviana* (Yungas; not illustrated): Differs from *fosteri* by thinner supercilium, and usually a larger orange breast patch. **Adult** ♀: Dusky upper mandible and notable yellow lower mandible. Fairly vivid olive above. Yellow below, brightest on the vent, and paler creamy-yellow on the belly. Grey wash on flanks is diagnostic. **Voice** Song is a fast, high-pitched chatter with a rattled quality, *sit-chchchchchchchchIT*, repeated several times, sometimes with a jumbled beginning. Also gives a rapid series of soft ticking notes. Calls include a *chit* and a very nasal *nwah*. [Saíra Dorada]

Blue Dacnis *Dacnis cayana* 13 cm

Paraná forest and gardens. **ID** Compact frugivorous tanager with a slender, slightly decurved bill. Mostly high-up in forest, but gregarious at garden feeders. **Adult** ♂: Cobalt-blue with a black back, tail, bib and centres of the wing-coverts. **Adult** ♀: Mostly vivid green, except for sky-blue cap and greyish throat. **Voice** Calls include very high-pitched, metallic, shrill notes *tzeet* or *zwreet*, in a loose series. [Saí Azul]

Chestnut-vented Conebill *Conirostrum speciosum* 11.5 cm

Yungas foothills to 1200 m, humid chaco and Paraná forest. **ID** Slender conical bill. Unobtrusive, usually in upper strata. Often in mixed flocks. **Adult** ♂: Blue-grey above, and pale grey below with a whiter belly and chestnut vent. **Adult** ♀: Bluish crown contrasts with green back, wings and tail. White below; variable combinations of buff wash on throat and vent, and/or yellow wash on breast sides and flanks. **Voice** Song is a short, fast, highly variable, series of 3–4 high-pitched notes, e.g. *tzi-twee-zip*. Also delivers a long, twittering, nasal warbled chatter. [Saí Celeste]

Bananaquit *Coereba flaveola* 11 cm

Misiones and ne. Corrientes. **ID** Small tanager with short, slender, decurved bill inhabiting forest clearings, parks and gardens. **Adult**: Sooty upperparts with long white supercilium and yellowish rump. Grey throat and yellow underparts. **Voice** Song is a shrill high-pitched, fast, variable twitter of sharp and buzzy notes. [Mielero]

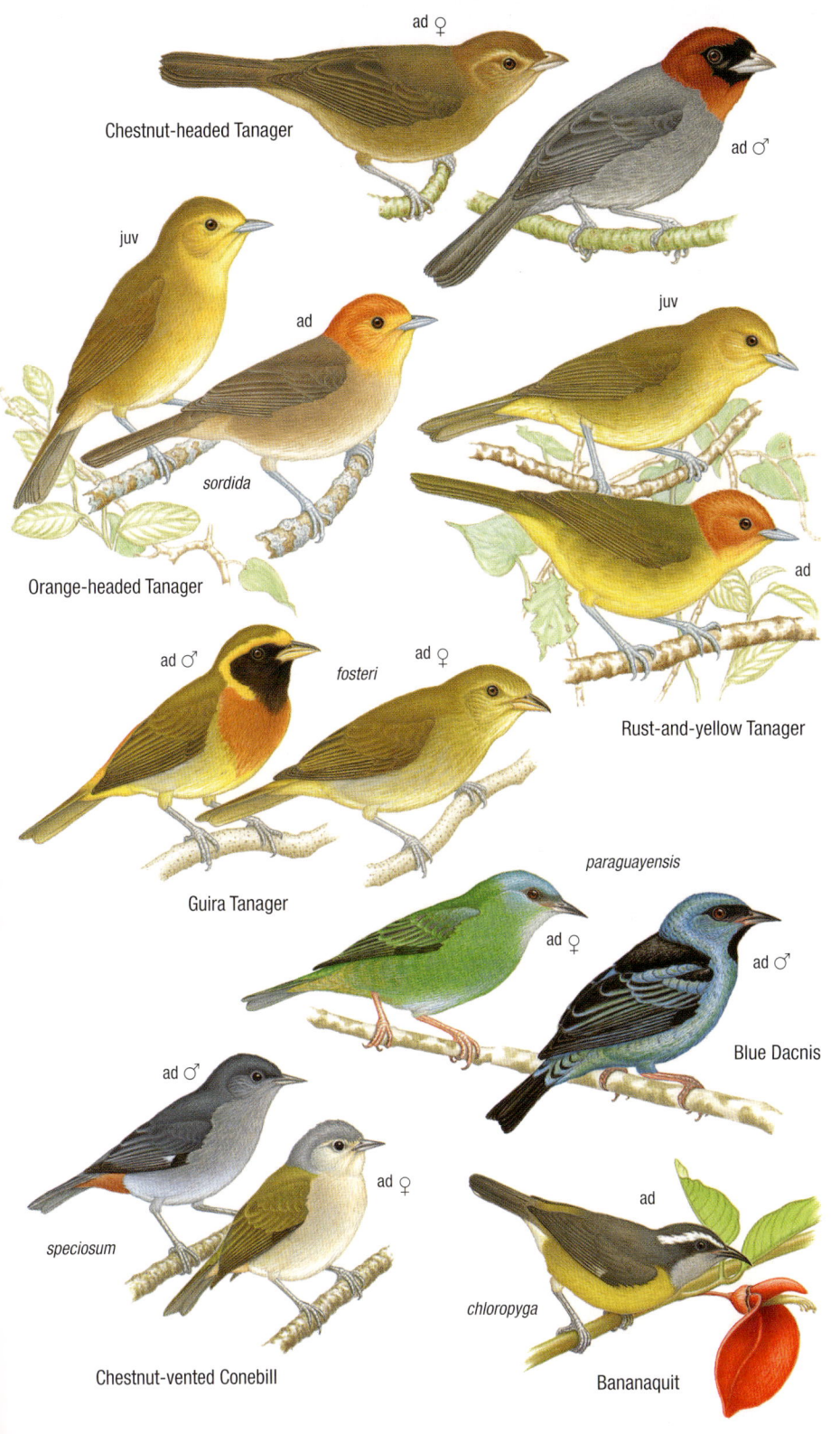

ad ♀

Chestnut-headed Tanager

ad ♂

juv

ad

sordida

Orange-headed Tanager

juv

ad

Rust-and-yellow Tanager

ad ♂

fosteri

ad ♀

Guira Tanager

paraguayensis

ad ♀

ad ♂

Blue Dacnis

ad ♂

ad ♀

speciosum

Chestnut-vented Conebill

ad

chloropyga

Bananaquit

Tachyphonus Relatively long-tailed tanagers with black males and warm brown females. Forage at all heights and give relatively simple repetitive songs. ***Ramphocelus*** Chunky tanagers of understorey and mid-strata with a very stout bill.

Ruby-crowned Tanager *Tachyphonus coronatus*   17.5 cm

Paraná forest. **ID** Adult ♂: Glossy blue-black with a concealed erectile scarlet coronal tuft. White underwing and inner lesser coverts, visible in flight. Adult ♀: Dull olive or greyish-olive crown and face. Whitish region around eye. Rufous-brown back and wing-coverts, becoming brighter rufous on rump and tail. Greyish-buff or whitish throat. Warm ochraceous below, becoming rufous on vent. Buff underwing-coverts. Overlaps with White-lined Tanager in e. Misiones and ne. Corrientes. **Voice** Song is a slow, medium-pitched, simple series of *schlup-CHIP* phrases, with the second note usually higher pitched, and an occasional third note-type interspersed; differs in general from similar White-lined by its lower pitch and slower cadence. Call is a very loud scolding *SCHIPP* note, repeated at 3–4 sec intervals. [Frutero Coronado]

White-lined Tanager *Tachyphonus rufus*   20 cm

Humid chaco, e. Misiones and Corrientes south to ne. Buenos Aires. **ID** Adult ♂: Larger-billed than Ruby-crowned Tanager. Mostly black with less blue gloss than Ruby-crowned, and more extensive white on lesser wing-coverts. Adult ♀: From Ruby-crowned by warm brown crown and face, contrasting dark chestnut tail, and whitish underwing-coverts. Overlaps with Ruby-crowned in e. Misiones and ne. Corrientes. **Voice** Song is very much like that of Ruby-crowned but generally higher-pitched and faster. Call is a repeated *chik* note. [Frutero Negro]

Brazilian Tanager *Ramphocelus bresilius*   c.18.5 cm

Hypothetical vagrant (escapee?) to Misiones. **ID** Adult ♂: Silvery-white base to black bill. Glistening crimson with black wings and tail. Adult ♀: Mostly brown with an orange-chestnut rump and belly. [Fueguero Escarlata]

Silver-beaked Tanager *Ramphocelus carbo*   c.18 cm

Not illustrated. Known from one (escapee?) Iguazú record. **ID** Adult: Bill as Brazilian Tanager. Mostly sooty-brown with a dark crimson throat and chest; belly and vent dark rosy. [Fueguero Oscuro]

Piranga Robust subcanopy and canopy tanagers with orange to red males and olive/yellow females. Olivaceous juveniles with streaky underparts quickly moult into female-like plumages. Immature males look like red-blotched females.

Hepatic Tanager *Piranga flava*   18.5 cm

Widespread. **ID** Adult ♂ *flava* (N lowlands and foothills to 1600 m): Dull reddish above with reddish-pink forehead, throat and breast. Adult ♀ *flava*: Olive above with a yellowish forehead, spectacles and throat, becoming olive-yellow below. Adult ♂ *saira* (n. Misiones; not illustrated): More crimson than ♂ *flava*. Adult ♀ *saira*: Golden forecrown. Brighter above than ♀ *flava*. Golden-yellow throat and breast, becoming paler below. **Voice** Rarely heard song is a slow, musical series of somewhat raspy mellow whistles and warbles, recalling a *Pheucticus* grosbeak or *Turdus* thrush. Contact calls comprise repeated *CHET* notes. [Fueguero]

Scarlet Tanager *Piranga olivacea*   c.17 cm

Not illustrated. Rare vagrant, known from two s. Buenos Aires records. **ID** Adult ♂: Striking deep scarlet with contrasting black wings and tail. Adult ♀: Olive above and yellow below with darker, drab-olive wings and tail. [Fueguero Boreal]

Habia Large, stout-billed passerine of forest understorey and mid-strata. Travels with mixed-species flocks.

Red-crowned Ant Tanager *Habia rubica*   19.5 cm

Paraná forest; scarce in s. Misiones. **ID** Adult ♂: Semi-concealed crimson coronal patch with dusky borders. Dull reddish-brown above. Rosy-red throat and breast, and dull pinkish-brown below. Pinkish-red undertail-coverts; often with a grey wash on the flanks. Adult ♀: Dull olive-brown above with a concealed cinnamon-orange coronal stripe. Bronze-olive below with a paler belly. **Voice** Dawn song is a slow, musical thrush-like series of loud, resounding notes. Groups are often detected by an excited long chatter of loud, steady, noisy rasps, that often develops into fast nasal 'skidding' notes *CH-CH-CH-CH-CH-CH-CH DJEEK DJEEK DJEEK DJEEK DIDIDJEEEK....* [Fueguero Morado]

Ruby-crowned Tanager
ad ♂
ad ♀

White-lined Tanager
ad ♂
ad ♀

ad ♀
dorsalis

ad ♂
Brazilian Tanager

ad ♀
saira

ad ♀
flava

ad ♂
Hepatic Tanager

ad ♀
ad ♂
Red-crowned Ant Tanager
rubica

Swallow Tanager *Tersina viridis* 15.5 cm

Paraná forest, especially in clearings; also various widespread erratic records. **ID** Large robust tanager found in all strata; often gregarious. Differs from all other tanagers by its relatively flattened bill, and by its cavity-nesting habits. **Adult** ♂: Glistening cobalt-blue with a black mask, barred flanks and white belly. **Adult** ♀: Vivid green with a brighter rump, yellowish centre of the abdomen and barred flanks. **Voice** Rarely heard song is a jumble of squeaky, nasal, ticking and high-pitched notes. Common call is a diagnostic, short, metallic upswept *dzeet*. [Tersina]

Diademed Tanager *Stephanophorus diadematus* 19 cm

n. and e. Misiones south along the Río Uruguay to ne. Buenos Aires. **ID** A very robust tanager with a bristle-like forehead and erectile coronal tuft. **Adult** ♂: Mostly deep blue with a black foreface, white crown and scarlet coronal tuft. Contrasting bright cobalt-blue shoulder and rump. **Adult** ♀ (not illustrated): Less white on the crown and much duller overall. **Juvenile** (not illustrated): Entirely brown. **Voice** Song is a rich, pleasant, steady warble of up to 3 secs, recalling Ultramarine Grosbeak and perhaps also Southern Yellowthroat. Calls include single or doubled, low-pitched *chwet* or *chup* notes. [Frutero Azul]

Fawn-breasted Tanager *Pipraeidea melanonota* 14.5 cm

Mainly Yungas and Paraná forest. **ID** A midstorey to subcanopy tanager with a dark mask. **Adult** ♂ *venezuelensis* (Yungas forest to 2400 m): Deep scarlet iris. Broad black mask. Sky-blue crown, nape, shoulder and rump. Rich buff below, tinged ochre on the breast. **Adult** ♀ *venezuelensis*: As ♂ but mask mostly dark brown, reduced blue on crown, shares bright blue rump, although pale buff below. **Adult** ♂ *melanonota* (Paraná forest; south to e. Buenos Aires in winter; not illustrated): Brown iris. Brighter than ♂ *venezuelensis*. **Adult** ♀ *melanonota*: Brown iris. Duller than ♀ *venezuelensis*; lacks blue on the crown, and has dull blue rump. **Voice** Rarely heard slow-song is a surprisingly varied string of imitations. Rapid-song is usually an extremely fast, very high-pitched repetition of shrill or pure notes, e.g. *zeezeezeezeezeezeezee*. Calls include an assortment of high-pitched notes. [Saíra de Antifaz]

Blue-and-yellow Tanager *Rauenia bonariensis* 17 cm

NE lowlands south to ne. Buenos Aires; a similar ssp. in the NW (to 3450 m) south to N Patagonia. **ID** Chunky tanager of midstorey to canopy. Pairs or mixed flocks. Generally in light woodland. **Adult** ♂ *bonariensis* (NE lowlands): Sky-blue head, throat, shoulder and fringes to wings and tail. Black lores and back. Orange rump and breast; yellow below. **Adult** ♀ *bonariensis*: Bicoloured bill. Brownish-olive above and buffy below. **Adult** ♀ *schulzei* (NW; not illustrated): Usually has a powdery-blue wash on the crown and ear-coverts, lacks olive tones above and is paler, and usually creamy below. **Voice** Song is a generally slow, long repetitive series of rich notes, *ziu-sweet ziu-sweet sut ziu-sweet…*, strongly recalling a warbling finch. Calls are single and disyllabic nasal notes. [Naranjero]

Sayaca Tanager *Thraupis sayaca* 17.5 cm

Widespread. **ID** Relatively short-tailed, midstorey to canopy tanager which readily joins mixed flocks. **Adult** *obscura* (NW Andes to 3400 m): Greyish-blue above; paler below especially on belly. Turquoise rump, shoulder and fringes to wings and tail. **Adult** *sayaca* (N lowlands south to n. Córdoba and n. Buenos Aires; not illustrated): Duller and darker overall. **Adult green morph**: Rare in both subspecies. Washed pale olive throughout, with a brighter rump. **Voice** Song is a jumbled series of rich, squeaky and nasal notes intermixed with pure drawn-out thin whistles, with much variation and sometimes consisting of mostly whistles, e.g. *swee-seee weee-you wee-see-you*. Calls include high-pitched *tsi* and *psiu* notes. [Celestino]

Palm Tanager *Thraupis palmarum* 19.5 cm

n. Misiones, often in towns and clearings where palms are present. **ID Adult**: Mostly olive. Dark brown wings and tail, fringed olive. Pale panel in primaries. **Voice** Recalls Sayaca Tanager but faster, higher-pitched and shriller. [Chogüí Oliváceo]

Swallow Tanager

ad ♀   *viridis*   ad ♂

Diademed Tanager   ad ♂

melanonota   ad ♀   ad ♀   *venezuelensis*   ad ♂   ad ♀   ad ♂   *bonariensis*

Fawn-breasted Tanager   Blue-and-yellow Tanager

green morph   ad   ad   *obscura*   ad

Sayaca Tanager   *palmarum*   Palm Tanager

Hooded Tanager *Nemosia pileata*　　　　　　　　12 cm

Local in N lowlands and foothills; mainly gallery forest. **ID** Small chunky tanager, mainly found in midstorey with mixed-species flocks. **Adult** ♂: Grey and white with a contrasting black hood, white lores, straw-coloured iris and bright orange tarsus. **Adult** ♀: Lacks ♂'s hood and shows a buff wash over the throat and breast, and contrasting creamy-yellow lower mandible. Compare with Black-capped Warbling Finch (Plate 182). **Voice** Long vocalisation is an unmusical, spluttered, shrill high-pitched trill. Calls include a scolding *chut* or *tek* note, sometimes accelerating into a short medium-pitched trill. Flight call is a sharp *sip sip.* [Frutero Cabeza Negra]

Magpie Tanager *Cissopis leverianus*　　　　　　　　29 cm

n. and c. Misiones; scarce in s. Misiones. **ID** A huge pied tanager with a long, graduated tail, as long as the body. Often gregarious. **Adult:** Unmistakable. Black-and-white plumage with a yellow iris. **Juvenile** (not illustrated): Brown replaces black plumage, and has a shorter tail. **Voice** Song is a fast, rhythmic series of springy, squeaky, high-pitched metallic notes followed by short or drawn-out whistles, e.g. *wich-wit-choo wich-wit-choo… warit-cheeeeu warit-chreeee warit-cheeeeu warit-chreeee…* Contact call is a sharp metallic *spick* note. [Frutero Overo]

Cinnamon Tanager *Schistochlamys ruficapillus*　　　　18 cm

Unconfirmed (sporadic?) in n. Misiones. **ID** Large chunky, tanager with stout, thickset bill, inhabiting forest edge and clearings. Sexes alike. **Adult:** Black loral mask contrasts with pinkish-cinnamon underparts. White centre to the belly. Olive-brown crown grades into blue-grey upperparts. [Frutero Canela]

Burnished-buff Tanager *Stilpnia cayana*　　　　　　15 cm

Local in gallery forest of ne. Corrientes; very rare in w. Misiones. **ID** **Adult** ♂: Mainly buff, tinged chestnut on the crown and with a contrasting black face and centre of throat, breast and belly. **Adult** ♀: Bronzy crown and greyish wash over face and throat, unlike ♀ Chestnut-backed Tanager. Fringing on wings much brighter than Chestnut-backed. **Voice** Song is a very fast phrase with alternating medium-pitched chipping notes and high-pitched, short, shrill whistles *ch-dip'zizi'ch-dip'zizi…* speeding into a jumble towards the end; may resemble a very fast Sayaca Tanager. Contact calls are thin, high-pitched *zeeu* and *tsi* notes. **Tax note 64**. [Saíra Pecho Negro]

Chestnut-backed Tanager *Stilpnia preciosa*　　　　　15 cm

Paraná forest chiefly in the *Araucaria* belt; some winter south along the Río Uruguay. **ID** Glistening chestnut head and mantle, opal wing-coverts and rump, and mainly turquoise below. **Adult** ♀: Drab olive above with a dull chestnut crown and blackish lores. Blue-green below with a creamy belly and obvious cinnamon undertail-coverts. Compare with Burnished-buff Tanager. **Voice** Common vocalisation from the canopy is a very high-pitched, shrill note *zreep*, usually slowly repeated once every 3–4 secs. [Saíra Castaña]

Black-backed Tanager *Stilpnia peruviana*

Not illustrated. Hypothetical historical reports from Misiones and Buenos Aires. **ID** **Adult** ♂: Differs from ♂ Chestnut-backed Tanager by its glossy black saddle. **Adult** ♀: Differs from adult♀ Chestnut-backed by a somewhat brighter crown and nape. [Saíra Espalda Negra]

Green-headed Tanager *Tangara seledon*　　　　　　14 cm

Paraná forest. **ID** **Adult** ♂: Unmistakable bright green tanager with blue breast, black throat and orange rump. **Adult** ♀ (not illustrated): Slightly duller. **Juvenile:** Dull green with a buffy centre to the abdomen and bright green wing fringing. **Voice** Foraging and contact call is an unmusical, nasal *tchwik* note, sometimes repeated continuously. Also delivers rapid metallic chatters and high-pitched calls. [Saíra Arcoíris]

Red-necked Tanager *Tangara cyanocephala*　　　　　12 cm

Hypothetical in Paraná forest. **ID** **Adult** ♂: Crimson scarf contrasts with blue crown, black mantle and bright green underparts. **Adult** ♀: Duller than ♂, lacking the black mantle and orange bar on the lesser coverts. [Saíra Militar]

ad ♀

ad ♂

caerulea

Hooded Tanager

ad

Magpie Tanager

major

Cinnamon Tanager

ad

ruficapillus

Chestnut-backed Tanager

ad ♀

chloroptera

ad ♂

ad ♀

ad ♂

Burnished-buff Tanager

ad ♂

juv

ad ♀

cyanocephala

ad ♂

Green-headed Tanager

Red-necked Tanager

Saltator Large passerines with a very stout bill of forest and lighter woodlands. Rarely in flocks, largely frugivorous, and accomplished songsters. Sexes differ marginally in some species.

Thick-billed Saltator *Saltator maxillosus*   21 cm

Rare in the sierras of ne. Misiones. **ID Adult** ♂: Huge bill mostly orange, variable in extent. Slate-grey above with narrow white supercilium and prominent black malar. Orange-buff throat and vent, otherwise dull ochraceous below, sometimes washed olive on the breast. **Adult** ♀: Blackish bill with a hint of orange at the base. Dark olive above. Underparts duller than ♂. Overlaps with Green-winged Saltator. **Voice** Rarely heard song is a sequence of groups of 2–4 high-pitched, short, metallic notes, *tsit zu ze zit... tsit zu... zrit zu tsi zit...* with variation. Most often delivers well-spaced calls, resembling notes in the song. [Pepitero Picudo]

Green-winged Saltator *Saltator similis*   21 cm

NE lowlands locally south to ne. Buenos Aires. **ID Adult:** Vivid olive above with long, narrow, white supercilium. White throat and upper breast, with black malar and dull buff wash below. **Juvenile:** Olive above with a white supercilium. The only saltator with ventral streaking; olive on breast and blurred on flanks. **Voice** Song is a fairly slow series of varied, loud, medium-pitched pure or slurred whistles that typically ascend upscale and finish on a lower-pitched note; sometimes a very short burry note can be added, e.g. *chu-che-chi-chíu-cheeo.* Lack of repeated notes distinguishes it from Greyish Saltator. Call is a high-pitched *tsit.* [Pepitero Verdoso]

Greyish Saltator *Saltator coerulescens*   22 cm

Widespread in Andean foothills and N lowlands except much of Misiones. Often in marshy habitats. **ID Adult:** Grey above with short white supercilium. White throat with black malar; greyish breast and buff below. **Juvenile:** Resembles adult Green-winged Saltator but with shorter, yellow supercilium, buffier throat, and only an indistinct malar. **Voice** Song includes repetition of 2–3 note-types and finishes in a long, trilling, burry note or in a descending whistle, e.g. *chuíchuí-dududu-breeee* or *fuifuifui-cheeeeo.* Rapid repetition of notes separates it from Green-winged Saltator. Also gives a noisy rapid series of squeaky, metallic notes (sometimes in duet), and a scolding *chink.* [Pepitero Gris]

Golden-billed Saltator *Saltator aurantiirostris*   20 cm

Widespread. **ID** The only saltator with black ear-coverts, pectoral band and long supercilium curving from eye to the side of the neck. ♂ has orange bill; ♀ has a blackish lower mandible. **Adult** *nasica* (Monte desert): Grey above with a white or buff supercilium. **Adult** *aurantiirostris* (N lowlands and NW foothills): Greyish-olive to olive-brown above. White supercilium. **Adult** *parkesi* (Entre Ríos): Dull bill. Buff supercilium and throat. No tail spots. **Juvenile** *aurantiirostris*: Black or brown bill. Brown ear-coverts. No pectoral band. **Voice** Song is a strident series of 3–5 fairly high-pitched, metallic whistles ending in an emphatic upward-inflected note, *pi-tsiú-SWEEP*, with much variation but general pattern is very consistent. Calls include metallic twittering and a high-pitched *tsip.* [Pepitero de Collar]

Black-throated Grosbeak *Saltator fuliginosus*   23.5 cm

Scarce in n. Misiones. **ID** Large, long-tailed forest grosbeak of midstorey and subcanopy. **Adult** ♂: Huge coral-red bill. Mainly dark slate with a black face, throat and breast. **Adult** ♀ (not illustrated): Grey throat and breast. **Voice** Song is a slow warble of two well-spaced alternating phrases of low-pitched, rich, powerful whistles, *CHUWÍ-chu-WÉU... diru-PI-chuwí-you.* Call is a very nasal, drawn-out, complaining *keuh.* [Pepitero Negro]

Many-coloured Chaco Finch *Saltatricula multicolor*   17 cm

Dry and sierran chaco; local in S monte desert and humid chaco. **ID** Large terrestrial and arboreal finch of xerophytic woodlands with a graduated tail. **Adult:** Stout yellow bill with a blackish culmen. Black face extending in a stripe onto sides of neck. Sandy-brown above with a broad white terminal band to black tail. Grey breast and apricot-pink sides of breast and flanks. **Voice** Song is a fast, slightly metallic *vira-vira* or *chiruviruví* repeated over, often for long sessions. Calls include a thin high-pitched *zeet* and a resounding, repeated *chi-dik* or *che-dek.* [Pepitero Chico]

Thick-billed Saltator

ad ♀

ad ♂

juv

ad

Green-winged Saltator

juv

ad ♂

ad ♂

Golden-billed Saltator

nasica

aurantiirostris

ad ♂

ad

ad ♂

juv

parkesi

aurantiirostris

Greyish Saltator

ad ♂

Black-throated Grosbeak

ad

Many-coloured Chaco Finch

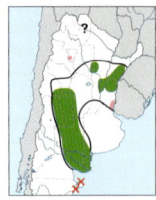

Yellow Cardinal *Gubernatrix cristata* 20 cm

Rare and local in Monte desert, Caldén and Espinal woodlands. **ID** Adult ♂: Mainly olive and yellow with black bib and crest. Adult ♀: White fore-supercilium and area surrounding bib. Grey cheeks and breast. Juvenile: Like ♀ but without the bib. **Voice** Song is an unhurried series of medium-pitched, rich, melodious whistles, often in short varied phrases of 3–6 notes, *pi-chu-pwi-tree… pwi-chew-tuee-chew…*, very much like that of Diuca Finch. Contact call, while perched and in flight, is a thin, high-pitched *sip* note, often repeated in succession. [Cardenal Amarillo]

Red-crested Cardinal *Paroaria coronata* 18.5 cm

Light woodlands in the N and C lowlands. **ID** Adult: Crimson foreface, throat and upstanding crest. Grey above and white below extending up sides of neck. Juvenile: Brown above with a warm brown or chestnut-brown face, throat and relictual crest. Sometimes occurs with Yellow and Yellow-billed Cardinals. **Voice** Song is a slowly delivered, rich melodious series of warbles, *chawa-WID cha-waro… chawee… cha-waro…* Also delivers a slow jumble of squeaks and whistles. Call is a nasal, rich *chwik*, resembling Ultramarine Grosbeak but sharper. [Cardenal Copete Rojo]

Yellow-billed Cardinal *Paroaria capitata* 17 cm

Gallery forest and marshes in the north; chiefly the NE. **ID** Adult: Pink or orange bill. Scarlet head. Glossy black back and bib. White below extending up sides of neck. Juvenile: Pale brown crown. Buffy ear-coverts and throat. Brown upperparts and white below. **Voice** Song consists of a steady repetition of 3–4 rising or falling notes in long phrases, *swi-chiú-wee-swi-chiú-wee…* or *chi-wit-chew-chu-chi-wit-chew-chu* repeated over and over, often recalling a warbling finch. Call is a nasal, metallic *tché*. [Cardenilla]

Black-backed Grosbeak *Pheucticus aureoventris* 21 cm

Yungas forest and C sierras; some winter across the chaco. **ID** A forest grosbeak with slight sexual dimorphism and huge bill. Adult ♂: Unmistakable with black upperparts, throat and breast; yellow below with striking white wing patches and tail corners. Adult ♀: Very similar but with blackish-brown upperparts. Juvenile: Pinkish lower mandible. Paler brown above than ♀; drabber yellow below with spotted throat and breast. **Voice** Song is a melodious, slow series of medium-pitched, varied warbles and drawn-out whistles, often ending in a repeated series of notes that switches between phrases, e.g. *chwe-chui-weeo-weeeu-didididid… chwe-chui-weeo-weeeu-twee-twee-twee…* Call is a brief, nasal *pink*. [Rey del Bosque]

Glaucous-blue Grosbeak *Cyanoloxia glaucocaerulea* 15 cm

NE gallery and Paraná forest. **ID** Adult ♂: Curved culmen and arched cutting edge; contrasting whitish lower mandible. Grey-blue to dull blue; brighter on forehead, cheek and shoulder. Brown flight feathers and tail, fringed dull blue. Adult ♀: Less contrasting whitish lower mandible. Brown above with darker brown wings and tail; sometimes warmer brown on rump. Warm brownish-buff below; often brighter on the flanks and vent. Overlaps with Blackish-blue Seedeater in winter (Plate 184). **Voice** Song is a very fast, erratic, somewhat scratchy warble lasting 2–3 secs. Contact call is an emphatic, low-pitched, metallic *SPINK*. [Reinamora Chica]

Ultramarine Grosbeak *Cyanoloxia brissonii* 16–17.5 cm

Widespread. **ID** Adult ♂ *argentina* (17.5 cm; N lowlands west of the Río Paraná, C sierras and NW Andes below 1600 m): Pale spot at base of blackish bill. Indigo with a brighter sky-blue forehead, cheek patch and shoulder. Adult ♀ *argentina*: Warm brown above; often rufescent on the rump. Dark brown wings and tail. Ochraceous-buff below. Adult ♂ *sterea* (16 cm; Paraná forest and gallery forest in Mesopotamia): Smaller-billed than *argentina* with a more rounded culmen; considerably darker plumage with reduced sky-blue forehead, cheek patch and shoulder. Adult ♀ *sterea*: Much darker brown above than *argentina* without rufescent tones. Darker tawny-brown below. **Voice** Song is a cheerful series of rather similarly-pitched warbles, with some more emphatic notes intermixed, *si-see-swee-si-sui-psíu-su-si-su-si-su*, often a little faster towards the end. Resembles Blackish-blue Seedeater but slower and lower-pitched. Call is a nasal *chwid* or *weend*. [Reinamora Grande]

Yellow Cardinal
ad ♀
ad ♂

Red-crested Cardinal
juv
ad

Yellow-billed Cardinal
juv
capitata
ad

ad ♂
juv
aureoventris
Black-backed Grosbeak

argentina
ad ♀
ad ♂
ad ♂
sterea
ad ♀

ad ♂
ad ♀
Glaucous-blue
Grosbeak

Ultramarine
Grosbeak

Atlapetes Large Andean understorey brushfinches, usually found in pairs. Sexes alike, but males are brighter. Song structure is surprisingly similar, but the ranges of the two species do not overlap.

Yellow-striped Brushfinch *Atlapetes citrinellus*　　　　16.5 cm

Endemic in Yungas forest from se. Jujuy to Catamarca (mainly 1200–2400m, lower in winter). **ID** **Adult**: Striking head pattern; black face, crossed by yellow supercilium and moustachial stripe, bordered by black malar streak. Mantle, wings and tail olive. Pale olive below with a yellowish centre to the abdomen. **Voice** Song is a fast 4–5 note, somewhat metallic, high-pitched phrase, *t-t-chu-TSIÚ-TSIÚ-TSIÚ*, sometimes alternating with a much faster phrase, *zee-jijijijijiji*. Calls include a sharp, high-pitched *tip* note, a scolding *tchac* and some wheezy notes. [Cerquero Amarillo]

Fulvous-headed Brushfinch *Atlapetes fulviceps*　　　　16 cm

Yungas forest of Jujuy and Salta (mainly 1400–2800m, lower in winter). **ID** **Adult**: Striking chestnut head with yellow loral spot and moustachial bordered by a chestnut malar streak. Plumage otherwise mostly olive with a yellowish centre to the abdomen. Compare with Rust-and-yellow Tanager (Plate 173). **Juvenile** (not illustrated): Differs chiefly by dull brown crown and ear-coverts with an indistinct yellowish supraloral spot and dark olive upperparts. **Voice** Song is a fast 4–5 note, somewhat metallic, high-pitched phrase, *chu-wi-TIÚ-TIÚ-TIÚ*, sometimes alternating with a much faster phrase, *chi-zidididi*. Calls include a sharp, high-pitched *tip* note, a buzzy *pit-ZEEUU*, and rarely some wheezy notes. [Cerquero Cabeza Castaña]

Arremon Understorey forest sparrows and brushfinches. High-pitched piercing songs.

Saffron-billed Sparrow *Arremon flavirostris*　　　　18 cm

Mainly humid chaco, n. and w. Corrientes, locally in w. Misiones and south along the Paraná to nw. Entre Ríos. **ID** **Adult**: Bright yellow bill. Black head and white supercilium that does not reach the bill. Grey back with contrasting olive wing-coverts. White below with a notable black pectoral band. **Juvenile**: Resembles adult but bill blackish, pectoral band narrower and brown, and throat and breast washed dull buff. **Voice** Song is a steady high-pitched, insect-like, piercing repetitive phrase with a faster beginning, *TSTSTS-TSI-TSI-TSI-tsi-tsi-tsi*. Calls include a very high-pitched shrill *zeet*. [Cerquero de Collar]

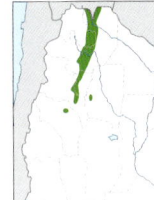

Moss-backed Sparrow *Arremon dorbignii*　　　　16 cm

Yungas forest to 1600 m. **ID** **Adult**: Recalls and overlaps with White-browed Brushfinch. Differs by its orange or yellow bill, grey upper back, and whiter underparts with a narrow black pectoral band. Note also the grey coronal stripe, full supercilium and narrow pectoral band compared to Saffron-billed Sparrow. **Voice** Song is a rhythmic series of varied high-pitched, shrill metallic notes, *tsit-tst-chr-tsit-zrr-tss*. Calls include very high-pitched *tsip* and *zwit* notes. Tax note 65. [Cerquero Musgoso]

White-browed Brushfinch *Arremon torquatus*　　　　17 cm

Yungas forest of Jujuy and Salta (mainly 1200–2000 m; lower in extreme NW and in winter). **ID** **Adult**: Stout black bill. Black head with grey coronal stripe and bold white supercilium. Olive back and white underparts, washed grey on the breast and dull olive-brown on the vent. Overlaps with Moss-backed Sparrow. **Juvenile**: Chestnut-brown crown and dusky face with a striking white moustachial spot and dusky malar. Long narrow cinnamon supercilium, and short coronal stripe. Drab bronze-olive above and mainly greyish below with a buffy throat. **Voice** Song alternates between two well-spaced phrases of high-pitched, thin, wheezy notes, e.g. *zip-zee-zirip-zeee... zwi-si-zeedoo-zi-dipzeee*; differs radically from Moss-backed Sparrow by its longer notes and more leisurely pace. Call is a very high-pitched, metallic *spit* or *sit*. [Cerquero Vientre Blanco]

Yellow-striped Brushfinch

ad

Fulvous-headed Brushfinch

ad

polionotus

ad

Moss-backed Sparrow

juv

ad

Saffron-billed Sparrow

ad

juv

borellii

White-browed Brushfinch

Rufous-collared Sparrow *Zonotrichia capensis* 15.5–16.5 cm

Very widespread; with 8 ssp. **ID** Ubiquitous sparrow with a permanent peaked crest. Gregarious, especially in winter. **Adult** *hypoleuca* (15.5 cm; NC lowlands): Grey crown with black lateral crown stripes, eye-stripe and line bordering ear-coverts. Contrasting rufous nape and sides of the breast with a black spot. Numerous similar subspecies. **Adult** *australis* (16.5 cm; S Patagonia, migrating to N lowlands and Andes): Larger. Lacks black head stripes. **Juvenile** *hypoleuca*: Lacks rufous on the nape and breast sides. Only slightly crested with less distinct brown or blackish head stripes than adult. White below, tinged buff on breast and finely spotted black. **Voice** Geographically variable. Common song-type consists of 2–3 musical whistles and a trill, *see-swee-siuuu-tritrtrtrtrtrtr*. In the Monte desert the trill is often replaced by a short rasp. Song of *australis* lacks the trill or rasp and only has whistles, *chi chiú chiú chu-chu-chu*. At night can give unpredictable, isolated complex songs. [Chingolo]

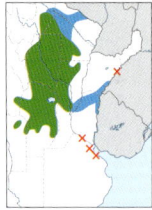

Chaco Sparrow *Rhynchospiza strigiceps* 15.5 cm

Endemic breeder; mainly in dry and sierran chaco to 1500 m. **ID** Long-tailed, stripe-headed sparrow with a stout bicoloured bill. Often in small flocks. **Adult**: Much smaller than Yungas Sparrow (no overlap). Fine black loral line. Rufous eye-stripe and lateral crown stripes with fine blackish crown streaks. **Voice** Song is composed of distinct musical phrases repeated at even intervals before switching to another phrase. Each phrase can be a simple trill or contain an introductory note followed by the repetition of a single note or of a multi-note pattern, e.g. *tching tching tch-tch-tch-tch or zee-zee didididdidi*. Calls include rapid liquid chatters, a metallic *tsink* or *chink*, and a high-pitched *tsuee*. [Alt: Stripe-capped Sparrow] [Cachilo Chaqueño]

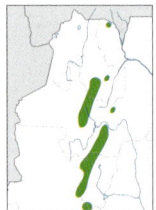

Yungas Sparrow *Rhynchospiza dabbenei* 19 cm

NW Andean brushland and forest borders to 1650 m; rarely higher. **ID** Much like Chaco Sparrow, but differs by larger size, loral pattern and chestnut tones. Often in small flocks. **Adult**: Grey coronal stripe, edged by chestnut lateral crown stripes. Black triangular loral patch. Broad grey supercilium and narrow chestnut eye-stripe. Streaked back and rufous shoulder. White moustachial bordered by black malar streak. Pale grey below with buff belly and flanks. **Voice** Song (mostly at dawn) is an unmusical long series of evenly-spaced, varied, short, sharp chirping notes that are repeated haphazardly, sometimes for minutes on end. Calls include rapid liquid chatters, a metallic *tsip* or *chiú* and a high-pitched *tseeeu*. **Tax note 66**. [Alt: Stripe-capped Sparrow] [Cachilo Yungueño]

Grassland Sparrow *Ammodramus humeralis* 13.5 cm

N and C lowlands, Andean foothills to 1550 m. **ID** Tiny inconspicuous open country sparrow. Not gregarious. **Adult**: Pale yellow lores and carpal visible at close range. Greyer above than juvenile. Unmarked greyish-white underparts. **Juvenile**: Brown crown, finely streaked black. Rather plain-faced with no appreciable markings. Brownish back, streaked black. Two narrow white wing bars. Buffy breast and white belly; lower throat and breast with diffuse spots or streaks. **Voice** Song has a short introduction followed by 1–3 insect-like, differently pitched, wheezy, raspy trills, e.g. *tip tsrree-schreeee* or *tsip-tsrreeee chi-chi-chi*. [Cachilo Ceja Amarilla]

Blue-black Grassquit *Volatinia jacarina* 11.5 cm

N lowlands (Sep–Apr). **ID** A small austral migrant grassquit of open country with glossy-plumaged ♂ and drab ♀. **Adult** ♂ breeding: Glossy blue-back. White underwing-coverts exposed during wing-flicking and short vertical display flight. **Adult** ♂ moulting: Variably mottled with brown above and grey below. **Adult** ♀: Brown above with slightly darker wings and tail. Whitish throat and dull buff below, becoming white in centre of belly; streaked brown on breast and flanks. **Voice** Song, during flight display, is a short metallic buzz followed by a lower-pitched trill, sometimes preceded by a rapid ticking series, *ti-ti-ti TZZ-TREE*. [Volatinero]

Rufous-collared Sparrow

ad

australis

ad

juv

hypoleuca

Chaco Sparrow

ad

ad

Yungas Sparrow

xanthorus

ad

juv

Grassland Sparrow

ad ♂

♂ display

breeding

ad ♂

ad ♀

moulting

jacarina

Blue-black Grassquit

Microspingus, *Poospiza* and *Poospizopsis* Small, restless finches with varying amounts of white in the outer tail feathers (tail spots). Songs are simple and monotonous or short scratchy phrases. All species join mixed flocks.

Grey-throated Warbling Finch *Microspingus cabanisi* 14 cm
Paraná forest and gallery forest in e. Corrientes, e. Entre Ríos and ne. Buenos Aires. **ID** Adult: Grey head with narrow white supercilium and weak moustachial. Reddish-orange flanks and bright rufous rump. Juvenile: Olive head with a yellowish supercilium and moustachial. Flanks and rump as adult. **Voice** Song has warbling finch quality, but is a relatively complex, hesitant, series of high-pitched whistles and shorter notes, e.g. *che-whitch-sip-it-spiú-slip-chd-seee*. Calls include a metallic *tsip* and a hasher *ZRiP*. [Monterita Litoraleña]

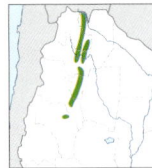

Rusty-browed Warbling Finch *Microspingus erythrophrys* 14 cm
Yungas forest to 3000 m. **ID** Adult: Grey crown and ear-coverts divided by narrow rusty supercilium. Broad white wing bar and wing panel. Mostly rich rufous below. Juvenile: Cinnamon supercilium and wing bar, and white wing panel. Cinnamon throat and breast, becoming paler below. **Voice** Song is a rich, fairly slow, medium-pitched, repetitive *chi CHUWEE siu*, sometimes faster and with double-notes, *swi CHA-CHA-SWEE*. Calls include a thin, high-pitched *szit*, and a sharp metallic *SPIT*. [Monterita Ceja Rojiza]

Black-and-rufous Warbling Finch *Poospiza nigrorufa* 15.5 cm
NE lowlands, often in marshy habitats. **ID** Adult ♂: Blackish ear-coverts bordered by a white moustachial and supercilium, turning rufous behind eye. Orange-rufous below with a white central abdomen. Adult ♀: Very similar; a little duller above and paler orange below. Juvenile: Blackish-brown to olive-brown above; blotchy blackish or brown streaking below. **Voice** Song is a simple and distinctive whistled phrase *pleased-to-meet-you*, repeated over and over. Calls include a high-pitched *see*. [Sietevestidos Pampeano]

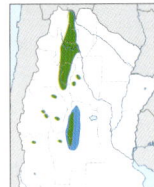

Black-and-chestnut Warbling Finch *Poospiza whitii* 14 cm
Scrub and light woodlands in Andean foothills and sierran chaco woodlands. **ID** Adult ♂: Resembles Black-and-rufous (no overlap) but deep maroon-chestnut below, and larger white tail spots. Adult ♀: Resembles Black-and-rufous, except for larger tail spots. Juvenile: Resembles juvenile Black-and-rufous although decidedly more olive above. **Voice** Song is a rhythmic, complex, 8–12 note, whistled phrase that includes couplets and trills, *choo we tip-tip sweet peer tweak trrree sweet peer*; abbreviated versions can resemble song of Black-and-rufous Warbling Finch. Calls include a high-pitched *zhee*. Tax note 67. [Sietevestidos Serrano]

Bolivian Warbling Finch *Poospiza boliviana* 14.5 cm
Very local in scrub and cultivation at 2650–3350 m mainly in n. Salta. Often found with Rufous-sided Warbling Finch. **ID** Adult: Long white supercilium and well-defined white throat. Contrasting orange-red pectoral band and flanks. White central abdomen and undertail-coverts. **Voice** Song is a sprightly, rich, simple, repetitive 2–3 note pattern, *we-CHUWEE we-CHUWEE...* or a slightly more complex *TWEEtsi TWEEtsi TWEEtsi Twi-twi-twi*. Call is a thin high-pitched *tsit*. [Monterita Boliviana]

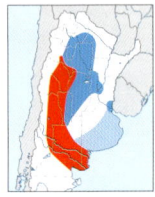

Cinnamon Warbling Finch *Poospiza ornata* 13.5 cm
Endemic breeder in Monte desert. **ID** Adult ♂: Grey crown and cinnamon supercilium. Maroon saddle and two broad buffy-white wing bars. Cinnamon-buff below with a deep chestnut pectoral band. Saddle and breast mottled in worn birds. Adult ♀: Grey-brown crown and brown mantle. Ochraceous below becoming buff on centre of belly. Immature: Biscuit-brown above with a cream supercilium, and buff or white wing bars. Buff below, streaked brown on breast. Juvenile: More extensive, thicker ventral streaking. **Voice** Song is a rapid, short, sharp, liquid phrase, e.g. *waCHÍCHUrípCHUrchip*, with much variation (some surprisingly fast), but often in couplets or triplets. Does not give trills, unlike Chaco Warbling Finch. Call includes thin, very high-pitched *sip* or *sit*. [Monterita Canela]

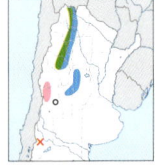

Rufous-sided Warbling Finch *Poospizopsis hypochondria* 16.5 cm
NW Andean scrub at 1700–4000 m; winter visitor in C Sierras. **ID** Adult: Dark grey crown and face with white supercilium and moustachial, bordering short blackish malar. Whitish throat, washed grey on breast or only at sides; buffy below with rufous flanks. Juvenile: Brown crown with buffy moustachial and shorter supercilium, and greyish malar. Greyish-brown wash on breast; cinnamon-buff below and pale rufous flanks. Tail spots much smaller than adult. **Voice** Song is a medium-pitched, repetitive, 2-note phrase *chup WI chup WI...* or *chrup WI chrup WI...* [Monterita Pecho Gris]

Grey-throated Warbling Finch

juv

ad

Rusty-browed Warbling Finch

erythrophrys

ad

juv

juv

Black-and-rufous Warbling Finch

ad ♂

Black-and-chestnut Warbling Finch

whitii

ad ♂

ad

imm

Cinnamon Warbling Finch

Bolivian Warbling Finch

ad ♀

ad

ad ♂

affinis

juv

Rufous-sided
Warbling Finch

Microspingus Small, restless finches with varying amounts of white in the outer tail feathers (tail spots). Songs are simple and monotonous or short scratchy phrases. All species join mixed flocks. See also Plate 181.

Black-capped Warbling Finch *Microspingus melanoleucus*   13 cm

N lowlands and foothills, except Misiones. **ID Adult**: Grey and white with a black cap, red iris and large white tail spots. **Juvenile**: Blackish mask; tinged buff below especially on flanks. Compare with Masked Gnatcatcher (Plate 163). **Voice** Song at dawn is a repetitive thin musical warble of 3–4 notes, e.g. *tew-chewí-tiú-tew-chewí-tiú...* Calls include a duetted, very fast jumble of scratchy and squeaky notes lasting up to 7 secs, and a sharp metallic *SPIT*. [Monterita Cabeza Negra]

Chaco Warbling Finch *Microspingus* [*torquatus*] *pectoralis*   13 cm

Endemic breeder? Dry and sierran chaco, and Monte desert. **ID Adult**: Grey and white with black ear-coverts and a white supercilium. Black pectoral band and chestnut undertail-coverts. **Juvenile**: Dark brown ear-coverts and buff pectoral band. Immatures show a brown or grey pectoral band. **Voice** Song at dawn is a rhythmic, fairy high-pitched, sweet series of clearly rising or falling whistles, always including a slow trill, e.g. s*wee-chwee-diuuu-swit-chiweep-trrr-sit*. Call is a thin *sip* note. **Tax note 68**. [Monterita de Collar]

Red Pileated Finch *Coryphospingus cucullatus*   13.5 cm

Widespread. **ID** Understorey finch with an erectile crest, white eye-ring and a stout conical whitish bill. **Adult ♂ *rubescens*** (forest and scrub in Mesopotamia): Crimson crest bordered by black lateral crown stripes. Dark brown above and deep red below. **Adult ♂ *fargoi*** (Chaco and Yungas forest to 1700 m): Mostly brown above and rosy-red below. **Adult ♀ *rubescens***: Similar to ♂ *fargoi* but without the head markings. **Adult ♀ *fargoi***: Biscuit-brown above with a chestnut rump and dark brown tail. Whitish lores. Buffy throat and pinkish-brown below. **Voice** Song is a medium-pitched, repeated phrase *wher cho wher cho...* or *wher chido wher chido....*, occasionally with extra notes thrown in; may resemble *Tachyphonus* tanagers. Call is a very high-pitched *tsi* or *pip*. [Brasita de Fuego]

Coal-crested Finch *Charitospiza eucosma*   11 cm

Single records from n. Misiones and Tucumán. **ID** Small arboreal and terrestrial finch with a stout bill and erectile crest. **Adult ♂**: Black foreface, crest, throat and centre of breast; bordered deep chestnut below. Large white oval cheek patch. Underparts mostly orange. Small white patch in wing, white rump and bases to outer tail form a striking pattern in flight. **Adult ♀**: Brown crest, whitish throat and sandy-brown below with buff undertail-coverts. Tail as ♂ but no wing patch. [Afrechero Canela]

Black-crested Finch *Lophospingus pusillus*   13 cm

Mainly dry chaco south in the C lowlands to n. San Luis. **ID** Plumage and white tail spots suggest a warbling finch but far more terrestrial and with an upstanding crest. **Adult ♂**: Black crest, thick eye-stripe and bib, highlighted by a broad white supercilium and broad moustachial. **Adult ♀**: Dark brown crest and eye-stripe. Unmarked white throat. **Juvenile** (not illustrated): Like ♀ but with fine breast and flank streaking. **Voice** Song is a melodious warble, recalling Many-coloured Chaco Finch. Calls include a soft *chip, chip-up*, long liquid chatters and a thin high-pitched *tsit* call. [Soldadito Chaqueño]

Diuca Finch *Diuca diuca*   15.5 cm

C lowlands and Patagonia (*minor*) with a similar but darker ssp. in the S Andes. See Plate 190 for high NW Andean ssp. (*crassirostris*). **ID** Widespread chunky finch with a short conical bill. Migrates northwards in winter. Often gregarious. **Adult *minor***: Essentially grey above with a peaked crown. Large white tail spots. White below with a broad grey pectoral band and small rusty flank patch. **Juvenile** (not illustrated): Browner above with an ill-defined pectoral band and buffy belly. **Voice** Song, mainly at dawn, is a mellow, measured *SCHup-CHI... SCHup CHIWol*, or *SCHUp-CHup-CHi.. SCHUp-CHoo* repeated over; often much like Yellow Cardinal. Calls include a variety of *schup, kwip* and *chip* notes. [Alt. Common Diuca Finch] [Diuca]

juv

ad

Chaco Warbling Finch

ad

juv

Black-capped Warbling Finch

fargoi

rubescens

ad ♂

ad ♂

fargoi

ad ♀

ad ♀

Red Pileated Finch

rubescens

ad ♀

ad ♂

Coal-crested Finch

ad ♀

ad ♂

ad

Black-crested Finch

minor

Diuca Finch

Long-tailed Reed Finch *Donacospiza albifrons* 13 cm

NE marshes. **ID** A slender marsh finch with a fairly long graduated tail with pointed tips. Perches upright. **Adult:** Grey face with a short buff or whitish supercilium. Brown above, streaked black on back. Ochraceous below, becoming buff on the belly. **Juvenile:** Mostly pale buff with fine streaking. **Voice** Song, is a medium-pitched, fast, warbling finch-like *tchu-TCHIP tchu-TCHIP tchu-TCHIP....* Calls include a buzzy *czep* or *djep* and a grating *gerp-gerp.* [Cachilo Canela]

Black-masked Finch *Coryphaspiza melanotis* 13 cm

Very rare in ne. Corrientes, ne. Santa Fe and sw. Misiones. **ID** Small reclusive finch of dry grasslands with a short, graduated tail and insect-like song. **Adult** ♂: Orange bill with a black upper mandible. Black head and narrow white supercilium. Back streaked dusky and buff. Broad white tail spots best seen in flight. Black spur on sides of breast and long black flank streaks. **Adult** ♀: Brown head, finely streaked buff. Short yellowish supercilium. Compare with Wedge-tailed and Lesser Grass Finches. **Voice** Song is a fast, high-pitched, metallic chatter followed by an insect-like buzz, *ZIZIZIZI-bzz* or *ZEZE-bzzbzz.* Call is a very high-pitched, insect-like *TZ* or *TZ-tz.* [Cachilo de Antifaz]

Great Pampa Finch *Embernagra platensis* 21 cm

Widespread. **ID** Large, conspicuous marsh finch with a very stout orange bill and long broad tail. **Adult** *olivascens* (NW Andes to 3150 m and W lowlands south to nw. Neuquén): Orange or reddish-orange bill with a curved culmen where black does not reach tip. Usually dull olive-brown above without dorsal streaking, and greyer ear-coverts. Flanks extensively washed pale cinnamon-brown. **Adult** *platensis* (NE lowlands to s. Buenos Aires): Orange bill with straighter black culmen. Grey face with blackish lores. Olive above with thick blackish dorsal streaking. Grey throat and breast, becoming whiter below. Pale cinnamon wash on flanks and vent. **Juvenile:** Buffy-olive above with thick blackish streaking. Pale yellowish supercilium and throat. Whitish below with black streaking on breast and flanks, and a cinnamon wash on the lower underparts. **Voice** Song is a fast, thin, short shrill warble, sometimes with metallic buzzes, e.g. *swi-zilip-zit-TZÍU* or *chu-zí-si-whizz,* with much geographic variation. Calls include diverse high-pitched notes and liquid metallic couplets. [Verdón]

Wedge-tailed Grass Finch *Emberizoides herbicola* 20 cm

NE marshes (overlaps with Lesser Grass Finch). **ID** Fairly large finch of humid grasslands with stout bill and long graduated tail. **Adult:** Orange bill. Broad whitish eye-ring. Brownish crown, ear-coverts and back; latter streaked black but rump unmarked. Brown tertials. White throat and buffy underparts, often paler in centre of belly. **Juvenile** (not illustrated): Differs by yellowish underparts and narrow yellowish supercilium ending just behind the eye. **Voice** Song is a rich, mellow series of 2–3 well-spaced, musical phrases, *di DU-LEEO... dji TWI-LEE... jew-i di-LEEE...,* alternating between high-pitched and low-pitched endings.. [Coludo Grande]

Lesser Grass Finch *Emberizoides ypiranganus* 19.5 cm

NE marshes (overlaps with Wedge-tailed Grass Finch). **ID Adult:** Slimmer build than Wedge-tailed. Orange-yellow bill. Narrow white eye-ring. Grey or olive crown and ear-coverts, finely streaked black. Fairly bright olive upperparts with crisper, thicker black streaking than Wedge-tailed, extending over rump. Black tertials unlike Wedge-tailed. Clean white underparts, with contrasting grey-brown flanks and undertail-coverts which are finely streaked black. **Voice** Song is a steady, harsh, grating unmusical series, sometimes with stuttering rhythmic endings, e.g. *dji-dj-dj-dj-dj-dji* or *ch-ch-ch-ch-ch´d-ch-d´d´d´.* Very diagnostic and repetitive, but may recall some phrases of Grass Wren. [Coludo Chico]

Long-tailed Reed Finch

ad

juv

ad ♂

Black-masked Finch

ad ♀

olivascens

ad

ad

juv

Great Pampa Finch

platensis

ad

ad

herbicola

Wedge-tailed Grass Finch

Lesser Grass Finch

Sporophila Large Paraná forest seedeaters. Nomadic Temminck's and Buffy-fronted have swollen bills and arched cutting edge with small upper mandible; they perch high, mostly in tall *Guadua* bamboo and are best detected by their loud, explosive and complex songs.

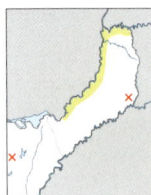

Temminck's Seedeater *Sporophila falcirostris*　　　　12 cm

Sporadic in Misiones; can be locally abundant. **ID Adult ♂**: Yellow bill. Dark grey above with brown flight feathers. Small white spot in primaries. Black-centred median-coverts. Pale grey below with a white centre to the belly and buff undertail-coverts. Can breed in a brown ♀-type plumage, but always shows yellow bill. **Adult ♀**: Grey bill. Olive to olive-brown above with a buffy or grey-brown face. No wing spot. Variable below, but with a paler belly and vent. **Voice** Song consists of two consecutive, very high-pitched, piercing/hissing phrases, *se-se-se-se-se-sisisisisisi*, sometimes followed by repetition of softer musical notes in simple or complex patterns. Calls include a loud, piercing, high-pitched trill, *TRRRRSSSS*, often preceding the song, and a high-pitched *TSEEU*. [Corbatita Picudo]

Buffy-fronted Seedeater *Sporophila frontalis*　　　　13 cm

Very rare and sporadic (unconfirmed) in Misiones. **ID Adult ♂**: Swollen orangey bill. Grey crown and face. White forehead and narrow postocular streak; sometimes lacking. Two whitish or buff wing bars. Small white spot in primaries. Whitish below, with a short dusky malar streak. Brown to olive wash on sides of breast and flanks. **Adult ♀**: Lacks head markings and wing spot of ♂. More olive above with two narrow buff wing bars. Yellowish throat, washed olive across breast and flanks. Creamy centre to belly. **Voice** Song is a unique, explosive, extremely loud, medium-pitched short series of usually 3–4 notes, *PITCHOCHÓ* or *CHOCHOCHOCHOTZI*, sometimes followed by softer musical notes audible at close range. Calls include loud single *CHIP* or *TEW* repeated. [Pichochó]

Blackish-blue Seedeater *Amaurospiza moesta*　　　　13 cm

Paraná forest. **ID** Forest seedeater with white underwing and a laterally broad bill. Feeds on bamboo shoots and leaves. **Adult ♂**: Glossy dark blue with a blackish forehead, lores, throat and breast. Dull blue below. **Adult ♀**: Ferruginous-brown above with a warmer rump. Brown wings and tail. Paler tawny below. Distinguished from ♀ Glaucous-blue Grosbeak by dark bill and from ♀ Ultramarine Grosbeak by smaller and notably wide bill. **Voice** Song is a melodious warble, recalling that of Ultramarine Grosbeak but generally shorter, faster and higher-pitched. [Reinamora Enana]

Sooty Grassquit *Asemospiza fuliginosa*　　　　11.5 cm

Sporadic in n. Misiones. **ID** A narrow-tailed grassquit with a broad-based, almost conical bill with only a shallow upper mandible and a bright orange or pink fleshy gape around the base of the bill. Perches high in *Guadua* bamboo or low in Paraná forest understorey. **Adult ♂**: Black bill. Pinkish-grey tarsus. Mostly blackish, tinged olive on the back. Dark brown wings and tail, and grey belly. **Adult ♀**: Dusky bill with variable yellow on lower mandible. Pinkish tarsus. Greyish-olive above becoming slightly paler on the rump. Contrasting brown flight feathers and tail, with olive fringes which are brighter than the back when fresh. Paler underparts becoming greyish-buff on the belly. Buff undertail-coverts, sometimes bordered white. **Immature ♂**: Dark grey bill with yellowish lower mandible. Greyish-olive above with a slightly paler rump. Dark brown wings and tail. Entirely brown above with a very slight olive tinge. Slightly darker throat and breast contrast with dull buff belly and vent. Central abdomen vaguely streaked brown. **Voice** Song is a fast, high-pitched, scratchy, insect-like metallic phrase, often followed by a stutter and rarely also an extremely high-pitched (barely audible) trill and a final whistle, *zir-slip-tr-ink-zidididi-zwwwwww-zréu*, lasting only 2 secs. [Espiguero Negro]

Uniform Finch *Haplospiza unicolor*　　　　13 cm

Paraná forest edge. **ID** Understorey finch with long conical bill; associated with bamboo seeding (mostly *Chusquea*) and grasses at forest edge. Often in mixed flocks. **Adult ♂**: Black bill and pinkish legs. Dull grey-blue above with a paler grey face and indistinct supercilium. Dark brown wings and tail. Pale ash-grey below, sometimes bluer on the breast. **Adult ♀**: Olive above with an indistinct paler supercilium. Grey-blue fringes to lesser and median coverts. Dull white to yellowish-buff below, with diffuse blurry brown breast streaking. **Immature ♂**: Pale grey head, otherwise mixed olive and grey. Indistinct supercilium. Bill and wings resemble ♀. Reduced ventral streaking. **Voice** Song is a somewhat explosive, very high-pitched, fast and brief metallic wheeze *zee-zr-zeee*, sometimes followed by high-pitched monotone trills or more varied shrill notes. [Afrechero Plomizo]

Temminck's Seedeater

ad ♂

ad ♀

ad ♂

ad ♀

Buffy-fronted Seedeater

ad ♂

ad ♀

Blackish-blue Seedeater

ad ♀

imm ♂

ad ♂

ad ♀

imm ♂

Sooty Grassquit

ad ♂

Uniform Finch

Sporophila Small migratory, sexually dimorphic tanager-finches. Bill colour and shape, and wing pattern can aid identification of drab ♀♀. The colourful 'capuchino' group on this plate has polymorphic males, a small white patch at the base of the primaries in the folded wing, and the females are largely indistinguishable. Vocalisations comprise measured, complex series of inflected whistles, chirps and churrs or rasps, which change over time. Widespread species exhibit vocal geographical variation. See also Plates 186 and 187.

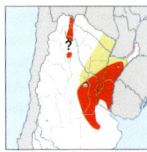

Dark-throated Seedeater *Sporophila ruficollis* 10.5 cm

Local in the N lowlands (Oct–Apr). Found in drier habitats. **ID Adult** ♂: Grey above with small white primary patch. Rufous rump and underparts. Contrasting black or dark face, throat and upper breast. Collared '*caraguata*' morph ♂ (Rare in e. Entre Ríos): Black collar and rufous back. **Adult** ♀: As ♀ Rufous-rumped Seedeater. [Capuchino Garganta Café]

Iberá Seedeater *Sporophila* sp. 11 cm

Iberá marshes, Corrientes (Oct–Mar). **ID Adult** ♂: Recalls a washed-out Dark-throated Seedeater, but note buff underparts and grey restricted to cap. **Adult** ♀: As ♀ Rufous-rumped Seedeater. **Tax note 69**. [Capuchino Iberá]

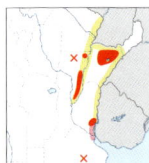

Rufous-rumped Seedeater *Sporophila hypochroma* 11 cm

NE marshes (Oct–Mar). **ID Adult** ♂: As Tawny-bellied Seedeater but rump and underparts deep rufous. **Adult** ♀: Olive-brown above and buff below with a small white primary patch. [Capuchino Castaño]

Tawny-bellied Seedeater *Sporophila hypoxantha* 11 cm

Chiefly in NE grasslands (Oct–Feb). **ID Adult** ♂: Grey above with rufous rump and small white primary patch. Rich cinnamon below. Many darker individuals are difficult to distinguish from Rufous-rumped Seedeater except by song. Pale-backed '*uruguaya*' morph (Rare; not illustrated): Nape and back concolorous with throat and belly. **Adult** ♀: As ♀ Rufous-rumped Seedeater. [Capuchino Canela]

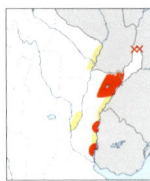

Chestnut Seedeater *Sporophila cinnamomea* 11 cm

Mainly Corrientes and ce. Entre Ríos (Oct–Apr). Usually in rolling terrain with sandy soil; to a lesser extent in marshes. **ID Adult** ♂: Rather bright chestnut with a grey cap and uppertail-coverts. Small white primary patch. **Adult** ♀: As ♀ Rufous-rumped Seedeater. [Capuchino Corona Gris]

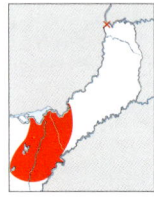

Pearly-bellied Seedeater *Sporophila pileata* 11 cm

Campos of s. Misiones to ne. Corrientes (Oct–Mar). **ID Adult** ♂: Black cap. Brown above with small white primary patch. White cheeks and underparts, often washed buff. Pale-backed '*andorinha*' morph (Rare; not illustrated): Nape and back concolorous with throat and belly. **Adult** ♀: From other ♀ seedeaters with a small white primary patch by whitish fringes to dark-centred wing-coverts and tertials. [Capuchino Boina Negra]

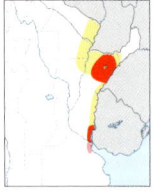

Marsh Seedeater *Sporophila palustris* 11 cm

NE marshes (Oct–Mar). **ID Adult** ♂: Grey crown, back and uppertail-coverts. White cheeks, throat, upper breast and small primary patch. Rufous rump and belly. Collared '*zelichi*' morph ♂ (Rare in Corrientes–e. Entre Ríos): From normal adult by white collar and rufous back. **Adult** ♀: As ♀ Rufous-rumped Seedeater. [Capuchino Pecho Blanco]

ad ♂

Dark-throated
Seedeater

collared
'*caraguata*'
morph

ad ♂

Iberá Seedeater

ad ♂

ad ♀

ad ♂

Tawny-bellied
Seedeater

Rufous-rumped
Seedeater

ad ♂

Chestnut
Seedeater

ad ♂

Pearly-bellied
Seedeater

ad ♀

ad ♂

ad ♂

collared
'*zelichi*'
morph

Marsh Seedeater

Sporophila Small migratory, sexually dimorphic tanager-finches, although larger, structurally different White-bellied Seedeater and Rusty-collared Seedeater are resident. Bill colour and shape, and wing pattern can aid identification of drab ♀♀. See also Plates 185 and 187. Chestnut-bellied is associated with forest marshes and riparian forests. ***Asemospiza*** Dull-coloured Grassquit is often confused with ♀ *Sporophila* seedeaters.

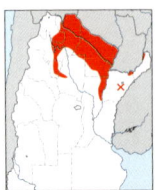

Lined Seedeater *Sporophila lineola* 11 cm

Mainly Andean foothills and chaco (Oct–Apr). **ID** Adult ♂: Glossy black above. White coronal patch and malar offset by black bib. Small white primary patch, narrow white rump and white underparts. Adult ♀ (not illustrated): Like ♀ Double-collared; perhaps separable by its slightly paler rump. **Voice** Song is a very rapidly delivered, uncomplicated, medium-pitched, bubbling or trilled sequence, *dididolodudududididiDIDÚ* in 2 secs, gradually rising in pitch and volume with emphatic final notes; sometimes followed by more complex rapid musical warbles. [Corbatita Overo]

Double-collared Seedeater *Sporophila caerulescens* 11 cm

Widespread in the N and C lowlands and foothills to 1500 m, locally higher (Oct–May). **ID** Adult ♂: Yellowish bill. Grey above with blackish lores, black bib and pectoral band. Adult ♀: Olive-brown above and buffy below; sometimes with a slight olive wash on the breast. **Voice** Song is a very rapid series of short, shrill whistles and scratchy notes, lasting 1.5–2 secs; generally descending in pitch and ending in emphatic notes, but highly variable. Calls include high-pitched and short nasal notes [Corbatita]

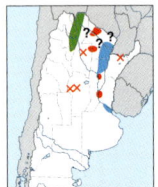

Dull-coloured Grassquit *Asemospiza obscurus* 11.5 cm

Mostly in the NW foothills and N chaco. **ID** A small finch-like thraupid of grassland clearings, often confused with ♀ *Sporophila* seedeaters. **Adult**: Brown above with narrow white eye-ring. Greyish cheeks, sometimes extending over the crown and nape suggesting a hood. Brownish throat, breast and flanks with a paler belly and vent. Note fleshy tarsus and bicoloured bill with fairly straight culmen; grey above and pink below. **Voice** Song is a fast, high-pitched, glassy, insect-like metallic phrase, repeated at short intervals, e.g. *za-zí-zlu-zer-dlree*. [Espiguero Pardo]

White-bellied Seedeater *Sporophila leucoptera* 13 cm

Resident in humid chaco; local in ne. Santa Fe and n. Corrientes. **ID** Adult ♂: Swollen pink bill. Slate-grey above with blackish wings and tail. Small white primary patch and white underparts. **Adult** ♀: Olive-brown above; paler buffy-brown below with a whiter belly. Primary patch and bill as ♂. **Voice** Song is a distinctive monotonous series of 5–20 well-spaced, upswept pure whistles, *fooEE-fooEE-fooEE-fooEE-fooEE-fooEE…* [Corbatita Blanco]

Chestnut-bellied Seed Finch *Sporophila angolensis* 14 cm

Clearings and edge of Paraná forest. **ID** Adult ♂: Glossy black head and upperparts. Small white speculum. Pure white underwing-coverts. Black throat and breast. Contrasting chestnut abdomen. **Adult** ♀: Recalls a small grosbeak. Ochre-brown to warm brown above; usually brighter on the rump. Underwing as ♂ but no white spot in closed wing. Overlaps with ♀♀ of Blackish-blue Seedeater (Plate 184), and Glaucous-blue and Ultramarine Grosbeaks (Plate 178), but pale buffy throat contrasts with cinnamon underparts. **Voice** Song is a pleasant, lazy warble with some more drawn-out notes, sometimes terminating in trills and churrs. [Curió]

Rusty-collared Seedeater *Sporophila collaris* 13.5 cm

Chiefly in NE marshes. **ID** Adult ♂: Stout black bill. Black cap and ear-coverts. White loral spot and subocular crescent, throat and primary patch. Rufous nape, rump and much of underparts with contrasting black back, wings, tail and pectoral band. Adult ♀: Large black bill. Brown and buff with a white throat and primary patch, and buff median-covert wing bar. **Voice** Song is an extremely complex, rapid series of, at times jumbled or repeating notes; frequently includes an extraordinary amount of imitation of other marshbirds. [Corbatita Dominó]

Lined Seedeater

ad ♂

Double-collared
Seedeater

ad ♂

ad ♀

ad

obscurus

Dull-coloured
Grassquit

ad ♂

Chestnut-bellied
Seed Finch

ad ♂

ad ♀

leucoptera

White-bellied
Seedeater

ad ♀

ad ♀

ad ♂

ad ♀

melanocephala

Rusty-collared
Seedeater

Sporophila Small vagrant, sexually dimorphic tanager-finches. Bill colour and shape, and wing pattern can aid identification of drab ♀♀. See also Plates 185 and 186.

Plumbeous Seedeater *Sporophila plumbea*   11 cm

Unconfirmed in Misiones grasslands. **ID** Adult ♂: Stout black bill; yellowish or dull-pink in winter. Blue-grey above and paler below. Contrasting blackish wings and tail. White throat, malar spot and small patch in primaries. Adult ♀: Black bill. Brown above with darker brown wings and tail; small white primary patch. Fawn below with a whitish centre to the belly. Some individuals show ruddy-brown ear-coverts and throat. [Corbatita Plomizo]

Yellow-bellied Seedeater *Sporophila nigricollis*   10.5 cm

Sparse records in n. Misiones. **ID** Adult ♂: Blackish hood and breast. Dull olive above; yellowish belly and vent. Adult ♀: Ochraceous olive-brown above with buffy ear-coverts and throat. Whitish-buff below, washed cinnamon on flanks and vent. [Corbatita Amarillo]

Black-bellied Seedeater *Sporophila melanogaster*   11 cm

Vagrant? to s. Misiones. **ID** Adult ♂: Grey crown, nape and flanks. Back and wings typically darker and browner. Small white primary patch. Throat and central underparts contrastingly black. Dark '*xumanxu*' morph (Rare): Differs by being entirely black below with a black nape and back. Adult ♀: As ♀ Rufous-rumped Seedeater. [Capuchino Vientre Negro]

Catamenia Andean 'seedeaters', larger than most lowland *Sporophila* (Plates 185–187), with stout, coloured bills showing a distinctly curved culmen and coloured undertail-coverts.

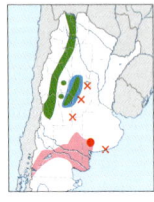

Band-tailed Seedeater *Catamenia analis*   12.5 cm

NW Andes south to N Patagonia and outlying sierras. **ID** Adult ♂: Yellow bill, bordered by black lores. Blue-grey above with white primary patch and band across undertail. Grey below with whitish belly. Chestnut undertail-coverts. In flight, white wing stripe and band across base of tail; similar in both sexes. Adult ♀: Yellow bill. Warm brown crown and mantle, with thick dark brown streaks. Greyish rump. White primary patch. White tail band slightly narrower than ♂. Buffy throat and breast, becoming white on belly; streaked brown on throat, breast and flanks. Cinnamon-buff undertail-coverts. Compare with Band-tailed Sierra Finch (Plate 188) which overlaps. **Voice** Song is composed of ringing, liquid, penetrating rattled phrases that eventually change to another type, *ZHIZHIZHIZHI... ZHIZHIZHIZHI... DRRRRRR...*, rarely becoming a complex series of whistles and churrs. [Piquitodeoro Chico]

Plain-coloured Seedeater *Catamenia inornata*   15 cm

N Andes, mainly above 2500 m south to Mendoza, and C Sierras at 700–2200 m. **ID** Adult ♂: Orange-pink bill. Ash-grey above, finely streaked black. Grey below with paler central abdomen and contrasting chestnut undertail-coverts. Adult ♀: Peach-coloured bill. Pale brown above, finely streaked on crown and thickly streaked dark brown on mantle. Unmarked greyish rump. Pale brown below with paler central abdomen and dull cinnamon undertail-coverts. **Voice** Song begins with 3–4 pure whistles followed by a slow series of 3–7 long churrs or wheezy trills that often slide downscale, e.g. *we-we-feeeee-CHRRRR-TWEEEERR-DUUURRR-ZHRRRII*. [Piquitodeoro Grande]

plumbea

ad ♀

Yellow-bellied Seedeater

ad ♂

ad ♂

ad ♀

nigricollis

Plumbeous Seedeater

ad ♂

ad ♂

Band-tailed Seedeater

ad ♀

Black-bellied Seedeater

analis

ad ♀

ad ♂

ad ♀

inornata

Plain-coloured Seedeater

Geospizopsis Small to medium Andean and sierran finches with small dull bills and dull tarsus. Males are grey while females and juveniles are streaked. *Rhopospina* Large robust finches of the Andes and Patagonia. Males are grey to black (with paler and streaked upperparts in fresh plumage) with yellow bills and tarsi; they perform characteristic display flights. Females and juveniles are streaked and have a yellow to pink tarsus.

Ash-breasted Sierra Finch   *Geospizopsis plebejus*   12.5 cm

High N Andes south to Mendoza, and sierras in Córdoba–San Luis. **ID Adult:** Short conical black bill. Short whitish supercilium and notable white eye-ring. Greyish-brown above with dark dorsal streaking, and unmarked grey rump. Grey below with whitish lower belly and vent. **Juvenile:** Much browner above than adult, thus unstreaked grey rump forms greater contrast. Note eye-ring and supercilium. Dull buff below with fine breast streaking. Compare with ♀ Plumbeous Sierra Finch **Voice** Song is a high-pitched buzz or trill followed by a few repeated couplets, e.g. *shrrrr chi-pid chi-pid chi-pid chi-pid* or *trrreee chee-chee-p chee-chee-p.* [Yal Chico]

Plumbeous Sierra Finch   *Geospizopsis unicolor*   15 cm

Throughout the Andes and C Sierras; 4 similar ssp. **ID Adult** ♂ *tucumanus* (N Andes): Uniform ash-grey above and paler grey below. **Adult** ♂ *ultimus* (S Andes ; not illustrated): Blue-grey above with a similar but much paler tinge below. **Adult** ♀: Brown above with thick dark brown streaking and unmarked greyish rump; buff below with thick blurry brown streaking mostly over breast and flanks. Dull bill and lack of a supercilium or wing bars help to distinguish from larger Band-tailed and smaller Ash-breasted Sierra Finches. **Voice** Song is a remarkably variable sprightly series of generally fast, modulated, rich, drawn-out whistles, e.g. *swee-duswee-duswee-duswee* or *sisisisisi-swee-su-sweeep.* [Yal Plomizo]

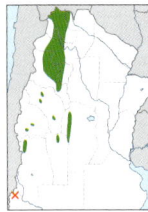

Band-tailed Sierra Finch   *Rhopospina alaudina*   16.5 cm

High NW Andes and C Sierras. **ID Adult** ♂: Bright yellow bill. Orange-yellow tarsus. Blackish lores. Generally blue-grey with fine black dorsal streaking and white central abdomen and vent. A white tail-band is visible in flight (in all plumages), as in Band-tailed Seedeater (Plate 187). **Adult** ♀: Yellow bill and tarsus distinguish from congeners. Buffy-brown with dark brown dorsal streaking. Buff wing bar on median-coverts. Pale buff below with variable streaking on throat, breast and flanks. **Immature** ♂/fresh adult: Brown back, streaked dusky. **Voice** Song is a high-pitched, rapid, shrill whistle repeated a few times, *zrizizi-zeeeu zrizizi-zeeeu…* or *zi zi-zeoo*, but with more variation in the Andes. During short parachuting display flight, with the tail fanned exposing the white band, song is more complex and incorporates a throaty, descending trill. [Yal Platero]

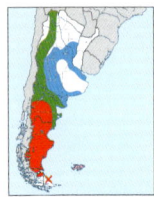

Mourning Sierra Finch   *Rhopospina fruticeti*   18 cm

Throughout the Andes and continental Patagonia. **ID Adult** ♂: Stout orange bill. Black foreface, throat and breast, heavily streaked above with 1–2 white wing bars, grey flanks and white belly and vent. Worn males can look black, with reduced wing bars, when best told from Carbonated Sierra Finch by shallower bill base and white vent. **Adult** ♀: Heavily streaked as congeners, but with chestnut ear-coverts, bordered by white moustachial and dusky malar. 1–2 distinctive white wing bars. **Voice** Song is a buzzy, icterid-like *WITCHEEEEEErrr Wiro* with some variation. More complex series in long parachuting flight, *dji-li-WIRRRO-RRROAAR WIRR-rororu.* Calls include high-pitched notes, a harsh nasal note (resembling Rufous-bellied Mountain Tanager in the NW) and a dry *tzec.* [Yal Negro]

Carbonated Sierra Finch   *Rhopospina carbonaria*   15 cm

Endemic. Local breeder in Monte desert of N Patagonia; winters erratically to the NW. **ID Adult** ♂: Bright yellow bill and orange-yellow tarsus. Sooty-black crown, face and underparts, mottled grey or whitish below when moulting. Darker above than Mourning Sierra Finch with poorly defined streaking, wing bars reduced or lacking, and broader-based, more conical bill. **Adult** ♀: Closest to Band-tailed (no overlap) but bill pink and tarsus orange; lacks grey on rump and buff median-covert bar is weaker. **Voice** Song, perched or in high circular display flight, is a high-pitched, fast, hissing *slizslizslizSLIZSLIZSLIZSLIZslizsliz*, louder in the middle portion. Calls include very high-pitched *tsi* notes. [Yal Carbonero]

ad

Ash-breasted Sierra Finch

Plumbeous Sierra Finch

tucumanus

ad ♀

plebejus

juv

ad ♂

imm ♂
fresh adult

ad ♂

ad ♀

venturii

Band-tailed Sierra Finch

ad ♂

ad ♂

fruticeti

ad ♀

ad ♀

Mourning Sierra Finch

Carbonated Sierra Finch

Phrygilus Large, colourful, robust sierra finches of the Andes and Patagonia. Sexually dimorphic to varying degrees; juveniles drabber, usually with malar streaks. Calls are similar among species, sounding like two pebbles struck together.

Grey-hooded Sierra Finch *Phrygilus gayi*   16.5 cm

Throughout the Andes and Patagonia. **ID** Adult ♂: Fairly bright olive back and rump with contrasting blue-grey hood and wing-coverts. Yellow below with white belly and vent, and greyish flanks. **Adult** ♀: Much drabber than ♂ with less clear-cut grey hood, olive-brown back, dusky throat streaking and dull apricot breast and flanks; paler on centre of belly and vent. **Juvenile**: Like ♀ but much paler throughout with a whitish belly and prominent dusky malar streak. **Voice** Song is an uninspiring series of medium-pitched, short metallic warbles and whistles, e.g. *slip slurrí slip slurrí slirríp slip slurrí...* [Comesebo Andino]

Patagonian Sierra Finch *Phrygilus patagonicus*   15 cm

Patagonian forest (some overlap with Grey-hooded). **ID** Adult ♂: From Grey-hooded by russet mantle and richer cadmium-yellow underparts with white restricted to the undertail-coverts. **Adult** ♀: Pale grey hood and olive mantle suggests ♂ Grey-hooded but wing-coverts grey-brown, and white restricted to undertail-coverts. **Juvenile**: Like ♀ but drabber still, with a dirty yellowish throat and brownish wing-coverts, often with two ochre wing bars. **Voice** Song is a highly variable, wheezy *chi-di chu-di churi-lit...* or *plip plu plip.. plip plu plip*, always higher-pitched, sweeter and more whistled than the similarly patterned Grey-hooded Sierra Finch. [Comesebo Patagónico]

Black-hooded Sierra Finch *Phrygilus atriceps*   17 cm

NW Andes above 2500 m south to nw. San Juan (overlaps with Grey-hooded). **ID** Adult ♂: Black hood and russet mantle; russet breast grading to cadmium-yellow with white centre to belly and vent. **Adult** ♀: Like ♂ but hood and wing-coverts sooty-brown. **Juvenile**: Paler grey-brown hood than ♀; also browner back, and duller and paler below. **Voice** Song is a modest series of tinny, monotonous warbles *klíri klíri...*, recalling Grey-hooded Sierra Finch but less varied. [Comesebo Cabeza Negra]

Melanodera Robust terrestrial bridled finches of the S Andes, Patagonia and the Falklands with stout conical bills. Found in isolated pairs in summer and roving flocks in winter. Streaky ♀♀ differ by wing and tail patterns.

White-bridled Finch *Melanodera melanodera*   16 cm

S Patagonia (very local) and the Falklands (common). **ID** Adult ♂ *melanodera* (Falklands endemic): Black lores and throat, surrounded by white. Contrasting yellow carpal and olive-yellow wing-coverts, fringes to flight feathers and outer tail. In flight, yellow-olive wings contrast with grey upperparts. **Adult** ♂ *princetoniana* (Grass-steppe of s. Santa Cruz and n. Tierra del Fuego, reaching s. Chubut in winter): From *melanodera* by bright yellow wing-coverts, primary fringes and tail fringes, and brighter yellow breast. In flight, striking contrast between grey upperparts and yellow wings. **Adult** ♀: Brown above with fine coronal and coarse dorsal streaking. Yellowish-fringed primaries and outer tail feathers. Unmarked buffy-white throat and streaked breast, broader at the sides. Essentially unmarked whitish abdomen. **Voice** Song is a variable series of leisurely whistles that carry well, even in strong wind, *tweedle twi susu* or *chido chi swreep*, with much variation. [Yal Austral]

Yellow-bridled Finch *Melanodera xanthogramma*   17.5 cm

S Andes. **ID** Adult ♂ *barrosi* grey morph (Andes above 1500 m from nw. Mendoza to sw. Santa Cruz): Black lores and bib surrounded by yellow, but note that some plumages show yellow spectacles and a white moustachial (also *xanthogramma*?). White belly, vent and tail fringes. **Adult** ♂ *barrosi* yellow morph (not illustrated): much as yellow morph *xanthogramma* but with white tail fringes and grey wing coverts. **Adult** ♂ *xanthogramma* yellow morph (Andes of Tierra del Fuego above 900 m, locally on Fuegian islands and at least formerly on the Falklands): Differs from *barrosi* by yellow belly, vent and tail fringes. In flight, little contrast, although the wing-coverts are mixed with olive. **Adult** ♂ *xanthogramma* grey morph (not illustrated): much as grey morph *barrosi* but with yellow tail fringes and olive wing-coverts. **Adult** ♀/immature *xanthogramma* yellow morph: Stouter-billed than White-bridled. Thicker crown steaking than ♀ White-bridled, and usually some facial streaking. Underparts also more coarsely streaked, extending onto flanks, over a pale yellow base colour. Flight feathers narrowly fringed white, with broad buff tertial fringes. **Adult** ♀ *xanthogramma* white morph: As yellow morph but with white ventral plumage; note described distribution of streaking, and broad tertial fringing. **Voice** Song is a pleasant repetitive series of leisurely whistles, *twidl tli choo* or *cli weeo wip chuwí*, with much variation. [Yal Andino]

ad ♂

juv

Patagonian Sierra Finch

caniceps

ad ♀

ad ♀

juv

ad ♂

Grey-hooded Sierra Finch

ad ♀

juv

ad ♂

Black-hooded Sierra Finch

White-bridled Finch

melanodera

princetoniana

ad ♂

ad ♀

ad ♂

ad ♂

ad ♂

melanodera

ad ♂
yellow morph

ad ♀
white morph

princetoniana

Yellow-bridled Finch

xanthogramma

ad ♀ / imm
yellow morph

ad ♂
yellow morph

xanthogramma

ad ♂
grey morph

xanthogramma

xanthogramma

barrosi

Idiopsar High altitude finches with a reddish iris, white throat and grey breast. All associate with cushion plants and nearby bogs and boulder piles.

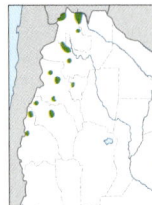

Red-backed Sierra Finch *Idiopsar dorsalis* 17 cm

NW Andes above 3600 m, south to nw. San Juan; always associates with cushion plants. **ID** Adult: Grey hood and breast with contrasting brick-red mantle. White or greyish throat, and white belly and vent. **Juvenile** (not illustrated): Much like adult, but with drabber back and browner-grey plumage. **Voice** Song is an unpatterned series of high-pitched, nasal, metallic notes, *tziú whí tze zi tziú tze whí*. Calls include a nasal *whí* or *wheenk*, and a very high-pitched *tz*. [Yal Altoandino]

Short-tailed Finch *Idiopsar brachyurus* 19 cm

Local in NW Andes above 2800 m, south to Tucumán. **ID** Adult: Lores and ear-coverts speckled white. Long, broad, grey bill with a paler mandible. Grey with dark brown wings and squarish tail. **Juvenile**: Much paler than adult and smaller-billed with a pink lower mandible. Dark iris. Pale pink tarsus. Ear-coverts grizzled white. **Voice** Song is a simple repetitive series of well-spaced, high-pitched notes, *zit psi tzi sip zit psi sip tzi…* Calls are similarly high-pitched isolated notes. [Yal Grande]

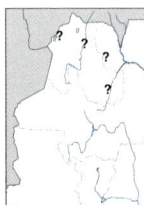

Glacier Finch *Idiopsar speculifer* 18.5 cm

Uncertain in Jujuy–Salta, above 3800 m. **ID** Adult: Fairly long black bill, with a straight culmen. Grey above and across breast. White subocular crescent, throat and large white wing panel. **Voice** All vocalisations are much like Red-backed Sierra Finch, although song has apparently never been recorded. [Alt: White-winged Diuca Finch] [Yal Glacial]

Diuca Finch *Diuca diuca* 18 cm

High Andes south to n. San Juan (*crassirostris*). See Plate 182 for lowland ssp. *minor*. **ID** Robust grey finch with white throat and stout bill. Sexes alike. **Adult** *crassirostris*: Large bill with an arched culmen. Mostly grey with a pinkish-white throat, grey pectoral band and flanks with a rusty patch at the rear; white central abdomen and vent. Large white tail spots, visible in flight. **Voice** Song, mainly at dawn, is a mellow, measured *SCHup-CHI… SCHup CHIWol*, or *SCHUp-CHup-CHi.. SCHUp-CHoo* repeated over; may recall a *Phrygilus* sierra finch. Calls include a variety of *schup, kwip* and *chip* notes. [Alt. Common Diuca Finch] [Diuca]

Grey-crested Finch *Lophospingus griseocristatus* 14.5 cm

Very local in arid Andean valleys of extreme n. Salta and Jujuy. **ID** Recalls a *Poospiza* warbling finch (Plate 181) but with an upstanding crest. **Adult** ♂: Ashy-grey with a blackish crest. Huge white tail spots. Pale grey below with a whitish belly and vent. **Adult** ♀ (not illustrated): Shorter crest and brown dorsal wash. **Juvenile**: Crest reduced or absent. Brown above with a white supercilium. Pale brown below with a white belly and vent. **Voice** Song is a measured series of medium-pitched short warbles (and some whistles), e.g. *chiro whiro khip whiro chiro whiro kip…* or *che pit chru chrr-brr chweet che…*, sounding like a deeper-toned warbling finch. Call is a metallic *zit*. [Soldadito Gris]

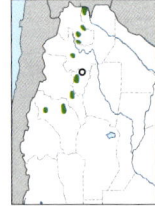

Tucumán Mountain Finch *Poospiza baeri* 17.5 cm

Virtually endemic in semi-isolated NW Andean ranges. **ID** An unobtrusive, chunky finch of secluded river gullies. Mostly terrestrial. Sexes alike. **Adult**: Grey with a rich rusty forecrown, supercilium, eye crescent, throat and upper breast. Olive-grey below with a chestnut vent. **Juvenile**: Olive-brown above; ochraceous-buff below with blurry breast streaking. **Voice** Song is a loud strident whistled *wer-chi, wer-CHI, …wer-CHI CHI.. wer-CHI-CHI CHIó*, with much variation and often quite patternless. Calls include a piercing *Szeet* and a thin *sip*. [Monterita Serrana]

ad

juv

Short-tailed Finch

ad

Red-backed Sierra Finch

ad

Glacier Finch

ad

crassirostris

Diuca Finch

juv

ad ♂

Grey-crested Finch

ad

juv

Tucumán Mountain Finch

Sicalis An often confusing group of sexually dimorphic finches ranging throughout the country from the coast to extremely high altitudes. Head pattern, bill shape, width of tertial fringes and primary projection can be important factors to distinguish some species. Fresh plumages show variable amounts of grey. Complex songs, often at dawn, can contain imitations of other birds, and only the diurnal songs are described here. Calls include soft *sweep* and ticking notes. See also Plate 192.

Citron-headed Yellow Finch *Sicalis luteocephala*   14 cm

Puna villages in n. Jujuy. **ID** Adult ♂: Cadmium-yellow face and yellow underparts contrast with grey hindcrown, back and extensive grey flanks. Olive leading edge of wing. Adult ♀: Resembles ♂ but with brown hindcrown and mantle. Overlaps with Bright-rumped, Puna and Greenish Yellow Finches. **Voice** Song is a distinctive raspy trill mixed with spluttered chatters. Ticking call is a gravelly *CHEK* or *CHOK*. [Jilguero Corona Gris]

Bright-rumped Yellow Finch *Sicalis uropygialis*   13.5 cm

Puna and high NW Andes. **ID** Adult ♂ fresh: Bronze-olive crown and contrasting grey cheeks. Yellow-olive rump. Bright yellow below with grey flanks. Adult ♀: Brown cheeks. Duller yellow underparts than ♂ with olive wash on breast. Whitish lower mandible. Pinched white eye-ring. Overlaps with Citron-headed, Puna and Greenish Yellow Finches. **Voice** Song is a semi-musical series, repeating up to 4 different note-types, *tre-tre-tre-tre-tre-chi-chi-chi-chi-pli-pli-pli*. [Jilguero Cara Gris]

Puna Yellow Finch *Sicalis lutea*   14 cm

Rocky Puna of Jujuy, Salta and Catamarca. **ID** Adult ♂ worn: Stout blackish bill unlike congeners. Fairly uniform sulphur-yellow with yellow-olive wing-coverts and bright rump and underparts. Adult ♂ fresh: variable amounts of grey on back; duller yellow on head and throat may impart a hooded appearance. Adult ♀: Duller than ♂, but still much brighter than ♂ Greenish Yellow Finch. Note yellow rump and dusky bill. Overlaps with Citron-headed, Bright-rumped and Greenish Yellow Finches. **Voice** Song is a sprightly series of mostly warbled couplets, *ti-ti-weeblo-weeblo-chwir-chwir-chirrup-chirrup-weechur-weechur-chidi-chidi*. [Jilguero Puneño]

Greenish Yellow Finch *Sicalis olivascens*   14 cm

NW Andes above 2150 m. **ID** Adult ♂ worn: Rather uniform olive with brighter yellow-olive rump and belly. Upperparts frequently mottled or streaked. Adult ♂ fresh: Note olive-yellow fringed wing-coverts and primary bases. See differences in tertial pattern of smaller-billed Monte Yellow Finch. Adult ♀: Mostly brown with an olive-tinged rump and yellow wash on the underparts. No yellow on the head. Bill stouter than Bright-rumped and culmen slightly curved unlike Greater Yellow Finch. Overlaps with Citron-headed, Bright-rumped, Puna, Greater and Monte Yellow Finches. **Voice** Song is a distinctive, yet uninspiring, monotone series of dry, grating trills or repetitive warbles, often descending and usually ending in a monotone rattle, *trrr-trrr-trrr-chrrr-chrrr-chrrr-chrrr-teeeerrrrrr*; unlike Monte Yellow Finch. [Jilguero Oliváceo]

Monte Yellow Finch *Sicalis mendozae*   13 cm

Endemic. Sandstone cliffs in Monte desert to 2150 m (locally higher in the NW). **ID** Adult ♂ worn: Bright olive-yellow above with little contrast between the back and rump; lacks dorsal streaks or mottling. Bright yellow underparts. Adult ♂ fresh: Little grey on the flanks. Bases of primaries fringed yellow or olive (not grey). Entire webs of tertials fringed grey versus narrow grey fringes to the outer webs of the tertials in Greenish Yellow Finch. Adult ♀: Intense sulphur-yellow patch covering the centre of the abdomen, sometimes reaching the throat. Olive rump and little or no olive on the lesser wing-coverts. Marginal overlap with Greenish in Salta/Tucumán. **Voice** Song is a harsh, metallic, fast-rolling series of ascending and descending syllables, unlike Greenish Yellow Finch. Ticking call is a dry *TEC*. **Tax note 70.** [Jilguero del Monte]

Citron-headed Yellow Finch

ad ♀

ad ♂

Bright-rumped Yellow Finch

ad ♀

uropygialis

ad ♂

ad ♂ fresh

ad ♂ worn

ad ♀

Puna Yellow Finch

Greenish Yellow Finch

olivascens

ad ♂ worn

ad ♀

ad ♂ fresh

ad ♂ fresh

ad ♂ worn

ad ♀

ad ♀

Monte Yellow Finch

Sicalis An often confusing group of sexually dimorphic finches ranging throughout the country from the coast to extremely high altitudes. Head pattern, bill shape, width of tertial fringes and primary projection can be important factors to distinguish some species. Fresh plumages with variable amounts of grey. Greater and Patagonian give complex songs, often at dawn, but only the diurnal songs are described here. Stripe-tailed and Grassland perform parachuting display flights. See also Plate 191.

Greater Yellow Finch *Sicalis auriventris*   15.5 cm

Andes from Catamarca to n. Santa Cruz; also Sierra de la Ventana, s. Buenos Aires. **ID** Bill shape, long primary projection and narrow tertial fringes clinch identification for both sexes. **Adult** ♂ **worn**: Straighter culmen and larger bill than congeners. Golden head and underparts. Grey cheek spot at close range. Olive back, with indistinct dusky streaking. Greyish-fringed wings. **Adult** ♂ **fresh**: From fresh Patagonian, Greenish and Monte Yellow Finches by combination of straight culmen and grey-fringed (not olive/yellow) wing-coverts. **Adult** ♀: Pale brown with an olive-washed rump and pale yellow centre of abdomen. Overlaps with Greenish and possibly Bright-rumped, Puna and Monte Yellow Finches. **Voice** Song is a mostly gravelly series of rather monotone, jumbled, metallic chattering mixed with some chirps. [Jilguero Grande]

Patagonian Yellow Finch *Sicalis lebruni*   14.5 cm

Patagonia; mostly in coastal regions. **ID Adult** ♂ **worn**: Olive crown and back, often mixed with grey. Vivid olive shoulder and rump. Broad grey tertial fringes. Rich yellow throat and central abdomen, washed grey on upper breast and flanks. Fairly notched tail as in worn Monte Yellow Finch. **Adult** ♂ **fresh**: Broad grey tertial fringes. Note extensive grey flanks. **Adult** ♀: Pale brown with an olive shoulder, yellow centre to belly, greyish flanks and white undertail-coverts. Broad buffy tertial fringes, brown rump and small bill. No known overlap with congeners. **Voice** Song is a fast, rollicking, semi-musical, staccato series, *chiri-dip chirip chiridip chidip...*, varied to a more musical *twi-dew di-dew...* [Jilguero Austral]

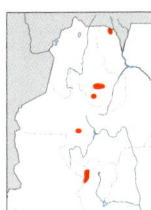

Stripe-tailed Yellow Finch *Sicalis citrina*   12 cm

Very local above 1400 m (mostly above 2000 m) in NW Andean meadows (Dec–May). **ID Adult** ♂: Conical bill. Fairly bright yellow crown, contrasting with olive back, streaked dusky. Rich cadmium-yellow underparts with an olive wash on the breast. In display flight, diagnostic white tail webs clearly visible. **Adult** ♀: Brown crown and mantle, densely streaked dark brown. Prominent buffy or white wing bars. Buff breast finely streaked brown and belly pale yellow. White inner webs of outer tail feathers visible from below. **Voice** Perched song has a buzz, then 2–3 drawn-out, wheezy, thin whistles *brzzzzz tzeeeeee TZiuuuuuuuuu-ít tzeeeee...*, second whistle slowly descending and with a distinctive ending. Flight song in parachuting display consists of a buzzy trill, a descending and decelerating chatter and often some pipit-like notes while descending with spread tail, *zzzrrr tetetetete-to-to-to-to to...* [Jilguero Cola Blanca]

Grassland Yellow Finch *Sicalis luteola*   12.5 cm

Widespread in lowlands and foothills (to 1350 m). **ID** Adult ♂: Streaky crown and mantle. Contrasting yellow supraloral, white spectacles and olive wash across breast. **Adult** ♀: Browner above than Stripe-tailed Yellow Finch with a brown wash across the breast. Vertical display flight (higher than Stripe-tailed) with parachuting glides. **Voice** Perched song is a rapid series of short, penetrating, metallic buzzy trills, often becoming longer towards the end. Flight song, in parachuting display, is a long series of rapid short, buzzy, ringing trills followed by successively longer notes, ending in very long churrs or buzzes while descending with spread tail, *czczczczczcz-shrshshrrshrrr-shrrreee-cheeeerrrrr-cheeeerrrr-cheeeeeerrr...* Common call is a high-pitched, disyllabic *P-TZÍU.* [Misto]

Saffron Finch *Sicalis flaveola*   12.5 cm

Widespread in the Andes and lowlands south to n. Chubut. **ID** Adult ♂: Mainly yellow with an olive back, indistinctly streaked. Diagnostic orange forecrown. **Adult** ♀: Pale brown above with dense streaking. Whitish below with brown-streaked breast and flanks, but not yellow like congeners. Olive wing panel and tail fringes. **Voice** Song is a highly variable series of rapid, high-pitched, chippers, whistles and trills, often quite inspired when songs can be longer, more piercing and twittering. Calls include single high-pitched *tzí* and soft *uip* or *swip* notes. [Jilguero Dorado]

ad ♀

Greater Yellow Finch

ad ♂ worn

ad ♂ fresh

Patagonian Yellow Finch

ad ♀

ad ♂ worn

ad ♂ fresh

♂ display

occidentalis

ad ♀

ad ♂

Stripe-tailed Yellow Finch

ad ♀

ad ♂

ad ♂

ad ♀

luteiventris

Grassland Yellow Finch

pelzelni

Saffron Finch

Spinus Small gregarious finches with conical bills, and a combination of black, yellow and often olive plumage. Bill shape, extent of hood, wing pattern and intensity of rump aid identification. Songs are fast, long and complex, often with repetitive syllables and at times include imitations of other birds. Calls comprise audible pure whistles and raspy notes, given while perched and in flight.

Black-chinned Siskin *Spinus barbatus* — 13 cm

Throughout the Patagonian Andes and steppe. Resident on the Falklands. **ID Adult ♂:** Black cap and bib contrast with yellow face. **Adult ♀:** Resembles ♀ Hooded Siskin, but note curved culmen, duller rump and broad yellowish rear supercilium, curving down to sides of neck. [Cabecitanegra Austral]

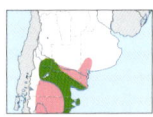

Undescribed endemic siskin *Spinus* sp. — 13 cm

Monte desert of n. Chubut, Neuquén, Río Negro, La Pampa and sw. Buenos Aires. **ID Adult ♂:** Paler yellow throughout and smaller-billed than Black-chinned Siskin. Black on head much reduced or lacking. **Adult ♀:** Distinctive, mostly grey plumage. Some overlap with Hooded Siskin. **Tax note 71.** [Silvestrín]

Hooded Siskin *Spinus magellanicus* — 11.5–13 cm

Widespread. **ID Adult ♂ *tucumanus*** (11.5 cm; N and C Andes above 800 m, rarely in the lowlands): Tiny delicate bill. Hood restricted to foreface and throat. Broad wing bar connects with wing panel. Dull yellow rump. Notably short-tailed. In flight, note mainly yellow underwing. Overlaps with Thick-billed Siskin. **Adult ♂ *magellanicus*** (13 cm; N and C lowlands and C sierras where it is the only siskin with a hood): Larger and with a stouter bill than *tucumanus*, and with a brighter rump, more extensive hood and longer tail. **Adult ♀ *magellanicus*:** Resembles adult ♂ without the hood. In flight, note yellow wing stripe and black tail band. Some overlap with the undescribed endemic taxon, especially in winter. [Cabecitanegra]

Thick-billed Siskin *Spinus crassirostris* — 13.5 cm

Rare above 2100 m in C Andes, and local above 2800 m in NW Andes; often in *Polylepis* woodlands. **ID Adult ♂:** Outsized blackish bill. Extensive black hood reaches upper breast. Narrow yellow wing bar along greater-coverts, touching wing panel. Fairly bright yellow rump. **Adult ♀:** Greyish-brown head and underparts, with an olive wash on the back. Contrasting yellow rump. Whitish undertail-coverts. Overlaps with Hooded Siskin (ssp. *tucumanus*), and with Black-chinned in winter in Mendoza. [Cabecitanegra Picudo]

Yellow-rumped Siskin *Spinus uropygialis* — 13 cm

Local above 1500 m in C Andes from San Juan to nw. Chubut; rare in the high NW Andes and Puna. **ID Adult ♂:** Blackish hood extends onto breast. Back appears black from a distance; streaked dull olive at close range. Extensive bright yellow rump, wing markings, tail base and underparts. **Adult ♀:** Resembles adult ♂ but with a brownish-olive hood and back; latter more obviously streaked. Overlaps with Black Siskin, chiefly in San Juan. [Cabecitanegra Andino]

Black Siskin *Spinus atrata* — 13 cm

NW Andes above 2000 m in Puna and Yungas forest tree-line. **ID Adult ♂:** Glossy black with yellow wing markings, belly and tail base. The only siskin with a black rump. **Adult ♀** (not illustrated): Virtually identical, lacking the gloss. Overlaps with Yellow-rumped Siskin, chiefly in San Juan. [Negrillo]

Black-chinned Siskin

ad ♀   ad ♂

Siskin sp.

ad ♀   ad ♂

ad ♀

ad ♂

ad ♀   ad ♂   ad ♂

Hooded Siskin

magellanicus   *magellanicus*   *tucumanus*

crassirostris

ad ♂

ad ♂

ad ♂

ad ♀

ad ♀

ad ♂

Thick-billed Siskin

Yelow-rumped Siskin

Black Siskin

Euphonia Small robust, mostly resident forest or woodland canopy passerines with stout bills and short tails. ♂♂ have dark glossy blue, purple or green dorsal plumage, and yellow or orange ventral plumage. ♀♀ are olive above; best distinguished by ventral plumage and bill size. *Chlorophonia* Small, chunky, *Euphonia*-like midstorey to subcanopy passerines with a small bill and sky-blue nuchal collar in ♂♂. Only adults are illustrated.

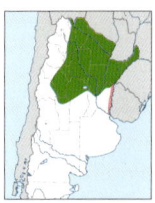

Purple-throated Euphonia *Euphonia chlorotica*   11 cm

Andean foothills to 1600 m, Monte desert, chaco and Paraná forest. Wanders widely. **ID** A benchmark species, overlapping with all congeners. **Adult** ♂: Broad yellow forehead and extensive dark throat. **Adult** ♀: Narrow yellowish forehead. Pale grey central underparts, bordered bronzy-yellow. **Voice** Song is a short, jumbled series of scratchy, nasal buzzy notes. Common calls are pure thin whistles, often doubled and evenly-pitched *fee-fee*, sometimes varied to versions with differently pitched whistles or a single drawn-out whistle (in Yungas forest may recall Buff-banded Tyrannulet). [Tangará Garganta Negra]

Violaceous Euphonia *Euphonia violacea*   11 cm

Paraná forest. **ID Adult** ♂: Dark purple above with a yellow-orange forehead and underparts. **Adult** ♀: Bronze forehead. Lacks white tail spots. Olive below, becoming yellowish on the abdomen. **Voice** Song is a highly variable, jumbled series of nasal buzzy notes, frequently including long strings of speeded-up imitations of other bird sounds. Calls are often doubled notes, including imitations. [Tangará Amarillo]

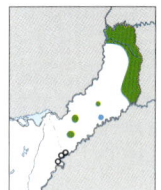

Green-throated Euphonia *Euphonia chalybea*   12.5 cm

Scarce in Paraná forest; movements poorly understood. **ID** Notably swollen bill with a white or pale base should distinguish from Purple-throated and Violaceous Euphonias. **Adult** ♂: Yellow of forehead reaches the front of the eye. Glossy black above with bottle-green reflections. Chin and upper throat usually look black (bottle-green reflections difficult to see). Lacks the white tail webs of Purple-throated and Violaceous. **Adult** ♀: Bronze forehead contrasts slightly with the vivid olive upperparts. Pale grey below with yellow-olive sides of breast, flanks and vent; usually more yellow on the face and throat. **Voice** Song recalls that of Purple-throated Euphonia but is lower-pitched, faster and gravelly. A distinctive gurgled, metallic, nasal chatter *gergergerger...*, with variations, is given repeatedly between song phrases. Calls include 2–4 rapid whistles *tiú-tiú-tiú*. [Tangará Picudo]

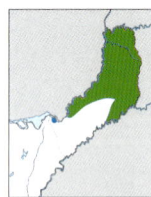

Chestnut-bellied Euphonia *Euphonia pectoralis*   12 cm

n. and c. Misiones. **ID Adult** ♂: Glossy blue-black upperparts and breast. Deep chestnut abdomen and golden-yellow pectoral tufts. **Adult** ♀: From all other euphonias by chestnut undertail-coverts. **Voice** Rarely heard song is a slow series of squeaky, nasal notes including drawn-out, thin, rising and falling whistles. Common call is a loud series of 2–8 gravelly, grating, relatively low-pitched notes, *drr-drr-drr...* or *shra-shra-shra...*, repeated over and over. [Tangará Alcalde]

Golden-rumped Chlorophonia *Chlorophonia cyanocephala*   11.5 cm

Yungas forest to 1600 m where resident, and Paraná and adjoining gallery forest, mostly in winter. **ID Adult** ♂: Black face and throat, sky-blue hood and collar, orange rump and underparts. **Adult** ♀: Chestnut forehead and sky-blue hood. Yungas birds (not illustrated) are yellower ventrally and may belong to an undescribed subspecies. **Voice** Song is an unbelievably fast, jumbled series of rich, throaty squeaks, tinkling whistles, nasal notes and rasps. Common calls include a disyllabic, metallic *chu-weenk*, and pure melancholic whistles, *peee* and *feeeu*. [Tangará Cabeza Celeste]

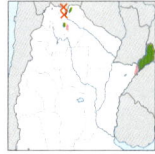

Blue-naped Chlorophonia *Chlorophonia cyanea*   11 cm

Paraná forest; sparsely in the n. Salta–se. Jujuy Yungas forest block. **ID Adult** ♂: Grass-green head and breast. Blue back, becoming brighter on rump. Bright yellow below. **Adult** ♀: Duller than ♂ with an olive back, more olive below and lacks the blue eye-ring. **Voice** Common vocalisations are pure, medium-pitched whistles, which can be fairly flat *peee* or descending *feeeu*, very similar to those of Golden-rumped Chlorophonia. [Tangará Bonito]

Purple-throated Euphonia

ad ♂

ad ♀

serrirostris

ad ♂

aurantiicollis

ad ♀

Violaceous Euphonia

ad ♀

ad ♂

Green-throated Euphonia

ad ♀

ad ♂

Chestnut-bellied Euphonia

ad ♀

aureata

ad ♂

Golden-rumped Chlorophonia

ad ♀

ad ♂

Blue-naped Chlorophonia

cyanea

Agelasticus Marsh icterids with slender bills. *Chrysomus* Gregarious icterid with a short thick conical bill. All are sexually dimorphic. Yellow-winged, Unicoloured and Chestnut-capped Blackbirds associate together.

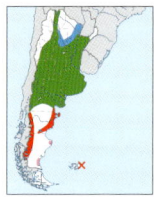

Yellow-winged Blackbird *Agelasticus thilius* 17.5 cm

Widespread. **ID Adult** ♂: Entirely black with yellow epaulets which are often concealed at rest. In flight, note striking yellow epaulets and axillaries. **Adult** ♂ **fresh**: Blackish head with broad buff supercilium. Brown- or chestnut-fringed upperparts, and buff-fringed underparts impart an overall mottled effect. **Adult** ♀: Much paler than fresh adult ♂ with buffy plumage and more distinct black streaking, especially below. Blackish malar and usually a pale coronal stripe. Yellow epaulets often concealed at rest. **Voice** Song has high-pitched twangy notes, some drawn-out, then a distinctive long note recalling a bandsaw cutting a large plank, *Twoo-laaaaaaay*, with much variation. Call is a simple *check*. [Varillero Ala Amarilla]

Unicoloured Blackbird *Agelasticus cyanopus* 19 cm

NE marshes and locally in the NW. **ID Adult** ♂: Long, slender, straight bill; culmen forms a straight line with the forehead. Plumage entirely glossy black. Compare with cowbirds and Chopi Blackbird (Plate 197), none of which inhabit dense marshes. **Adult** ♀: Brown above with contrasting chestnut-fringed wings and chestnut mantle, streaked black. Yellow or ochre-yellow below with blackish streaks on the flanks. **Immature** ♂: Resembles ♀ but with a black face and throat. **Voice** Song comprises a large variety of well-spaced liquid trills, gurgled chatters and repeated whistles, *grrlllllll...... trrreeewwww.. ... tiú-tiú-tiú-tíu. ...sieu sieuu sieuu*. Call is a soft, raspy *djep*. [Varillero Negro]

Chestnut-capped Blackbird *Chrysomus ruficapillus* ♂ 19 cm; ♀ 17.5 cm

N and C lowlands and foothills. **ID Adult** ♂: Dark chestnut crown and large bib (often difficult to see), otherwise entirely black. **Adult** ♀: Drab brown-olive above with blurry streaking on the back. Dull olive below with a yellowish-buff throat. **Juvenile**: Resembles ♀ but with blurry streaking below. **Immature** ♂: Intermediate between ♂ and ♀ with a streaked back and variable scattering of black spots below, usually with an indistinct chestnut throat. **Voice** Song is complex, but often includes 1–2 flat, high-pitched whistles followed by a long nasal rasp and short, rapid whistles, *feee feee laaaaaaay titítitítí*; the first part may recall Yellow-winged Blackbird and the second part Grey-breasted Crake. Call in flight is a liquid *pwik*. [Varillero Congo]

Icterus Slender-bodied icterid with comparatively long, slender tail and bill.

Orange-backed Troupial *Icterus croconotus* 25 cm

Rare in dry and humid chaco. **ID Adult**: Orange and black, with a large white wing flash. Black face mask with long, pointed bill and yellow eye. **Voice** Song is a slow series of rich whistles at different pitches, recalling a rusty swing, *fueee fueee weeee wooo*. Complex vocal repertoire includes mimicking of other birds. [Matico]

petersi

ad ♂

ad ♀

ad ♂

Yellow-winged Blackbird

ad ♂
fresh

ad ♂

imm ♂

ad ♀

cyanopus

Unicoloured Blackbird

imm ♂

ruficapillus

ad ♀

juv

Chestnut-capped
Blackbird

ad ♂

stictifrons

ad

Orange-backed Troupial

Pseudoleistes Chunky brown-and-yellow marsh icterids with stout conical bills. Sexes alike. *Xanthopsar* Savanna icterid with a short straight bill; found in boggy swales, often feeding in nearby crops. Saffron-cowled Blackbird associates with Yellow-rumped or Brown-and-yellow Marshbirds, depending on location. *Amblyramphus* Large conspicuous marsh icterid with a long conical bill. Often gregarious.

Yellow-rumped Marshbird *Pseudoleistes guirahuro* 24 cm

Mainly s. Misiones and n. Corrientes. **ID Adult**: Blackish foreface, throat and breast, otherwise brown above and yellow below with a large yellow rump. In flight, conspicuous yellow rump, epaulets and underwing-coverts. **Juvenile** (not illustrated): Resembles juvenile Brown-and-yellow Marshbird but lacks brown flanks and shows a dull yellow rump. Often associates with Brown-and-yellow Marshbird and Saffron-cowled Blackbird. **Voice** Song is very complex and often given by groups. A repetitive pattern, *ti-waaaah-chew... te-de-chew* frequently overlaps with low-pitched churrs and upswept whistles. Calls include a musical *ch-KLY... KLÚY?...*, patterned like a wheezy version of Brown-and-yellow Marshbird. [Pecho Amarillo Grande]

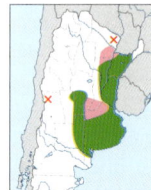

Brown-and-yellow Marshbird *Pseudoleistes virescens* 24.5 cm

Widespread in the NE. **ID Adult**: Brown head, throat, breast and upperparts. Yellow belly with extensive brown flanks. In flight, yellow epaulets and underwing-coverts. **Juvenile**: Pale brown and yellow with variable brown streaking over the breast and a brown malar streak. Sometimes found with Yellow-rumped Marshbird and Saffron-cowled Blackbird. **Voice** Distinctive far-carrying vocalisations in flight include harsh, gravelly chatters and pure whistles, *tarrrrreee tarrrreeeee* or *chrr-dwee...chrr-dwee...* [Pecho Amarillo]

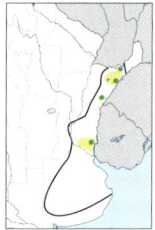

Saffron-cowled Blackbird *Xanthopsar flavus* 18.5 cm

Very local in NE savanna. **ID Adult** ♂: Flame-yellow foreface. Yellow throat, breast and belly. Black hindcrown and upperparts. In flight, contrasting yellow epaulets and small square rump patch. **Adult** ♀: Drab brown above, with blurry streaking on the back. Dull yellow rump and small yellow epaulets. Yellow supercilium and face with a narrow brown eye-stripe and black lores. Dull orange-yellow throat and breast becoming yellow below and white on the vent. **Voice** Song is a short phrase including brief high-pitched, twangy whistles and a harsh nasal rasp or churr, *trillwraaah-chidú*, with much variation and recalling other marsh blackbirds. Calls include a low *chwup* and a buzzy *zeeu*. [Tordo Amarillo]

Scarlet-headed Blackbird *Amblyramphus holosericeus* ♂ 26.5 cm; ♀ 23.5 cm

NE marshes and locally in NW. **ID Adult** ♂: Black with scarlet head, neck and thighs. **Adult** ♀ (not illustrated): Somewhat smaller, more matt black, and scarlet areas have a slight orange tone. **Immature**: Dirty red throat and breast, and scattered red spots on the head. **Voice** Sings in duet, with one sex giving a gargled, ringing *kiuw-ki-di-di* while the second gives a low-pitched whistle followed by a trill, *peee-ou kwewewewe*. Also, flat whistles and rapid liquid trills. Flight call is 4–5 short whistles, *dew dew-dew-dew*. [Federal]

Yellow-rumped
Marshbird

ad

Brown-and-yellow
Marshbird

ad

ad ♂

juv

ad ♀

♂

ad ♂

imm

Saffron-cowled Blackbird

Scarlet-headed Blackbird

Agelaioides Gregarious icterid with a stout pointed bill. Nests in abandoned or active nests of other birds, chiefly furnariids. Parasitised by Screaming Cowbird with which it often occurs. *Molothrus* Robust gregarious icterids, which lay their eggs in other birds' nests. Screaming Cowbird almost exclusively parasitises Greyish Baywing which the juvenile closely resembles.

Greyish Baywing *Agelaioides badius*   18.5 cm

N and C lowlands and Andes. **ID Adult:** Pale brown with chestnut wings. Bill more slender and longer than juvenile Screaming Baywing, lores blacker, primary projection notably short, and grey underwing-coverts in flight. **Voice** Song is a steady series of rather flat whistles and metallic trills that move up and down the scale in semitones, like an orchestra tuning instruments. Common call is a geographically variable *chac-ziii.* [Tordo Músico]

Screaming Cowbird *Molothrus rufoaxillaris*   20 cm

N and C lowlands and Andes. Often found with Greyish Baywing. **ID Adult:** Stout bill, shorter and more conical than Shiny Cowbird. Entirely black with little or no gloss. Chestnut iris and dark chestnut axillaries difficult to see. Flocks readily identified from Shiny Cowbird by lack of brown ♀♀. **Juvenile:** Pale brown with chestnut wings. Very similar to Greyish Baywing but bill shaped as adult, long primary projection, indistinct blackish lores, and black underwing-coverts visible in flight. **Juvenile moulting:** Irregular scattering of black spots and blotches. **Voice** Song is a disyllabic, metallic whistle, *ZU-KLEE*. Call is a shrill, rising, metallic churring rasp, *SCHRRREE* or *schr-SCHRR-SCHRRREE.* [Tordo Pico Corto]

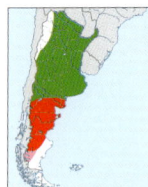

Shiny Cowbird *Molothrus bonariensis*   ♂ 20.5 cm; ♀ 18.5 cm

Widespread. **ID Adult ♂:** Fairly long bill with straightish culmen. Black with a distinct purple-blue gloss in favourable light. **Adult ♀:** Uniform brown above and greyish-brown below with a paler throat. **Adult ♀ dark morph** (Only in Misiones and e. Formosa): Entirely dark brown. **Juvenile:** Light biscuit-brown with indistinct streaking below. **Voice** Song during ruffed-neck display is a low gurgle followed by piercing liquid notes. Song while perched or in flight is a canary-like musical series of rapid, high-pitched notes and trills. ♀♀ give a diagnostic, slow cackling chatter. [Tordo Renegrido]

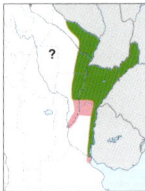

Chopi Blackbird *Gnorimopsar chopi*   23 cm

Light woodlands and palm savanna in the NE. **ID** Chunky, gregarious icterid and colonial nester with comparatively short squarish tail. At close range, fine pointed nuchal feathers are visible. **Adult:** Stout, broad-based bill with slightly curved culmen and grooves on the base of the lower mandible, visible only at close range. Entirely glossy black. **Voice** Musical song is interspersed with *Cho-Pí* phrases and churrs. Also delivers rich single whistles on different pitches. [Chopí]

Austral Blackbird *Curaeus curaeus*   26.5 cm

Patagonian forest and shrub-steppe. **ID** Fairly large gregarious forest icterid with perfectly straight culmen and gonys. **Adult *curaeus*** (Patagonia): Entirely glossy black. No other icterid in range with such a long straight bill. **Adult *reynoldsi*** (Tierra del Fuego; not illustrated): even longer and deeper-based bill. **Voice** Common song theme includes the nasal flight call, *ch-de-dew*, often followed by a whistle and a variable ringing chatter, e.g. *ch-de-dew wee drreelll*, sometimes in long series. [Tordo Patagónico]

Giant Cowbird *Molothrus oryzivorus*   ♂ 35 cm; ♀ 30 cm

Forest clearings in extreme N lowlands and foothills. **ID Adult ♂:** Proportionately small head. Glossy purple-black with a large black bill and a curved culmen. **Adult ♀:** Less glossy than ♂ but with some purple on the breast. Smaller-billed than ♂ but bill still very stout. Head is proportional to the body, unlike ♂. Iris variable. **Voice** Song is a piercing, shrill, long whistle, *ZHEEEEEEEEH* or a similar series of shorter notes. Call is a muffled *shack* or *chwit*, often in flight with audible wingbeats. [Tordo Gigante]

Greyish Baywing

badius

ad

juv moulting

Screaming Cowbird

juv

ad

ad ♀
dark morph

bonariensis

ad ♀

juv

ad ♂

Shiny Cowbird

ad

chopi

Chopi Blackbird

ad

curaeus

ad ♂

oryzivorus

Austral Blackbird

ad ♀

Giant Cowbird

Icterus Slim-bodied icterid with comparatively long slender tail and bill. Usually in pairs or with mixed-species flocks but more gregarious in winter.

Variable Oriole *Icterus pyrrhopterus*   19.5 cm

Forest and secondary growth in N lowlands. **ID** Adult *pyrrhopterus* (widespread): Black with chestnut shoulders, sometimes difficult to see. Silvery-white underwing often visible in flight. Adult *tibialis* (rare in n. Misiones): Golden-yellow epaulets. Birds with orange-brown epaulets also recorded. Nest on Plate 202. **Voice** Song comprises a complex series of melodic whistles, chuckles, trills and rasps, e.g. *swee-siú-see-see-suwée-chrr-shhh*. Calls include wheezy notes, rasps and often imitations of raptors and woodpeckers. [Boyerito]

Cacicus Noisy, black icterids with pale conical bills. Nests are long pendulous sacs.

Golden-winged Cacique *Cacicus chrysopterus*   19.5 cm

Yungas forest and NE lowlands. **ID** Adult: Whitish bill and iris. Black with golden rump and stripe across wing-coverts. Nest on Plate 202. **Voice** Song is a soft, low, liquid warble followed by powerful slurred whistles delivered in a combination of brief and well-spaced phrases, with much variation, e.g. *gu-gu-kreeeel-WREEU-TZEEwí*. Common call is a wheezy, nasal, descending *wheyyh*, often repeated. [Boyero Ala Amarilla]

Solitary Cacique *Cacicus solitarius*   29 cm

Chaco woodlands and gallery forest south to ne. Buenos Aires. **ID** Adult: Entirely black with a long ivory or straw-coloured bill. Nest on Plate 202. **Voice** Song is a low-pitched *kuk-kuk-kuk-kuk* followed by a resonant, melodic, ventriloquial *ki-woo ki-WHAA* or *k-cho kwo*, sometimes squeaky with repeated couplets ending in a gurgled trill, with much variation. Common call is a nasal *wheerr*, more raspy and less wheezy than Golden-winged Cacique. [Boyero Negro]

Red-rumped Cacique *Cacicus haemorrhous*   28.5 cm

Paraná forest. **ID** Gregarious colonial nester. Adult ♂: Black with a large yellowish bill and contrasting red rump. Adult ♀ (not illustrated): Averages smaller with browner underparts. Nest on Plate 202. **Voice** Song is a ringing gurgle *grrleew*, often followed by audible wing-flapping with twittering notes. Noisy groups give loud parrot-like series intermixed with harsh, nasal *SHWRRAP* notes or whistles. [Boyero Cacique]

Psarocolius Large, long-tailed forest icterid with long, broad-based bill. Nests are spectacular pendulous sacs.

Crested Oropendola *Psarocolius decumanus*   ♂ 46 cm; ♀ 35cm

Yungas foothill forest; local in gallery forest in the NE. **ID** Adult ♂: Black with chestnut rump and vent and contrasting yellow outer tail feathers. In flight, striking yellow tail. Adult ♀ (not illustrated): Obviously smaller, with browner plumage and reduced crest. Nest on Plate 202. **Voice** Song is a complex series beginning with soft knocking notes, then a gurgled trill and loud descending whistles, followed by rustling wing sounds with body thrust forward, e.g. *k-k-k-grrleeeeeWOOOOOOW*. Call is a dry, harsh *SCRAK* or *CHAK*. [Yapú]

Variable Oriole

pyrrhopterus

ad

ad

Golden-winged Cacique

ad

tibialis

ad

Solitary Cacique

ad ♂

Red-rumped Cacique

affinis

ad ♂

maculosus

ad

Crested Oropendola

Bobolink *Dolichonyx oryzivorus* 17 cm

Marshes in the N lowlands, especially in *Cyperus* rushes. **ID** Finch-like non-breeding boreal migrant with a short spiky tail and short, stout conical bill. Many males arrive and depart in breeding plumage. **Adult ♂ breeding**: Mainly black, scalloped with buff and with a contrasting orange-buff nuchal patch. **Adult ♀ / adult ♂ non-breeding**: Pinkish bill with black culmen. Buff coronal stripe bordered by black lateral crown stripes. Striking buff scapular braces. Buff below with prominent black flank streaking. **Voice** Song is a rapid series of squeaks, short whistles and nasal notes, simplified to *bob-o-link*. Calls include a nasal *pink* (often given in flight) and a harsh *tjed*. [Charlatán]

Leistes Sexually dimorphic icterids of natural grassland and crops with variable amounts of red on the underparts and a white or buff supercilium. Tail and bill length varies between species. Underwing coloration is diagnostic. Gregarious.

White-browed Blackbird *Leistes superciliaris* ♂ 17.5 cm; ♀ 16 cm

N and C lowlands. **ID Adult ♂ breeding**: Black with a pure white rear supercilium and crimson throat and breast. In flight, black underwing at all ages and in both sexes. **Adult ♂ non-breeding**: Rich buff supercilium. Black upperparts and belly, thickly scalloped with buff. Reddish throat and breast. **Adult ♀**: Resembles ♀ Bobolink (note habitat differences) with a less well-defined coronal stripe and more prominent supercilium. Lacks scapular braces and tail is comparatively shorter, barred and square-ended (not spiky). Diagnostic pinkish blotches on the abdomen. Flanks spotted instead of streaked. **Voice** Song during display flight consists of short whistles, a long buzz and a stutter while descending, e.g. *ZI-ZI-CHZZZZZ-TEE-chuck-chak-cha…cha*. Perched song is a series of thin, musical whistles and short rasps. Calls include a scolding, metallic rasp *tzeek*. [Pecho Colorado]

Pampas Meadowlark *Leistes defilippii* 21 cm

Chiefly restricted to sw. Buenos Aires; very rare in NE and C lowlands. **ID** Adult ♂ breeding: Red loral spot unlike White-browed Blackbird. From Long-tailed Meadowlark by deeper crimson throat and breast with a narrower black border and a rounded lower edge; also comparatively shorter, straighter bill and shorter tail. In flight, black underwing in all plumages, unlike Long-tailed and Sierran Meadowlarks. Makes a parachuting song display flight unlike Long-tailed. **Adult ♂ non-breeding**: More heavily fringed with buff above than breeding birds, duller red breast and pale scalloping on the belly. Best distinguished from Long-tailed by buff (not white) supercilium and rounded lower edge of the darker breast. **Adult ♀**: From ♀ Long-tailed by its smaller and buffy (not white) throat patch without a clearly defined border, and short indistinct buffy malar. Breast usually heavily spotted. **Voice** Song during display flight is a series of short penetrating whistles followed by a buzz and a chatter while parachuting down, *zizizi-BZZZZZ-ch-d-d-d*. Perched song has 2–3 brief whistles followed immediately by a rising wheezy rasp, *zi-li-leee-zeeee*, much simpler and shorter than Long-tailed Meadowlark. Call is a soft *chwup*. [Loica Pampeana]

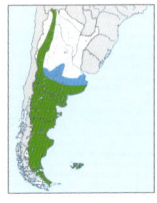

Long-tailed Meadowlark *Leistes loyca* 25.5 cm

Almost throughout, including NW Andes (to 3100 m) and Falklands, with similar ssp. **ID** Adult ♂: Longer-billed with a slightly curved upper mandible, and longer-tailed than Pampas and Sierran Meadowlarks; also paler and browner above and on the flanks. Paler red throat and breast terminating in a jagged point on the belly. Sings from a perch not during a display flight. In flight, silvery-white underwing distinguishes all plumages from White-browed Blackbird and Pampas Meadowlark. **Adult ♀**: Buffy-white supercilium. Clear-cut white or buffy-white throat, demarcated by blackish flecking around edges. Reddish-pink centre of abdomen. **Voice** Perched songs include rapidly repeated notes followed by a long, metallic, grating buzz, *twi-twi-chichichi-TZZZRRREA*, often modified into a broken, falling then rising buzz, *twi-chwi-ZRREO TZREEA*. Call is a soft *chwup*. [Loica]

Sierran Meadowlark *Leistes [loyca] obscurus* 24.5 cm

Endemic to the sierras of Córdoba–San Luis. **ID** Adult ♂: Slightly smaller and considerably darker throughout than Long-tailed Meadowlark with blacker upperparts and belly, and with deeper, richer red throat and breast. No overlap with Pampas Meadowlark, but has silvery-white underwing like Long-tailed (not black as in Pampas). Conducts song display flights like Pampas, completely unlike Long-tailed. **Adult ♀**: Unmistakable. More extensive and redder patch on abdomen than ♀ Long-tailed, more extensive black streaking around pinkish-white throat, and more extensive black on belly, being generally darker overall. **Voice** Perched songs include short introductory notes and two long metallic grating buzzes, often separated by a whistle, e.g. *chi dew ZRREEEO weee WHEERRR* or *zik-zik-ZRRREW ch-wee DZZZZZ*. Call is a soft *chwup*. **Tax note 72**. [Loica Serrana]

Bobolink

ad ♂ br

ad ♀ /
ad ♂ non-br

White-browed Blackbird

ad ♀

ad ♂ non-br

ad ♂

ad ♂ br

ad ♀

ad ♂ br

ad ♂

ad ♂ non-br

ad ♂

Pampas
Meadowlark

ad ♀

Sierran
Meadowlark

ad ♂

loyca

ad ♂

ad ♀

ad ♂

Long-tailed Meadowlark

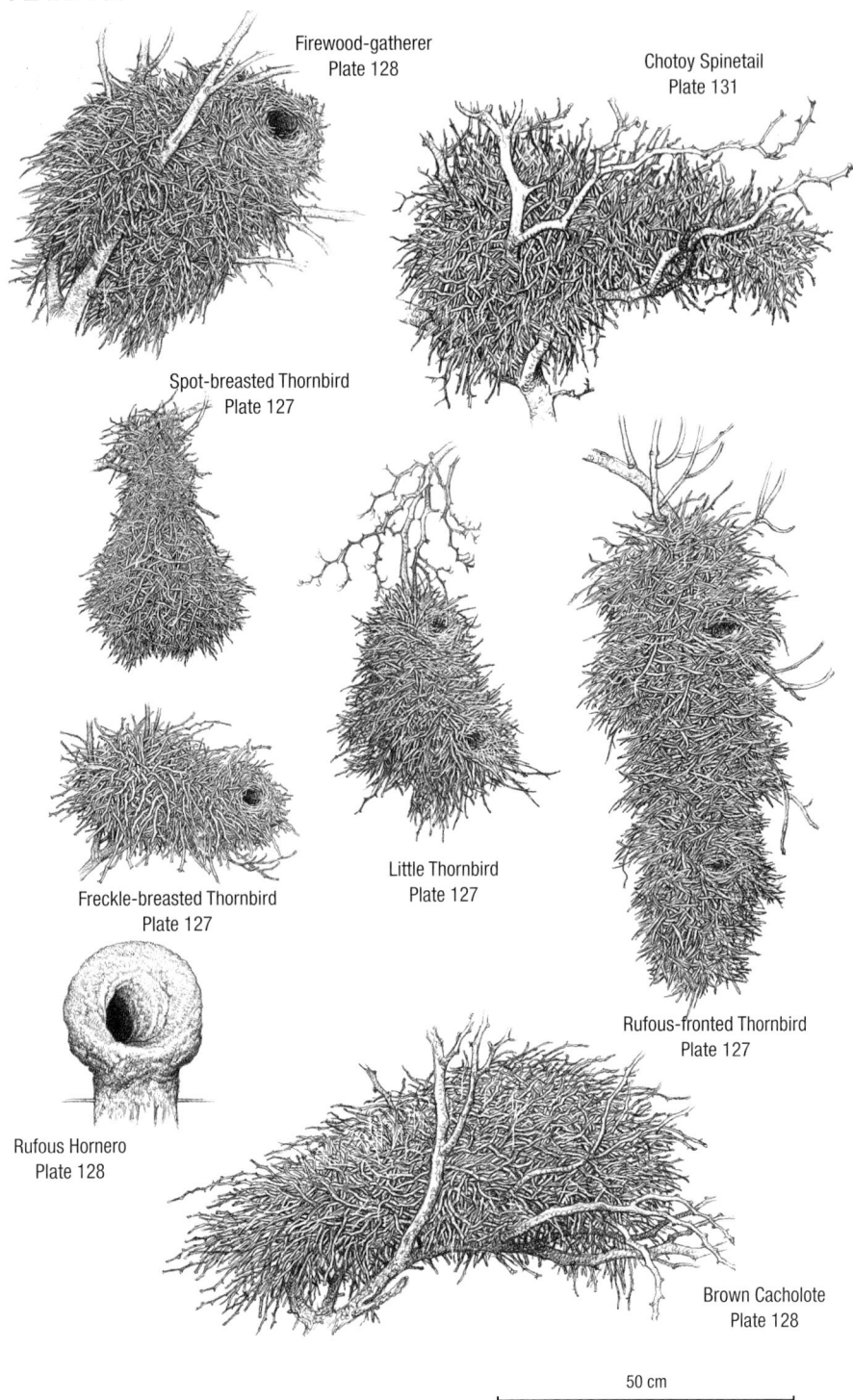

Firewood-gatherer
Plate 128

Chotoy Spinetail
Plate 131

Spot-breasted Thornbird
Plate 127

Little Thornbird
Plate 127

Freckle-breasted Thornbird
Plate 127

Rufous-fronted Thornbird
Plate 127

Rufous Hornero
Plate 128

Brown Cacholote
Plate 128

50 cm

PLATE 201: FURNARIID NESTS II

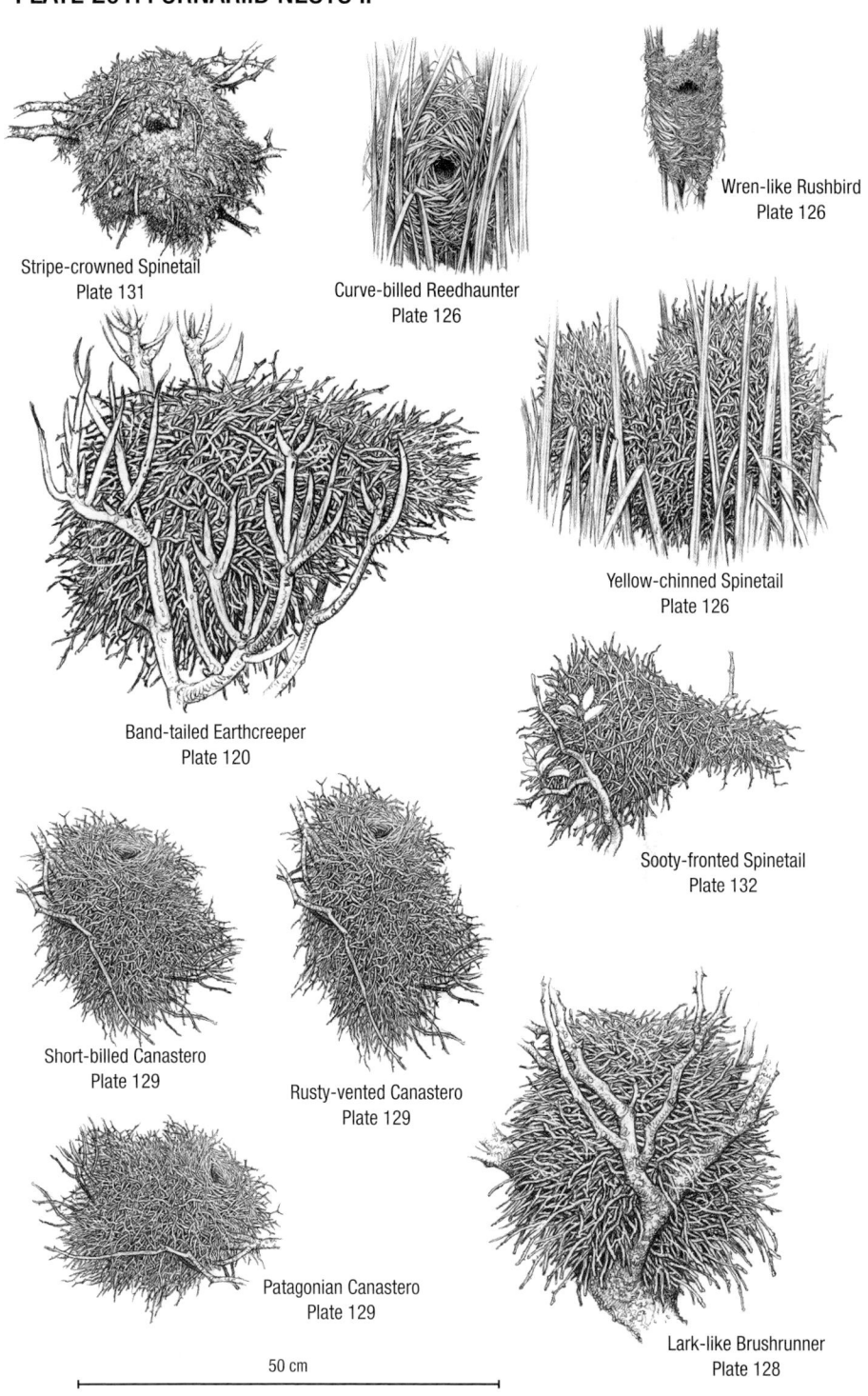

Stripe-crowned Spinetail
Plate 131

Curve-billed Reedhaunter
Plate 126

Wren-like Rushbird
Plate 126

Band-tailed Earthcreeper
Plate 120

Yellow-chinned Spinetail
Plate 126

Sooty-fronted Spinetail
Plate 132

Short-billed Canastero
Plate 129

Rusty-vented Canastero
Plate 129

Patagonian Canastero
Plate 129

Lark-like Brushrunner
Plate 128

50 cm

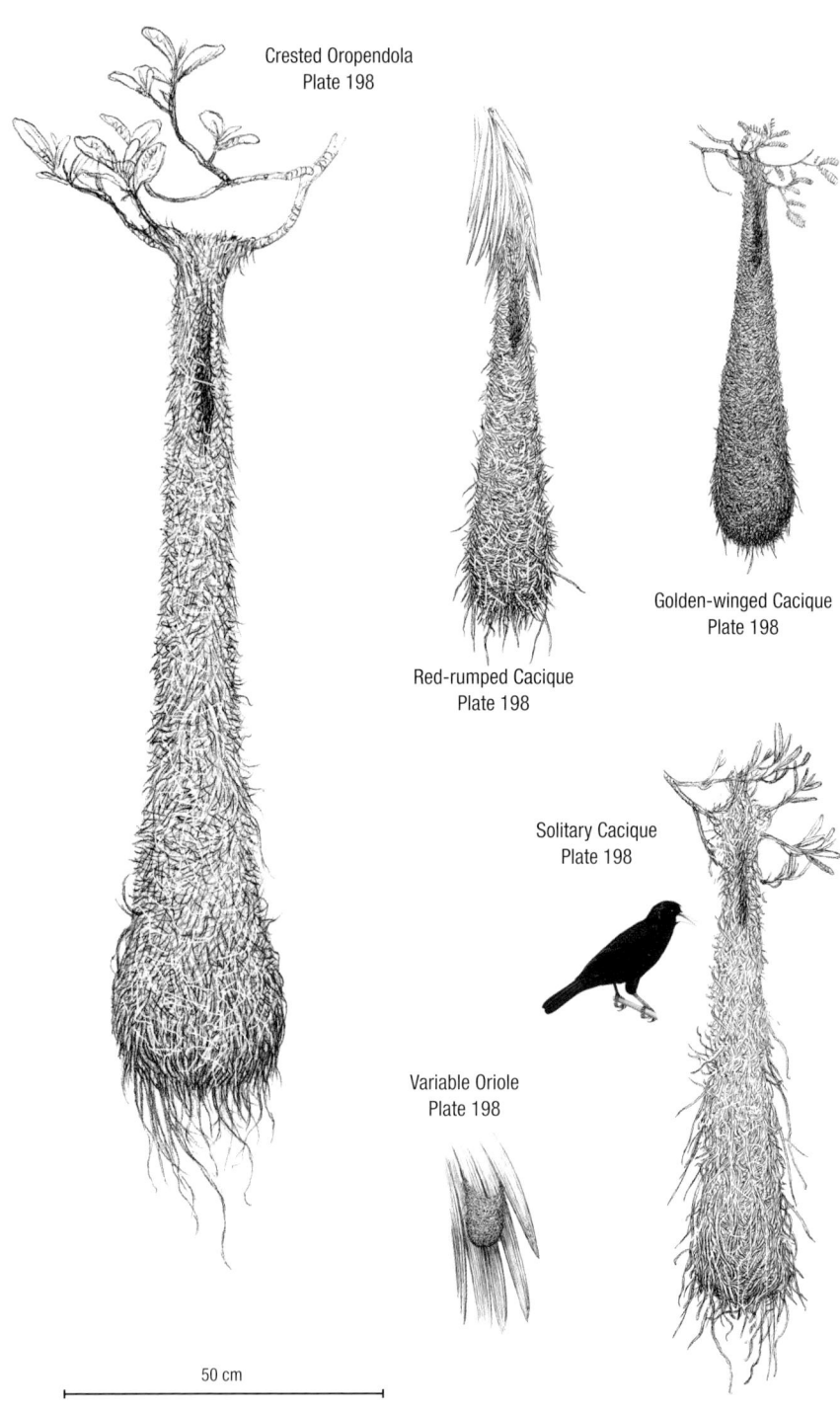

Crested Oropendola
Plate 198

Red-rumped Cacique
Plate 198

Golden-winged Cacique
Plate 198

Solitary Cacique
Plate 198

Variable Oriole
Plate 198

50 cm

APPENDIX 1: INTRODUCED SPECIES

Greylag Goose *Anser anser*
77 cm

Naturalised in e. Entre Ríos, coastal Buenos Aires, w. Santa Cruz and the Falklands. **ID** Gregarious, thick-necked goose with grey, white or intermediate plumage. **Adult** Note orange bill and legs, usually with much white in the wing (sometimes just the primaries) and outer tail-feathers. **Voice** Cackling calls and some honking. [Ganso Común]

California Quail *Callipepla californica*
28 cm

Brush-steppe of CW Andean foothills (chiefly Neuquén, w. Río Negro and Chubut; to 1200 m), expanding into the lowlands. **ID** Terrestrial and arboreal. **Adult** ♂: Striking recurved bulbous antenna-like crest. Black foreface outlined in white. Chestnut crown, bordered white. Remainder of head, nape and breast blue-grey. Buffy belly, and notable white or buff flank streaking and ventral spotting. **Adult** ♀: Whitish throat, short coronal tuft; otherwise brown with similar flank and belly markings as ♂. **Voice** Song is a far-carrying rhythmic *KUK WY-CU,.. KUK WY-CU...* Calls include a sudden, medium-pitched questioning *KUU* note, rapid spluttered, nasal chattering and loud clicks. [Codorniz de California]

Silver Pheasant *Lophura nycthemera*
♂ 90 cm; ♀ 63 cm

Patagonian forest and dense exotic forest on Isla Victoria, w. Río Negro. **ID Adult** ♂: Mostly silvery-white with a black crown and underparts, red facial skin, lappets and legs, downwardly-arched white tail. **Adult** ♀: Mostly brown with some white vermiculations, bare orange facial skin, and shorter brown and black tail than ♂. [Faisán Plateado]

Feral Pigeon *Columba livia*
35 cm

Ubiquitous, mostly in towns and cities; highly gregarious. **ID** Adult: Plumage highly variable, often grey with a violet/green nape and two black bands on the wing-coverts; otherwise with various amounts of white. **Voice** Deep cooing and gurgled sounds. [Paloma Doméstica]

Crested Myna *Acridotheres cristatellus*
26 cm

Urban areas, plantations and various light woodlands from La Plata, Buenos Aires south to ec. Chubut and sparsely in the west; gregarious. **ID** Adult: Black with a notable frontal crest (shorter in ♀), square white patch in base of primaries, and rounded tail with terminal white band. [Estornino Crestado]

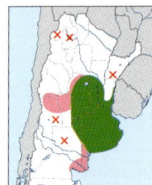

Common Starling *Sturnus vulgaris*   22 cm

Widespread in urban areas and the Pampas, expanding south to Río Negro and west to Mendoza; one Falklands record. **ID Adult:** Breeding birds have a pointed, slender yellow bill, and black plumage glossed blue-green; non-breeders are heavily spotted with buff and have a dark bill. Wings are triangular-shaped in flight. **Voice** Imitates all sorts of sounds. Common call is a shrill *tzhreee*. [Estornino Pinto]

European Greenfinch *Chloris chloris*   15 cm

Widespread in e. Buenos Aires; often in pine forest. **ID Adult** ♂: Stout pinkish bill. Mostly rich olive with a yellow primary flash, visible on the folded wing. Notched tail with yellow bases to the outer tail-feathers. **Adult ♀:** Browner and duller. Compare with ♀ Hooded Siskin (Plate 193). **Juvenile:** Fine dorsal and ventral streaking on greyish white underparts. **Voice** Song comprises fast or slow monotone trills. Call is a long rasp, *SCHRREEEEU* or *ZREEE.* [Verderón]

European Goldfinch *Carduelis carduelis*   12.5 cm

Local in e. Buenos Aires and ce. Entre Ríos. **ID Adult:** Striking red foreface bordered by white, with black running from the crown to the nape. Brownish mantle and white rump. Black wings with a broad golden wing-stripe and white tertial spots. Notched black tail with a white subterminal band. Mostly white below with pale brown flanks and breast spurs. **Juvenile:** Plain greyish head, fine dorsal and breast streaking. Best identified by striking wing pattern as adult. **Voice** Song is complex, with siskin- or wren-like series of trills, twittering and metallic tinkling notes. [Cardelino]

House Sparrow *Passer domesticus*   15 cm

Widespread in towns throughout, also caldén and sometimes espinal woodland. **ID Adult** ♂: Grey crown and cheeks bordered by a brown hindcrown and nape. Striking black bib and upper breast. Brown above, streaked black. Greyish below with a short white median-covert wing bar. **Adult ♀:** Rather nondescript. Drab brown crown and narrow eyestripe, bordering a conspicuous buffy rear supercilium. Greyish face and underparts. Upperparts like ♂ with a narrow whitish median-covert wing bar. **Voice** Noisy; gives rather unmusical, strident series of chirps or short metallic warbles. [Gorrión]

APPENDIX 2: KEYS TO PRION IDENTIFICATION
(Illustrations on Plate 23)

a) Rapid aid to prion (*Pachyptila*) identification in the region

	Bill width	Bill length	Tail-band width	M on upperwing
Broad-billed *P. vittata*	very broad	long	medium	fairly bold
Antarctic *P. desolata*	broad	long	medium	bold
Slender-billed *P. belcheri*	narrow	fairly long	narrow	indistinct
Fairy *P. turtur*	narrow	short	broad	bold

b) Biometric measurements of adult prions (*Pachyptila*) in the region

Bill width is given at the widest point. Figures in parentheses are from breeding sites outside the region. Measurements are in millimetres.

	Bill length	Bill width	Wing	Tail
Broad-billed *P. vittata*	>30	17–25	191–229	94–120
Antarctic *P. desolata*	26.4–28.8	13.3–15.7	185–199	85–98
Slender-billed *P. belcheri*	23.4–27.6	9.0–11.5	175–191	81–96
Fairy *P. turtur*	21 (19.8–26.0)	(10.0–12.5)	185 (168–191)	96 (78–99)

Knowledge of prion bill morphology can also aid their identification. In the hand, the interramal space on the underside of the bill is feathered in Slender-billed and Fairy Prions, whereas Antarctic and Broad-billed have bare skin between the rami, and in the latter species there is a fleshy distensible pouch.

APPENDIX 3: KEYS TO PIPIT IDENTIFICATION

(Illustrations on Plates 168–169)

a) Field identification of pipits

	Short-billed	Puna	Yellowish	Pampas	Hellmayr's	Paramo	Correndera	Ochre-breasted
Hindclaw	short	medium	long	short	fairly long	long	fairly long	very long
Malar streak	narrow	narrow	indistinct	none	none	indistinct	narrow	notable
Dorsal braces	none	none	none	none	none	none	distinctive	distinctive
Fresh breast	buff	ochre	sulphur	dull buff	pale buff	strong buff	buff	ochre-yellow
Fresh belly	greyish-white	greyish-white	sulphur white	greyish-white	greyish-buff	greyish-white	greyish-white	yellowish-
Fresh breast markings	dense spots	spots	fine streaks	sparse streaks	sparse streaks	small streaks	thick spots	sparse streaks
Flank markings in fresh plumage	sparse streaks	sparse streak	none	long streaks	broad streaks	long streaks	oval spot-streaks	long streaks
Worn breast and belly	white – pale buff	white – pale buff	white – grey-white	white ?	white – grey-white	pale buff ?	whitish	buff – white
Worn breast markings	small spots	small spots or streaks	sparse streaks	?	sparse streaks in some	?	small spots or streaks	very sparse streaks

b) In-hand identification of pipits by tail pattern

	outermost rectrix r1	penultimate rectrix r2	third outermost rectrix r3
Short-billed	mostly white	mostly white	no pattern
Puna	mostly white and tip to inner vane	mostly white	no pattern
Yellowish	mostly white	white tip	no pattern
Pampas	white	white?	no pattern
Hellmayr's *brasilianus* and *dabbenei*	mostly white	white distal half	no pattern
hellmayri	dull buff	dull buff tip	no pattern
Paramo	cream	cream tip	no pattern
Correndera *correndera*	mostly white	mostly white	no pattern
calcaratus	buffy-white	buffy-white	no pattern
catamarcae, chilensis and *grayi*	mostly white	partially white	no pattern
Ochre-breasted	mostly white	mostly white	white tip

APPENDIX 4: SONOGRAMS

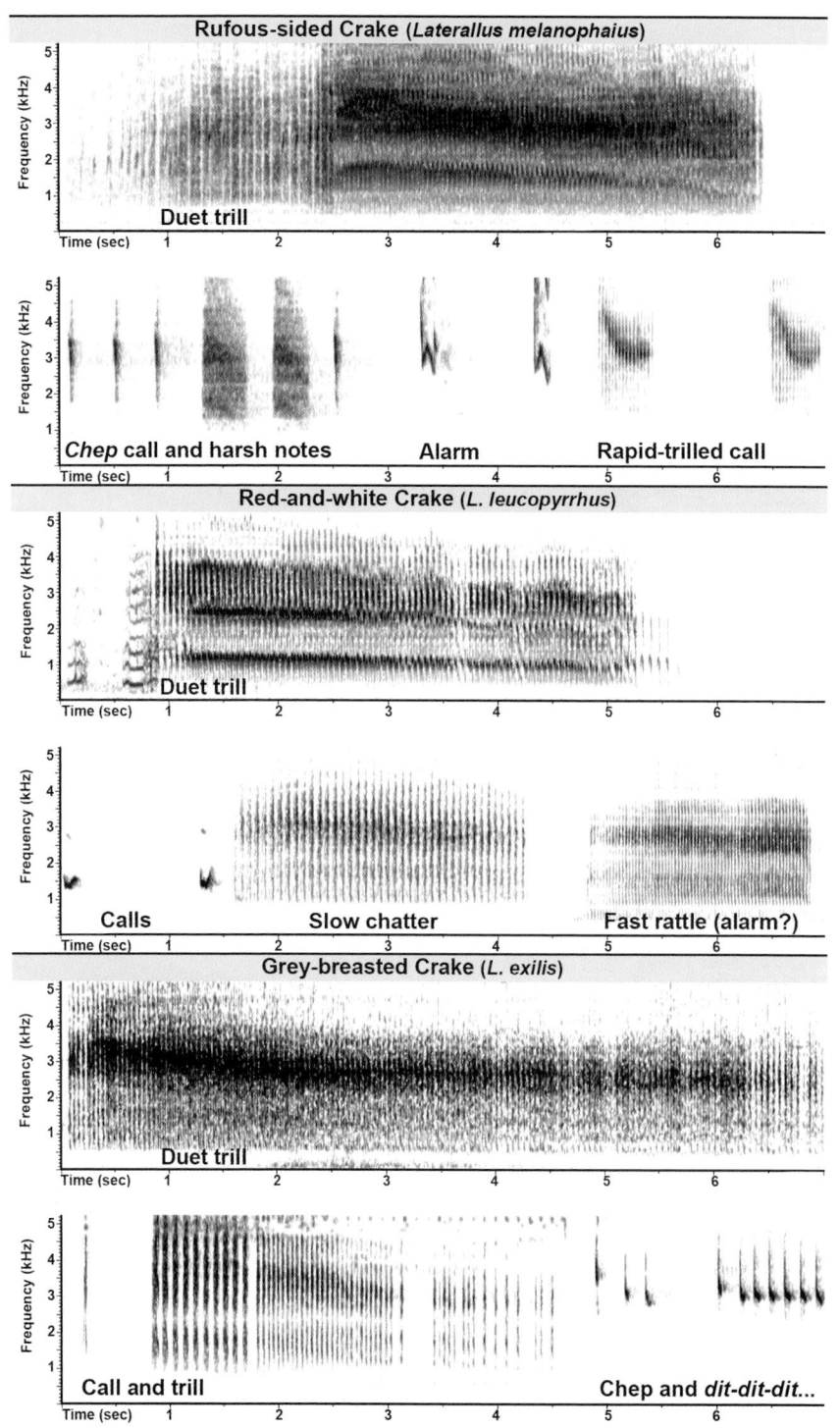

Rufous-sided Crake (*Laterallus melanophaius*)

Frequency (kHz) / Time (sec)

Duet trill

Chep call and harsh notes · Alarm · Rapid-trilled call

Red-and-white Crake (*L. leucopyrrhus*)

Duet trill

Calls · Slow chatter · Fast rattle (alarm?)

Grey-breasted Crake (*L. exilis*)

Duet trill

Call and trill · Chep and *dit-dit-dit...*

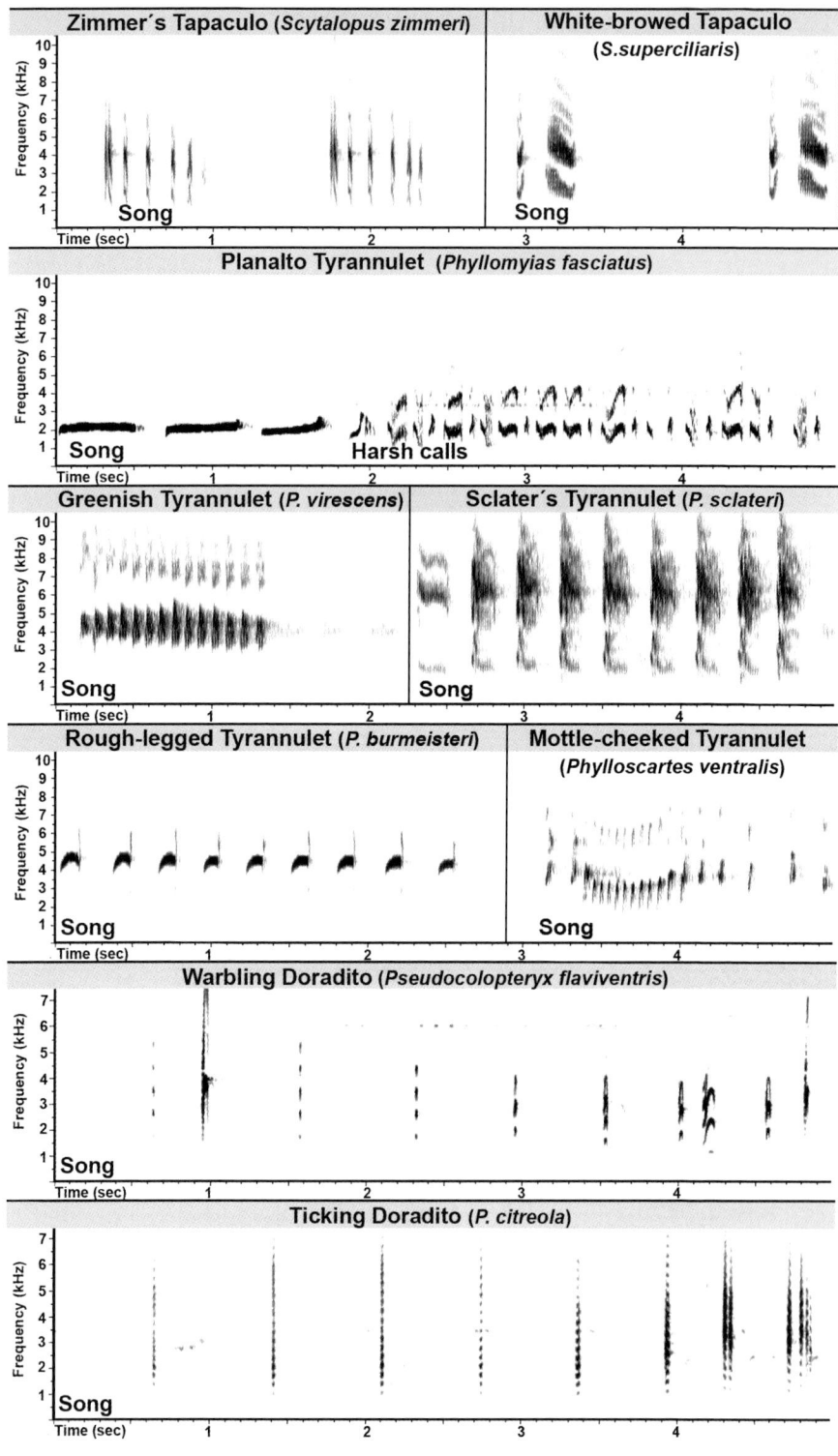

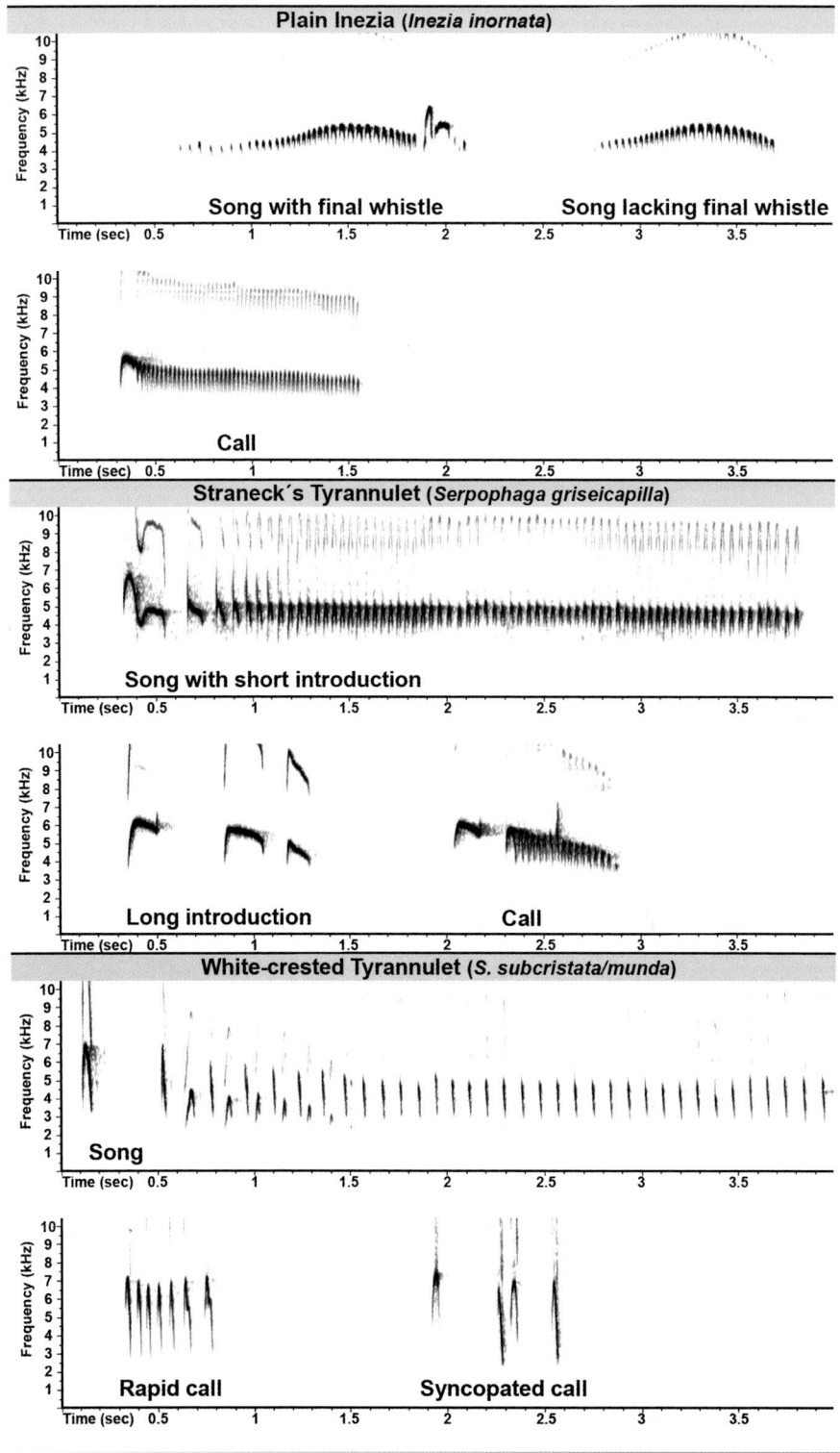

Plain Inezia (*Inezia inornata*)

Song with final whistle

Song lacking final whistle

Call

Straneck's Tyrannulet (*Serpophaga griseicapilla*)

Song with short introduction

Long introduction

Call

White-crested Tyrannulet (*S. subcristata/munda*)

Song

Rapid call

Syncopated call

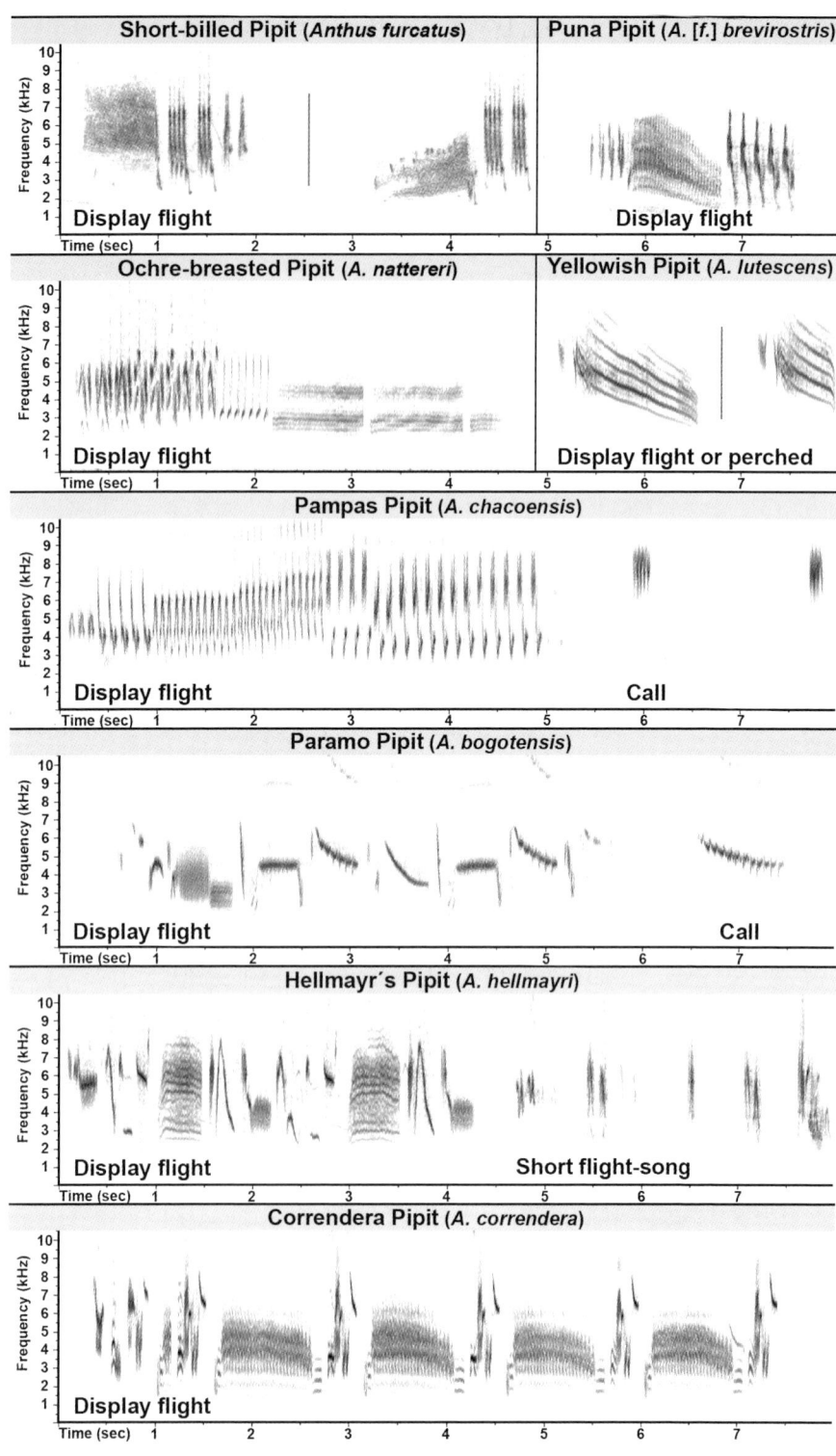

APPENDIX 5: LIST OF ILLUSTRATED FLORA

This appendix identifies 58 species of plants illustrated on the plates and a further 14 genera, listed in alphabetical order with scientific name synonyms in brackets. The most commonly used or available vernacular names are provided, together with some other alternative names in common use, general group names and 'translated names' in parentheses where deemed useful. As far as possible, the flora were chosen to show typical habitats for the bird species concerned and plants are illustrated to scale with the birds.

Scientific plant name [synonym]	Vernacular plant name (alternatives, useful 'translations')	Plate number	English bird name * mostly found on this plant ** almost exclusive to this plant
Acacia aroma	Tusca	182	Chaco Warbling Finch
Acacia caven	Espinillo	132	Austral Spinetail
		129	Short-billed Canastero
		127	Little Thornbird
		180	Chaco Sparrow
Allophylus edulis	Cocú (Chal chal)	156	Wing-barred Piprites
Alnus acuminata [A. jorullensis]	Aliso del Cerro (alder)	5	Red-faced Guan *
		107	Dot-fronted Woodpecker *
Aloysia cf. gratissima	Cedrón de Monte	155	White-tipped Plantcutter
Araucaria angustifolia	Pino Paraná (Paraná Pine, Monkey Puzzle)	125	Araucaria Tit-Spinetail **
		158	Azure Jay *
Azorella sp.	Yareta (cushion plant)	190	Red-backed Sierra Finch **
		190	Glacier Finch **
Baccharis salicifolia	Chilca	143	Dinelli's Doradito *
Bauhinia microstachya	Caí-escalera (monkey ladder)	123	Plain Xenops
Berberis sp.	Barberry	125	Plain-mantled Tit-Spinetail
		130	Austral Canastero *
Brassavola tuberculata	Niña de la Noche (epiphytic orchid)	110	Helmeted Woodpecker
Bulbophyllum napellii	Orquídea (tree orchid)	108	Yellow-browed Woodpecker
Campylocentrum grisebachii	Orquídea áfila (leafless orchid)	158	Black-tailed Tityra
Cecropia pachystachya	Ambay (Ambaí)	5	Rusty-margined Guan
		107	White-spotted Woodpecker
		107	Yellow-fronted Woodpecker
Celtis chichape [C. pallida]	Tala churqui	178	Ultramarine Grosbeak
Celtis ehrenbergiana [C. tala]	Tala	106	White-barred Piculet
		131	Chotoy Spinetail
		182	Black-capped Warbling Finch
		192	Saffron Finch
Cercidium praecox	Brea	182	Black-crested Finch
Chusquea culeou	Coligüe (bamboo)	125	Des Murs's Wiretail *
		114	Magellanic Tapaculo
		114	Ochre-flanked Tapaculo

Scientific plant name [synonym]	Vernacular plant name (alternatives, useful 'translations')	Plate number	English bird name * mostly found on this plant ** almost exclusive to this plant
		116	Rufous Gnateater *
Chusquea ramosissima	Tacuarembó (bamboo)	111	Tufted Antshrike
		132	Grey-bellied Spinetail *
Condalia microphylla	Piquillín	188	Mourning Sierra Finch
Copernicia alba	Caranday (caranday palm)	108	Pale-crested Woodpecker *
Cortaderia selloana	Cortadera (Pampas grass)	183	Great Pampa Finch *
Discaria chacaye	Chacay	130	Sharp-billed Canastero
Eichhornia azurea	Aguapey (water hyacinth)	54	Yellow-breasted Crake *
		56	Giant Wood Rail
Eichhornia crassipes	Camalote (water hyacinth)	57	Red-fronted Coot
Eryngium sp.	Serrucheta (saw grass)	126	Straight-billed Reedhaunter **
		187	Band-tailed Seedeater
		196	Saffron-cowled Blackbird **
Erythrina crista-galli	Ceibo (Seibo)	198	Solitary Cacique
Festuca sp.	Festuca (Fescue grass)	130	Scribble-tailed Canastero **
Foeniculum vulgare	Hinojo (Fennel)	199	White-browed Blackbird
Geoffroea decorticans	Chañar	106	Checkered Woodpecker
		120	Chaco Earthcreeper
Guadua trinii [Bambusa trinii]	Yatevó (bamboo)	77	Purple-winged Ground Dove *
		106	Mottled Piculet *
		111	White-bearded Antshrike **
		184	Blackish-blue Seedeater *
Guadua chacoensis	Tacuaruzú (bamboo)	184	Temminck's Seedeater *
		184	Sooty Grassquit *
Larrea cuneifolia	Jarilla Macho (creosote)	182	Diuca Finch
Larrea divaricata	Jarilla Hembra (creosote)	129	Patagonian Canastero
Larrea nitida	Jarilla Crespa (creosote)	188	Carbonated Sierra Finch
Ludwigia sp.	Cruz de Malta	126	Yellow-chinned Spinetail
Merostachys claussenii	Tacuapí (bamboo)	114	Spotted Bamboowren *
		124	Canebrake Groundcreeper *
Microgramma sp.	Suelda Consuelda (epiphytic fern)	5	Bare-faced Curassow
Mimosa bonplandii	Rama Negra	112	Rufous-capped Antshrike
Monttea aphylla	Mata Sebo	189	Grey-hooded Sierra Finch
Myrcianthes cisplatensis	Mato	182	Red Pileated Finch
Nothofagus antarctica	Ñire	109	Chilean Flicker
		122	Patagonian Forest Earthcreeper **
		193	Black-chinned Siskin
Nothofagus betuloides	Guindo	189	Patagonian Sierra Finch
Nothofagus dombeyi	Coihue	94	Rufous-legged Owl
		131	Thorn-tailed Rayadito

Scientific plant name [synonym]	Vernacular plant name (alternatives, useful 'translations')	Plate number	English bird name * mostly found on this plant ** almost exclusive to this plant
Nothofagus pumilio	Lenga	47	Rufous-tailed Hawk
		76	Chilean Pigeon
Opuntia sp.	Tuna (cactus)	118	Scimitar-billed Woodcreeper
Peperomia sp.		94	Spectacled Owl
		117	Black-banded Woodcreeper
		118	Olivaceous Woodcreeper
Pistia stratiotes	Repollo de Agua (water lettuce)	57	Red-fronted Coot
Podocarpus parlatorei	Pino del Cerro	76	Band-tailed Pigeon
		92	Yungas Pygmy Owl
Polylepis sp.	Queñoa (Tabaquillo)	125	Tawny Tit-Spinetail **
		187	Plain-coloured Seedeater
		190	Tucumán Mountain Finch
		193	Thick-billed Siskin *
Porlieria microphylla	Cucharera (Chucupí)	177	Many-coloured Chaco Finch
Portulaca sp.	Flor de Seda	128	Crested Hornero
Prosopis alpataco	Alpataco	181	Cinnamon Warbling Finch
Prosopis caldenia	Caldén	155	White-tipped Plantcutter
Prosopis nigra	Algarrobo negro	107	White-fronted Woodpecker
		110	Black-bodied Woodpecker
Schinus fasciculata	Moradillo (Molle)	132	Sooty-fronted Spinetail
Schinus johnstonii	Molle Patagónico	192	Patagonian Yellow Finch
Schinus sp.	Molle	177	Golden-billed Saltator
Schoenoplectus californicus	Junco (California sedge)	126	Wren-like Rushbird *
		195	Yellow-winged Blackbird
Scirpus giganteus	Paja Brava (sedge)	126	Curve-billed Reedhaunter *
Sebastiania commersoniana	Blanquillo (Palo de leche)	181	Black-and-chestnut Warbling Finch
Senecio crithmoides		193	Yellow-rumped Siskin
Sesbania punicea	Acacia Mansa	181	Black-and-rufous Warbling Finch
Solanum granulosum-leprosum	Fumo Bravo	177	Thick-billed Saltator *
Sophronitis cernua	Orquídea epífita (epiphytic orchid)	117	White-throated Woodcreeper
Thalia sp.	Pehuajó	196	Scarlet-headed Blackbird
Tillandsia aëranthos	Clavel del Aire Azul (airplant)	125	Tufted Tit-Spinetail
Tillandsia recurvata	Clavel del Aire Chico (airplant)	182	Black-capped Warbling Finch
Trichocereus atacamensis	Pasacana (cactus)	40	Variable Hawk
Typha sp.	Totora (Bulrush)	54	Red-and-white Crake
Zanthoxylum sp. [Fagara sp.]	Tembetarí (Teta de Perra)	111	Great Antshrike
		177	Green-winged Saltator
Ziziphus mistol	Mistol	178	Ultramarine Grosbeak

APPENDIX 6: TAXONOMIC NOTES

1. **Yungas Guan *Penelope bridgesi*** Recently split from Dusky-legged Guan *P. obscura* on account of morphological differences (Evangelista-Vargas & Silveira 2018). Vocal differentiation and further morphological and morphometric differences support the split (del Hoyo & Motis 2004).

2. **Comb Duck *Sarkidiornis sylvicola*** The Old World Knob-billed Duck *S. melanotos* and the New World Comb Duck *S. sylvicola* were until recently considered to be subspecies. The lack of evidence of interbeeding between these widely allopatric forms, and differences in colour of flanks, size and possibly vocalisations all suggest that these are better considered as two separate species, although genetic and behavioural studies are needed to better understand this question (Remsen *et al.* 2020).

3. **Inca Teal *Anas [flavirostris] oxyptera*** Nominate Speckled Teal *A. f. flavirostris* and the high-altitude subspecies *oxyptera* exhibit diagnostic plumage differences. Genetic studies show a recent divergence of 0.5–1 million years, while genomic analyses indicate that populations of *flavirostris* and *oxyptera* are genetically distinct, despite low levels of genomic divergence (Graham *et al.* 2017). Genes related to life at high altitude in *oxyptera* differ markedly from *flavirostris*, suggesting that differences in these genes might underlie their population divergence, and could be responsible for occasional asymmetric introgression in evolutionary times, with genetic flow from *flavirostris* to *oxyptera* but not *vice versa* (Graham *et al.* 2017). Both taxa breed syntopically at least sporadically without any apparent hybridisation at Dique La Angostura in Tucumán province (Argentina), although overlap occurs mostly during winter at intermediate altitudes in the Andes of NW Argentina, where *flavirostris* is a winter visitor and *oxyptera* an altitudinal migrant. Collectively, these data suggest that *oxyptera* might be better recognised as a separate species from *flavirostris*.

4. **Albatrosses *Diomedea* and *Thalassarche*** Species limits in both of these genera are contentious with different taxonomists arguing from one extreme to the other; that each and every insular island breeding population should be awarded species status, or that genetic differences are so slim (in many cases less than 1% divergence) suggest that a return to a more traditional taxonomy should be afforded. We take the middle ground and follow the majority of recent authorities, while understanding that new changes can be expected.

5. **Fuegian Storm Petrel *Oceanites [oceanicus] chilensis*** This distinctive subspecies is in need of further study, but may merit recognition as a different species (Howell & Zufelt 2019).

6. **Pincoya Storm Petrel *Oceanites pincoyae*** Recently described as a distinctive species (Harrison *et al.* 2013). Breeding grounds are unknown. Its species status has recently been questioned (Howell & Schmitt 2016), but more evidence is needed to disprove the specific distinctiveness of this taxon.

7. **Imperial Shag *Phalacrocorax atriceps*** The subspecies *bransfieldensis* of Antarctica has sometimes been elevated to species status ('Antarctic Shag') originally based on osteological differences and purported sympatry with *atriceps* after the excavation of *bransfieldensis* bones on Tierra del Fuego (Siegel-Causey & Lefevre 1989). While nothing is currently known of *bransfieldensis* in Tierra del Fuego, osteological differences can be clearly expected in insular populations, and experienced observers report them to be difficult or impossible to separate in the field. Comparative DNA studies suggest that *bransfieldensis* belongs to a group including several subantarctic and antarctic island taxa often afforded full species status (*melanogenis*, *nivalis*, *purpurascens*, and *verrucosus*) which would be sister to a recently diverged group including *atriceps*, *georgianus* and *albiventer* (Kennedy & Spencer 2014). However, while all these taxa are clearly related, phylogenetic relationships are uncertain and further studies are needed.

8. **Azara's Bittern *Ixobrychus [exilis] erythromelas*** Vocalisations of South American breeders differ markedly from nominate Least Bittern *I. e. exilis* (Behrstock 1996). These differences are maintained across vast areas, indicating that South American birds should be recognised as a different species. We are currently finishing work on their taxonomy.

9. **Turkey Vulture *Cathartes aura*** Subspecies *jota* and *falklandica* differ in structure, plumage and head colour from *ruficollis*. Further continent-wide studies are needed to understand their distribution, morphology and taxonomy.

10. **Roadside Hawk *Rupornis magnirostris*** Two subspecies groups can be recognised: northern *magnirostris* (smaller, slimmer, paler grey hood and back) and southern *pucherani* (larger, stockier, darker hood and back), which also seem to differ in vocalisations. Argentine taxa belong to the *pucherani* group. When looking at the extremes, the differences suggest the existence of at least two species in South America, but more studies on phenotypic and genetic variation are needed to assess the degree of intermediacy among the numerous forms.

11. **Chilean Hawk *Accipiter [bicolor] chilensis*** This temperate forest hawk differs markedly in vocalisations and plumage from Bicoloured Hawk *A. bicolor*. Although often regarded as a separate species (e.g., Ferguson-Lees & Christie 2005), no formal studies have been published.

12. **Rufous-thighed Hawk *Accipiter [striatus] erythronemius*** Up to four allopatric species have been recognised within the widely ranging Sharp-shinned Hawk *A. striatus* (Ferguson-Lees & Christie 2005, Remsen *et al.* 2020), but evidence for splitting them is weak, and based purely on plumage and size characters (Storer 1952, Ferguson-Lees & Christie

2001, Remsen *et al.* 2020). Here we opt to highlight *erythronemius* as a distinctive taxon that may merit species-level treatment pending further study.

13. Crested Caracara *Caracara plancus* On the basis of several plumage features, Dove & Banks (1999) considered *C. plancus* to comprise three biological species with birds of northern South America assigned to *C. cheriway* and an extinct species which occurred on Guadalupe Island, Mexico as *C. lutosus*. This treatment has been adopted almost universally without scrupulous critique. However, it is noteworthy that *plancus* and *cheriway* grade into one another over a large area of Amazonia with 21 intermediate specimens noted (Dove & Banks 1999). We believe that the data suggest a subspecific relationship at most, while clinal variation in plumage traits needs evaluation. For these reasons, and the lack of any other evidence supporting species-level differences, we treat *cheriway* as a subspecies of *C. plancus* under the original name Crested Caracara.

14. Black-necked Stilt *Himantopus mexicanus* Subspecies *mexicanus*, usually together with *knudseni* of Hawaii, is sometimes considered a single species, in which case *H. melanurus* (of Argentina and s. S. America) is known as 'White-backed Stilt', but there is no published evidence to warrant this treatment. The two forms were reported to intergrade in ne. Peru and Minas Gerais, Brazil (Hellmayr & Conover 1948, Blake 1977) and intermediate specimens have been described from ne. Brazil (Short 1975). Vocal differences between *knudseni* + *mexicanus* and *melanurus* have been reported, but no formal studies have been published (Jaramillo 2003).

15. Southern Lapwing *Vanellus chilensis* Two distinct groups can be recognised: widespread *lampronotus* together with *cayenensis* of NW South America; and Patagonian *fretensis* with nominate *chilensis* of the C Andes of Argentina and much of C and S Chile. These groups differ in plumage, by the lack of a crest and long-distance migration in southern birds (*chilensis*), and especially by the marked distinctions in their voices. However, in areas where they meet in e. Río Negro and ne. Chubut, there are specimens showing mixed characters which cannot be attributed to either *lampronotus* or *fretensis*, in addition to an intermediate *lampronotus* x *chilensis* specimen from San Juan (Navas & Bó 1986). Our preliminary studies reveal birds with one plumage producing vocalisations of the other in the intergradation zone, and others producing somewhat intermediate voices. More study is required.

16. Magellanic Snipe *Gallinago magellanica* Recently split from South American Snipe *G. paraguaiae* based mainly on the marked differences in vocalisations and mechanical sounds made by the tail during aerial displays (Miller *et al.* 2019).

17. Willet *Tringa semipalmata* Morphological, vocal and genetic differences suggest that two species should possibly be recognised: Eastern Willet *T. semipalmata* and Western Willet *T. inornata* (Douglas 1998, Oswald *et al.* 2016).

18. Sandwich Tern *Sterna sandvicensis* The problematic taxonomy of Sandwich Terns is in need of rigorous work. The subspecies *eurygnathus* has frequently been afforded specific status as 'Cayenne Tern', while subspecies *acuflavidus* has less often been separated as 'Cabot's Tern'. Recent genetic data suggest that nominate *sandvicensis* is sister to a clade composed of conspecific *eurygnathus* and *acuflavidus* which are seemingly more closely related to Elegant Tern *T. elegans* (Efe *et al.* 2009). Genetic divergence in mitochondrial DNA between *eurygnatha* and *acuflavidus* is reduced (0.25–0.29%; Bridge *et al.* 2005, Efe *et al.* 2009), being greater between them and *elegans* (*c.*1%), and *sandvicensis* (*c.*2.7%) (Efe *et al.* 2009). However, genetic sampling might be insufficient to justify elevation of *eurygnathus*/*acuflavidus* as a separate species (Chesser *et al.* 2013). To complicate matters further, two populations included within *eurygnathus* that breed in Argentina and Brazil exhibit consistent differences in size, breeding and moult schedules, and may deserve taxonomic treatment as separate from the Caribbean breeding *eurygnathus* (Voous 1968, Escalante 1970, 1973, Schoch & Azpiroz 2016). Birds with yellow tips and black bills in South America could be visiting North American *acuflavidus* or little-known variants/stages ('*acuflavidus*') of birds fledged in South America (Schoch & Azpiroz 2016). Further study is obviously needed.

19. Black Skimmer *Rynchops niger* Notable differences in plumage, biometric measurements and migratory behaviour of the mostly Amazonian breeder *cinerascens* compared with the resident South American *intercedens*, and to the African *R. flavirostris* and Indian *R. albicollis* species, suggest that *cinerascens* may have diverged to species level but more study is required, since all taxa in the genus appear closely related.

20. Band-tailed Pigeon *Patagioenas fasciata* Two subspecies groups, northern *fasciata* from northern Nicaragua northwards, and southern *albilinea* from northern Costa Rica southwards, can be recognised based on bill and plumage colour, and differences in vocalisations. Further studies are needed, but evidence suggests that two separate species are involved.

21. Cliff Parakeet *Myiopsitta* [*monachus*] *luchsi* Many authorities consider the Cliff Parakeet to represent a distinct species from Monk Parakeet *M. monachus*, based primarily on its non-communal stick nests built on cliffs in loose colonies versus massive communal stick nests placed in trees in *monachus*. Apart from some notable plumage differences, there are also vocal distinctions although these have yet to be published. The two are allopatric and come within 175 km of one another, and are also segregated by altitude, with *luchsi* mostly above 1400 m in Bolivia and above 2900 m in NW Argentina. In a study of mitochondrial DNA, no haplotypes of *luchsi* were shared with *monachus* (Russello *et al.* 2008). The differences strongly suggest that *luchsi* is a distinct species, but further DNA studies and a published vocal analysis are needed to fully endorse this stance.

22. **Patagonian Nightjar** *Systellura [longirostris] bifasciatus* and **Siku Nightjar** *Systellura [longirostris] atripunctatus* These subspecies differ markedly in vocalisations from nominate Band-winged Nightjar *S. l. longirostris* and from other subspecies usually placed within this broadly conceived species. We are currently working on a broad-scale genetic and vocal analysis to elucidate the taxonomy of these taxa (Areta, Valqui, Pearman *et al.* in prep.).

23. **Magellanic Horned Owl** *Bubo [virginianus] magellanicus* The subspecies *magellanicus* and *nacurutu* differ diagnostically in vocalisations, morphology and distribution, being parapatric in CW and NW Argentina (e.g., Salta, Córdoba, San Luis) with *magellanicus* in open areas in highlands and *nacurutu* in forested lowlands without apparent hybridisation (pers. obs., Traylor 1958, König *et al.* 1996). The situation is less clear regarding vocalisations and segregation from Peru northwards (Traylor 1958, Schulenberg *et al.* 2007, López-Lanús 2015). Further vocal, morphological and genetic studies of all subspecies in this group are needed to elucidate species limits.

24. **American Barn Owl** *Tyto furcata* Genetic evidence suggests that the cosmopolitan Barn Owl *T. alba*, can be subdivided into at least three species: American Barn Owl *T. furcata*, Western Barn Owl *T. alba* and Eastern Barn Owl *T. javanica* (Alibadian *et al.* 2016, Uva *et al.* 2018). Besides the morphological and genetic differences, American Barn Owl also differs vocally from nominate *alba* (Robb & The Sound Approach 2015).

25. **Yungas Speckled Hummingbird** *Adelomyia [melanogenys] inornata* Genetic evidence and plumage differences suggest that multiple species are involved in the Speckled Hummingbird, with the southernmost birds being basal to other taxa (Chaves *et al.* 2011).

26. **Yungas White-bellied Hummingbird** *Elliotomyia [chionogaster] hypoleuca* Differences in undertail pattern and vocalisations between southern subspecies *hypoleuca* and northern *chionogaster* suggest that two species might be involved, especially because vocalisations of *chionogaster* are apparently more similar to those of Green-and-white Hummingbird *E. viridicauda* than to *hypoleuca* (Robbins *et al.* 2013). The genus *Elliotomyia* has been recently described (Stiles & Remsen 2019).

27. **Purple-crowned Plovercrest** *Stephanoxis loddigesii* Recently afforded specific status and separated from Green-crowned Plovercrest *S. lalandi*, from which it differs notably in crest colour and ventral coloration in males (Cavarzere *et al.* 2014).

28. **Black-throated Trogon** *Trogon [rufus] chrysochloros* The Atlantic Forest subspecies *chrysochloros* differs markedly in plumage and song from other subspecies in this complex. Genetic and formal taxonomic studies are needed to elucidate how many species should be recognised in this currently polytypic species (see Dickens 2015).

29. **Bolivian Woodpecker** *Veniliornis [lignarius] puncticeps* This largely neglected taxon, restricted to the inter-Andean dry valleys of Bolivia and Argentina, has often been merged with nominate Striped Woodpecker *V. l. lignarius* (Short 1982). However, it is vocally more similar to White-spotted Woodpecker *V. spilogaster* than to either Striped or Checkered Woodpeckers *V. mixtus*. Our ongoing research indicates that Checkered, Striped and Bolivian may represent three or just a single species (Areta, Pearman & Burgos in prep.).

30. **Golden-olive Woodpecker** *Colaptes rubiginosus* The southern taxa differ in plumage and vocalisations from nominate *rubiginosus* (and other taxa) from northern South America, most notably in their calls. Further studies are required and at least two species appear to be involved.

31. **Golden-breasted Woodpecker** *Colaptes [melanochloros] melanolaimus* The vocally, ecologically and morphologically divergent *melanolaimus* group was formerly considered a distinct species 'Golden-breasted Woodpecker', but interbreeding with nominate *melanochloros* has been shown to be extensive along the Paraná and Uruguay Rivers (notably in the province of Corrientes; Short 1972), thus most subsequent authors regard these forms as conspecific. Although this is not a clear-cut case, we afford a distinct status to the *melanolaimus* group to emphasise the evident differences from *melanochloros*, which greatly exceed those expected for conventional subspecies.

32. **Olive-crowned Crescentchest** *Melanopareia maximiliani* The sierran subspecies *argentina* and *maximiliani* differ notably in vocalisations from the lowland *pallida*; however, a thorough analysis of vocal, plumage and genetic variation is underway to quantify their divergence and clarify whether variation is clinal or discrete among these groups (Toledo, Pearman & Areta in prep.).

33. **Olivaceous Woodcreeper** *Sittasomus griseicapillus* Morphological and vocal differences between nominate *griseicapillus* and *sylviellus* are striking, providing strong evidence that there are two different species. However, this highly polytypic species is in need of detailed research as more than two species are likely involved. Until formal comparative studies are published, we provisionally treat them as part of the same species.

34. **Scaled Woodcreeper** *Lepidocolaptes squamatus* The subspecies *falcinellus* is frequently afforded full species status as 'Scalloped Woodcreeper', but morphological evidence suggests this treatment is weak (Silva & Straube 1996). Variation in plumage appears to be clinal, linking *squamatus* and *falcinellus*, and vocal differentiation seems tenuous at best (pers. obs.). Here we treat them as a single species until more evidence clarifies the situation.

35. **Common Miner** *Geositta cunicularia* Numerous subspecies are recognised, and limits and specific status of many remain contentious (Remsen 2003). Nominate *cunicularia* seems to differ vocally from *titicacae* but probably not from

other Argentine taxa, suggesting the existence of at least two separate species (pers. obs., Fjeldså & Krabbe 1990). Further studies across the wide range of this miner are needed to fully understand species limits.

36. Buzzing Miner *Geositta* [*rufipennis*] *rufipennis* and Trilling Miner *Geositta* [*rufipennis*] *fasciata* Plumage and vocalisations are clearly different among the *rufipennis* and *fasciata* groups of the 'Rufous-webbed Miner'. Our research shows that taxa from these groups are narrowly parapatric, at least in Mendoza province but possibly also in Neuquén and San Juan, with *fasciata* occurring at higher altitude than *rufipennis*, whilst playback experiments evidence complete discrimination among these groups (Areta, Pearman & Jaramillo in prep.).

37. Sharp-tailed Streamcreeper *Lochmias nematura* Nominate Atlantic Forest *nematura* and Andean *obscuratus* differ dramatically in size and plumage, while other subspecies in South and Central America differ from them to different extents (Remsen 2003). Vocalisations seem rather similar, but more studies are needed.

38. Grey-flanked Cinclodes *Cinclodes oustaleti* and Cream-winged Cinclodes *C. albiventris* Olrog's Cinclodes *C. olrogi* has been a controversial 'endemic' Argentine species ever since its description (Nores & Yzurieta 1979). Immediately, Olrog (1979) treated *olrogi* as a subspecies of *oustaleti*, while Nores (1986) subsequently considered *olrogi* to be a subspecies closely related to *riojanus* (currently within *C. albiventris*). A morphological study of the *oustaleti* complex demonstrated that *olrogi* should be regarded as a subspecies of *oustaleti* (Navas & Bó 1987). A phylogenetic analysis of mitochondrial DNA (complete COII and ND3 genes) showed a divergence of 0.5% between *olrogi* and *oustaleti*, and suggested they were conspecific (Chesser 2004). Using the same genes and an expanded dataset, Sanín *et al.* (2009) recovered a sister relationship of *olrogi* and *oustaleti*, but interestingly a haplotype of *C. albiventris tucumanus* was more similar to *oustaleti* than to *olrogi*. They suggested *C. albiventris* should comprise five subspecies: *albiventris*, *tucumanus*, *yzurietae*, *riojanus* and *rufus*. The issue of whether *albiventris* is conspecific or not with *olrogi-oustaleti* is a lingering one. Interestingly, the taxa *yzurietae-riojanus-tucumanus-rufus* are but reddish versions of *oustaleti-olrogi*, while *albiventris* is a browner version sharing the same structural plan. Preliminary analyses indicate that vocalisations of *oustaleti*, *olrogi* and *albiventris* are indistinguishable (Areta, Pearman & Krabbe unpubl.). We follow previous authors in treating *olrogi* as conspecific with *oustaleti*, but refrain from merging *albiventris* with them until our ongoing work clarifies the situation.

39. Tussacbird *Cinclodes* [*antarcticus*] *antarcticus* and Fuegian Cinclodes *Cinclodes* [*antarcticus*] *maculirostris* Plumage, morphology and ecology of these forms differ radically to the extent that it seems likely that they are different species as originally described. Genetic data (COI) found no differences among a single sample of *maculirostris* in comparison with five samples of *antarcticus*, suggesting a very recent divergence (Campagna *et al.* 2012). Vocal data of *maculirostris* is difficult to obtain, but might hold the key to the taxonomy of these birds. When treated conspecifically, the two forms are known as Blackish Cinclodes.

40. Patagonian Forest Earthcreeper *Upucerthia saturatior* Recently split from the narrowly parapatric Scale-throated Earthcreeper *U. dumetaria*, based on differences in plumage, morphometrics, vocalisations and habitat (Areta & Pearman 2009).

41. Buff-breasted Earthcreeper *Upucerthia validirostris* Subspecies *validirostris* was formerly considered an Argentine endemic species distinct from 'Plain-breasted Earthcreeper' *U. jelskii*. Vocalisations do not differ between them, with each readily responding to playback of forms located >1000 km apart, while plumage and morphology has been shown to be clinal (Areta & Pearman 2013).

42. Buff-browed Foliage-gleaner *Syndactyla rufosuperciliata* Two subspecies groups, Andean *cabanisi* (including *similis* and *oleaginous*) and Atlantic forest *rufosuperciliata* (including *acrita*) have recently been afforded species status based on their degree of genetic differentiation (Cabanne *et al.* 2019). However, the Argentine taxa *oleagineus* and *acrita* are vocally similar and further quantitative studies are warranted to justify their split.

43. Puna Tit-Spinetail *Leptasthenura* [*aegithaloides*] *berlepschi* Multiple species differing in plumage and vocalisations are currently included within a broadly defined Plain-mantled Tit-Spinetail *L. aegithaloides* (Jaramillo 2003). Vocalisations of *berlepschi* differ markedly from those of *pallida* and *aegithaloides*, and they do not respond to reciprocal playback, whereas vocal differences between the latter two are perhaps insufficient to distinguish them at the species level. We are currently researching this (Jaramillo & Areta in prep.).

44. Puna Canastero *Asthenes sclateri* Recognition of Puna Canastero as specifically distinct from Stripe-backed Canastero *A. wyatti* is a matter of controversy, and subspecies have been variously allocated to different species (Navas & Bó 1982, Ridgely & Tudor 1994, Vuilleumier 1997, Krabbe 2000). Vocalisations are similar, differences in plumage are slight and there is some evidence of intergradation between the northernmost subspecies *punensis* of Puna and the southernmost subspecies *graminicola* of Stripe-backed in the Lake Titicaca region (Fjeldså & Krabbe 1990, Vuilleumier 1997, Krabbe 2000, Remsen 2003). A narrowly circumscribed 'Córdoba Canastero' has been recognised as a separate species for taxa in the Sierras de Córdoba (*sclateri*) and Sierras de San Luis (*brunnescens*) (Fjeldså & Krabbe 1990, Vuilleumier 1997). It seems likely that Puna (including 'Córdoba') are better merged with Stripe-backed Canastero, but further study is required.

45. Ochre-cheeked Spinetail *Synallaxis scutata* Geographic variation in vocalisations of nominate *scutata* (SW to NE

Brazil) and *whitii* (mostly NW Argentina, SE Bolivia, SW Brazil and locally in NW Paraguay) is not well understood. However, vocal differences seem marked and at least two species might be involved. An isolated population in SE Peru (Puno) merits further study as well, but seems vocally more similar to *whitii.*

46. **Austral Spinetail** *Synallaxis* [*albescens*] *australis* The migratory subspecies *australis* differs notably from other subspecies of Pale-breasted Spinetail *S. albescens* in vocalisations, suggesting that at least two species should be recognised (pers. obs., Ridgely & Tudor 1994, Remsen 2003). The presence of nominate *albescens* as a breeder in Misiones and Corrientes provinces needs corroboration. We are actively working on these issues.

47. **White-crested Elaenia** *Elaenia albiceps* Based on molecular phylogenetics and vocal analyses, it has been proposed that the southern taxon *chilensis* either constitutes a separate species, 'Chilean Elaenia' (Rheindt *et al.* 2009) or that it is conspecific with Sierran Elaenia *E. pallatangae*, and only distantly related to nominate *E. albiceps* (Chattopadhyay *et al.* 2017, Tang *et al.* 2018). We acknowledge that the taxonomy of *chilensis* is in need of revision. However, the patterns of introgression and level of vocal differentiation in these taxa are not clear, and we provisionally retain *chilensis* under *E. albiceps* pending further studies.

48. **Straneck's Tyrannulet** *Serpophaga griseicapilla* Described as a cryptic species (Straneck 2007). Known for some time with the name Grey-crowned Tyrannulet *S. griseiceps* (Straneck 1993) which was subsequently shown to apply to the juvenile of subspecies *munda* of White-crested Tyrannulet (Herzog & Mazar Barnett 2004).

49. **White-crested Tyrannulet** *Serpophaga subcristata* The subspecies *munda* has often been regarded as a separate species, 'White-bellied Tyrannulet' (Ridgely & Tudor 1994, Remsen *et al.* 2020). However, intermediate specimens occur in C Argentina (Bó 1969), vocalisations are identical, and they readily respond to reciprocal playback experiments (Straneck 1993). Although concerns have been raised concerning Straneck's work (Herzog 2001), massive amounts of newly available material confirm their merging as a single species (Mazar Barnett, Pearman & Areta unpubl. data).

50. **Yellow-olive Flycatcher** *Tolmomyias sulphurescens* Multiple species are likely included within this species as currently delineated (Fitzpatrick 2004). In Argentina, two vocal groups can be recognised, nominate *sulphurescens* in Paraná Forest with raspy whistles, and *grisescens* (Humid chaco) and *pallescens* (Yungas) with pure whistles. Geographic variation needs study, as populations might be interconnected and clinal variation may occur.

51. **Whistling Fuscous Flycatcher** *Cnemotriccus* [*fuscatus*] '*bimaculatus*' The nominate subspecies *fuscatus* of Fuscous Flycatcher from the Atlantic Forest of Brazil differs vocally and morphologically from what has been historically regarded as the widespread lowland taxon *bimaculatus*. Additionally, these forms do not respond to playback of each other (Belton 1976), providing strong evidence of the existence of two different species. However, it is possible that the name *bimaculatus* applies to another vocally distinct population in NW Bolivia (Herzog *et al.* 2016) and SE Peru. Further clarification should await formal studies of the genus *Cnemotriccus*, and here we choose to express uncertainty on the name of the taxon in Argentina as '*bimaculatus*'.

52. **Ticking Doradito** *Pseudocolopteryx citreola* Cryptic species, recently separated from Warbling Doradito *P. flaviventris* based on differences in vocalisations and results from playback experiments that show complete heterospecific discrimination (Ábalos & Areta 2009).

53. **Plumbeous Tyrant** *Knipolegus cabanisi* Formerly considered a subspecies of Jelski's Black Tyrant *K. signatus* from which it was recently split based on remarkable differences in plumage of both sexes and deep genetic divergence (Hosner & Moyle 2012).

54. **Cliff Flycatcher** *Hirundinea ferruginea* The northern subspecies group (*sclateri* and *ferruginea*) differs in plumage and vocalisations from the southern subspecies group (*bellicosa* and *pallidior*) and may constitute two separate species.

55. **Solitary Flycatcher** *Myiodynastes* [*maculatus*] *solitarius* The long-distance migratory subspecies *solitarius* of Streaked Flycatcher has been considered a separate species (Hellmayr 1927). Possible hybrids with nominate *maculatus* have been reported from the right bank of the Río Xingú and elsewhere in Brazil, but the geographic origin of these is unknown and the issue merits further analysis (Zimmer 1937). Plumage and vocalisations of *solitarius* differ markedly from other subspecies including calls, song and dawn song, but thorough geographic sampling is needed to test for clinality. We are currently researching this issue.

56. **Grass Wren** *Cistothorus platensis* Recent genetic and vocal studies have suggested that as many as eight species could be embraced within the present species (Traylor 1988, Robbins & Nyári 2014). In our region, the subspecies *platensis*, *hornensis* (together with *falklandicus*), and *tucumanus* were elevated to species rank (Robbins & Nyári 2014). However, we feel that this treatment, while perhaps correct, needs to be tested by thorough analysis of vocal and geographic variation in plumage and genetics.

57. **Chivi Vireo** *Vireo chivi* Split from Red-eyed Vireo *V. olivaceus* with which it was long considered conspecific. Recent genetic studies show that South American taxa in the *chivi* group are distantly related to nominate *olivaceus* from North America (Slager *et al.* 2014) and sister to Black-whiskered Vireo *V. altiloquus*, with some evidence of reduced asymmetric hybridisation (Battey & Klicka 2017). Differences in vocalisations, migratory behaviour and plumage support the split.

58. Chiguanco Thrush *Turdus chiguanco* The subspecies *anthracinus* (blackish plumage, orange bill and legs, with eye-ring) differs notably from nominate *chiguanco* (brownish plumage, yellow bill and legs, lacking eye-ring). They approach each other in northern Chile, and although there is no definite evidence of hybridisation, an intermediate specimen has been reported near their possible contact zone (Fjeldså & Krabbe 1990, Jaramillo 2003). Genetic data show that they are closely related and recently diverged (Batista *et al.* 2020), but further elucidation of their relationships to Great Thrush *T. fuscater*, coupled with vocal analyses and surveys at their possible contact zones are needed.

59. Sclater's Nightingale-Thrush *Catharus maculatus* All Neotropical forms of this nightingale-thrush were until recently united under the name Spotted Nightingale-Thrush *C. dryas*, but this now applies to Central American forms having been split on the basis of genetic, morphometric and vocal data (Halley *et al.* 2018).

60. Puna Pipit *Anthus [furcatus] brevirostris* Recently recognised as a distinct species on account of genetic divergence from Short-billed Pipit *A. furcatus* and diagnostic song (Van Els & Norambuena 2018). Although better geographic sampling is needed to fully understand the genetics of these pipits (Remsen *et al.* 2020), we tentatively afford species status to the high-altitude *brevirostris* based mostly on its peculiar vocalisations and genetic divergence which resemble differences between Paramo *A. bogotensis* and Hellmayr's Pipits *A. hellmayri* (Van Els & Norambuena 2018).

61. Southern Yellowthroat *Geothlypis [aequinoctialis] velata* The three South American forms *aequinoctialis*, *velata* and *auricularis* may constitute separate species that differ in plumage and genetics as much as congeners do (Escalante-Pliego 1992, Curson *et al.* 1994, Escalante *et al.* 2009). Detailed vocal analyses and more extensive geographic sampling might be needed to better support these likely splits (Remsen *et al.* 2020).

62. Yellow Warbler *Setophaga [petechia] aestiva* Three subspecies groups have been recognised which differ in habitat use, migratory behaviour and plumage (Curson *et al.* 1994). Although three to four species may exist in the complex, more data are needed (Klein & Brown 1994, Remsen *et al.* 2020).

63. Golden-crowned Warbler *Basileuterus culicivorus* Three subspecies groups have been recognised as different species: Middle American *culicivorus* 'Stripe-crowned Warbler', Colombian-Venezuelan *cabanisi* 'Cabanis's Warbler', and the widespread South American *auricapillus* 'Golden-crowned Warbler' to which Argentine birds pertain (Curson *et al.* 1994). They differ in plumage and to some extent in vocalisations, but genetic studies show that different, complicated, possible subdivisions of the many taxa in this complex are in need of a critical assessment (Vilaça & Santos 2010).

64. Burnished-buff Tanager *Stilpnia cayana* Seven subspecies are often recognised in this species and divided in two groups: southeastern *flava* group (typically with dark masks extending into a central throat and belly stripe) and northwestern *cayana* group (typically with black restricted to a mask). Extremes of variation are very different (*chloroptera* in the *flava* group, and *cayana* in its group), but geographic variation is not well understood, and some subspecies near the contact zones of both groups seem to exhibit intermediate characters (Isler & Isler 1987, Ridgely & Tudor 1989).

65. Moss-backed Sparrow *Arremon dorbignii* Recently split from Saffron-billed Sparrow *A. flavirostris* based on differences in plumage, genetics and vocalisations (Buainain *et al.* 2016, Trujillo-Arias *et al.* 2017).

66. Yungas Sparrow *Rhynchospiza dabbenei* Recently split from the parapatric Chaco Sparrow *R. strigiceps* from which it differs in plumage, size, proportions, habitat and vocalisations (Areta *et al.* 2019). When treated conspecifically, the two forms were known as Stripe-capped Sparrow.

67. Black-and-chestnut Warbling Finch *Poospiza whitii* Separated from Black-and-rufous Warbling Finch *P. nigrorufa* from which it differs genetically by *c.* 2.5% (Lougheed *et al.* 2000, Shultz & Burns 2013), plumage, vocalisations, morphology and habitat (Jordan *et al.* 2017). Reciprocal playback experiments confirm that both taxa are able to discriminate songs of each other (Jordan *et al.* 2017).

68. Chaco Warbling Finch *Microspingus [torquatus] pectoralis* The endemic Bolivian form *torquatus*, generally known as Ringed Warbling Finch, is restricted to the Bolivian highlands and differs by its larger size, discrete plumage differences and song. One old study indicated a far-removed genetic footprint (Lougheed *et al.* 2000), but more study is required to reaffirm all of these differences that seemingly confirm specific status. If confirmed, this would add another endemic breeding species to the Argentine avifauna (Pearman & Areta in prep.)

69. Iberá Seedeater *Sporophila* sp. This 'recently discovered' seedeater has had a tumultuous taxonomic history and is here regarded provisionally as a valid species with an uncertain name (Areta *et al.* 2016). Although two new specific names have been proposed (López-Lanús '2015', Di Giacomo & Kopuchian 2016), the purportedly first-published name was regarded as unavailable due to uncertainties with the publication process and date (Remsen *et al.* 2020). Regardless of this, neither of the new names might be required, as pre-existing names may apply to this taxon. For details discussing the alternative names and preliminary clarifications on its discovery, see Areta *et al.* (2016). Further work in progress will clarify the correct name of this taxon.

70. Monte Yellow Finch *Sicalis mendozae* This endemic species has been recently split from Greenish Yellow Finch *S. olivascens*, based on differences in plumage, morphometrics, vocalisations and habitat (Areta *et al.* 2012).

71. **Undescribed endemic siskin *Spinus* sp.** The status of this peculiar siskin form (Wetmore 1926, Areta *et al.* 2011) is being researched by a team (Areta, Delhey, Soto & Mahler in prep.). Our preliminary findings show that genetic divergence from Black-chinned Siskin *S. barbatus* is relatively low, as seems to be commonplace in the siskin radiation (Beckman & Witt 2015).

72. **Sierran Meadowlark *Sturnella* [*loyca*] *obscurus*** The resident endemic taxon of the Sierras de Córdoba and Sierras de San Luis differs dramatically in plumage from all forms of Long-tailed Meadowlark *S. loyca*, while differences in its aerial display are more akin to Pampas Meadowlark. Vocal and morphological differences are also noteworthy in regarding *obscurus* as a full species (Pearman, Areta & Fraga in prep.).

REFERENCES

Ábalos, R. & Areta, J. I. (2009) Historia natural y vocalizaciones del Doradito Limón (*Pseudocolopteryx* cf. *citreola*) en Argentina. *Ornitologia Neotropical* 20: 215–230.

Alibadian, M., Alaei-Kakhki, N., Mirshamsi, O., Nijman, V. & Roulin, A. (2016) Phylogeny, biogeography, and diversification of barn owls (Aves: Strigiformes). *Biological Journal of the Linnean Society* 119: 904–918.

Areta, J. I., Depino, E. A., Salvador, S.A., Cardiff, S. W., Epperly, K. & Holzmann, I. (2019) Species limits and biogeography of *Rhynchospiza* sparrows. *Journal of Ornithology* 160: 973–991.

Areta, J. I., Hernández, I. & Prieto, J. (2011) Birds of sand: birding the central Monte Desert. *Neotropical Birding* 8: 52–58.

Areta, J. I. & Pearman, M. (2009) Natural history, morphology, evolution, and taxonomic status of the earthcreeper *Upucerthia saturatior* (Furnariidae) from the Patagonian forests of South America. *Condor* 111: 135–149.

Areta, J. I. & Pearman, M. (2013) Species limits and clinal variation in a widespread High Andean furnariid: the Buff-breasted Earthcreeper (*Upucerthia validirostris*). *Condor* 113: 131–142.

Areta, J. I., Pearman, M. & Ábalos, R. (2012) Taxonomy and biogeography of the Monte Yellow-Finch (*Sicalis mendozae*): understanding the endemic avifauna of Argentina's Monte Desert. *Condor* 114: 654–671.

Areta, J. I., Piacentini, V.Q., Haring, E., Gamauf, A., Silveira, L. F., Machado & Kirwan, G. M. (2016) Tiny bird, huge mystery— the Possibly Extinct Hooded Seedeater (*Sporophila melanops*) is a capuchino with a melanistic cap. *PLoS ONE* 11: e0154231.

Batista, R., Olsson, U., Andermann, T., Aleixo, A., Ribas, C. C. & Antonelli, A. (2020) Phylogenomics and biogeography of the world's thrushes (Aves, *Turdus*): New evidence for a more parsimonious evolutionary history. *Proceedings of the Royal Society* B 287: 20192400.

Battey, C. J. & Klicka, J. (2017) Cryptic speciation and gene flow in a migratory songbird species complex: Insights from the Red-Eyed Vireo (*Vireo olivaceus*). *Molecular Phylogenetics and Evolution* 113: 67–75.

Beckman, E. J. & Witt, C. C. (2015) Phylogeny and biogeography of the New World siskins and goldfinches: Rapid, recent diversification in the Central Andes. *Molecular Phylogenetics and Evolution* 87: 28–45.

Behrstock, R. A. (1996) Voices of Stripe-backed Bittern *Ixobrychus involucris*, Least Bittern *I. exilis*, and Zigzag Heron *Zebrilus undulatus*, with notes on distribution. *Cotinga* 5: 55–61.

Belton, W. (1976) Taxonomy of certain bird species from Rio Grande do Sul, Brazil. *National Geographic Society Research Reports* 17: 183–188.

Blake, E. R. (1977) *Manual of Neotropical birds.* Volume 1. University of Chicago Press, Chicago, Illinois.

Bó, N. (1969) Acerca de la afinidad de dos formas geográficas de *Serpophaga*. *Neotrópica* 15: 54–58.

Bridge, E. S., Jones, A. W. & Baker, A. J. (2005) A phylogenetic framework for the terns (Sternini) inferred from mtDNA sequences: implications for taxonomy and plumage evolution. *Molecular Phylogenetics and Evolution* 35: 459–469.

Buainain, N. R., Brito, G. R., Firme, D. F., Figueira, D. M., Raposo, M. A. & Assis, C. P. (2016) Taxonomic revision of Saffron-billed Sparrow *Arremon flavirostris* Swainson, 1838 (Aves: Passerellidae) with comments on its holotype and type locality. *Zootaxa* 4178: 547–567.

Cabanne, G. S., Campagna, L., Trujillo-Arias, N., Naoki, K., Gómez, I., Miyaki, C. Y., Santos, F. R., Dantas, G. P. M., Aleixo, A., Claramunt, S., Rocha, A., Caparroz, R., Lovette, I. J. & Tubaro, P. L. (2019) Phylogeographic variation within the Buff-browed Foliage-gleaner (Aves: Furnariidae: *Syndactyla rufosuperciliata*) supports an Andean-Atlantic forests connection via the Cerrado. *Molecular Phylogenetics and Evolution* 133: 198–213.

Campagna, L., St Clair, J. J. H., Lougheed, S. C., Woods, R. W., Imberti, S. & Tubaro, P. L. (2012) Divergence between passerine populations from the Malvinas – Falkland Islands and their continental counterparts: a comparative phylogeographical study. *Biological Journal Linnean Society* 106: 865–879.

Cavarzere, V., Silveira, L. F., Vasconcelos, M. F., Grantsau, R. & Straube, F. C. (2014) Taxonomy and biogeography of *Stephanoxis* Simon, 1897 (Aves: Trochilidae). *Papéis Avulsos de Zoologia* 54: 69–79.

Chattopadhyay, B., Garg, K. M., Gwee, C. Y., Edwards, S. V. & Rheindt, F. E. (2017) Gene flow during glacial habitat shifts facilitates character displacement in a Neotropical flycatcher radiation. *BMC Evolutionary Biology* 17: 210.

Chaves, J. A., Weir, J. T. & Smith, T. B. (2011) Diversification in *Adelomyia* hummingbirds follows Andean uplift. *Molecular Ecology* 20: 4564–4576.

Chesser, R. T. (2004) Systematics, evolution, and biogeography of the South American ovenbird genus *Cinclodes. Auk* 121: 752–766.

Chesser, R. T., Banks, R. C., Barker, F. K., Cicero, C., Dunn, J. L., Kratter, A. W., Lovette, I. J., Rasmussen, P. C., Remsen Jr, J. V., Rising, J. A., Stotz, D. F. & Winker, K. (2013) Fifty-fourth supplement to the American Ornithologists' Union Check-list of North American Birds. *Auk* 130: 558–571.

Curson, J., Quinn, D. & Beadle, D. (1994) Warblers of the Americas: an identification guide. Houghton Mifflin, Boston.

del Hoyo, J. & Motis, A. Update Chapter. Pp 322–476 in Delacour, J. & Amadon, D. (2004) *Curassows and Related Birds.* Second edition. Lynx Edicions and The National Museum of Natural History, Barcelona and New York.

Di Giacomo, A. S. & Kopuchian, C. (2016) Una nueva especie de capuchino (*Sporophila:* Thraupidae) de los Esteros del Iberá, Corrientes, Argentina. *Nuestras Aves* 61: 3–5.

Dickens, J. K. (2015) Taxonomy of *Trogon rufus* (Gmelin, 1788) and Amazonian ring-shaped clinal variation. MSc Thesis, Universidade de São Paulo.

Douglas, H. D. (1998) Response of Eastern Willets (*Catoptrophorus s. semipalmatus*) to vocalisations of Eastern and Western (*C. s. inornatus*) Willets. *Auk* 115: 514–518.

Dove, C. J. & Banks, R. C. (1999) A taxonomic study of Crested Carcaras (Falconidae). *Wilson Bulletin* 111: 330–339.

Efe, M. A., Tavares, E. S., Baker, A. J. & Bonatto, S. L. (2009) Multigene phylogeny and DNA barcoding indicate that Sandwich Tern complex (*Thalasseus sandvicensis*, Laridae, Sternini) comprises two species. *Molecular Phylogenetics and Evolution* 52: 263–267.

Escalante, P., Márquez-Valdelamar, L., de la Torre, P., Laclette, J. P., Klicka, J. (2009) Evolutionary history of a prominent North American warbler clade: the *Oporornis–Geothlypis* complex. *Molecular Phylogenetics and Evolution* 53: 668–678.

Escalante-Pliego, B.P. (1992) Genetic differentiation in yellowthroats (Parulinae: *Geothlypis*). *Acta XX Congressus Internationalis Ornithologici.* 333–341.

Escalante, R. (1970) Notes on the Cayenne Tern in Uruguay. *Condor* 72: 89–94.

Escalante, R. (1973) The Cayenne Tern in Brazil. *Condor* 75: 470–472.

Evangelista-Vargas, D. O. & Silveira, L. F. (2018) Morphological evidence for the taxonomic status of the Bridges's Guan, *Penelope bridgesi,* with comments on the validity of *P. obscura bronzina* (Aves: Cracidae). *Zoologia* 35: e12993.

Ferguson-Lees, J. & Christie, D. A. (2001) *Raptors of the World.* Houghton Mifflin, Boston.

Ferguson-Lees, J. & Christie, D. A. (2005) *Raptors of the World.* Princeton University Press, NJ.

Fitzpatrick, J. W. (2004) Family Tyrannidae (tyrant flycatchers). Pp. 170–462 in del Hoyo, J., Elliott, A. & Sargatal, J. (eds). *Handbook of the Birds of the World. Volume 9: Cotingas to pipits and wagtails.* Lynx Edicions, Barcelona, Spain.

Fjeldså, J. & Krabbe, N. K. (1990) *Birds of the High Andes.* Zoological Museum, University of Copenhagen, and Apollo Books, Svendborg, Denmark.

Graham, A. M., Lavretsky, P., Muñoz-Fuentes, V., Green, A. J., Wilson, R. E. & McCracken, K. G. (2017) Migration-selection balance drives geographic differentiation in genes associated with high-altitude function in the Speckled Teal (*Anas flavirostris*) in the Andes. *Genome Biology and Evolution* 10: 14–32.

Halley, M. R., Klicka, J. C., Sesink Clee, P. R. & Weckstein, J. D. (2018) Restoring the species status of *Catharus maculatus* (Aves: Turdidae), a secretive Andean thrush, with a critique of the yardstick approach to species delimitation. *Zootaxa* 4276: 387–404.

Harrison, P., Sallaberry, M., Gaskin, C. P., Baird, K. A., Jaramillo, A., Metz, S. M., Pearman, M., O'Keeffe, M., Dowdall, J., Enright, S., Fahy, K., Gilligan, J. & Lillie, G. (2013) A new storm petrel species from Chile. *Auk* 130: 180–191.

Hellmayr, C. E. (1927) *Catalogue of birds of the Americas and the adjacent islands.* Field Museum of Natural History, Zoological Series. Volume 13, part 5.

Hellmayr, C. E. & Conover, B. (1948) *Catalogue of birds of the Americas.* Field Museum of Natural History, Zoological Series. Volume 13, part 1, number 3.

Herzog, S. K. (2001) A re-evaluation of Straneck's (1993) data on the taxonomic status of *Serpophaga subcristata* and *S. munda* (Passeriformes: Tyrannidae): conspecifics or semispecies? *Bulletin of the British Ornithologists' Club* 121: 273–277.

Herzog, S. K. & Mazar-Barnett, J. (2004) On the validity and confused identity of *Serpophaga griseiceps* Berlioz 1959 (Tyrannidae). *Auk* 121: 415–421.

Herzog, S. K., Terrill, R. S., Jahn, A. E., Remsen Jr, J. V., Maillard, Z. O., García-Solíz, V. H., MacLeod, R., Maccormick, A. & Vidoz, J. Q. 2016. *Birds of Bolivia: Field Guide.* Asociación Armonía, Santa Cruz de la Sierra, Bolivia.

Hosner, P. A. & Moyle, R. G. (2012) A molecular phylogeny of black-tyrants (Tyrannidae: *Knipolegus*) reveals strong geographic patterns and homoplasy in plumage and display behaviour. *Auk* 129: 156–167.

Howell, S. N. G. & Schmitt, F. (2016) Pincoya Storm Petrel: comments on identification and plumage variation. *Dutch Birding* 38: 384–388.

Howell, S. N. G. & Zufelt, K. (2019) *Oceanic Birds of the World.* Princeton University Press, NJ.

Jaramillo, A. (2003) *Birds of Chile.* Princeton University Press, NJ.

Jordan, E. A., Areta, J. I. & Holzmann, I. (2017) Mate recognition systems and species limits in a warbling-finch complex (*Poospiza nigrorufa/whitii*). *Emu* 117: 344–358.

Isler, M. L. & Isler, P. R. (1987) *The Tanagers.* Smithsonian Institution Press, Washington DC.

Kennedy, M. & Spencer, H. G. (2014) Classification of the cormorants of the world. *Molecular Phylogenetics and Evolution* 79: 249–257.

Klein, N. K. & Brown, W. M. (1994) Intraspecific molecular phylogeny in the Yellow Warbler (*Dendroica petechia*) and implications for avian biogeography in the West Indies. *Evolution* 48: 1914–1932.

König, C., Heidrich, P. & Wink, M. (1996) Zur Taxonomie der Uhus (*Bubo* ssp.) im südlichen Südamerika. Stuttgart. *Beitrage für Naturkunde* Series A 540: 1–9.

Krabbe, N. K. (2000) Rediscovery of *Asthenes wyatti azuay* (Chapman 1923) with notes on its plumage variation and the taxonomy of the *Asthenes anthoides* superspecies. *Bulletin of the British Ornithologists' Club* 120: 149–153.

López-Lanús B. (2015) Análisis comparativo de las vocalizaciones de distintos taxa del género *Bubo* en América. *Hornero* 30: 69–88.

López-Lanús B. ('2015') Una nueva especie de capuchino (Emberizidae: *Sporophila*) de los pastizales anegados del Iberá, Corrientes, Argentina. Pp. 473–489 in López-Lanús, B. *Guía Audiornis de las aves de Argentina, fotos y sonidos; identificación por características contrapuestas y marcas sobre imágenes.* Primera edición. Audiornis Producciones, Buenos Aires, Argentina.

Lougheed, S. C., Freeland, J. R., Hanford, P. & Boag, P. T. (2000) A molecular phylogeny of Warbling-Finches (*Poospiza*) paraphyly in a Neotropical Emberizid genus. *Molecular Phylogenetics and Evolution* 17: 367–378.

Miller, E. H., Areta, J. I., Jaramillo, A., Imberti, S. & Matus, R. (2019) Snipe taxonomy based on vocal and non-vocal sound displays: the South American Snipe is two species. *Ibis.* https://doi.org/10.1111/ibi.12795

Navas, J. R. & Bó, N. A. (1986) Revisión de las subespecies Argentinas de *Vanellus chilensis* (Aves: Charadriidae). *Neotrópica* 32: 157–165.

Navas, J.R. & Bó, N. A. (1982) La posición taxonómica de *Thripophaga sclateri* y *T. punensis* (Aves, Furnariidae). *Comunicaciones del Museo Argentino de Ciencias Naturales 'Bernardino Rivadavia'* 4: 85–93.

Navas, J. R. & Bó, N. A. (1987) Notas sobre Furnariidae argentinos (Aves, Passeriformes). *Revista del Museo Argentino de Ciencias Naturales 'Bernardino Rivadavia'* 14: 55–86.

Nores, M. (1986) Diez nuevas subespecies de aves provenientes de islas ecológicas argentinas. *Hornero* 12: 262–273.

Nores, M. & Yzurieta, D. (1979) Una nueva especie y dos nuevas subespecies de aves (Passeriformes). *Academia Nacional de Ciencias de Córdoba Miscelánea* 61: 4–8.

Olrog, C. C. (1979) Nueva lista de la Avifauna Argentina. *Opera Lilloana* 27: 1–324.

Oswald, J. A., Harvey, M. G., Remsen, R. C., Foxworth, D. U., Cardiff, S. W., Dittmann, D. L., Megna, L. C., Carling, M. D. & Brumfield, R. T. (2016) Willet be one species or two? A genomic view of the evolutionary history of *Tringa semipalmata.* *Auk* 133: 593–614.

Remsen Jr, J. V. (2003) Family Furnariidae (ovenbirds). Pp. 162–357 in del Hoyo, J., Elliott, A. & Christie, D. A., eds. *Handbook of the Birds of the World, Volume 8. Broadbills to Tapaculos.* Lynx Edicions, Barcelona.

Remsen Jr, J. V., Areta, J. I., Bonaccorso, E., Claramunt, S., Jaramillo, A., Pacheco, J. F., Robbins, M. B., Stiles, F. G., Stotz, D. F. & Zimmer, K. J. (2020). A classification of the bird species of South America. American Ornithological Society, <http://www.museum.lsu.edu/~Remsen/SACCBaseline.htm>

Rheindt, F. E., Christidis, L. & Norman, J. A. (2009) Genetic introgression, incomplete lineage sorting and faulty taxonomy create multiple cases of polyphyly in a montane clade of tyrant-flycatchers (*Elaenia,* Tyrannidae). *Zoologica Scripta* 38: 143-153.

Ridgely, R. S. & Tudor, G. (1989) *The Birds of South America: The Oscine Passerines.* Volume 1. University of Texas Press, Austin.

Ridgely, R. S. & Tudor, G. (1994) *The Birds of South America: The Suboscine Passerines.* Volume 2. University of Texas Press, Austin.

Robb, M. & The Sound Approach (2015) *Undiscovered Owls: A Sound Approach Guide.* The Sound Approach, Poole, UK.

Robbins, M. B., Schulenberg, T. S., Lane, D. F., Cuervo, A. M., Binford, L. C., Nyári, A. S., Combe, M., Arbeláez-Cortés, E., Wehtje, W. & Lira-Noriega, A. (2013) Abra Maruncunca, dpto. Puno, Peru, revisited: vegetation cover and avifauna changes over a 30-year period. *Bulletin of the British Ornithologists' Club* 133: 31–51.

Robbins, M. B. & Nyári, A. S. (2014) Canada to Tierra del Fuego: species limits and historical biogeography of the Sedge Wren (*Cistothorus platensis*). *Wilson Journal of Ornithology* 126: 649–662.

Russello, M. A., Avery, M. L. & Wright, T. F. (2008) Genetic evidence links invasive monk parakeet populations in the United States to the international pet trade. *BMC Evolutionary Biology* 8: 217.

Sanín, C., Cadena, C. D., Maley, J. M., Lijtmaer, D. A., Tubaro, P. L. & Chesser, R. T. (2009) Paraphyly of *Cinclodes fuscus* (Aves: Passeriformes: Furnariidae): implications for taxonomy and biogeography. *Molecular Phylogenetics and Evolution* 53: 547–555.

Schoch, D. T. & Azpiroz, A. B. (2016) Revisiting Escalante's terns: provenance of Sandwich (Cayenne) Terns on the Uruguayan coast. *Neotropical Birding* 19: 44–55.

Schulenberg, T. S., Stotz, D. F., Lane, D. F., O'Neill, J. P. & Parker III, T. A. (2007) *Birds of Peru.* Princeton Univ. Press, Princeton, New Jersey.

Shultz, A. J. & Burns, K. J. (2013) Plumage evolution in relation to light environment in a novel clade of Neotropical tanagers. *Molecular Phylogenetics and Evolution* 66: 112–125.

Short Jr, L. L. (1972) Systematics and behaviour of South American flickers (Aves, *Colaptes*). *Bulletin of the American Museum of Natural History* 149: 1–109.

Short Jr, L. L. (1975) A zoogeographic analysis of the South American Chaco avifauna. *Bulletin of the American Museum of Natural History* 154: 163–352.

Short Jr, L. L. (1982) *Woodpeckers of the World.* Delaware Museum of Natural History, Greenville, Delaware.

Siegel-Causey, D. & Lefevre, C. (1989) Holocene records of the Antarctic Shag (*Phalacrocorax* [*Notocarbo*] *bransfieldensis*) in Fuegian waters. *Condor* 91: 408–415.

Silva, J. M. C. & Straube, F. C. (1996) Systematics and biogeography of Scaled Woodcreepers (Aves, Dendrocolaptidae). *Studies in Neotropical Fauna and Environment* 31: 3–10.

Slager, D. L., Battey, C. J., Bryson Jr, R. W., Voelker, G. & Klicka, J. (2014) A multilocus phylogeny of a major new world avian radiation: the Vireonidae. *Molecular Phylogenetics and Evolution* 80: 95–104.

Stiles, F. G. & Remsen Jr, J. V. (2019) The generic nomenclature of the Trochilini: a correction. *Zootaxa* 4691: 195–196.

Storer, R. W. (1952) Variation in the resident Sharp-shinned Hawks of Mexico. *Condor* 54: 283–289.

Straneck, R. (1993) Aportes para la unificación de *Serpophaga subcristata* y *Serpophaga munda*, y la revalidación de *Serpophaga griseiceps* (Aves: Tyrannidae) *Revista del Museo Argentino de Ciencias Naturales 'Bernardino Rivadavia', Zoología* 16: 51-63.

Straneck, R. (2007) Una nueva especie de *Serpophaga* (Aves: Tyrannidae). *Revista FAVE - Ciencias Veterinarias* 6: 31–42.

Tang, Q., Edwards, S. V. & Rheindt, F. E. (2018) Rapid diversification and hybridisation have shaped the dynamic history of the genus *Elaenia. Molecular Phylogenetics and Evolution* 127: 522–533.

Traylor Jr, M. A. (1958) Variation in South American Great Horned Owls. *Auk* 75: 143-149.

Traylor Jr, M. A. (1988) Geographic variation and evolution in South American *Cistothorus platensis* (Aves: Troglodytidae). *Fieldiana Zoology* New Series 48: 1–35.

Trujillo-Arias, N., Dantas, G. P. M., Arbeláez-Cortés, E., Naoki, K., Gómez, M. I., Santos, F. R., Miyaki, C. Y., Aleixo, A., Tubaro, P. L. & Cabanne, G. S. (2017) The niche and phylogeography of a passerine reveal the history of biological diversification between the Andean and the Atlantic forests. *Molecular Phylogenetics and Evolution* 112: 107–121.

Uva, V., Päckert, M., Cibois, A., Fumagalli, L. & Roulin, A. (2018) Comprehensive molecular phylogeny of barn owls and relatives (Family: Tytonidae), and their six major Pleistocene radiations. *Molecular Phylogenetics and Evolution* 125: 127–137.

Van Els, P. & Norambuena, H. V. (2018) A revision of species limits in Neotropical pipits *Anthus* based on multilocus genetic and vocal data. *Ibis* 160: 158–172.

Vilaça, S. T. & Santos, F. R. (2010) Biogeographic history of the species complex *Basileuterus culicivorus* (Aves, Parulidae). *Molecular Phylogenetics Evolution* 57: 585–597.

Voous, K. H. (1968) Geographical variation in the Cayenne Tern. *Ardea* 56: 184–187.

Vuilleumier, F. (1997) Status and distribution of *Asthenes anthoides* (Furnariidae), a species endemic to Fuego-Patagonia, with notes on its systematic relationships and conservation. *Ornithological Monographs* 48: 791–808.

Wetmore, A. (1926) Observations on the birds of Argentina, Paraguay, Uruguay and Chile. *Bulletin of the U. S. National Museum* 133: 1–448.

Zimmer, J. T. (1937) Studies of Peruvian birds 28. Notes on the genera *Myiodynastes, Conopias, Myiozetetes*, and *Pitangus. American Museum Novitates* 963: 1–28.

INDEX

Chaco Owl *Strix chacoensis* (*Aldo Chiappe*).

KEY TO THE DISTRIBUTION MAPS

Year-round resident. Birds that breed and stay year-round in the same area.

Spring–summer resident. Birds that breed and then wholly or mostly vacate the breeding grounds. Note that a small proportion may overwinter.

Seabirds close to their breeding range during spring–summer.

Winter visitor.

Scarce winter visitor (seabirds only).

Seasonal non-breeding visitor. Used mainly for spring–summer boreal migrant landbirds, and for seabirds breeding elsewhere and visiting the South-west Atlantic mostly during spring–summer.

Scarce seasonal non-breeding visitor (seabirds only).

Passage migrant. Birds that are found only while on passage. Also used for **nomadic breeders** which may be present/absent for many years without any seasonal pattern (mostly bamboo seed-eating birds). It is also used for Saffron-cowled Blackbird whose colonies are seldom in the same places, thereby suggesting some degree of inter-seasonal movements.

Year-round non-breeding range at sea or along coasts.

Sparse occurrence. Frequently used for vagrants that have occurred repeatedly and which are likely to occur again. Also used for some birds that are in expansion and, in a few cases, for areas for which information did not allow us to decide whether a bird breeds, overwinters or is merely on passage. Overall, pink indicates situations of low-density, low chances of seeing the bird and uncertainty on status.

Bold lines indicate the former area occupied by birds that have experienced drastic range retractions (e.g., Yellow Cardinal), or that are currently considered extinct in the study area (e.g., Glaucous Macaw).

X   Accidental, unique or extremely sparse records.

o   Confirmed specimen record when other evidence is lacking, or a species has a restricted range or is rare in the study area.

?   Uncertain or inconclusive records.

Small arrows are used to point to small distribution areas and seabird breeding colonies. Longer or double-ended arrows indicate migration routes.